Bloodstain Pattern Analysis with an Introduction to Crime Scene Reconstruction

Bloodstain Pattern Analysis with an Introduction to Crime Scene Reconstruction, Fourth Edition, provides criminal investigators and forensic scientists with a complete and comprehensive handbook on Bloodstain Pattern Analysis.

Bringing over 90 combined years of practical experience – and thousands of cases worked collectively – the authors explain the complex mechanics of blood spatter analysis, including anatomical issues relative to Bloodstain Pattern Analysis, a discussion of blood and the circulatory system, and the nature of bleeding associated with various traumatic and non-traumatic injuries. The book also details specific methodologies for crime scene analysis and reconstruction, explaining the proven methodology involved in the process. Such methodology is built upon the scientific method and provides focus and structure to the analyst as they evaluate evidence and conduct their work and the investigation.

Finally, all chapters are fully revised and updated to address the latest taxonomy and terminology, with over 400 full-color photographs included to illustrate the dynamics of Bloodstain Pattern Analysis.

Key features:

- Presents a specific and detailed taxonomy of bloodstain pattern characteristics, outlining recently updated NIST's OSAC terminology
- Includes two full-color fold-outs Decision Map to guide analysts through the classification process
- Details the theory, principles, and methodology for crime scene reconstruction
- Expands the bloodstain on clothing chapter, to include new developments in the understanding of bloodstain features and characteristics on cloth

Offering practical advice and tips for new and experienced professionals alike, *Bloodstain Pattern Analysis with an Introduction to Crime Scene Reconstruction, Fourth Edition*, offers readers the necessary tools to guide and focus any investigative effort.

CRC SERIES IN
PRACTICAL ASPECTS OF CRIMINAL AND FORENSIC INVESTIGATIONS
VERNON J. GEBERTH, BBA, MPS, FBINA *Series Editor*

Informants, Cooperating Witnesses, and Undercover Investigations: A Practical Guide to Law, Policy, and Procedure, Second Edition
Dennis G. Fitzgerald

Practical Military Ordnance Identification
Tom Gersbeck

Practical Cold Case Homicide Investigations Procedural Manual
Richard H. Walton

Autoerotic Deaths: Practical Forensic and Investigative Perspectives
Anny Sauvageau and Vernon J. Geberth

Practical Crime Scene Processing and Investigation, Second Edition
Ross M. Gardner

The Counterterrorism Handbook: Tactics, Procedures, and Techniques, Fourth Edition
Frank Bolz, Jr., Kenneth J. Dudonis, and David P. Schulz

Practical Forensic Digital Imaging: Applications and Techniques
Patrick Jones

DiMaio's Forensic Pathology, Third Edition
Vincent J.M. DiMaio and Kimberley Molina

Bloodstain Pattern Analysis with an Introduction to Crime Scene Reconstruction, Fourth Edition
Tom Bevel, Ross Gardner, and Tom J. Griffin

Bloodstain Pattern Analysis with an Introduction to Crime Scene Reconstruction

Fourth Edition

Tom Bevel, Ross M. Gardner, and Tom J. Griffin

CRC Press
Taylor & Francis Group
Boca Raton London New York

CRC Press is an imprint of the
Taylor & Francis Group, an **informa** business

Fourth edition published 2025
by CRC Press
2385 NW Executive Center Drive, Suite 320, Boca Raton FL 33431

and by CRC Press
4 Park Square, Milton Park, Abingdon, Oxon, OX14 4RN

CRC Press is an imprint of Taylor & Francis Group, LLC

First edition published by CRC Press 1997
Third edition published by CRC Press 2008

Library of Congress Cataloging-in-Publication Data
Names: Bevel, Tom, author. | Gardner, Ross M., author. | Griffin, Tom J., author.
Title: Bloodstain pattern analysis / Tom Bevel, Ross M. Gardner and Tom J. Griffin.
Description: Fourth edition. | Boca Raton, FL : CRC Press, 2025. | Series: Practical aspects of criminal and forensic investigations series | Includes bibliographical references and index.
Identifiers: LCCN 2024041177 (print) | LCCN 2024041178 (ebook) | ISBN 9781032528502 (hardback) | ISBN 9781032528441 (paperback) | ISBN 9781003408734 (ebook)
Subjects: LCSH: Bloodstains. | Forensic hematology. | Crime scene searches.
Classification: LCC HV8077.5.B56 B48 2025 (print) | LCC HV8077.5.B56 (ebook) | DDC 363.25/62--dc23/eng/20241211
LC record available at https://lccn.loc.gov/2024041177
LC ebook record available at https://lccn.loc.gov/2024041178

ISBN: 9781032528502 (hbk)
ISBN: 9781032528441 (pbk)
ISBN: 9781003408734 (ebk)

DOI: 10.4324/9781003408734

Typeset in Minion
by Deanta Global Publishing Services, Chennai, India

Bloodstain Pattern Analysis with an Introduction to Crime Scene Reconstruction

Fourth Edition

Tom Bevel, Ross M. Gardner, and Tom J. Griffin

CRC Press
Taylor & Francis Group
Boca Raton London New York

CRC Press is an imprint of the
Taylor & Francis Group, an **informa** business

Fourth edition published 2025
by CRC Press
2385 NW Executive Center Drive, Suite 320, Boca Raton FL 33431

and by CRC Press
4 Park Square, Milton Park, Abingdon, Oxon, OX14 4RN

CRC Press is an imprint of Taylor & Francis Group, LLC

First edition published by CRC Press 1997
Third edition published by CRC Press 2008

Library of Congress Cataloging-in-Publication Data
Names: Bevel, Tom, author. | Gardner, Ross M., author. | Griffin, Tom J., author.
Title: Bloodstain pattern analysis / Tom Bevel, Ross M. Gardner and Tom J. Griffin.
Description: Fourth edition. | Boca Raton, FL : CRC Press, 2025. | Series: Practical aspects of criminal and forensic investigations series | Includes bibliographical references and index.
Identifiers: LCCN 2024041177 (print) | LCCN 2024041178 (ebook) | ISBN 9781032528502 (hardback) | ISBN 9781032528441 (paperback) | ISBN 9781003408734 (ebook)
Subjects: LCSH: Bloodstains. | Forensic hematology. | Crime scene searches.
Classification: LCC HV8077.5.B56 B48 2025 (print) | LCC HV8077.5.B56 (ebook) | DDC 363.25/62--dc23/eng/20241211
LC record available at https://lccn.loc.gov/2024041177
LC ebook record available at https://lccn.loc.gov/2024041178

ISBN: 9781032528502 (hbk)
ISBN: 9781032528441 (pbk)
ISBN: 9781003408734 (ebk)

DOI: 10.4324/9781003408734

Typeset in Minion
by Deanta Global Publishing Services, Chennai, India

Our love and thanks go out to our families for their support of this project. In particular, we wish to thank our wives, Elizabeth, Karen, and Nancy, for putting up with us during the long periods we spent behind the keyboard.

Editor's Note

This textbook is part of a series entitled "Practical Aspects of Criminal and Forensic Investigations." This series was created by Vernon J. Geberth, New York City Police Department Lieutenant Commander (Ret.), who is an author, educator, and consultant on homicide and forensic investigations.

This series has been designed to provide contemporary, comprehensive, and pragmatic information to the practitioner involved in criminal and forensic investigations by authors who are nationally recognized experts in their respective fields.

Contents

Foreword

Bloodstain Pattern Analysis with an Introduction to Crime Scene Reconstruction, Fourth Edition, in full-color by Ross M. Gardner, Tom Bevel – and newly added author Tom "Grif" Griffin – has been completely updated from the last edition of this seminal text published in 2008. It has been several years since any new Bloodstain Pattern Analysis (BPA) book has been released, and, as the recognized BPA standard, this new edition has been entirely revised and improved. It is concisely written to maintain its practicality and is more in line with current OSAC BPA standards. It is the most complete and comprehensive handbook to date from the perspective of the criminal investigator and forensic scientist on the subject of Bloodstain Pattern Analysis.

The three authors have been active in Bloodstain Pattern Analysis, crime scene reconstruction, and instruction and have over 100 years' combined practical experience as they regularly attend conferences and lecture on the topics covered in this edition. Each author is very qualified to write on this subject matter with their respective backgrounds in actual criminal investigations.

Tom Bevel, my friend and colleague for many years, is a retired police Captain from Oklahoma City, Oklahoma. He spent 27 years there working homicide and major cases. Tom continued his BPA work as the President and creator of Bevel, Gardner and Associates Inc., a forensic education and consulting company. He has an extensive experience in Bloodstain Pattern Analysis. Ross Gardner, whom I consider a personal and professional colleague, served for the United States Army Criminal Investigation Command (USACIDC) for over 24 years as a felony criminal investigator. Ross also served four years as a Chief of Police for a small suburban Atlanta Police Department. Ross also has an extensive experience in Bloodstain Pattern Analysis as he continued his BPA work. Ross first brought the idea of taxonomy to the attention of the SWGSTAIN GROUP in 2002. Tom "Grif" Griffin has an extensive experience in Bloodstain Pattern Analysis. Tom served 27 years with the Colorado Bureau of Investigation (CBI) as a criminal investigator and laboratory agent where he provided analysis and testified as an expert witness in Colorado district courts and/or federal courts in BPA, crime scene reconstruction (CSR), shooting incident reconstruction (SIR), and crime scene investigation (CSI). Please see "The Authors" section in the book for further information.

Most BPA experts agree that an updated edition is warranted based on the vast amount of research and knowledge that has amassed in this field, especially the updated terminology, since the last edition.

This new edition is based on a true taxonomy. The future of Bloodstain Pattern Analysis will be based on descriptive characteristics. Building on well-established classification ideas in Bloodstain Pattern Analysis, the authors have refined an objective classification system based on a taxonomic approach. *Bloodstain Pattern Analysis with an Introduction to Crime Scene Reconstruction,* Fourth Edition, is the basis for standardizing Bloodstain Pattern

Analysis. It establishes the need for universal rules defining this discipline by stressing the underlying scientific basis and how to apply this knowledge to cases in the field objectively.

Bloodstain Pattern Analysis with an Introduction to Crime Scene Reconstruction, Fourth Edition, provides specific details on crime scene analysis/reconstruction in explaining a proven methodology involved in the process. This methodology is built upon the scientific method and provides focus and structure to the analyst as they conduct the analysis.

The major additions are that the first few chapters dealing with *Bloodstain Pattern Analysis: Its Function and Historical Perspective, Bloodstain Pattern Terminology, and Bloodstain Classifications* in the Third Edition have been enhanced, taking the information and making those connections clearer.

This new revised edition connects the dots between the concepts of science and the BPA discipline. The authors illustrate and explain the foundational theory of blood as fluid and how it responds to the forces in a predictable way. The authors elucidate the three subtheories Pattern Diversity, Stain Shape and Vector Correlation, and the Physically Altered Bloodstain.

The authors address updates on virtual methods used to determine Area of Origin and explain the basis of the Physically Alter Bloodstain Principle and the limitations imposed by this principle. The authors also include an expanded discussion of the various issues relating to bloodstains on fabrics and how these issues impose limitations on the analysis.

Together with the revised chapters, the most significant improvement in this Fourth Edition is the inclusion of almost 400 photographs, 370 of which are in full-color. These visual aids graphically illustrate the dynamics of Bloodstain Pattern Analysis, enhancing the reader's understanding and retention of the material.

The authors have included two additional appendices. The first explains the application of the Bevel, Gardner Bloodstain Classification Decision Map, the current version is included, developed and revised through years of practical application. The second outlines the OSAC Bloodstain Classification Process Map and the various issues relating to its application.

In my seminal textbook *Practical Homicide Investigation: Tactics, Procedures, and Forensic Techniques, Fifth Edition,* I point out:

> Solving homicides, especially those without witnesses are extremely more difficult to solve because your main witness, the deceased, is dead. One must develop the ability to "absorb" the crime scene and be able to read the uncollectible nuances of the event.

The classification and analysis of bloodstain patterns within the crime scene oftentimes provide the investigator with the critical information to reconstruct the crime. Used properly, Bloodstain Pattern Analysis can help establish specific events associated with the crime.

I personally believe that without practical scene experience, there is a deficiency in crime scene reconstruction. Seasoned practice necessitates that the practitioner have that ability to "absorb" the crime scene, and be able to read the uncollectible nuances of the event. This is what we refer to as "scene experience" as opposed to a strict "laboratory" mentality. Tom Bevel, Ross Gardner, and Tom Griffin all have this "scene experience" as well as the necessary knowledge to evaluate and apply the scientific methodology to the reconstruction process.

Bloodstain Pattern Analysis with an Introduction to Crime Scene Reconstruction, Fourth Edition, is a masterful blend of "Practice and Theory" with practical crime scene knowledge and the application of scientific methodology to the process of crime scene reconstruction. The new edition follows a logical path throughout the text, highlighted with excellent color examples making it easy for the reader to navigate and comprehend the content.

Vernon J. Geberth, M.S., M.P.S.
Author of Practical Homicide Investigation
Series Editor, Practical Aspects of Criminal & Forensic Investigations

Preface

The goal of forensics is to seek the truth, and analysts should have no other agenda. In pursuing this end, analysts revisit what we hope is a not too distant past and attempt to understand better the events that unfolded. This task is anything but simple, and the tools employed are all from the forensic disciplines.

Each area of forensics provides insight and a glimpse back into this past. Each has its place in evaluating the aftermath of crime – the physical evidence. In the most classical sense, most forensic disciplines provide knowledge of the "who" of crime. Fingerprints, DNA, trace, and fiber evidence all allow us to associate people or objects with a crime scene. Forensic pathology, on the other hand, has always been a primary link to the "what" of crime, providing insight into some of the events that occurred during the incident.

Bloodstain Pattern Analysis (BPA) is a scientific discipline that significantly contributes to answering the question of "what" happened. When appropriate, BPA helps establish specific actions associated with violent crimes. In this capacity, BPA acts as a critical bridge between classical forensics and crime scene reconstruction.

Bloodstain Pattern Analysis is anything but a young discipline. It predates the development of even fingerprints as a forensic discipline. That history is a consistent history, what the pioneers of BPA believed, we still recognize as predictable aspects of the discipline. Yet we still see aggressive discussions between analysts over what they can or cannot opine from a specific stain. Often, these arguments consume our objectivity. One of the discipline's greatest concerns is returning to objective analysis, employing empirical data, rather than "my experience" as a foundation. Arguments based on "experience" alone lead us to a darker side of forensics, where subjective analysis reigns. To fight this tendency, the continuing goal of the discipline must be to understand its underlying scientific basis, and where and how the predictability present in bloodstains can be directed objectively in understanding crime scenes. In this edition, we have striven to provide a clearer picture of where that predictability is for BPA conclusions.

The authors of this book come to you from distinctly different backgrounds, and all have a high level of experience in "on-scene" crime scene evaluations. One is a retired law enforcement leader, one a retired Army CID Special Agent, and the other a retired forensic scientist. All are nationally and internationally respected in their fields. Different roads led them to the same destination. For two authors, those roads crossed outside the city of London, at the Metropolitan Police Detective Training School. Although several years apart, two of your authors attended the Scenes of Crimes Officer (SOCO) Course.

The British approach to scenes of crime is, at the very least, one of the most methodical in the world. The SOCO course teaches students to understand and incorporate all forensic evidence in evaluating crime. It places responsibility for understanding the interrelationship of that evidence on none other than a generalist, the crime scene investigator.

Perhaps the SOCO course serves as the wellspring of the authors' shared passion and belief in conducting crime scene evaluations using a holistic approach. Inherent in this thought is that case resolution critically depends on proper Crime Scene Reconstruction (CSR). However, case resolution is not just a matter of proving someone guilty. The competent investigator seeks to establish the truth, no matter what it may be. This demands a consideration of all evidence available, and the correlation of such evidence to identify reasons for contradictory results when they happen to occur. Both BPA and CSR as disciplines offer an avenue to this goal. Practiced objectively, BPA provides a clearer understanding of the aftermath of bloodshed associated with violent crimes. As a class characteristic evidence, BPA provides the means to corroborate or refute specific claims about bloodshed events. CSR (also referred to as Crime Scene Analysis) provides proven methodologies that allow objective snapshots of the crime to be established, and in many instances, sequenced. Although it does not define an absolute truth, this information is always effective for helping the court define its concept of what that "truth" may be.

In the criminal justice system, it is not uncommon to encounter a lawyer who adamantly believes, no matter what the nature of the testimony, that the analyst/investigator established in their own mind the innocence or guilt of a subject before completing the investigation. It appears incomprehensible to counsel that the investigator can take the oft-subjective information reported and conduct an objective investigation. Such a reaction should not surprise us because the idea of objectivity is relatively foreign to trial law. No matter the underlying truth, lawyers (both the prosecution and defense) highlight the information that best serves their position and attempt to diminish or ignore what works against them. Law professors have effectively described science as a kind of smorgasbord, where the lawyers step up to the table and take from science that which they want. Anything that looks unappetizing is ignored or dismissed, a classic form of confirmation bias. This is the actual mindset of those who claim criminal investigators are subjective!

The bloodstain pattern/crime scene analyst can ill afford to pursue their end with the same mindset. Choosing what evidence one will or will not consider in the analysis is heresy in science. Unfortunately, that trap is far too easy to fall into.

Within the scene is the evidence, which, if properly analyzed, provides everyone with the ability to define specific facts and infer others. Based on the totality of this information, it is possible to determine the most probable events surrounding the incident in question. Even if the overall incident is not defined, proper analysis allows for eliminating certain events, which still adds clarity to a situation.

No single forensic discipline has the potential to provide as much clarity regarding the occurrences at a crime scene as does BPA. However, that cannot lead to an expectation that the bloodstain evidence will stand alone. Yet in the right hands, BPA is a highly effective tool for defining the truth.

Bloodstain Pattern Analysis and crime scene analysis are not for the casual investigator who intends only to dip their toe in the water, find a quick answer, and move on. They are not one and the same as they operate on different theoretical bases and follow their own distinct methodologies.

But the bloodstain pattern analyst truly reconstructs crime scenes from the BPA aspect. As such, they must understand the nuances of their discipline and what other forensic disciplines bring to the table. The analyst must be able to objectively apply each category of evidence to the situation, inferring as little as possible, but recognizing the whole. The bloodstain pattern analyst must also understand and apply proven crime scene analysis methods to accomplish this task. The evidence establishes a knowledge base from which the analyst reaches a glimpse at the "truth."

In this edition, we have striven to provide a clearer picture of where that predictability exists; a melding of science, theory, qualitative observation, and experimentation as a basis for the BPA conclusions. We hope the student and practitioner find this book a good source document that can aid in teaching these disciplines or in maintaining or enhancing the practitioner's skills.

As we stated in previous editions, investigations and forensics are about defining truth as effectively and objectively as possible. In part, the oath of office for a US Army Criminal Investigation Special Agent states: "I shall at all time seek diligently to discover the truth, deterred neither by fear nor prejudice." We dedicate this newest edition to everyone, analysts and investigators alike, who recognizes and understands the importance of their roles as objective truth seekers.

Acknowledgments

In considering a project of this nature, the effort of the authors alone rarely ensures success. We would like to offer thanks and acknowledgment to the following individuals for their support and efforts:

Our editor, Vernon Geberth, for his continued support over the years and editions.

Philippe Esperança, for his significant contributions in formulating and refining the decision map explained. Philippe's background in entomology, as well as his understanding of Bloodstain Pattern Analysis, was instrumental in our efforts to build a functional decision map.

Rudi Jaehser of Frankfurt Germany, for his untiring efforts in leading Ross through the bowels of the Frankfurt University library those many late evenings.

Regina Dearborn, for her expert translations of many of the articles discussed in Chapter 2. Without her skill in German and her ability as a legal translator, many of these articles would have remained unknown.

Toby Wolson of Miami, Florida, for his support in providing excellent examples of the Roadmapping concept.

Sgt. Ron Wortham and Dr. Jon Loffi with the Oklahoma City Police Department, who designed and constructed equipment used in bloodstain pattern research, the results of which appear in this text.

Jouni Kiviranta of the Helsinki Police, whose outstanding photography skills produced many of the enclosed figures.

Richard Wintory, formerly with the Oklahoma County District Attorney's Office, for his support and working to get both Bloodstain Pattern Analysis and Crime Scene Reconstruction through vicious voir dire hearing and accepted in Oklahoma Courts.

Dr. Martin Fackler of Florida, for his support and critical eye. Dr. Fackler readily shared his knowledge and research with us.

Leonard Conn, formerly with the University of Oklahoma Police Department, who assisted in building equipment for both the stop-motion photography of blood impacting at various angles and the high-speed photographs of a bullet impact into a bloody target. Examples of both are included in this book.

Dr. Alfred Carter of Carleton University, Ottawa, for remaining patient and calm in the face of Ross' repetitive need for physics instruction, as well as for his permission for using images from the FCO software Backtrack Images, Tracks, and Backtrack/Win. Special thanks from the authors go out to Dr. Carter for no reason other than no other single scientist or member of academia has given as much to this discipline as he has.

Dr. Kenneth Beard of the University of Illinois at Urbana-Champaign, whose original research led to our five-year search to better understand droplet impacts.

Don Coffey, of the US Army Criminal Investigation Command, for sharing his approach and knowledge on dealing with pattern transfers.

David Baldwin, Ph.D., Former Director of the Midwest Forensic Resource Center, Iowa State University and Todd Zdorkowski, Assistant Director, for their support of the numerous high-speed digital film experiments.

Dr. Ronald O. Gilcher and Arlene Wilson with the Oklahoma Blood Institute, who over many years have provided much information about blood components and the processing involved in blood products provided for research.

The members of the Metropolitan Police Forensic Crime Laboratory, London, England, who allowed the use of still images from their outstanding video "Blood in Slow Motion."

To all the students from the bloodstain pattern schools conducted by Tom, Ross, and Grif. They have constantly demanded more and more knowledge that caused their instructors to continue to grow within this discipline. The same is true of all of the students of the Basic Crime Scene Reconstruction course. From every class, we have gained knowledge and learned new techniques.

Lt. Travis Witcher (Ret.) Oklahoma City Police Department that allowed Tom to attend training under MacDonell and Bunker. Lt. Witcher required at the end of this training that Tom start his own research and training to pass on the knowledge gained to OCPD officers. We can all benefit from the insight in his comment, "Teaching will force you to really learn the discipline." His wisdom is more than evident and serves as a challenge for all of us. Lt. Witcher next secured funding from outside sources to send Tom to study at Hendon Police College, London Medical College, and as a Lab Liaison for the Met and Forensic Laboratory in the UK.

William Turner, formerly of the USACIDC, who sparked a flame in Ross by sharing the knowledge he gained from a Corning Institute bloodstain pattern course. Bill probably has no idea of the events he set in motion, but we thank him anyway.

Richard Hopkins, formerly of the Arapahoe County Sheriff's Office, Colorado, for being a mentor and bloodstain analyst partner for Grif.

The many instructors of the Scenes of Crime Officer Course, Metropolitan Police Detective Training School, Hendon, England. Their insight and knowledge are responsible in a large part for forging Tom and Ross' shared passion.

Three units of the Oklahoma City Police Department and their respective members through the years: Technical Investigations, the Forensic Lab, and Homicide. All of them helped focus this discipline and develop Tom's "scene sense." Thank you for your never-ending support.

Martin Eversdijk of the Amsterdam Police Laboratory, for his excellent photographs.

Kim Duddy (Ret.) Washington State Patrol Crime Laboratory, for her review of Chapter 12.

Mary Schleicher, Unified Forensic Laboratory Englewood, Colorado, for her assistance resolving DNA questions.

The following agencies and individuals for supplying various photographs in support of the project: Henry Muse, E-Systems; Lt. William D. Gifford, Anchorage Police Department, Alaska; our late friend Donald R. Schuessler, Department of Public Safety, Eugene, Oregon; John Graham, Arvada Police Department, Colorado; Helena Komuleinen, curator of the Finnish National Gallery; Assistant

Chief Johnny Kuhlman, Oklahoma City Police Department; the Edmond Police Department, Oklahoma; Ray Clark, the Oklahoma City Police Department; Heikke Majamma, National Bureau of Investigation, Helsinki, Finland; *The Journal of Forensic Identification*; T. Daniel Gilliam, Larimer County Sheriff's Department, Ft. Collins, Colorado; David Stiles, Texarkana Police Department, Arkansas; Louis Laurel, State of Texas Attorney General's Office; Don Blake and Bill May, Norman Police Department, Oklahoma; Mike McGuffey, Covington Police Department, Kentucky; Claire Dawson-Brown, Travis County District Attorney's Office, Texas; Iris Dalley, OSBI, Oklahoma City, Oklahoma; Steve Chancellor, USACIDC; Sandra M. Roberts, Eugene Police Department, Oregon; Lt. Steve Kohne, Tippecanoe County Sheriff's Office, Indiana; Michael Taylor, Institute of Environmental Science and Research, New Zealand; Matt Noedel, Tacoma, Washington; Malcolm Fletcher, UK; Herbert MacDonell, Corning, New York; Jeff Svobda, Ft. Collins Police Department, Colorado; Gillian Leak, Principal Forensic Services, UK.

The International Association of Bloodstain Pattern Analysts, the Rocky Mountain Association of Bloodstain Pattern Analysts, and the Association for Crime Scene Reconstruction, for establishing an environment in which we can all grow.

There are many more individuals who may not be named here. All have contributed to this discipline in some fashion, which in turn assisted in the writing of this book. We salute them for their individual effort, insightfulness, and resulting research in support of Bloodstain Pattern Analysis and crime scene reconstruction.

Introduction

Bloodstain Pattern Analysis and crime scene reconstruction are two distinct forensic disciplines. Bloodstain Pattern Analysis seeks to define "what" caused the bloodstains in the scene. In pursuing this end, bloodstain pattern analysts cannot help but reconstruct the bloodstain aspects of a crime or incident. Crime scene reconstruction, also referred to as crime scene analysis, seeks to establish objective actions associated with the crime or incident in question and, when possible, to order those actions. Thus, crime scene reconstruction is also concerned with the "what" of crime.

Crime scene reconstruction demands that we evaluate all physical evidence to derive some conclusion as to what occurred. As Geberth reminds us for homicide investigations, case resolution hinges on "*careful and intelligent* [italics added] examination of the scene."[1] Much as a reporter does, the crime scene analyst seeks to define a crime's who, what, when, where, and how to understand what really happened. Interestingly enough, as we will discover, the "why" of crime lies outside the realm of the crime scene analyst. An unfortunate problem is associated with this process: there is no standard by which we can test our ultimate conclusions. An archeologist once made an analogy about this difficulty – discussing a dig and its findings, he said, "It's something like putting together a jigsaw puzzle without having access to the box top. You really don't know what the picture is supposed to look like."[2]

The crime scene analyst shares the archeologist's dilemma because the analyst's box top is unavailable. Despite this limitation, crime scene analysis attempts to define the nature of actions that are so dynamic that even if we had a videotape of the incident, they might not be fully understood. It should not be surprising that we share in the archeologist's dilemma, for, as we will explain, archeology and crime scene analysis share common purposes and methodologies. We will call upon some of the established principles of archeology to help define crime scene reconstruction methodologies.

The crime scene analyst's primary goal is to identify the actions that make up the crime or incident being evaluated and the order of those events. Thus, identifying specific actions associated with the crime, i.e., the "what" of a crime, is of significant concern.

In their most classic use, most forensic disciplines provide the investigator with information regarding the "who" of the crime. DNA results and fingerprint evidence help us decide who was present at the scene. In this concept of who, we also include forensic chemistry, biology, geology, and trace evidence examinations, as they too help associate items with our players and the scene. This ultimately serves the function of defining the "who" of the reconstruction.

Our answers to "what" actions occurred during the crime are sought quite often through the application of forensic pathology. As Dr. James Luke stated:

> From the standpoint of forensic pathology, the two major parameters that form the basis of any case investigation are (1) identification and documentation of the postmortem findings present, and (2) interpretation of those findings in the context of the circumstances of death.[3]

In the past few decades, the scientific discipline of Bloodstain Pattern Analysis has reawakened to its role in documenting these circumstances. Bloodstain Pattern Analysis allows the investigator to better define actions that could or could not have occurred during bloodshed. Once identified, these facts are considered in light of all other evidence to corroborate or refute statements, confessions, or investigative theories.

Bloodstain Pattern Analysis, in many ways, mirrors the role of forensic pathology. Once again, quoting Dr. Luke, "It is the responsibility of the forensic pathologist finally to construct a scaffolding of factual information against which witnesses' and suspects' statements can be evaluated."[4] Bloodstain evidence in this role (acting as a scaffolding) cannot stand apart from other evidence.[5] Reconstruction demands that we consider all evidence. Viewed from a holistic approach, all the evidence available should preferably lead those who view it to a similar conclusion.

This concept of a generally agreed-upon conclusion should not be a foreign thought, particularly when considering bloodstain patterns. They are, after all, graphically oriented. For example, in describing an unknown pattern as consistent with a known pattern type, any analyst should be able to point to some physical characteristic present in the unknown stain that exists in the known pattern type. Having done so, another analyst cannot simply ignore this information. If a stain is observable and reproducible, it is difficult for two analysts to rationally argue their beliefs about the nature of the stain from mutually exclusive positions. When this occurs, it is very likely the result of subjective analysis on the part of one or both. Unfortunately, subjective analysis in both bloodstain pattern and crime scene analysis is a fact of life. To help preclude subjectivity, the analyst should attempt to achieve several things.

First, the analyst must understand all areas of forensic science and have been directly exposed to crime scenes. Author Griffin often remarks that analysts need a "scene sense." This sense gives the analyst a more rounded perception, considering the many subtleties and interrelationships found at scenes. Evidence viewed from the confines of a white-walled laboratory is far too sterile. It leaves the viewer lacking a realistic perspective of crimes and crime scenes. It is this perspective that makes up "scene sense." As the SOCO course at the Hendon Detective Training School teaches, we gain much from viewing evidence in situ. This is true, not only from a case perspective, but also for the long-term professional development of the analyst.

One should refrain from construing the necessity of understanding these disciplines as meaning the analyst is an expert in all of them. Far from it, crime scene analysts are generalists. They have the knowledge, much like a manager, to combine the experience and expertise of the other team members. This process is at the heart of crime scene reconstruction.

Second, analysts must understand their discipline. Bloodstain Pattern Analysis, in particular, is far from being a static field. Our understanding changes due to research efforts. But as we base Bloodstain Pattern Analysis on applying physical laws to blood, there are specific rules we can apply. These rules are as true today as when bloodstains were first considered in forensics.

In this edition, we have expanded the discussion of where predictability exists in BPA. The foundational theory that blood as fluid responds to forces in a predictable manner is better understood by looking at the three sub-theories. These are Pattern Diversity, Stain Shape and Vector Correlation, and the Physically Altered Bloodstain. It is in these sub-theories that the "predictability" of BPA can be specifically defined. Understanding, applying, and articulating this knowledge provides the discipline with a major step forward in removing the claims of "subjectivity" leveled at Bloodstain Pattern Analysis. The methodology described in Bloodstain Pattern Analysis follows the scientific method, creating a consistent protocol every analyst can utilize. Once again, understanding, applying, and articulating this methodology should functionally aid every analyst employing it to help fend off the "subjectivity" attack in court. The practice of forensics demands we find objective ways to pursue our discipline; each aspect serves this function and attempts to pull Bloodstain Pattern Analysis up by its bootstraps into the twenty-first century.

As with Bloodstain Pattern Analysis, common themes throughout the history of crime scene reconstruction guide the analysts in their purpose. Understanding these universal themes, and then seeking those areas that require further study should be the lifetime goal of every analyst. Since 1997, we have taught and refined the methodology known as Event Analysis, which incorporates these basic themes.

The interrelationship of these two disciplines places an added burden on the practitioner. Because crime scene reconstruction so often relies on the information developed by the Bloodstain Pattern Analysis, it is imperative that the crime scene analyst have a solid understanding of bloodstain patterns. As bloodstain pattern analysts are engaged in creating a partial reconstruction, they must understand the basic tenets and methodologies of crime scene reconstruction to be effective.

Neither of these disciplines can be taken lightly. The analyst must approach them with a willingness to immerse themselves in the history and methodology of both. Neither discipline is for the faint of heart, both demand objectivity, dedication, and significant study. But when these disciplines are applied appropriately and acceptably, they provide functional and objective data that will aid juries and judges in seeking justice.

This text will explore the tenets of Bloodstain Pattern Analysis in depth – what it is, how it is used, and practical methodologies to employ to achieve defensible results. We will also explore the role of crime scene analysis and reconstruction and its relationship to Bloodstain Pattern Analysis.

We do not pretend to offer the "only" methodologies and techniques for pursuing Bloodstain Pattern Analysis and crime scene analysis. We do provide proven methodologies and techniques. Applied appropriately, this book's ideas and practical approaches will guide and focus any investigative effort.

References

1. Geberth, V.J., *Practical Homicide Investigation*, 2nd ed., CRC Press, Boca Raton, FL, 1996, p. 12.
2. Anonymous. Our apologies to the original author. We heard him some years ago on a public television program. We cannot identify him.
3. Ressler, R.K. et al., *Sexual Homicide: Patterns and Motives*, Lexington Books, Lexington, MA, 1988, p. 155.
4. Ressler, R.K. et al., *Sexual Homicide: Patterns and Motives*, Lexington Books, Lexington, MA, 1988, p. 162.
5. Bevel, T. and Gardner, R.M., *Bloodstain Pattern Analysis Theory and Practice – A Laboratory Manual*, TBI, Inc, Oklahoma City, OK, 1990, p. 1.

The Authors

Captain Tom Bevel (Ret.) is President and creator of Bevel, Gardner and Associates Inc., a forensic education and consulting company. He was an associate professor in the Masters of Forensic Science program at the University of Central Oklahoma, Edmond, OK. He retired after 27 years with the Oklahoma City Police Department. His last assignment was the Commander of the Homicide, Robbery, Missing Persons, and the Major Unsolved Case units.

Captain Bevel holds a Master of Arts degree in Administration of Criminal Justice from the University of Central Oklahoma. He is a graduate of the Scenes of Crime Course, Hendon, England; the Technical Investigations Course, Central US Police Institute, Oklahoma State University at Oklahoma City; the FBI National Academy; and the Post-Graduate Medical-Legal Course, London Medical College, London, England. He is a doctoral candidate in Forensic Science (DFS) at Oklahoma State University. He has been inducted into the Oklahoma Law Enforcement Museum and Hall of Fame with the Oklahoma Historical Society.

Captain Bevel was a charter member of the FBI Scientific Workgroup on Bloodstain Pattern Analysis (SWGSTAIN) and served on the Board of Directors; he is Fellow of the Association for Crime Scene Reconstruction (ACSR); Distinguished Member of the International Association of Bloodstain Pattern Analysts (IABPA); on the Editorial Board for the *Journal of Forensic Identification*; member of the Southwestern Association of Forensic Scientists (SWAFS); and Member of the American Academy of Forensic Sciences (AAFS); and he served on the Board of Advisers for the Master of Forensic Science program for the University of Central Oklahoma. He is coauthor of *Practical Crime Scene Analysis and Reconstruction.*

Captain Bevel has served as a crime scene consultant in 49 US states and 9 foreign countries. He has been qualified as an expert in crime scene reconstruction and Bloodstain Pattern Analysis in US state and federal courts and countries outside the United States.

Special Agent Ross M. Gardner (Ret.) is Vice President of Bevel, Gardner and Associates. He worked for the US Army Criminal Investigation Command (USACIDC) as a felony criminal investigator for nearly 20 years. He retired as a Command Sergeant Major and Special Agent in 1999 after serving a total of 24 years in the US Army law enforcement. Mr. Gardner subsequently served four years as the Chief of Police for Lake City, Georgia, a small suburban Atlanta police department. He is now retired and active in independent consulting.

Mr. Gardner holds a Master's degree in Computer and Information Systems Management from Webster University, a Bachelor's degree in Criminal Justice from Wayland Baptist University, and an Associate's degree in Police Science from Central Texas College. He graduated first in his class at the Scenes of Crime Officers Course, New Scotland Yard, Hendon, England, in 1985, and between 1988 and 1996 served as an adjunct professor of the police science program at Central Texas College. He is a former president of the Rocky Mountain Association of Bloodstain Pattern Analysts (RMABPA), as well as the Association for Crime Scene Reconstruction (ACSR), and has served as the chairman of the Education Committee for both the RMABPA and the International Association of Bloodstain Pattern Analysts (IABPA). Mr. Gardner was recognized both as Fellow and Distinguished Member of ACSR. He is a distinguished member of the IABPA. He was a charter member of the FBI Scientific Workgroup on Bloodstain Pattern Analysis (SWGSTAIN) and served as chairman of the Taxonomy and Terminology subcommittee for 11 years.

Mr. Gardner was certified by the International Association for Identification as a Senior Crime Scene Analyst. He is an active instructor and consultant throughout the United States in crime scene analysis, Bloodstain Pattern Analysis, and crime scene investigation, teaching to a variety of groups ranging from police, forensic laboratory, and investigative organizations to trial counsel professional development groups. He is also the coauthor of *Practical Crime Scene Processing and Investigation* and *Practical Crime Scene Analysis and Reconstruction* and editor of Michael Maloney's *Crime Scene Investigation Procedural Guide.*

Criminal Investigator Tom J. "Grif" Griffin (Ret.) is the owner of Griffin Forensic Solutions & Consulting, LLC, and a partner in Bevel, Gardner and Associates. Mr. Griffin holds a Bachelor's degree in Chemistry from the Colorado College and did graduate work in chemistry at the University of Northern Colorado where he designed and instructed a course in forensic chemistry. He worked as a criminalist and CSI at the Greeley, Colorado Police Department for four years before going to the Denver office of the Colorado Bureau of Investigation as a forensic chemist, crime scene investigator, and bloodstain pattern analyst. He retired from CBI after 27 years to become a private consultant and instructor.

Mr. Griffin was a charter member of the FBI Scientific Working Group on Bloodstain Pattern Analysis (SWGSTAIN) and was vice chair for the lifetime of SWGSTAIN. He is a past president of the International Association of Bloodstain Pattern Analysts (IABPA) and Distinguished Member; he is a co-founder and past president of the Rocky Mountain Association of Bloodstain Pattern Analysts (RMABPA); he served on both the Bloodstain Pattern Analyst Certification Board and the Crime Scene Certification Board of the International Association for Identification (IAI) and currently certified as Senior Crime Scene Analyst and Bloodstain Pattern Analyst; he also received IAI's Distinguished Member status and the IAI's Good of the Association Award; he is also a past president of the Association for Crime Scene Reconstruction (ACSR) and the Rocky Mountain Division of the IAI (RMDIAI). Mr. Griffin is a co-founder and past president of NecroSearch International, a private group that has assisted law enforcement in approximately 45 states and several foreign countries for over 35 years in the detection of clandestine graves and the recovery of human remains. Mr. Griffin is a member of additional forensic associations. He is a coauthor of the self-published *Mass Crime Scene Handbook* based on the experiences and lessons learned in preparing for and processing the Columbine High School shooting in 1999. In addition to consulting on case work, Mr. Griffin is an instructor in Bloodstain Pattern Analysis, crime scene reconstruction, and crime scene processing, fields in which he has testified as an expert witness as well as forensic chemistry.

Bloodstain Pattern Analysis
Its Function and a Historical Perspective

1

> And Cain quarreled with Abel his brother: and it came to pass, when they were in the field, that Cain rose up against Abel his brother, and slew him. And the Lord said unto Cain, Where is Abel, thy brother? And he said, I know not: Am I my brother's keeper?
>
> And the Lord said, What hast thou done? The voice of thy brother's blood crieth unto me from the ground. And now art thou cursed from the earth, which hath opened her mouth to receive thy brother's blood from thy hand. When thou tillest the ground, it shall not henceforth yield unto thee her strength; and a vagabond shalt thou be on the earth.[1]
>
> **Genesis 4:10**

Reading these passages, one might wonder was God really unaware of what transpired between Cain and Abel? Was He then convinced of the crime only by the presence of Abel's blood? This seems unlikely given the nature of the Judeo-Christian God. Rather, it would appear that God was reminding Cain that no matter what Cain's denial, the physical evidence of the deed spoke in as strong, if not a more convincing voice. In these few short paragraphs, we find what may be the first recorded use of Bloodstain Pattern Analysis in a judicial setting.

The Function of Bloodstain Pattern Analysis

What is the function of Bloodstain Pattern Analysis? Like any forensic discipline, Bloodstain Pattern Analysis seeks to define the facts surrounding some incident in question. The examination of the physical nature of bloodstains provides information that may be of significance in understanding the events that occurred during the incident.

We often refer to what the analyst evaluates as the "static aftermath" of an event.[2] Dispersion, shape characteristics, volume, pattern, the number and size of bloodstains, and their relationship to the surrounding scene are part of this aftermath. This information provides the investigator with a window into the past, helping define an objective history for a given incident. Clarity is not a guarantee, for it is possible the information present in the bloodstains will fail to illuminate any of the issues in question. Often, however, the analyst finds direct and convincing information that makes the role of the fact finder much easier.

Bloodstain Pattern Analysis is based on a very simple theory: blood as a fluid (a colloidal fluid, but a fluid nonetheless) will react to external forces in a predictable fashion. The cohesive forces of surface tension and viscosity, the various external forces (e.g., whatever the inertial force disrupting the blood) as well as gravity and air resistance all act together to produce similar results (patterns) under generally similar conditions. Thus, bloodstain patterns are reproducible phenomena. The physical characteristics of the bloodstain pattern help define the general nature of the event that created it. "General nature" implies

DOI: 10.4324/9781003408734-1

some basic force acting on the blood, not a crime scene–specific mechanism (e.g., a blow by a 2x4). Only after recognition of this basic force can an analyst consider the context of a unique scene and perhaps arrive at a conclusion as to the specific source of the bloodstain. To be clear, BPA conclusions demand objective recognition of patterns through specific physical characteristics – criteria that must be clear and evident. The basic force and event types that can be differentiated by these characteristics include the following:

- Blood dispersed from a point/area source by a force (e.g., impact patterns)
- Blood ejected over time from an object in motion (e.g., cast-off patterns)
- Blood ejected in volume in a streaming ejection (e.g., spurt and gush patterns)
- Blood dispersed as a function of gravity (e.g., drip, drip trails)
- Blood that accumulates and or flows on a surface (e.g., pools and flows)
- Blood that is deposited through contact transfer (e.g., smears and pattern transfers)

Additional information we are likely to discover through an examination of the bloodstain patterns includes the following:

- The direction in which a spatter stain was traveling when deposited.
- The angle at which a spatter stain impacted a surface.
- The area of origin (in a three-dimensional space) for impact patterns.
- The direction from which a force was applied or the direction of deposition.
- The nature of the object(s) involved in creating the pattern.
- In some instances, the approximate number of blows struck during an incident.
- The relative position(s) in the scene of the suspect, victim, or other related objects during the incident.
- Movement of individuals and/or objects during and after bloodshed.
- Sequencing of multiple events associated with an incident.

The end products of a Bloodstain Pattern Analysis are a series of defined actions that occurred during the incident; that is to say, the bloodstain patterns help us understand not only the nature of the actions that created them, but also where in the scene these actions occurred, any direction or motion associated with these actions, and in some instances what order the actions occurred. The ultimate conclusions offered by the analyst are intended to put this information into a context for the incident in question, associating them when possible to specific acts or actions (e.g. associating an impact pattern to a gunshot) by involved parties.

A Historical Perspective of Bloodstain Pattern Evidence

In seeking an understanding of Bloodstain Pattern Analysis, it is important to look back at its evolution. In his book, *Blood Evidence*, Craig Lewis wrote: "The science of bloodstain pattern analysis, a field in which the only textbook in existence was written by MacDonell, was little known."[3] This viewpoint, although completely inaccurate, is not uncommon. Out of ignorance, authors, judges, counsel, and investigators continue to proclaim the discipline as something "new" and without supporting research.

To those who spend time at scenes of violent crime, it seems unimaginable to ignore the presence of blood or not ask some simple questions as to its relationship to the crime.

As we will discover, Bloodstain Pattern Analysis has a nearly 170-year history in formal forensic science that predates many modern forensic disciplines. This history is a consistent history. Although study and research has increased our overall understanding of cause and effect in Bloodstain Pattern Analysis, the basic ideas that early pioneers claimed they could do, we can still accomplish. Therefore, it is helpful to understand the genesis of Bloodstain Pattern Analysis.

Right or wrong, the mere presence of blood, excluding any formal analysis, has long been held as evidence of foul play. For example, in the development of early Germanic law, tribal code (private, as compared to state) was often created. One of the surviving documents of this nature is the Sachsenspeigel. Compiled as a record of Saxon custom, a knight named Ritter Repkow wrote the document during the period 1220–1235. Common Law II, 63, Article 1-3 dealt in part with the raising of the "hue and cry" and the necessity of proving one's innocence for acting against a criminal caught in the act. He who had acted was required to prove an incontestability of the criminal's deed.[4] Examples of this proof, although not quite beyond a reasonable doubt, included "a criminal caught red handed."[5] The Sachsenspeigel, with its detailed illustrations, makes reference to the criminal having been caught with blood on his hand.

The work of William Shakespeare, although not the typical reference material of the criminalist, is filled with references to blood and the prejudicial nature by which its discovery is viewed. Shakespeare's various works, written between 1582 and 1616, like all authors, reflect the perceptions of his time. For example, in Act II, Scene II of Macbeth, after stabbing King Duncan, Macbeth says:

> "What hands are here! Ha! they pluck out mine eyes. Will all great Neptune's ocean wash this blood Clean from my hand?"

The Sachsenspeigel and Shakespeare's plays may not be of significant concern to the modern-day bloodstain pattern analyst, but they do reflect the consideration of bloodstains as a basic issue, preceding the formal development of forensic science. They also reflect the prejudicial manner in which bloodstains are often viewed. An interesting artistic example of the prejudicial nature of bloodstains hangs in the Central Art Archives of Finland. The work is entitled *Fratricide*, by Akseli Gallen-Kallela (circa 1897). It shows a young man with a bloodied sword and clothing as his mother points to numerous stains on his clothing (see Figure 1.1).

Another author whose writings reflect man's inquisitive nature with blood evidence is Sir Arthur Conan Doyle. In 1887, Doyle wrote *A Study in Scarlet*, introducing the world to his brilliant character Sherlock Holmes. In this story, the master detective concerns himself not only with discovering a reliable reagent test for blood, but also with the examination of "gouts and splashes of blood, which lay all around."[6] Again, we see artistic indications that those concerned with the early investigation of crime considered the relationship of bloodstains to such crime.

Early Scientific References

Having considered literary references, it seems appropriate to turn attention to the scientific evaluation of bloodstain patterns. Your authors intend to discuss some of the more critical and insightful research conducted over the last century; however, there are many

Figure 1.1 *Fratricide* by Akseli Gellen-Kallela (1897), a painting owned by the Anteneum, the Antell Collection, and the Central Art Archives of Finland, photographed by Hannu Aaltonen. Beyond the issue of his bloody sword, the pattern transfer and blood spots on the young man's garments speak to his mother as to his guilt. (Photo courtesy of the National Art Archives, Helsinki, Finland.)

references that may not be mentioned or discussed in detail. The analyst should recognize this and consider seeking these historical references for greater enlightenment.

Europe was the primary location in which Bloodstain Pattern Analysis found its initial footing in science. The primary periodical through which many of the continental authors expressed themselves was the *Viertaljahresschrift fur gerichtliche Medizin* (Quarter-Year Writings for Forensic Medicine), which was published in Germany. This was *the* scientific journal of the time. Although much of the concern between the period of 1850 and 1940 dealt with the identification of blood, scattered throughout these discussions are references to patterns observed at scenes.

Before proceeding, it is important to note that in German the term "blutspritzen" was widely used by various authors. This term is translated most often as "blood sprinkles."[7] There is, however, no absolute translation and the terms sprinkle, spatter, splash, or spurt are all in acceptable usage.[8] It becomes difficult, then, to always understand the specific context intended by any author. A caution is in order when considering the translations offered here.

In 1856, J.B. Lassaigne wrote *Neue Untersuchungen zur Erkennung von Blutflecken auf Eisen und Stahl* (New Examination to Differentiate Bloodspots from Iron and Steel). In the latter section of his paper, Lassaigne discussed marks that appeared similar to bloodstains, but were caused by insects.[9]

Although Lassaigne made it clear he believed "crushing" of dead flies created such stains, he implied having found such stains at scenes and associated their presence with the presence of flies. Based on the description, these stains seem similar to what we might refer to as insect or fly specks. Unfortunately, the translation and overall description of Lassaigne cannot satisfy the issue. Had Lassaigne found "fly specks"? Was he then establishing a causal connection without considering the regurgitation of blood by flies at bloodstained scenes? Whatever the case, Lassaigne's observations establish his attention to detail and his concern for differentiating such stains from other types of bloodstain patterns.

In 1863, John Beck and Theodric Beck wrote *Elements of Medical Jurisprudence*. The book, as reviewed by Herbert MacDonell, discussed various cases in which Bloodstain Pattern Analysis was utilized. Specific references were made to the "situation" of wounds and, as MacDonell noted, the authors also used the term "blood sprinkles." The latter term tends to indicate a German influence, although the book originated in Philadelphia.[10]

In 1880, Dr. Henry Faulds published *On Skin-Furrows of the Hand*, describing bloody fingerprints a common form of pattern transfer and their likely usage to identify the criminal.[11]

In 1882, Professor Charles M. Tidy of London published *Legal Medicine*, in which he stated:

> Few things hold so important a place as, or involve investigations of a greater nicety, than determining the precise nature of various [blood] spots or stains found on fabrics, instruments.[12]

Perhaps one of the most impressive treatises written on the subject of Bloodstain Pattern Analysis is *Ueber Entstehung, Form, Richtung und Ausbreitung der Blutspuren nach Hiebwunden des Kopfes* (Concerning Origin, Shape, Direction and Distribution of Bloodstains Following Blow Injuries to the Head), written by Eduard Piotrowski in 1895 at the University of Vienna.

Dr. Piotrowski's application of the scientific method to his observations and evaluations of bloodstains is unequaled in the known writings of the time. He reconstructed his scenes to model those in question, controlled and adjusted various variables to determine their specific effect on his experiments, and went so far as to use live rabbits. See Figure 1.2. Although the latter issue would not sit well with many modern groups, Dr. Piotrowski recognized the dynamic nature of what he was studying and felt the use of live subjects assisted his understanding.

Dr. Piotrowski considered this dynamic nature and used it as one of his variables. Calling the concept a "complicirtem morde," he used multiple methods of attack against his study subjects (rabbits), thereby ensuring to himself that he was considering all possibilities and their effects.[13]

In considering this complicirtem morde, Dr. Piotrowski properly recognized bodily reactions of his subjects when struck. In the following example, he observed small stains "similar to dots" which radiated out from the rabbit's head, and concluded:

> As far as their position is concerned, they spread out in a radiating pattern. The center of this outward radiating droplet [pattern] was the nostrils and mouth from which the accumulated blood was forced out … [expectorate blood].[14]

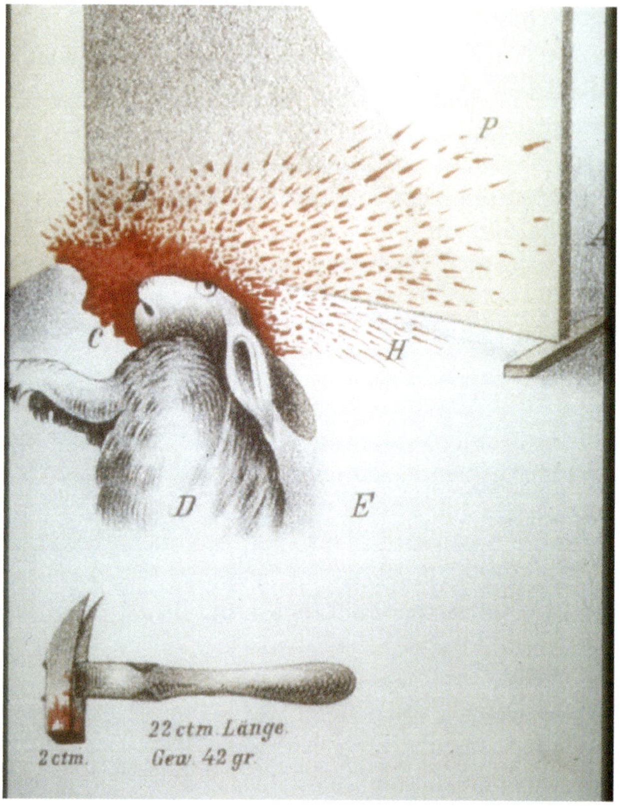

Figure 1.2 In addition to his written description, Piotrowski created detailed drawings of his various experiments and their results. (Photograph courtesy of Herbert MacDonell, Laboratory of Forensic Science, Corning, NY.)

Evident to Piotrowski was the correlation between the location of the stain's tail and the direction the droplet was traveling at impact. He also recognized the causative factors of cast-off stains. He included in his evaluation of the first, the effect of a parabolic arc on the resulting stain. In the latter, he isolated not only the fact that blood flung from a weapon would create a specific stain, he also correlated this with stain directionality. This gave him an ability to define the direction in which the weapon was being moved.[15] In this work's concluding statement, Dr. Piotrowski commented: "the formation, shape, and distribution of bloodstains follow specific rules and that these, allowing for many modifications considering the nature of the case, are not to be underestimated and are of great value in the judgment."[16]

In 1901, Jurgen Thorwald, writing about the efforts of Professor A. Florence, stated:

> [Florence] had worked out a whole system for classifying bloodstains caused by dripping, splashing, spurting, or grazing contact. Round stains, or roundish jagged stains, for example, indicate that blood fell vertically; oblong stains result from impact at various angles.[17]

Thorwald based his discussion on Florence's article "Les Taches de Sang au Laboratoire De Medicine Legale De Lyon." This article originally appeared in the *Archives De Anthropolgie Criminelle* in 1901.

In 1902, Dr. John Glaister discussed the role of bloodstains in his book *Medical Jurisprudence Toxicology and Public Health.* As other authors of his time, Glaister's primary concern was the issue of whether various stains were or were not human blood, but he also commented on patterns of blood and in particular the basic pattern types:

> As has already been pointed out, in every case in which a dead body with wounds upon it is examined *in situ*, examinations should be carefully made for the presence of blood-stains and their incidence upon the body and in its vicinity. The examiner must expect to meet every possible variety of stains, both in respect to character, incidence and magnitude: as (a) sprays, spurts, or jets; (b) smears of various forms; or (c) pools of blood.[18]

Dr. Glaister did not further refine his ideas of what these different patterns looked like, but it is clear investigators at the time were considering not only the presence of blood, but also the nature of the bloodstains found.

In 1904, Hans Gross wrote *Handbook fur Untersuchnungsrichter Als System Der Kriminalistik.* Gross is perhaps best remembered as the father of the modern criminal investigation, and particularly the concepts we now consider viable crime scene investigation. In his book, he provided a detailed discussion of not only the evaluation of bloodstains, but also their collection and documentation.[19] Gross felt bloodstains were of critical importance in the investigation and he devoted some 30 pages of the book to his considerations of blood and bloodstain patterns in the investigative process (see Figure 1.3).

Figure 1.3 One of many drawings from Gross' translated text. This drawing depicts variations in the shape of a stain as a function of how the drop impacts the surface (the impact angle), and the correlation of satellite stains, scallops, and tails produced, which provide information about the directionality of the stain.

This book was considered *the* reference of its time and in 1924 with Gross' permission, it was translated into English and republished by J. Collyer Adam as *Criminal Investigation.*

In 1912, Russian Professor R.A. Reiss discussed the use of bloodstain patterns in an article entitled "Scientific Techniques of Criminal Investigation." His primary considerations were shape, but also included a discussion of arterial patterns and concern over the difference between stains on an absorbent target compared to those on a nonabsorbent target.[20]

In 1914, Haberda wrote *Eine besondere Form von Blutspritzen* (A Special Form of Bloodstain), discussing a specific pattern observed in airway injuries. He described such stains as droplets of various shapes, which contained small air bubbles mixed with the blood. Beyond his consideration for this particular stain, Haberda offers many insightful lessons that are still applicable. Consider the following:

> The discovery of quantity, spread, form, and arrangement of bloodstains at a blood spattered crime scene can be of high importance. Evaluation requires many years of experience, usually learned little by little through practice, but never from books. Nevertheless, experts are often not careful enough when it comes to the necessary evaluation.
>
> Forced by precise questions of police, jurisdictional or governmental authorities, the experts sometimes answer too exclusively and draw the wrong conclusions about the bloodstains on a corpse or in the surrounding crime scene."[21]

We would agree wholeheartedly that oftentimes subjective analysts make conclusions that are far too exclusive.

Haberda also made reference to bloody fingerprints, clothing pattern impressions, and his particular stain: the foamy bloodstain. In describing the shape of stains that might be encountered, Haberda said:

> Even though the distance of the fall, or the angle with which the blood hits the ground influences the shape of bloodstains, which are for example more or less round, bear paw like, club or bottle like.[22]

How similar to modern descriptions are Haberda's? Round and bear paw are still adjectives used to describe stains, while elliptical replaced terms like club or bottle-like. There can be no doubt Haberda was observing and classifying the very things we still look for today.

Ernest Ziemke is another author of interest. His work is found in *Gerichtsarztliche und polizieartzliche Technik*, a book written by Theodor Lochte in 1914. Chapter 7, entitled "Die Untersuchnung von Blutspuren" (The Examination of Blood Tracks), was written by Dr. Ziemke. The work includes 14 pages of text with numerous pictures; it details various stains and the information represented by those stains. Ziemke dealt with a wide range of issues affecting Bloodstain Pattern Analysis.

In the chapter preamble, Dr. Ziemke states:

> Blood tracks [blood effects or evidence] are the most important tracks that stay behind after a crime is committed. Very often they alone have been of significant enough importance for the conviction of the suspect, and have been the focal point in a trial with only circumstantial evidence… Their evaluation should be efficient, because during an investigation it may be hard to foresee what might be of importance later during a court trial.[23]

In discussing the search for blood, both for serological and bloodstain pattern evidence, Dr. Ziemke offers that:

> The suspect, his clothes, all items he carries, or which are in his pockets should be carefully examined for bloodstains.[24]

This is sound advice which some of the best bloodstain pattern analysts in the country continually stress and teach today. Far too often such minute traces are simply overlooked by investigators.

In discussing pattern analysis and the conclusion drawn, Dr. Ziemke comments:

> Very important conclusions can be drawn from the arrangement, location, size, and form of bloodstains ... Based on our own experience and experiments we want to point out that it is necessary to be cautious and not draw conclusions from a single or very few blood tracks. This happens quite often ... Only when a large number of bloodstains are examined and compared, is it possible to exclude errors [in logic].[25]

Dr. Ziemke provided detailed descriptions of various types of bloodstains. His terminology followed that of Piotrowski, but he added terms such as "thorn apple shaped" to refer to the bear claw–shaped stain. He acknowledged that secondary droplets (satellite stains/wave cast-off stains) in these instances would assist in understanding the direction of motion of the blood droplet. He included figure examples of cast-off blood and drip patterns in which he described how to define directionality. Figure 1.4 is an example of one of these illustrations. In closing his discussion of bloodstain patterns, he said:

> When the examination of bloodstains is done efficiently and carefully, and if all possibilities are exhausted, shape, location, and the site of the bloodstain can give important details about the circumstances of the deed, eventually even be of importance for the conviction of the suspect.[26]

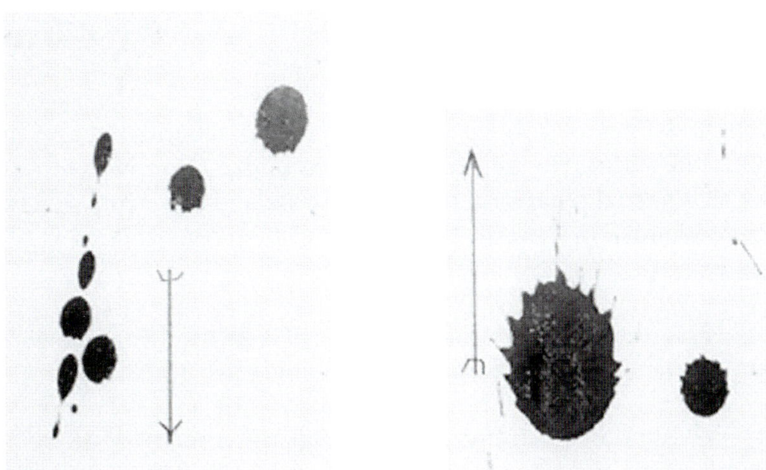

Figure 1.4 Illustrations from Dr. Ziemke's chapter on Bloodstain Pattern Analysis. The two figures (Ziemke's Figures 42 and 43) demonstrate a drip effect from walking and running. Ziemke added the arrows to show the directionality evident in the stains.

Dr. W.F. Hesselink considered various issues related to bloodstain analysis in his article "Blutspuren in der Kriminalistchen Praxis," written in 1931. For example, Hesselink considered the subsequent condition of the dried stains on clothing as a manner of differentiating whether such clothing had been worn or in use since the deposit of the stains. He also considered a method of dissolving the bloodstain to determine its relative age when compared to some other stain. Hesselink eventually concluded this technique was inaccurate, indicating that the manner in which each droplet dried and coagulated would impact on the results.

He also considered whether a suspect would or would not be stained in a violent murder, making the following observations:

> The answer depends upon two main circumstances: one, if many bloodstains are found at the scene, and two, what instrument was used. Regarding this I did several experiments. When using the hammer, the liquid squirts in all four directions, and will also cover the suspect. If a stick is used, the blood squirts left and right, and the suspect might not show any blood marks. When a forward bent file is used, the blood squirts only to the front. The foreground [of an illustration] shows many spatter while only a few went towards the subject. While the suspect of a bloody murder, performed with a hammer, usually will be bloodstained, the suspect, in case of death caused by hitting with a bent file might have few or no bloodstains on the clothes.[27]

This issue is one that analysts routinely face. Presence at the scene does not always create significant evidence in the form of bloodstains. It is only by considering the scene context, where were spatter-type bloodstains dispersed, that any conclusion is possible. When spatter is dispersed in a large radiating fashion (its dispersion cone), one in which the subject must have been present to accomplish the attack, some conclusion may be possible. In instances where there is a small more directed dispersion cone, nothing demands the attacker will be stained.

Hesselink went on to describe the use of a pattern transfer he discovered, stating:

> In one murder, I found a blood spot on the bedroom floor. First, it did not look too informative, but when I examined it later with a very strong light, I found remarkable prints of about 14 shoe nails.[28]

Hesselink sought to define not only the nature of the stains found, but also the manner of the event creating the stains. He then correlated the information with other forensic evidence, resulting in a partial reconstruction of his scene and the crime. His recognition of the importance of Bloodstain Pattern Analysis for scene reconstruction is reflected in his closing statement:

> Therefore, when examining blood, the blood investigation [to identify a substance as blood] itself is not as important as the clarification of surrounding circumstances.[29]

In 1931, Henry T.F. Rhodes, a criminologist in England, a member of the International Academy of Criminology, and an admirer of the great criminologist Edmund Locard, published his book *Some Persons Unknown*. The book was a compilation of case studies with a history of the scientific development of criminology. Although Rhodes did not discuss the use of bloodstain patterns in and of itself, he discussed in depth the development of the identification of blood in forensic investigation. He commented on the ever-increasing discrimination required by the court in terms of making identification of blood. Rhodes stated that this manifested itself in the change of accepted testimony from "it looks like blood," to "yes, it is blood," to "yes it is blood and it's human." At the time of publication, the four categories of blood, which ultimately became known as the ABH/ABO grouping system, had just been discovered. Rhodes recognized this advancement as a significant investigative tool with which to primarily exclude individuals as suspects.[30]

Rhodes followed his first book with a second entitled *Clues and Crime*. Published in 1933, the theme of the book was consideration of how science assisted in the detection of crime. Although Rhodes did not comment on his personal use of bloodstain patterns, he gave specific mention to the efforts of Hans Gross in the area of Bloodstain Pattern Analysis.[31]

In their writings, it is obvious that Hesselink and Rhodes clearly viewed blood in different perspectives. Rhodes understood what Gross was pursuing in the area of bloodstain patterns, but still felt the scientific individualization through the blood grouping systems of the time was of greater significance. Hesselink, on the other hand, felt the individualization and identification was secondary and that the patterns themselves were of greater significance. Today we recognize that both were correct. The individualization of bloodstains without consideration of the pattern itself or vice versa is tantamount to reading a book but skipping chapters.

In 1939, at the XXII Congress of Forensic Medicine, one of the most insightful papers relating to Bloodstain Pattern Analysis was presented by Dr. Victor Balthazard, R. Piedelievre, Henri Desolille, and L. Derobert. Entitled *Etude Des Goutes De Sang Projecte,* the proposed purpose of the research was to pinpoint characteristic elements of a bloodstain which might "give decisive hints" as to its origin.[32]

Balthazard et al. felt it necessary not only to examine the resulting spatter and stains, but also to understand the manner in which blood exits wounds, the trajectories of such blood, and the manner in which the blood droplet changed to become the resulting stain. As with Piotrowski, Balthazard recognized the importance of the dynamics involved in bloodshed, and realized bloodshed could not always be mimicked under laboratory conditions. To provide more realistic data, Balthazard utilized rabbits to produce actual bleeding injuries.[33]

Balthazard's consideration of the length-to-width ratio of a stain as a function of impact angle is one of the most important contributions of this work. His recognition of this empirical relationship between shape and impact angle is foundational to understanding the concept of Area of Origin. Analysts repeat his experiment in nearly every 40-hour Bloodstain Pattern Analysis course taught today. Professor Balthazard offered a caution in considering impact angle estimations, however, stating:

> In practical application, it [impact angle evaluations] should not be looked after for an illusionary accuracy. Nevertheless, this curve permits an estimate of the impact angle with an acceptable accuracy, sufficient for practical purposes.[34]

Balthazard furthered the cause and process for making Area of Origin estimations. His group's efforts set the stage for current "stringing" techniques. Yet in considering the issue of the unknown parabola of the droplet, Balthazard stated:

> Practically these methods can only be applied in a limited manner. [For instance] sometimes it is necessary to find out whether a victim, at the moment he was injured, stood on his feet or was lying.[35]

Thus, Balthazard recognized absolute points were unlikely to be identified from analyzing impact angles, but rather such effort would establish limits of possible and impossible. Finally, Balthazard's group considered the nature of the target on which a droplet fell. They found that many deformations would be possible, given various target characteristics. In summary, Balthazard stated:

> These modifications [deformations caused by the target] are sometimes obvious, but we must constantly remember that on an apparently homogeneous target, uneven areas can occur which may cause slight disfigurations.[36]

It is likely that world events in Europe prevented further effort in developing the research of Balthazard and his fellow researchers. Nevertheless, their efforts were not completely lost or without some impact. In 1941, John Gohringer, a US citizen studying for his medical degree at the Institute of Forensic Medicine, University of Heidelberg, wrote as his dissertation "Kann Aus Dem Bild Auf Verscheidenen Unterlagagen Die Fallohe Und Richtung Des Gefallenen Tropfens Ermittelt Wreden?" (Is it possible to establish the falling height and directionality of a dropped down drop of blood from the appearance of the stain on various targets?). Much of Dr. Gohringer's inspiration developed out of Balthazard's research. As an interesting historical note, Dr. Gohringer completed his studies in December of 1941 and was subsequently held as an internee by the German government for seven months before he was released to Vienna.[37] He later returned to the United States, where he left forensics and began practicing medicine. Dr. Gohringer had occasion in 1992 to speak before the International Association of Bloodstain Pattern Analysts (IABPA) training conference in Colorado Springs sharing some of his thoughts on the history of Bloodstain Pattern Analysis.

Dr. Paul Leland Kirk, of Berkeley, California, also added immeasurably to the knowledge of Bloodstain Pattern Analysis. Dr. Kirk, Professor of Criminalistics and Biochemistry, was active in assisting law enforcement organizations in the United States between 1935 and 1967. Kirk's book, *Crime Investigation*, published in 1953, included a chapter entitled "Blood: Physical Investigation." In that chapter, Kirk discussed the general application of Bloodstain Pattern Analysis to criminal investigations.

Another source for Kirk's beliefs on the subject is evident in his affidavit filed in the Court of Common Pleas, Criminal Branch, in the case of *State of Ohio v. Samuel H. Sheppard*. This document provides immense insight on his approach to bloodstain analysis.

In the Sheppard case, Dr. Kirk considered the drying times of blood and the evaluation of blood trails as evidence. He specifically sought to evaluate plausible causes for such trails. Further, he identified a void in the bedroom of Mrs. Sheppard that others missed, which established the most likely position of the attacker. He then correlated the cast-off patterns found in the room to the position of this void. Dr. Kirk clearly utilized a whole-scene approach.[38]

In 1955, Svensson, Wendel, Sodermann, and Stenlund published *Rikospaikka-Tutkimus* (Techniques of Criminal Investigation) in Finland. The text included 12 pages depicting various aspects of Bloodstain Pattern Analysis and describing its role in the criminal investigation. Although the primary direction of the text was to explain what kind of evidence to look for and how to collect it, they described basic types of patterns, discussed directionality, and recognized Bloodstain Pattern Analysis was an integral part of violent crime investigation. The book was later published in English by the American Elsevier Publishing Co. in 1965.

In 1960, Dr. Jozef Radzicki of Warsaw, Poland, published *Slady Krwi w Praktyce Sledczej* (Bloodstain Prints in the Practice of Technology). Although a full translation is not available, portions were translated for inclusion in the IABPA library. In this work, Dr. Radzicki established three basic groups of bloodstains, basing them upon the mechanisms of creation. These three groupings were:

1. Bloodstains resulting directly from extravasations. This grouping included drops, gushes, and pools of blood.
2. Bloodstains resulting from the application of various instruments. This grouping included spatter, cast-offs, and patterns resulting from direct contact.
3. Bloodstains resulting from the wiping or removal of blood.[39]

As with previous authors, Radziki cautioned that a critical consideration was the characteristics of the target surface, as bloodstains created under similar circumstances but falling to different targets might well result in slightly dissimilar stains. Several other interesting points made by Dr. Radziki included his consideration that "how" an instrument was employed would clearly affect the nature and distribution of the resulting spatter much as Hesselink described. Dr. Radziki also discussed arterial bloodstain patterns in depth.[40]

One of the most interesting arguments coming from modern-day legal antagonists of forensic science is that only research accomplished in this day and age has scientific value, and further that Bloodstain Pattern Analysis developed primarily within practical environments (e.g., the police and forensic services). Further is the claim that there was no academic development of the discipline and thus scientifically valid effort is lacking. Such a claim is irrational on its face. One doesn't throw out the foundations of individuals like Newton, simply because he did not accomplish his research under the full scrutiny of today's modern scientific mindsets. Many of the authors of the published research identified had significant scientific backgrounds. Evaluation of their work clearly establishes they had training in and used the scientific method in the development of their ideas about bloodstain patterns. Piotrowski, for example, was a medical doctor employed by the Institute for Forensic Medicine in Crakow. Victor Balthazard was a professor of forensic medicine at the Sorbonne University in Paris. The majority of the treatises described were either published in the peer-reviewed scientific journals of the day or were published in academic environments. There is no question that the early roots of Bloodstain Pattern Analysis are firmly planted in science and the scientific method.

Modern Works in Bloodstain Pattern Analysis

Having dealt with a "historical" view of Bloodstain Pattern Analysis, we arrive at a more recent history. Following Dr. Kirk's efforts through the 1960s was what might be considered the modern renaissance of the discipline at least in the United States. The number of authors writing on this subject increased dramatically, professional associations related to the field were established, and the discipline as a whole took on a far more accepted status in court.

Many proclaim Herbert MacDonell as the father of modern Bloodstain Pattern Analysis. Whether this is accurate depends upon your individual perspective. What cannot be denied is that Herbert MacDonell brought about a distinct reawakening of this discipline.

In 1970, after conducting extensive research, MacDonell and Lorraine Bialousz coauthored *Flight Characteristics and Stain Patterns of Human Blood*. As the two were working under a Law Enforcement Assistance Administration (LEAA) grant, the LEAA published the report. The LEAA document remained available for nearly 12 years before going out of print. In 1982, MacDonell revised the original work and self published it as *Bloodstain Pattern Interpretation* under his name alone. MacDonell subsequently completed a third work on the subject in 1993, entitled *Bloodstain Patterns*. In the latter, he further expounded on the original works ideas.

In 1983, Dr. Henry Lee, Peter Deforest, and Dr. R.E. Gaensslen wrote *Forensic Science: An Introduction to Criminalistics*. Included in this work is a 12-page section dedicated to explaining bloodstain patterns.

In 1983, the IABPA was formed. The association's stated purpose is to promote the general knowledge, techniques, and understanding of bloodstain pattern evidence.

In 1986, the *Journal of Forensic Science* published two papers by Peter Pizzola, Steven Roth, and Peter Deforest: entitled "Blood Droplet Dynamics I and II." The group sought to examine the dynamics of liquid droplets in flight and photographed droplet impacts, providing a more accurate understanding of the action that is often called "wave cast-off." In "Blood Droplet Dynamics II," Pizzola et al. clearly demonstrated that motion of the target could mimic characteristics of an impact at a greater angle than that which occurred. Their findings are important in our efforts to understand dynamic scenes of crime.[41]

In 1989, William Eckert and Stuart James published *Interpretation of Bloodstain Evidence at Crime Scenes*. Although the book received a very critical review in the *Journal of Forensic Science*, it was the first attempt in almost seven years to tackle Bloodstain Pattern Analysis as a single reference text.

The following year, the International Association for Identification accepted Bloodstain Pattern Analysis as a discipline, removing it as a subcategory of crime scene analysis. That same year, your authors (Bevel and Gardner) wrote *Bloodstain Pattern Analysis: Theory and Practice: A Laboratory Manual*.

In 1998, James and Eckert published their second edition of *Interpretation of Bloodstain Evidence at Crime Scenes* with CRC Press, significantly adding to its value. Prior omissions were corrected and the text remains a valuable reference.

James followed the second edition with *Scientific and Legal Applications of Bloodstain Patterns Interpretation*. James put together an impressive group of contributors, including Dr. Alfred Carter, William Fischer, Carol Henderson, Paul Erwin Kish, Maria Saccoccio, and T. Paulette Sutton. The text was directed toward trial counsel in an effort to help lawyers better understand how to apply Bloodstain Pattern Analysis in criminal trials.

In 2001, Anita Wonder published her book *Bloodstain Dynamics*. Although this work from an overall viewpoint is a significant resource, Wonder included issues that were to say the least controversial. This included the claim that she could differentiate the nature of spatter based on a single stain from any spatter pattern. Despite some of these more controversial issues, the book remains an excellent resource for the discipline.[42] Wonder followed this initial effort with a second book entitled *Bloodstain Pattern Evidence* in 2007.

Wonder's efforts in seeking an objective manner of classifying basic pattern types set the stage for your authors' general approach to a taxonomic classification system.

In 2002, the Federal Bureau of Investigation formed the Scientific Working Group on Bloodstain Pattern Analysis (SWGSTAIN). Calling on expertise from police, laboratory, and independent consultants worldwide, the purpose and focus of this group was to explore and define functional guidelines for the discipline. The group remained intact and active until 2012.

In 2005, Stuart James along with coauthors Paulette Sutton and Paul Kish published *Principles of Bloodstain Pattern Analysis: Theory and Practice* with CRC Press. This work was a far more comprehensive document than James' previous texts.

Following in SWGSTAIN's footsteps, the Organization of Scientific Area Committees (OSAC) for Forensic Science was created by the National Institute of Standards and Technology (NIST) in 2014. Its purpose was to help develop discipline-specific scientific standards for the various forensic disciplines. It has a wide-ranging number of Forensic Science Standard Boards that direct such effort, including those for physics/pattern interpretation. One of these is directed at Bloodstain Pattern Analysis.

Summary

Whatever we might say of Bloodstain Pattern Analysis, it has a rich and consistent history: one that indicates the consideration of bloodstains in solving crime predates modern forensics. As to the issue of being a "new" discipline, the examination and consideration of bloodstain patterns and their historical acceptance in forensics is well documented for anyone bothering to look. Bloodstain Pattern Analysis has a nearly 170-year documented history; and as important as the length of that history is, just as important is the fact that it is a consistent history. Like all sciences, early ideas have been scrutinized and reevaluated. Much of what the early pioneers thought possible remains valid today and ideas that were subsequently refuted were abandoned. Like all sciences, ideas about BPA have been refined and better explained over that 170-year period. And as with all valid sciences, when considered in light of modern-day scientific philosophies, BPA as a discipline fits neatly into such mindsets and when applied appropriately can meet all scientific rigors. In Chapter 2, we will discuss these ideas in detail and demonstrate this effectively.

Long recognized for its ability to support the evaluation of scenes of crime, Bloodstain Pattern Analysis serves the investigator and the court by illuminating "what did not happen," as much as "what happened." Keep in mind that Bloodstain Pattern Analysis is a class characteristic evidence, and as such it is most effective in refuting claims. It cannot tell us in all cases "who" or "exactly what," but as Piotrowski and Hesselink discussed, the ability to define the "situation" or "circumstances" of the crime is often just as important.

Working with an understanding of all areas of forensics and with experience in evaluating crime scenes, the investigator can often use Bloodstain Pattern Analysis to reconstruct the events surrounding a given incident. More and more, this process of reconstructing the incident is being tested in our judicial system; it is only through the application of quality objective analyses that the discipline can hope to serve its intended function.

As analysts, our purpose must always be to guard against subjective analysis. A failure to understand the theoretical basis, the failure to use criteria-based analysis, the failure to follow functional methods, standards, and controls all lead to "subjective" analysis. The single greatest detractor in BPA today lies in the training of those who are often allowed to testify in court on BPA issues. Courts across the United States, all the while screaming about a "subjective" discipline, allow people with limited training and those who have ignored appropriate methods, standards, and controls to testify as "experts." Despite the common argument of the legal profession that the discipline itself is subjective, the true problem is not the scientific basis of the discipline. The problem lies in subjective application, by incompetent or inadequately trained personnel. As described, the courts themselves (both counsel and judges) are primarily guilty for failing to prevent such subjective analysis.

As we wrote in 1990, each day, blood from scenes of a crime cries out to investigators. The use of proper Bloodstain Pattern Analysis simply enhances the criminalist's ability to be an active listener to this very vocal witness.

Notes

1. Anonymous, The Bible, Genesis, 4:10.
2. MacDonell, Herbert L., *Bloodstain Pattern Interpretation*, Laboratory of Forensic Science, Corning, NY, 1982, p. 4.
3. Lewis, Craig A., *Blood Evidence*, Berkeley Books, New York, 1992, p. 168.
4. Hinckeldey, Christopher, *Criminal Justice Through the Ages*, Druckerei Shulis, Heilbron, Germany, 1981, p. 55.
5. Hinckeldey, Christopher, *Criminal Justice Through the Ages*, Druckerei Shulis, Heilbron, Germany, 1981, p. 74.
6. Doyle, Arthur C., *The Original Illustrated Strand Sherlock Holmes*, Wordsworth Editions, Hertforshire, England, 1990, p. 21.
7. MacDonell, Herbert L., Segments of History: The Literature of Bloodstain Pattern Interpretation, *International Association of Bloodstain Pattern Analyst News*, Volume 8, Number 1, March 1992 Dearborn, Regina, Personal communications, January 1993.
8. Dearborn, Regina, Personal communications, January 1993.
9. Lassainge, J.B., Neue Untersuchnung zur Erkennung von Blutflecken auf Eisen und Stahl, Viereljahresschrift fur gerichtliche und offentliche Medicin, 1856, pp. 285–289.
10. MacDonell, Herbert L., Segments of History: The Literature of Bloodstain Pattern Interpretation, *International Association of Bloodstain Pattern Analyst News*, Volume 8, Number 1, March 1992.
11. MacDonell, Herbert L., Segments of History: The Literature of Bloodstain Pattern Interpretation, *International Association of Bloodstain Pattern Analyst News*, Volume 8, Number 1, March 1992.
12. MacDonell, Herbert L., Segments of History: The Literature of Bloodstain Pattern Interpretation, *International Association of Bloodstain Pattern Analyst News*, Volume 8, Number 1, March 1992.
13. Piotrowski, Eduard, *Ueber Entstehung, Form, Richtung u. Ausbreitung der Blutspuren nach Hiebwunden des Kopfes*, Golos Printing, Elmira, NY, 1992, p. 8 (German version).
14. Piotrowski, Eduard, *Ueber Entstehung, Form, Richtung u. Ausbreitung der Blutspuren nach Hiebwunden des Kopfes*, Golos Printing, Elmira, NY, 1992, p. 18 (German version).
15. Piotrowski, Eduard, *Ueber Entstehung, Form, Richtung u. Ausbreitung der Blutspuren nach Hiebwunden des Kopfes*, Golos Printing, Elmira, NY, 1992, p. 16 (German version).
16. Piotrowski, Eduard, *Ueber Entstehung, Form, Richtung u. Ausbreitung der Blutspuren nach Hiebwunden des Kopfes*, Golos Printing, Elmira, NY, 1992, p. 32 (English version).

17. As quoted by MacDonell, Herbert L., Segments of History: The Literature of Bloodstain Pattern Interpretation, *International Association of Bloodstain Pattern Analyst News*, Volume 8, Number 4, December 1992.
18. Glaister, John, *Medical Jurisprudence Toxicology and Public Health*, E.S. Livingstone, Edinburgh, UK, 1902, p. 234.
19. Gross, Hans Dr., *Handbuch fur Untersuchnungsrichter Als System Der Kriminalistik*, J. Swietzer Verlag, Munchen, 1904.
20. Reiss, R.A., Scientific Techniques of Criminal Investigation, Russian Ministry of Justice Magazine, St Petersburg, Mar 1912.
21. Haberda, H., Eine besondere Form von Blutspritzern, *Vierteljahresschrift fur gerichtliche Medizin*, Volume 47, Supplement IX, 1914, pp. 380–381.
22. Haberda, H., Eine besondere Form von Blutspritzern, *Vierteljahresschrift fur gerichtliche Medizin*, Volume 47, Supplement IX, 1914, p. 381.
23. Lochte, Theodor and Ziemke, Ernest, *Gerichtsarztliche und polizieartzliche Technik*, J.F. Bergmann, Wiesbaden, 1914, p. 152.
24. Lochte, Theodor and Ziemke, Ernest, *Gerichtsarztliche und polizieartzliche Technik*, J.F. Bergmann, Wiesbaden, 1914, p. 153.
25. Lochte, Theodor and Ziemke, Ernest, *Gerichtsarztliche und polizieartzliche Technik*, J.F. Bergmann, Wiesbaden, 1914, pp. 157–158.
26. Lochte, Theodor and Ziemke, Ernest, *Gerichtsarztliche und polizieartzliche Technik*, J.F. Bergmann, Wiesbaden, 1914, p. 164.
27. Hesselink, W.F., Blutspuren in der kriminalistischen Praxis, *Zeitschrift fur die angewandte Chemie*, Volume 44, Number 31, 1931, pp. 653–654.
28. Hesselink, W.F., Blutspuren in der kriminalistischen Praxis, *Zeitschrift fur die angewandte Chemie*, Volume 44, Number 31, 1931, p. 655.
29. Hesselink, W.F., Blutspuren in der kriminalistischen Praxis, *Zeitschrift fur die angewandte Chemie*, Volume 44, Number 31, 1931, p. 655.
30. Rhodes, Henry T.F., *Some Persons Unknown*, John Murray, London, 1931, pp. 193–208.
31. Rhodes, Henry T.F., *Clues and Crime*, John Murray, London, 1933, p. 58.
32. Balthazard, Victor, Piedelievre, R., Desoille, Henri, and Derobert, L., ƒtude des Gouttes De Sang Projecte, XXII Congres De Medicine Legale, Paris, France, June 1939, p. 1.
33. Balthazard, Victor, Piedelievre, R., Desoille, Henri, and Derobert, L., ƒtude des Gouttes De Sang Projecte, XXII Congres De Medicine Legale, Paris, France, June 1939, p. 6.
34. Balthazard, Victor, Piedelievre, R., Desoille, Henri, and Derobert, L., ƒtude des Gouttes De Sang Projecte, XXII Congres De Medicine Legale, Paris, France, June 1939, p. 8.
35. Balthazard, Victor, Piedelievre, R., Desoille, Henri, and Derobert, L., ƒtude des Gouttes De Sang Projecte, XXII Congres De Medicine Legale, Paris, France, June 1939, p. 25.
36. Balthazard, Victor, Piedelievre, R., Desoille, Henri, and Derobert, L., ƒtude des Gouttes De Sang Projecte, XXII Congres De Medicine Legale, Paris, France, June 1939, p. 27.
37. MacDonell, Herbert L., Euology for Dr. Gohringer, presented by H. MacDonell, to the International Association of Bloodstain Pattern Analysts Training Conference, Colorado Springs, CO, November 12, 1999.
38. Kirk, Paul L., Affadavit Regarding the State of Ohio vs. Samuel H. Sheppard, Court of Common Pleas, Criminal Branch, No. 64571, 26 April 1955.
39. Radziki, Jozef, SladyKrwi w Praktyce Sledczej, Wydawnictwo Zakladu Kryminalistyki, Komendy Glowenj MO, Warszawa, 1960.
40. Radziki, Jozef, SladyKrwi w Praktyce Sledczej, Wydawnictwo Zakladu Kryminalistyki, Komendy Glowenj MO, Warszawa, 1960.
41. Pizzola, P.A., Roth, S., and Deforest, P.R., Blood Droplet Dynamics Ñ II, *Journal of Forensic Science*, JFSCA, Volumber 31, Number 1, January 1986, p. 58.
42. Wonder, Anita, *Blood Dynamics*, Academic Press, San Diego, CA, 2001.

Bloodstain Pattern Analysis's Place as a Scientific Discipline

2

Science as we know it is under constant attack in this day and age. Anti-vaxers, GMO issues, climate-deniers, you name it and someone somewhere is standing up and arguing that their opinion trumps the current state of established science. Nowhere is this more evident than in forensic science, where the law profession, in particular, constantly challenges even the most basic established concepts. See Example 1. This antagonistic environment is very much the case for the discipline of Bloodstain Pattern Analysis (BPA). We repeatedly hear how BPA is untested, without foundational research, and very much subjective. Of course, many of these arguments ignore the difference between subjective analysis by a human being, which is certainly at issue, and the objectivity of the underlying science. Yet these opinions continue to gain ground.

Example: 1 While testifying in a trial in Federal Court in New York, author Gardner commented on a pattern on a normal smooth substrate that was clearly some form of spatter. In particular it seemed associated with some streaming ejection. The judge who had already shown a preference to the prosecution, interrupted testimony and in front of the jury stated: You can't say that. That's contact. He went on to criticize the author, detailing his obviously rich and significant knowledge of Bloodstain Pattern Analysis.

As we'll discuss recognizing those stains on a typical surface produced as a function of free flight through the air (spatter) and those produced through contact is one of the simplest and most basic concepts involved in BPA. Yet ignorant of this knowledge and believing his own opinion to be of greater value, the author was released by the judge who said: "You may be an expert, but you don't have anything to tell my jury." This of course, after he allowed a prosecution witness to testify about BPA. For those who testify routinely, such arrogance often appears inside the courtroom.

Now to be clear, should science be challenged? The obvious answer is yes. The scientific method itself is about refutation of beliefs, not so much about absolute conclusions. Good science is constantly moving forward, challenging itself, correcting itself, finding better more complete answers, and often self-correcting itself. So challenging scientific beliefs will always be appropriate, but the manner in which science is challenged becomes important. Mere opinion and beliefs based solely on anecdotal information rarely has a place in any true challenge of science. Even when such opinions are expressed and ultimately found to be correct, the only process by which they can truly be "found correct" relies on scientific investigation.

Forensic science dutifully accepts its role in the legal system. Analysts understand their conclusions will be challenged in the courtroom. But ignorance of basic concepts and the continuous representation of unsubstantiated ideas are not useful to the administration of true justice, nor, frankly speaking, are they particularly ethical. Yet it would be rare to encounter an analyst of any nature who has not been presented with ludicrous arguments by those in the legal profession. The nature of criminal defense in some ways makes

DOI: 10.4324/9781003408734-2

some of these arguments at least palatable. Presented with a guilty client and overwhelming evidence, a defense counsel has a duty to offer other scenarios or theories to the jury. Unfortunately, this tactic is no longer reserved for defense. Prosecution teams now engage in the same business. See Example 2. To the authors this is troubling at best, as a prosecutor represents the community and truth, whatever it may be. Thus truth should be at the heart of any arguments put forth. Sadly, this is the current environment that forensic science in general finds itself and certainly the one Bloodstain Pattern Analysis finds itself in.

Example: 2: In a trial associated to a husband who shot his wife through a back seat as they were travelling down the road, the author was cross examined by an excellent and capable prosecutor whom he had previously worked for. Although not truly an aspect of the prosecution's theory, present on the seat back was a circular pattern of gray material, dispersed and driven into the leather. Centered in the pattern was the bullet hole. As the weapon had been fired through a light plastic bag, the investigators had not sampled or even tested the pattern. The concern was cited that the testing would not define the range of distance the weapon was fired from. In one aspect that was correct, but the presence of the obvious stippling pattern in the seat leather certainly defined the minimum distance the weapon could have been fired from.

Testifying with regard to crime scene analysis issues and specifically to its presence in conjunction with a ballistic examiner's evaluation of the minimum distance, the facts were relatively straight-forward. Whatever occurred the weapon was not placed against the seat back when fired and was so testified to by the author.

Under cross examination the prosecutor followed with this line of questioning:

P: Are you aware that prior to arriving at the location where the shooting occurred, the three parties had been travelling on a dirt road?

A: No I was not.

P: Then how do you know that this pattern wasn't just dust that was simply clinging to the seat back.

The analyst was to say the least, a little taken back by the argument presented. It doesn't require any major scientific understanding to know dust from stippling; but more to the point, assuming dust was present (and none was present anywhere else in the vehicle), one would have to make an independent assumption that somehow, a circular area that just happened to be centered around a bullet hole had become electrostatically charged while the remaining seat back did not. Then one simply need contrast the force of the electrostatic charge holding the dust in place against the force of the 800+ fps muzzle blast inches away. "Hmmm, I wonder what will happen to the dust?" Not exactly rocket science.

With a straight face the prosecutor argued this wasn't really a stippling effect, but rather just some seriously sticky dust. Considering this argument was not particularly germane to the Prosecution theory, what purpose did it serve? Being frank, it is nothing more than the classic legal profession's position, that nothing the other side says is true. For no other reason, an otherwise capable and respected Prosecutor was willing to put forth this ludicrous argument to his jury.

It is not uncommon that unsubstantiated arguments are heard with specific regard to Bloodstain Pattern Analysis. Regardless of the environment that the legal profession chooses to operate, BPA must stand in defense of its basic tenets and constantly move forward to refine those tenets. As this chapter and other chapters will illustrate, the science of BPA is simple. The foundational theory is that blood as a fluid responds to forces in a predictable manner.[1] Three underlying principles or sub-theories define the range of this predictability, as well as setting the limitations of appropriate Bloodstain Pattern Analysis:

1. Pattern Diversity
2. Stain Shape and Vector Correlation
3. The Physically Altered Bloodstain[2]

This theory and these concepts are discussed in detail throughout this book. Suffice it to say that it is easy to claim one has a theory, or even state a theory of some sorts. If, however, one is to accept this theory and the associated tenets as valid, first and foremost one must ask and be confident that BPA meets the expectations of true science. That BPA is not some subjective form of "voodoo science" or a "I'll know it when I see it" circumstance. Simply put, the question becomes: Can BPA meet the rigors of scientific explanation? Answering that question demands consideration of the nature of science itself.

The Science behind the Science

A critical question for anyone interested in understanding science is first and foremost what is science?

Merriam-Webster offers the following definition: "a department of systematized knowledge as an object of study."[3]

Dictionary.com offers the following definition: "systematic knowledge of the physical or material world gained through observation and experimentation."[4]

The authors preferred definition of science is as follows: the observation, experimental investigation, and theoretical explanation of some phenomena with what is hopefully a practical application.

Definitions are always nice, but what is entailed in this explanation of phenomena? First and foremost, all science in general hopes to define the following:

- What a thing is.
- How it works.
- What predictive value does that knowledge represent.[5]

As Pogge explained, science gives humans a framework in which "to confront what we don't know." It demands certain rigors and it relies on verifiable data. That data may be quantitative, derived from experiment and measurement of a thing; or qualitative derived from observation of a thing.[6] This is true of any scientific effort, in any discipline. The idea of "systematic" however forces the rigorous mindset. Ghost hunters may well be pursing the aspects of observation, investigation, and explanation, but unless they pursue such effort following rigorous scientific mindsets, their effort is unlikely to be considered scientific.

There are clearly different philosophical approaches to how science should proceed. The standard approaches of Deductive/Nomological (DN) and Inductive-Statistical/Nomological (IS or IN) mindset are perhaps what immediately comes to mind. This approach is based on the idea that any scientific explanation is rooted in some covering law or laws (e.g., a specific scientific generalization). To be "lawful" assumed certain characteristics of the law; as Woodward stated: "they [laws] are exceptionalness generalization meeting various other familiar conditions – they must contain qualitative predicates [that] support certain counterfactuals."[7] Thus, under this typical approach, a certain scientific law or laws are applied to phenomena as a means of understanding how they work, or, as Salmon put it, an "exportability on the basis of lawful connections."[8]

An interesting aspect of these models is dealing with various explanations that *cannot* be precisely explained through specific lawfulness. In accepting such explanations, Hempel made the argument that in such situations, we may be confident in making some relevant uniform connection even though we may not be able to state them precisely, and thus we are willing to take them for granted without explicit mention.[9] The idea is that "laws" are involved, but the exact nature or articulation of said law(s) is unnecessary. Woodward referred to this concept as the "hidden structure." Various DN theses rely on the idea that certain complex systems are still systems whose behavior is governed by fundamental physical laws. In this sense, it is uncontroversial that there is some nomothetic structure "underlying" their behavior. Nonetheless, it is also uncontroversial that we cannot understand these systems by describing the characteristics of each micro-physical constituent and then writing down and solving equations governing all of their interactions.[10] In other words, even the DN mindset accepts that we can understand certain causal explanations and accept them as scientific even though we cannot define the exact laws that prescribe why the explanation is true.

This "hidden structure" idea is critical as we explore Bloodstain Pattern Analysis with other scientific philosophies in mind. As Woodward explains, these ideas regarding the hidden structure "suggests that there must be some marks or criteria of successful explanation… that are specifiable or recognizable independently of facts about fundamental physics and other 'underlying', but inaccessible theories."[11] Acceptance of Woodward's idea, specifically as applied to the special sciences (which Bloodstain Pattern Analysis is), functionally refutes the argument by antagonists who demand that only someone versed in fluid dynamics and physics can adequately explain bloodstain patterns or even suggest such a detailed theoretical explanation is possible.

The DN model is often referred to as the Hempel–Popper model. Karl Popper was very much a proponent of the DN mindset (e.g., empirical explanation based on laws); but Popper was also proponent of the idea of falsification in science. Falsification is another rigor within normal scientific approaches. With regard to falsification, Popper felt that rather than accepting some hypothesis, science should seek to falsify or refute it by every possible means. He believed that only when a theory or hypothesis had withstood such a rigorous critical examination should it be accepted. As discussed previously, science has the duty to constantly challenge any theory or conclusion when some valid new means of falsification is presented.

In typical Popper mindset, first a theory or hypothesis must have a means to be tested, and second only after critical evaluation that sought to refute the belief and the belief withstood such effort could the belief be considered good science. These ideas are integral to current thought of what is a good scientific method. When presented with a problem, one must put forth hypotheses (both that which one believes and an appropriate null hypothesis). Then, actively gather appropriate data and use that data to test the belief. Thus, Popper's ideas are rooted in the "methodical/systematic" examination of the belief; experimental effort directed at refutation, as well as a belief in some scientific generalization that formed the foundation of the belief.

An excellent example of this process was evident with regard to Einstein's Theory of General Relativity. After proposal, this theory was generally accepted. But Einstein's theory demanded a simple prediction – in the presence of a large mass, light will bend due to curvature of space. Sir Arthur Eddington's famous experiment in 1919 during a solar eclipse set out to test if this prediction was accurate. Although Eddington believed Einstein's theory to be correct, had his results been different, it would have demanded revisiting Einstein's theory itself.

So, let's return to the discussion on a scientific "law." Popper certainly believed in the DN and lawful mindset. But just how "exception-less" are they really? By current definition, a scientific law is a "statement based on repeated experimental observations that describes some aspect of the world. A scientific law always applies under the same conditions and implies that there is a causal relationship involving its elements."[12] This definition alone argues against the "exception-less" mindset of the classic Popper/Hempel DN thinking.

In 1990, James Woodward challenged the basis of the DN approach. Woodward argued that "explanatory generalizations in the special sciences do meet, or somehow serve as stand-ins, for generalizations that meet the standard condition of law-hood."[13] He argued for his concept of Invariance Under Intervention, which has since become known as "manipulationist" or "interventionists" account of explanation and causation. The basic concept of this mindset is that scientific generalizations based on the manipulation of certain variables that change other variables can be used as scientific explanation, even when such relationships hold over a certain limited domain. In other words, they are not scientific laws that apply at all times and all places, but do apply in some range of circumstances.[14] His basic belief, now generally accepted within the scientific community, was that based on the various examples that argue against exacting DN thinking, not all science and specifically the special sciences can operate based on strict "scientific laws." Generalizations that provide explanation of causal relationship, even in specific limited domains, are still scientific. As Woodward argued, these relationships are not laws but are relationships that are invariant under a given set of circumstances.

Woodward's ideas are significant in the consideration of Bloodstain Pattern Analysis, certainly a discipline that would be considered a "special science." As will be discussed, the associated generalizations (the predictable aspects) of BPA do not meet the strict DN definition of "laws," predictions that can be counted on in every single instance and circumstance. There are domains (ranges of circumstance involving known variable conditions) in which a given concept of BPA can be applied and other domains (outside of that predictable range) in which analysts recognize changes in the associated variables prevent utilization of these generalizations. As Woodward and others have effectively argued for the last 30 years, this does not invalidate their value in *that* specific domain. They are still valid scientific explanation. In forthcoming chapters, the authors will demonstrate specific examples of how this mindset applies to BPA.

Understanding Causality in Modern Scientific Mindset

As described, a basic tenet of science is in identifying for some thing/phenomenon "how does it work." This idea is the heart of science – causality; understanding what causes some ultimate effect and differentiating it from a mere correlation. Thus, causality has always been an area of great philosophical discussion with the scientific community. How do we know that some *Cause* (C) actually produces some *Effect (E)* and what definition or standard did we use to achieve that belief?

Over the years, various science philosophers have presented definitions of what defines "causality." Often these definitions become convoluted, if not circular. One astute aspect of Woodward's effort was in creating a particularly clear definition of causation, but his effort is more than a mere definition; it is effectively a standard of how to describe a causation relationship.

To believe that *C* causes *E*, Woodward set out his causation idea as follows. Let's assume we believe that some variable X acting on Y produces an effect Z. Let's also assume there are other associated variables Q, R, S. To be confident that this causation relationship is a true function of causation and not merely a correlation, the following must be true:

Keeping X invariant (unchanged), all the while changing the associated variables Q, R, and S. Z should remain invariant.
But when X is changed, while keeping the associated variables Q, R, and S remain stable (invariant), then a corresponding change should also occur in Z.

Let's put that in a simple example for Bloodstain Pattern Analysis:

X – is the angle of impact, where:
X1 is 10 degrees
X2 is 45 degrees
X3 is 90 degrees
Y – is the collapse of a stable droplet upon impact.
Z – is the resulting width to length ratio of the stain produced.
Q – is the substrate on which the collapse occurs.
R – is the mass of the stable droplet involved.
S – is the distance the droplet was projected before collapse.

Let's imagine we set X to X1 (keeping it invariant throughout each iteration of the experiment). In a series of experimental iterations, Q is altered, to include a variety of relatively smooth surfaces (e.g., tile, gypsum, glass). R is altered in a second series by changing the volume of the droplets produced. Finally, in a third series, S is altered, by moving the target surface varying distances from the source. Under those circumstances, the resulting width to length ratio (Z) will remain consistent.

As described by Woodward, that is only half of the required effort. In a new series of iterations, we must leave Q, R, and S invariant throughout, while changing X to X2. Under those circumstances, Z will change and in a consistent way. Such effort establishes a clear scientific causation relationship between the angle of impact and the resulting width/length ratio of the stain versus a mere correlation.

Recall however, Invariance Under Intervention accepts that not every causation relationship is exception-less. We recognize based on our understanding that alterations to some variables may lead to a breakdown of this causation relationship. In this instance, the substrate (variable Q) is the most obvious. Should we change Q to an absorbent or highly textured surface, this will affect the symmetrical collapse of the droplet. This in turn will affect Z, altering the width to length ratio, irrespective of X. Under such circumstances, Z is no longer "predictable" as a function of X.

Recognizing the range in which a causation relationship is valid is a significant factor in applying any discipline's foundational beliefs. But, the fact that these relationships are not exception-less does not invalidate their value under a "predictable" range of variables.

In each of the chapters associated with the BPA sub-theories, the authors will provide an example of how these concepts describe the causation relationships associated with bloodstain pattern predictability.

Primary, Secondary, Tertiary, or Quaternary Effects – Why Should We Care?

As previously discussed, there are certainly those antagonists of BPA who argue that lacking a complete understanding of the intricacies of fluid dynamics, or the ability to explain in exacting quantitative terms a given stain, analysts are not qualified to offer opinions. To be clear, the creation of bloodstain patterns is obviously attributable to "fluid dynamics." Primary physics effects right down to the quantum level define what is happening. But that's true of everything. Thus, *is* it also true that only a scientist with a clear understanding of the detailed aspects of quantum mechanics could or should apply BPA?

That argument fails on its face for one primary reason. The details of many variables associated with the creation of bloodstains in any given crime scene are effectively unknown! Only the primary variables are effectively definable and utilized in arriving at valid conclusions. The specific viscosity of the blood involved, the specifics of the exact manner (angle, velocity, etc.) in which a force is applied to disperse the blood, the exact temperature of the blood, substrate, and environment are unknown. Statements like "I'm a fluid dynamic scientist and you can't understand the details of these interactions" are hyperbole. First, it fails to recognize what information valid BPA is actually based on. Second, no one in their right mind can claim they can define these variables from a crime scene situation sufficiently to apply detailed physics equations in any attempt to provide a more quantitative conclusion.

BPA analysts are looking singularly at primary effects (obvious qualitative physical relationships) of the physics involved in the creation of a given bloodstain. Secondary, tertiary, and certainly deeper effects are beyond anyone's true comprehension when applied to a unique crime scene. A simple example of this issue is the relationship of the diameter of a given droplet falling to a surface. Dropping a given sized droplet from the same height, might on its face, presume a resulting droplet of a given size each and every time. But that is not the case and the most obvious explanation (although not fully understood) is the viscosity of the fluid. When the fluid is less viscous, as the droplet collapses against the surface, the fluid flows more easily during collapse. This results in a larger stain than that created when the viscosity is greater. Thus, we understand viscosity certainly appears to be a secondary affect altering an aspect of the resulting stain; yet no one can claim to know the specific viscosity of any given droplet in any given scene. Viscosity ranges widely from human to human and often over time can vary widely within the same human. There is no manner to establish that unique variable. This effect, however, in no way alters any valid predictable claim associated with BPA.

A knowledgeable bloodstain analyst is or should be aware of limitations imposed in the application of the predictable aspects of BPA. For example, target surface concerns are taught from the onset of an analyst's training. As such, analysts recognize the need to self-limit opinions when presented with asymmetrical, irregular-shaped stains or stains where characteristics are ambiguous or muted. They are aware of the fundamental aspects being evaluated.

In fact, as is often demonstrated when academic scientists begin discussing Bloodstain Pattern Analysis, they attempt to apply quantitative conclusions to the behavior of blood but fail to understand or demonstrate an understanding of the more obvious fundamental aspects. See Example 3. Thus, you see interesting discussions, but ones that cannot be applied in any practical sense. It is not that academic research into BPA may not have some subsequent value at some time in the future. But claims that fail to consider the fundamental ideas and limitations of the discipline have no practical value.

Example 3. At a conference in Korea, the author had occasion to discuss the work of a scientist employed by the military in developing fabrics that were fluid resistant. This gentleman's knowledge of fluid dynamics is without question. His research ultimately led to applying his knowledge to bloodstain patterns on fabrics. He had specifically concluded that impact angle, to include basic shape recognition (circle or ellipse) was not possible.

To be clear, BPA has long recognized that impact angle is not functionally determined on fabrics. At best, BPA recognizes that one can define if a droplet struck at an acute or perpendicular angle (based on the resulting shape). Even then only in some instances. Some fabric substrates are recognized to affect the collapse of the droplet, and thus the shape.

Thus, it was discussed between the two individuals, why shape recognition was not possible in *any* circumstance. A series of photos were presented in which, his research had shown that absorbency and thread pattern clearly affected the resulting shape. But the examples (droplets falling at 90 degrees, but resulting in elliptical shapes) demonstrated obvious abnormal ellipses, completely atypical to those expected of a "well formed" stain. When questioned as to the nature of the experiment that produced the results presented, the scientist explained his procedure. Droplets were simply dropped onto the targets from a distance of 3 feet.

Given the procedure used, the specific results shown were anything but unexpected. The experiment failed to achieve a force for collapse close to terminal velocity, let alone the more typical forces associated with impact spatter. The collapse of the droplet (the flow of blood across the surface) is significant in defining the ultimate shape. If there is insufficient energy to cause that flow and collapse, then flow will obviously not occur. Thus absorbency and thread orientation become the more critical factors in defining the resulting shape. The conclusion he offered might well be valid for similar circumstances as provided in his experimental procedure, but one atypical of those encountered by an analyst. In BPA Fabric courses, the correlation between impact angle and shape in impact spatter is consistently demonstrated across a wide range of fabrics. As was already understood, there are fabrics in which this is not true. But analysts understand when presented with a pattern that fails to follow typical impact shapes and pattern orientations, tests of the fabric are necessary before offering any conclusion. Once again, the predictable aspects of BPA are not exception-less. But the range in which they are invariant, are still valid.

Class and Individual Characteristic Evidence

A basic idea within forensic science is the difference between individual and class characteristic evidence. Both types offer valid information, but their application must be understood and applied correctly. Thus, it is important to understand that BPA is clearly a class type of evidence.

By definition, class characteristic evidence has traits that allow the item/evidence to be compared to some group of items/evidence. Based on the presence or absence of various traits, the item can then be included or excluded from that group. Trace evidence examinations (e.g., fibers, soils, glass) are a typical form of class characteristic evidence.

Class type evidence is distinct from individual characteristic evidence; the latter allows the item/evidence to be compared to a specific object/individual. Individual characteristic evidence examination may result in "identification" and/or a "match" to that which it is compared against. Obvious forms of individual characteristic evidence are DNA and fingerprints.[15]

The value of class characteristic evidence is primarily one of refutation. Once an item of evidence is excluded from a given group, any subsequent claim that the item's explanation is related to that group is scientifically excluded. Of course, inclusion is helpful as well, as this indicates a potential explanation for the item. The Pattern Diversity principle (the classification of bloodstains) allows a scene stain to be compared with the basic forms of known bloodstains. When valid criteria-based exclusions occur (e.g., this stain does not have characteristic a or b), this effectively refutes any offered investigative theory that demands inclusion. For example, a single spatter stain cannot be explained as occurring from some form of incidental contact, as in Example 1.

Inclusion of a given stain to a given class/group of stains demands identifying physical stain characteristics that are consistent with that group. But in practical work, this may present a situation in which multiple possible distinct mechanisms associated with a scene could have produced the stain/pattern involved. Using the example of the single spatter stain above, based on the unique scene context, there may be various spatter mechanisms that have produced the stain. Nothing about patterns, in and of themselves, demand a specific scene unique mechanism. They are not individual in nature and the classes that can be identified may not limit the stain's source sufficiently for the court. Once classification occurs, a stain/pattern is considered against the unique scene context in which it is found. Then and only then can some further defined opinion become possible regarding the specific nature of what caused the stain. As the Pattern Diversity chapter will discuss, a pattern – any pattern – simply defines the basic nature of the force(s) involved in its creation, not the absolute crime scene mechanism.

Any failure to understand and apply this concept of the class nature of classification of bloodstains effectively creates issues (e.g., claiming a specificity that is not valid.) Ultimate specific opinions as to the source of a stain are possible, but they are based first on the general classification and then consideration of that classification in the unique scene context presented.

The Foundational Theory and Sub-theories of Bloodstain Pattern Analysis

As discussed, scientific explanation demands a foundational theory. Although historically not articulated effectively by prior authors, in 2009, two of your authors described these beliefs in a more succinct fashion. The basic foundational theory of BPA is as follows:

Blood as a fluid responds to forces in a predictable manner.

This predictability lies in the three sub-theories or principles applied in BPA. The first is Pattern Diversity, which can be articulated as follows:

> *Pattern Diversity – The variations in combinations of blood volumes and forces acting on those volumes lead to recognizable classes of patterns.*[16]

When presented with patterns produced by unknown scene mechanisms, analysts are often able to recognize (classify) a scene pattern and therefore relate it to a basic event mechanism. As a comparative science, BPA utilizes physical characteristics of known patterns, comparing those present in the unknown scene pattern. These characteristics vary but generally include size, shape, volume, appearance, and orientation of the stains in the pattern. For inclusion to a group/class of patterns, sufficient articulable characteristics must be present. Lacking said characteristics, the pattern is generally excluded as belonging to a given pattern group. There are always exceptions, as when presented with a highly ambiguous scene stain where no conclusion is possible beyond "it's blood," situations where the class characteristics are shared by several pattern types (e.g., linear spatter) or when presented with a complex pattern as described in the Taxonomy chapter.

In considering Pattern Diversity, it is imperative to differentiate a "classification" with a "conclusion." Classification of a pattern is based solely on the pattern – what physical characteristics are present? Done properly, classification is an objective statement. It is based on the primary physics and resulting qualitative observations resulting from the interactions between the dispersing force and the blood mass. Classification is the starting point of any BPA effort. It is difficult to imagine a situation in which an analyst offers some BPA opinion, without first considering classification. A conclusion, on the other hand, is a statement/opinion describing the ultimate scene unique source of the pattern. How did it get there? Any conclusion offered, starts with a valid classification and then must consider the context in which that unique scene pattern was found.

Conclusions are obviously what the court desires. Can one explain to the jury how this pattern came to be in a given crime scene? That question is answered by applying the full methodology of BPA. At times, a scientifically valid conclusion may be quite specific, assuming a specific scene context. (see example below). At other times, that answer may be quite ambiguous. But everything begins with proper classification, correlating a scene pattern to a particular pattern group.

> **Scene Context Example:** Presented with stain(s) that have characteristics consistent with spatter, there may be multiple potential source explanations within the scene. This might include impact spatter and an expiration pattern/spatter. Given a consideration of expiration, a context is necessary for verification. The autopsy report should find blood present in the lungs, trachea, mouth, nose, etc. If there is a lack of blood in these areas, then spatter produced by expiration can be eliminated. Were blood to be found at autopsy in these areas, then expiration has to be included as possible.

Classification is merely a statement of inclusion or exclusion to the basic bloodstain pattern types. It is a class characteristic statement and nothing more. Classification is not a statement about specific crime scene mechanisms. It defines a basic creation force/mechanism, which are described in detail in a subsequent chapter.

The second sub-theory/principle in BPA is Stain Shape and Vector Correlation, which is articulated as follows:

> *Stain Shape and Vector Correlation – The shape of certain bloodstains provide indicators as to the direction of deposition as well as to the spatial origin of the blood.*[17]

There are two critical principles supporting Stain Shape and Vector Correlation. These include the concept of Impact Angle and Directional Angle; both are discussed in detail in later chapters. Generally, when a relatively stable droplet strikes a surface, the collapse creates a circular or elliptical bloodstain. Spatter stains that are well-formed (e.g., symmetrical) have an identifiable major and a minor axis. There is an empirical relationship between the ratio of the length and width of a stain and the angle at which the droplet struck that surface. This is the stain's impact angle. The long axis of such well-formed stains in combination with the tapered tails and scalloping produced during the collapse effectively defines the direction the droplet was moving at the moment of collapse. This is the stain's directional angle.

Impact and directional angle are significant aspects of Stain Shape and Vector Correlation, but they are not the end-all. Other types of stains also provide additional indicators as to direction of deposition and issues of motion. Examples include wipes, flows, and drip trails. Understanding all of these assist BPA analysts in understanding where some source event was occurring.

The final sub-theory/principle in BPA is known as the Physically Altered Bloodstain (PAB or PABS), and is articulated as follows:

> *Physically Altered Bloodstains – Exposed blood will react to environmental conditions (e.g., air flow, temperature, humidity, variations of surface or disruption) in a relatively predictable manner.*[18]

Once exposed to environmental conditions, blood dries, coagulates, or wicks into permeable surfaces. Certain levels of predictability exist within this response to the environmental conditions. These effects may allow the bloodstain analyst the ability to define minimum periods of time since the blood was first exposed or perhaps estimate the volume of blood involved. The predictability associated with the PAB principle is described in detail in later chapters. Disruptions of a preexisting stain(s) are effectively an alteration, but in this instance there is a specific classification (e.g., wipe) for such occurrences.

It is important to recognize that the PAB principle is perhaps the most significant Achilles' Heel in BPA. Specific scene environmental conditions are the primary variables that define the nature of this predictability. Knowledge of these variables and case-based experimentation are routinely required to offer any opinion. Additionally, to date, specific research in coagulation has been limited. But, as with any science, as additional effort is directed, PAB issues are constantly being reevaluated.

These are the theoretical foundations of the science of Bloodstain Pattern Analysis. They are well-defined, having been discussed in scientific journals for over 170 years. That history as described is consistent. Much of what early pioneers in BPA observed and believed has withstood the test of time and scientific scrutiny. As a science, BPA continues to explore and reevaluate these ideas, but when applied appropriately, they remain scientifically valid.

Conclusion

Bloodstain Pattern Analysis is a defined special science, operating clearly with in the philosophical concepts of Invariance Under Intervention as described by Woodward. BPA has a clear foundational theory, further defined by the three sub-theories, which define specifically where BPA predictability lies. It is understood that where such predictability exists, it does so within a given range of circumstance/variables. Applied appropriately, analysts frame opinions within those limitations.

That predictability is based primarily in qualitative (observable/describable) cause and effect relationships. The source of these qualitative relationships is based in the primary physical interactions of forces acting on blood masses. Secondary, tertiary, even quantum effects obviously are in play in these interactions. These concerns are validly beyond practical application to BPA, due to the various unknown variables involved in crime scenes. The accepted ideas of "hidden structure" associated with the philosophical approaches to science, specifically the deductive/nomological mindset, establish that ignoring these unknowable aspects is not *inappropriate*. Granted, if continued research ultimately yields some viable theoretical knowledge that can be applied, BPA as a discipline will adjust and as with all science reevaluate its tenets.

Notes

1. Gardner, R. and Griffin, T., Foundations of the Discipline of Bloodstain Pattern Analysis, A Response to the Report of the National Academy of Sciences, *Journal of Forensic Identification*, Volume 60, Number 4, 2010, p. 478.
2. Gardner, R. and Griffin, T., Foundations of the Discipline of Bloodstain Pattern Analysis, A Response to the Report of the National Academy of Sciences, *Journal of Forensic Identification*, Volume 60, Number 4, 2010, p. 479.
3. Merriam Webster On-Line Dictionary, https://www.merriam-webster.com/dictionary/science, dtd. 1 Feb 2019.
4. Dictionary.com, https://www.dictionary.com/browse/science, dtd. 1 Feb 2019.
5. Pogge, Richard, Astronomy 141, Lesson 1, Ohio State University, 23 Sept 2009.
6. Pogge, Richard, Astronomy 141, Lesson 1, Ohio State University, 23 Sept 2009.
7. Woodward, J., Explanation and Invariance in the Special Sciences, *British Journal for the Philosophy of Science,* 2000, pp. 197–198.
8. Woodward, J., *Making Things Happen*, Oxford University Press, Oxford, NY, 2003, p. 606.
9. Hempel, C., Aspects of Scientific Explanation and Other Essays in the Philosophy of Science, New York Free Press, New York, NY, 1965, pp. 360–361.
10. Woodward, J., *Making Things Happen*, Oxford University Press, Oxford, NY, 2003, p. 641.
11. Woodward, J., *Making Things Happen*, Oxford University Press, Oxford, NY, 2003, p. 643.
12. Definitions.net, https://www.definitions.net/definition/scientific%20law, 1 Feb 2019.
13. Woodward, James, Explanation and Invariances in the Special Sciences, *British Journal for the Philosophy of Science,* #51, 2000, p. 198.
14. Woodward, James, Explanation and Invariances in the Special Sciences, *British Journal for the Philosophy of Science,* #51, 2000, p. 198.
15. Gardner, R.M. and Krouskup, D., *Practical Crime Scene Processing and Investigation*, 3rd Edition, CRC Press, Boca Raton, FL, 2019, p. 20.
16. Gardner, R. and Griffin, T., Foundations of the Discipline of Bloodstain Pattern Analysis, A Response to the Report of the National Academy of Sciences, *Journal of Forensic Identification,* Volume 60, Number 4, 2010, p. 480.

17. Gardner, R. and Griffin, T., Foundations of the Discipline of Bloodstain Pattern Analysis, A Response to the Report of the National Academy of Sciences, *Journal of Forensic Identification,* Volume 60, Number 4, 2010, p. 484.
18. Gardner, R. and Griffin, T., Foundations of the Discipline of Bloodstain Pattern Analysis, A Response to the Report of the National Academy of Sciences, *Journal of Forensic Identification,* Volume 60, Number 4, 2010, p. 486.

Bloodstain Pattern Terminology

3

Terminology in Bloodstain Pattern Analysis has and remains a significant issue. The terms associated with BPA have developed over the entire history of the discipline and there are multiple terms applied to the same concepts. To highlight just how distinct this problem is, during the period when the Scientific Working Group on Bloodstain Pattern Analysis (SWGSTAIN) was active, Carolyn Gannett independently produced a consolidated document detailing the varying terms in use in BPA. That list included over 500 different words. In this chapter we will attempt to provide an overview of current terminology; this includes terms associated with BPA, specific pattern names, and some important characteristics evaluated in BPA. Where current AAFS Standards Board (ASB) adopted terminology falls short, we will continue to include those terms we feel are necessary to the conversation.

There has been and remains a constant theme in recent years to adopt a single set of terms across the entire discipline of Bloodstain Pattern Analysis. The thought being that all analysts must use these terms if the discipline is to be considered "scientific." While this is a positive thought, on its face it fails to address a basic issue associated with the discipline. BPA is practiced worldwide and the "suggested" terms presented are, of course, defined in English. Analysts operate in effectively every language. Anyone adept at multiple languages understands that literal translations are not always possible. For example, an analyst in South America recently contacted one of the authors asking for clarification on the terms relating to spatter (back and forward spatter), writing:

> What I can deduce is precisely that in English-speaking linguistics, the antagonistic concepts "back" and "forward" do not conflict with the complement "spatter". However, in Spanish, when translating it literally, the first of them, generate conflicts."[1]

This issue becomes even more distinct when a given language is not based on a Latin alphabet. We think it imperative to say that a consolidated, generally accepted set of terms has value for the discipline; but it does not in and of itself establish the validity of the science, nor is it the most crucial concern.

Another critical aspect of terminology lies in the labels assigned to the various classifications of patterns. As with many other authors and organizations, your authors continue to use various mechanistic terms as labels for specific stains. These mechanistic terms (e.g., expiration, impact, smear) create certain inferences to lay juries that add to the potential confusion associated with BPA terminology. Your authors offer a potential solution in Chapter 4, for future consideration in labeling of "patterns."

Wherever the terminology in this chapter conflicts with current suggested terminology, we will ensure as effectively as possible to recognize those variations.

DOI: 10.4324/9781003408734-3

Referring to the Discipline

Heard from analysts and in court are multiple terms for the discipline itself. Bloodstain Pattern Analysis may at any given time be referred to as follows:

- Bloodstain pattern interpretation
- Blood spatter analysis
- Bloodstain spatter analysis
- Blood splatter analysis/interpretation
- Bloodstain Pattern Analysis

Perhaps the most appropriate term is the last, Bloodstain Pattern Analysis. This statement is based on several issues. "Analysis" implies a methodical, structured approach to any given effort. Analysis implies examinations to determine the nature of that examined, and determine any interrelationships present.[2]

The reason for using the word "Pattern" versus "Spatter" or "Splatter" is less subjective. Bloodstain pattern analysts evaluate all forms of bloodstains, which includes "spatter" stains as well as "non-spatter" patterns. Spatter is a specific form of bloodstain; stains that have been in free flight. Non-spatter includes contact stains, accumulations, as well as gush and drip stains.

As for the terms "spatter" and "splatter," the latter is a verb. As analysts are discussing a thing, some pattern, splatter is technically incorrect. Spatter, as described, refers to a specific form of bloodstain pattern and is not inclusive of all recognized forms of bloodstains.

Terms Associated with Bloodstain Pattern Analysis

There are various terms in use worldwide when referring to concepts associated with BPA. The most recent Organization of Scientific Area Committees for Forensic Science (OSAC) adopted terminology, the source of which is the ASB, have been identified and included.[3] Older terms often encountered are included (as synonyms) as well for consideration.

Accompanying Drop

(OSAC/ASB definition) A small blood drop produced as a by-product of drop formation.

Synonyms: Follow-on Drop

As a drop forms particularly under the force of gravity and begins to release from any given surface, a long narrow column of fluid connecting the droplet mass to the dripping source forms to interconnect the two. This columnar fluid structure or spindle is technically known as a jet in fluid dynamics. When gravity or any other force is sufficient to allow detachment of the drop, this spindle itself is not always drawn back into either the source blood mass or the droplet. Minor perturbations in this column of blood lead to its breakup through what is known as Rayleigh Plateau Instability.[4] This breakup leads to small masses that then assume spheroid shapes and fall along with the parent stain to the target surface.

Altered Stain

***(OSAC/ASB definition):** A bloodstain with characteristics that indicate a physical change has occurred.*

Synonyms: Physically Altered Bloodstain, PAB, PABS

Both prior and after creation of a bloodstain, certain alterations may occur that manifest themselves in the outward appearance of the bloodstain. The nature of those manifestations may suggest or indicate the nature of that alteration. Examples include dilution, or even disruption of a coagulated blood mass (e.g., altered impact pattern). Most alterations are not a "class" of pattern, with the exception of the wipe. Alterations can occur in any given pattern. These alterations (e.g., dried, flaked) are merely adjectives used to fully describe the pattern, when present.

Angle of Impact

(OSAC/ASB definition): The angle (alpha), relative to the plane of a target, at which a blood drop strikes the target.

The alpha angle is the acute angle of a droplet's intercept with a target as viewed from the side (the XZ plane) (see Figure 3.1.).

Synonyms: Impact Angle, Alpha Angle

As a droplet in flight strikes a target and collapses, it does so at some angle. This is the stain's angle of impact (alpha angle) and is always relative to the target. Droplets striking a horizontal or vertical target perpendicularly each have an impact angle of 90 degrees. This angle is a representation of the droplet's vector relative to the target as viewed from the side. It is distinct from the directional (gamma) angle. This angle is always represented as an acute angle (90 degrees or less). Defining this angle is a major aspect of the Stain Shape and Vector Analysis principle.

Area of Convergence

(OSAC/ASB definition): The space in two dimensions to which the directionalities of spatter stains can be retraced to determine the location of the spatter producing event.

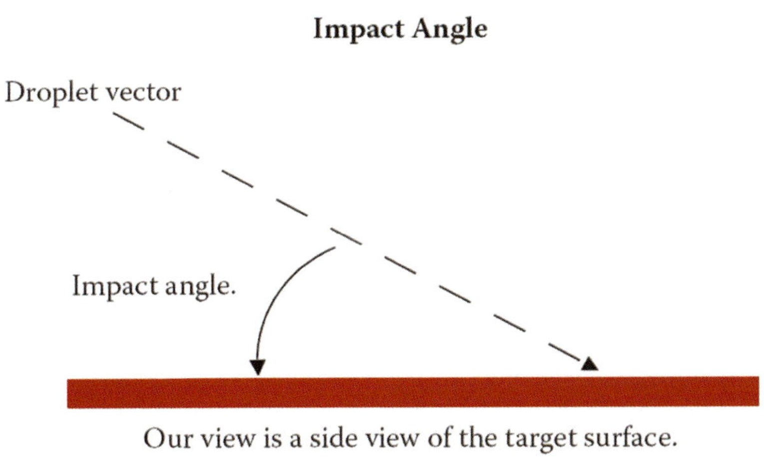

Figure 3.1 Impact angle: the acute angle (less than 90°) at which a droplet strikes a given target.

Synonyms: Convergence Point, Point of Convergence

When a spatter stain of any nature has a sufficiently elliptical shape, the long or major axis of that ellipse defines the droplet's directional movement on the target during collapse. Obviously, that line or axis represents two potential directions. The shape of the ellipse and presence of any associated spines and satellite stains (which are found primarily on the far side of the stain) help isolate the stain's directional component. When such spatter stains are found in a radiating pattern on a given surface, tracing a line (either virtually or physically) back in the opposite direction of the directional component for several well-formed stains will lead to an area in which the lines converge on that surface. This is the area of convergence. The three-dimensional source (the area of origin) of the spatter stains will correlate to this two-dimensional convergence.

In the early days of BPA, this area was often referred to as a "point" of convergence. Due to considerations of minor error associated when defining the true direction of motion on collapse, as well as the nature of the spatter producing event (spatter does not always originate from a singular XYZ point), these lines rarely produce a "point" of convergence Thus "area of convergence " is the most appropriate term.

Area of Origin

> *(OSAC/ASB definition):* The space in three dimensions to which the trajectories of spatter stains can be utilized to determine the location of the spatter producing event.

Synonyms: Point of Origin

Where the directional angles help isolate a point in two-dimensions (the area of convergence) for a group of radiating spatter stains, combining the information provided by both the directional and impact angles provides indications as to the actual source of the spatter in three-dimensions. The lines for the convergence are extended (again either virtually or physically) individually back toward the source based on the directional angle but at the calculated impact angle. In many instances, these lines will converge in three-dimensional space (in instances of highly perpendicular impacts, no such convergence may occur). This convergence of lines limits the possible area of origin for the event. This technique is referred to as "stringing" and can be accomplished either physically or virtually. This technique is fully explained in later chapters.

In the early days of BPA, this three-dimensional area was often referred to as a "point" of origin. It is imperative to recognize these lines do not define an absolute XYZ point in space. Thus, "area of origin" is the most appropriate term.

Back Spatter Pattern

> *(OSAC/ASB definition):* A bloodstain pattern resulting from blood drops which can be produced when a projectile creates an entrance wound.

Synonyms: Back Spatter

The passage of a bullet into the body produces an entry wound. Ejection of back spatter from the entry is common and believed to be primarily a product of the temporary wound cavity (produced by the passage of the bullet in the flesh) collapsing back to the damaged tract. The radiating spatter are ejected from the entry wound generally perpendicular to the wound surface, which is generally back toward the shooter and weapon.

Bloodstain

(OSAC/ASB definition): A deposit of blood on a surface.

Bloodstain Pattern

(OSAC/ASB definition): A grouping or distribution of bloodstains that indicates through regular or repetitive form, order, or arrangement the way the pattern was deposited.

Synonyms: Pattern

The Pattern Diversity principle recognizes that various basic force mechanisms in combination with variations of blood masses produce stains with consistent characteristic patterns. These patterns allow for inclusion and exclusion of an unknown stain into the various classes/groupings, which further identifies the potential source event (e.g., the manner of deposition). Note that even in the ASB definition, patterns are defined *based on physical characteristics* (e.g., regular or repetitive form, order, or arrangement). *Patterns are defined not by context, but by recognition of physical characteristics.* Conclusions associated with the identified patterns are defined first by the objective classification of the "pattern" and then through consideration of this pattern in the unique scene context found.

Blood Clot

(OSAC/ASB definition): A gelatinous mass formed by a complex mechanism involving red blood cells, fibrinogen, platelets, and other clotting factors (see Figure 3.2).

Synonyms: Clot

Clots occur as a function of the fibrin cascade, a chemical process that reduces or stops blood loss upon injury. The various clotting factors present in liquid blood lead to the fluid becoming a gelatinous mass. The clot is made up of insoluble fibrin, which effectively

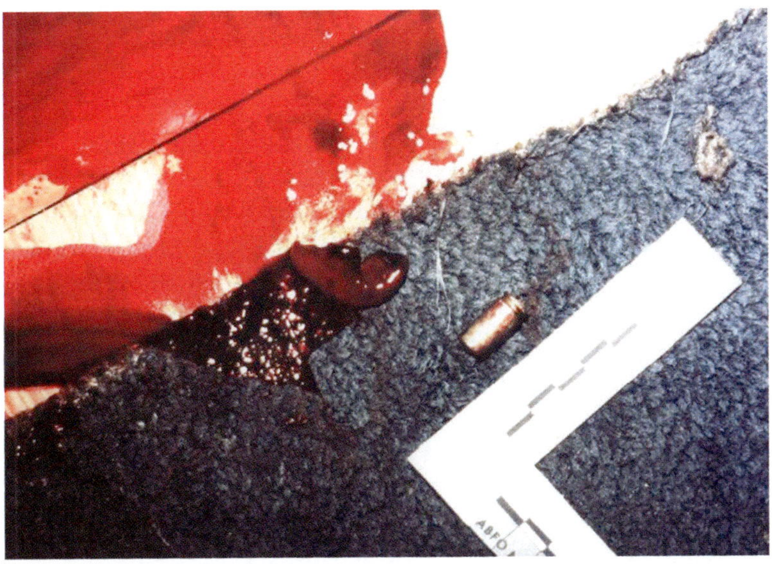

Figure 3.2 A clotted mass on a carpet, lying adjacent to a fired cartridge case.

captures most of the cellular material in the blood. Often associated with the clot is a straw or pink colored fluid, the plasma.

Beyond existing independent of some pattern (a clot may be distinct having been disrupted and disassociated with its source), clots or clotting may be present in nearly any accumulation of blood. Clots and clotting are not a specific form of classification in the taxonomy approach. Instead, a clot or clotting is simply an adjective to describe a condition found within some pattern or accumulation of blood.

Blowback Effect

A circumstance in which blood is deposited inside the barrel of a weapon after discharge.

Blowback is a distinct, but older term associated with the concept of blood ejected back toward the weapon. When the muzzle of a weapon is held in close or near contact with gunshot wounds, blood and tissue may be projected back into the muzzle itself. Bloodstains of this nature are referred to as "blowback." If the stains in the muzzle consist solely of small circular or elliptical stains, then properly the term blowback is not appropriate. In that instance, the term spatter is more correct, although such staining might well be referred to as blowback staining. There has always been some question as to the process by which blowback occurs. As for the process in which blowback is created, when presented with only spatter stains the analyst will understand this is simply a form of back spatter. But in those instances where stains other than spatter and/or tissue are present, explanations have included a belief that over-pressure in the wound coupled with a change in pressure in the muzzle draws these contaminants into the barrel. There are documented situations in which blowback consisting of both blood and tissue have been found as far as 10 inches up a barrel. There may be a correlation between the distance the muzzle was from the wound and the distance of penetration into the barrel; however, this is not fully understood.[5]

Bubble Ring

(OSAC/ASB definition): An outline within a bloodstain resulting from air in the blood.

Synonyms: Vacuole

When air and blood are mixed and droplets subsequently created, small bubbles of air may be present within the ejected droplet. As the droplet collapses on a surface, these bubbles routinely (but not always) collapse leaving behind a small voided area within the resulting stain. Bubble rings/vacuoles are a typical characteristic found in expiration patterns. They may occur in other limited circumstances, for example, when an extended beating occurs involving a volume of blood.

Capillary Action

A condition exhibited by the attraction of a liquid to surfaces it is in contact with and its own surface tension.

Capillary action does not strictly follow the scientific definition of capillary attraction. The latter results from surface tension and adhesion of liquids when in contact with solids, particularly tubes. The term capillary action is used to describe two observed actions related to capillary attraction.

When objects are in contact and blood is present between the two, upon separation the blood mass attempts to remain a cohesive mass. The cohesive force of the blood draws the blood up in between the separating surfaces. The adhesive qualities of the blood keep the blood connected to the objects. The separation will eventually overcome the cohesive force, and these connecting structures shear and are drawn back to the objects. When drawn back to the surface, these small connecting aspects of the blood often result linear features in the contact stain that have no relationship to the target surface. Figure 3.3 demonstrates this action, while Figure 3.4 shows the result.

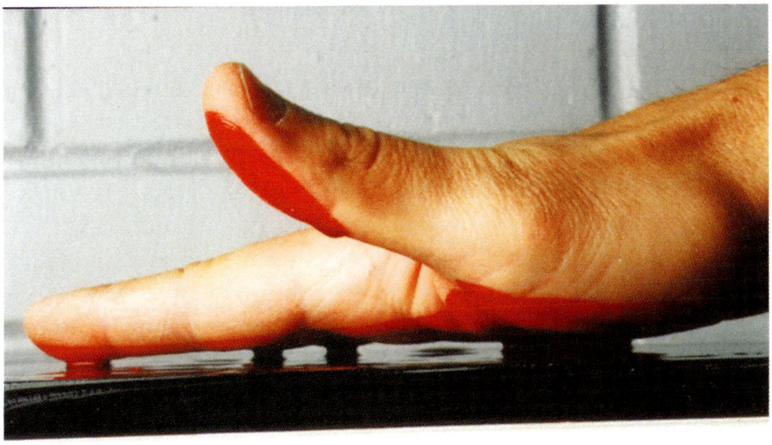

Figure 3.3 An example of capillary action. As the hand is withdrawn, the surface tension of the liquid in conjunction with its adhesive quality causes a bridging effect. (Photograph courtesy of Jouni Kiviranta, National Bureau of Investigation, Helsinki, Finland.)

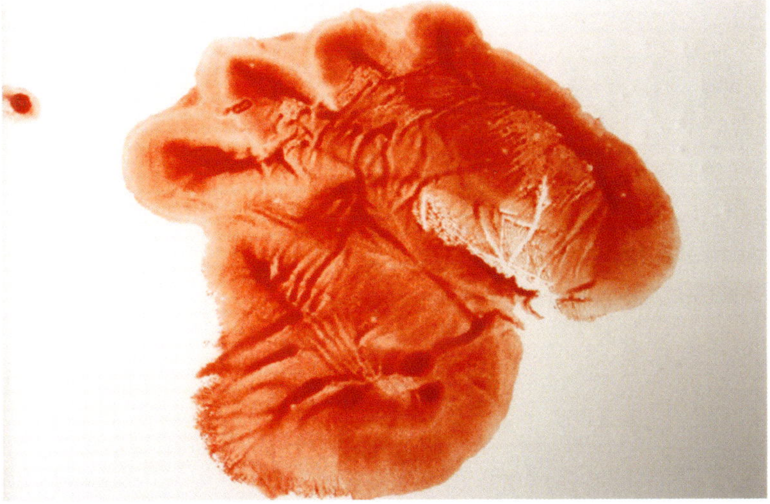

Figure 3.4 The result of Figure 3.3. Distinct linear features are present in the pattern because of the capillary action. (Photograph courtesy of Jouni Kiviranta, National Bureau of Investigation, Helsinki, Finland.)

Capillary action may also produce odd flow characteristics. For example: a flow is expected to follow the affect of gravity, but the adhesive quality may cause the flow to follow a surface contour. Blood flowing on a horizontal surface, but in contact with the victim's arm may follow the arm contour. Whether on a horizontal or vertical surface, this behavior may present what appears as an unnatural blood flow.

Cast-off Pattern

(OSAC/ASB definition): A bloodstain pattern resulting from blood drops released from an object due to its motion.

Synonyms: Swing Cast-off

Cast-off patterns are a classification of patterns described in Chapter 4. They result from blood adhering to a surface that is then accelerated. When a bloodied object is accelerated in an arcing motion, the ability of the blood to remain a mass on the object is overcome by that acceleration. At times, the blood may be projected directly from the accelerating surface as individual spatter, or it may flow off the object as a spindle, which then breaks up into individual spatter. In either case, these blood droplets are released from the object at distinct points and follow distinct paths to the surrounding surfaces. As they are released at various points along the object's path, they are deposited in linear orientations relative to that path (see Figure 3.5). Cast-off patterns are produced by any number of objects, this includes weapons (e.g., a club or knife) and/or the victim or subject (e.g., hands, hair). Cast-off can reflect both offensive and defensive activities. In many instances, cast-off may

Figure 3.5 Two cast-off patterns. The two distinct linear patterns on the cabinet are cast-off, produced by the bloody object that was swung right to left.

identify the minimum number of swings made and will almost always provide information as to the direction of motion involved in their creation.

There are no viable quantitative criteria for the size of the droplets and thus the stains produced by cast-off. The resulting spatter size is a function of several factors, including the surface area of the object, the volume of blood adhering to the object, and the speed associated with the swing of the object. Any given weapon may produce different size spatter when swung under different situations.

A second form of spatter produced by accelerated motion is cessation cast-off. Cessation stains are certainly the result of an object's accelerated motion, but differ from typical swing cast-off. They have no discerning characteristics. When the object ceases motion in a specific direction, the blood adhering to the object continues in that direction. The adhesive forces holding the blood to the object are often insufficient and the blood continues in the original direction. These droplets produced by the cessation of the object's motion may be ejected from the entire bloody surface. Cessation cast-off can appear somewhat linear, or they may radiate. Lacking specific reliable characteristics to associate with all cessation cast-off, this form of bloodstain is not included in the taxonomic classification of patterns. It is generally recognized only by the unique scene context.

Contact Stain

> Any stain or pattern created by the transfer of blood from one object to another through physical contact.

Contact stains entail a broad classification of non-spatter stains. They include smears, wipes, swipes, and pattern transfers. A secondary purpose in the taxonomy approach for the "contact" classification is to act as the parent stain of all wipes and swipes. If a stain cannot be clearly distinguished as either wipe or swipe, it is classified as contact. See Chapter 4 for a full discussion of the various stains produced by contact.

Directionality

> *(OSAC/ASB definition):* The characteristic of a bloodstain that indicates the direction blood was moving at the time of deposition.

One important piece of information evident in the Stain Shape and Vector Correlation principle is the directionality of the stain in question. Directionality relates to the path the droplet was following relative to the XY plane of the target it struck (e.g., as viewed looking down on that target).

As a droplet intersects a target, the collapse of the droplet produces either a circular or an elliptical stain. The long axis reflects the droplet's movement as it collapsed. The long axis, of course, has two potential orientations, but the droplet's collapse generally produces tails, scallops, or satellite stains that are deposited primarily on the side opposite its movement on impact. Shape and these associated features allow analysts to recognize the direction the droplet was moving at the time of deposition. Figure 3.6 shows spatter with two distinct directional angles. The directionality of a stain can be described in various ways: either as cardinal directions (e.g., W-E), relative to the target (e.g., L-R) or as a specific angle (e.g., gamma angle of 90 degrees).

Figure 3.6 Spatter on a floor exhibiting two different directions of travel of the spatter droplets (as indicated by the red arrows).

Directional information (motion) is also present in other types of stains beyond spatter. This includes wipes, swipes, and flows, but this is somewhat distinct from the typical usage of the word directionality.

Directional Angle

(OSAC/ASB definition): The angle (gamma) between the long axis of a spatter stain and a defined reference line on the target (see Figure 3.7).

Synonyms: Gamma Angle

Directional angle is a specific angle measured for the directionality of the stain relative to a surface and a known reference point (e.g., generally up or north). This angle can describe directionality in an unambiguous manner (e.g., the stain's directional angle was measured as 210 degrees) or used for defining the flight path with Area of Origin software. Gamma angle can be described in several ways, either as a specific angle of 1–360 degrees or as in the case of some software, as an angle of 1–180 degrees with the additional indicator of a minus sign. Angles oriented to the right are reflected as simply a number (1–180), whereas angles oriented to the left are depicted with the minus sign.

Drip Pattern

(OSAC/ASB definition): A bloodstain pattern resulting from a liquid that dripped into another liquid, at least one of which was blood.

Synonyms: Blood into Blood

A drip pattern is a specific class of pattern. As problematic as the original term (blood into blood) was, the ASB term Drip Pattern is itself confusing. It is another

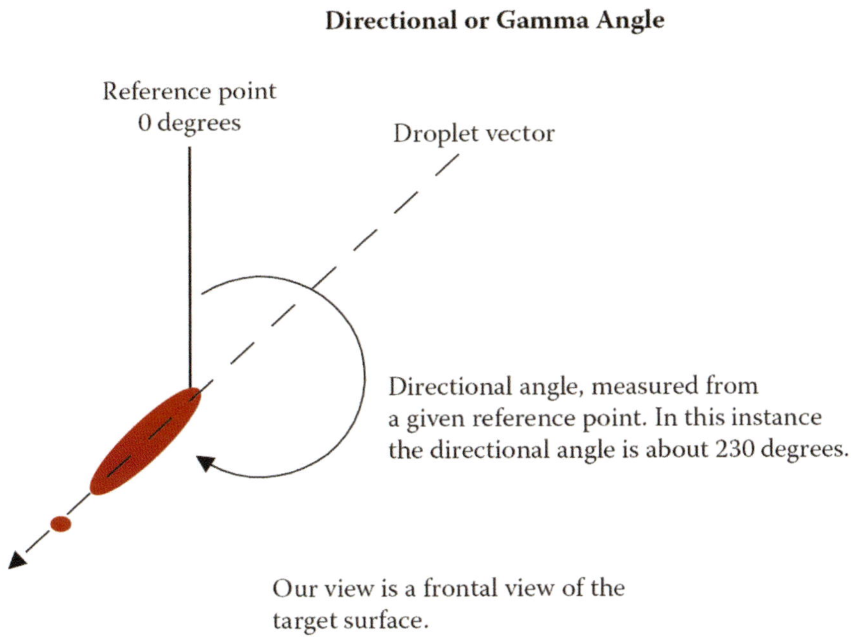

Figure 3.7 Directional angle. This angle describes the directionality of the droplet, but does so from a standard reference point, thus allowing it to be used by forensic software.

term developing from the "Likes" and "Dislikes" of the SWGSTAIN group (see a more detailed discussion of this process under the projected pattern definition). As a term for a specific pattern type, it is functional, but also has the potential to confuse a lay jury. Drips and drip trails are additional labels associated with specific classes of patterns. All represent patterns produced from an underlying gravity-induced force, dripping. When discussing these three patterns (drip, drip trail, and blood into blood), it is common to hear analysts refer to them all as "drip patterns." As heard by the jury, an analyst discussing and showing a specific "Drip Pattern" (e.g., blood into blood) may at a later point in testimony refer to distinctly different patterns (drips and drip trails) as drip patterns as well.

Drip patterns occur on both horizontal and vertical surfaces. On a horizontal surface, the drip pattern causes random, somewhat irregular satellite stains around the immediate margin of the irregular shaped primary stain. Further out from the immediate margin, these satellites will appear similar to impact spatter. As these satellites take on a more typical spatter form further out from the primary stain, it may be difficult to differentiate between them from typical spatter. When a drip pattern occurs in proximity to a vertical surface such as a wall, the satellite spatter aspects of the pattern are deposited on the vertical surface. Within this vertical area, a voided area may be found in the pattern. The vertical pattern often resembles a funnel or V-shape (see Figure 3.8).

Drip Stain

(OSAC/ASB definition): A bloodstain resulting from a falling drop that formed due to gravity.

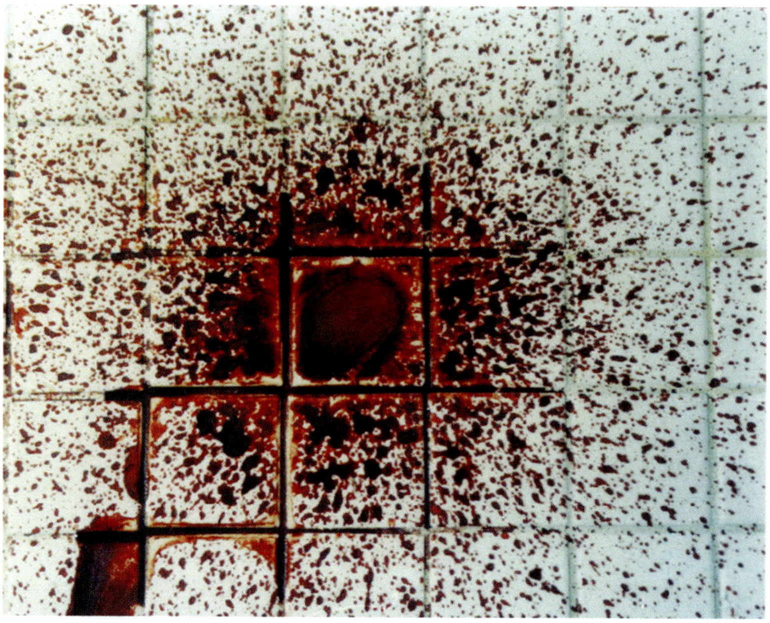

Figure 3.8 A drip pattern, also known as blood into blood pattern. These patterns are produced by a series of drips landing in the same area.

Synonyms: Drip

The drip stain represents another specific class of pattern found in bloody scenes. As a blood mass collects on an object, gravity draws that mass downward. As this occurs, a spindle is produced between the blood on the object and the forming droplet. At some point, the force of gravity overcomes the adhesive and cohesive forces holding the droplet, spindle, and source together. This shears the spindle, releasing the droplet. The specific volume necessary to produce a single drip is approximately 0.08 ml, which allows a quantitative value for the smallest drip stain that can be produced. The largest value for the primary drip stain is generally understood, but is based on anecdotal information. Gravity-induced drops falling downward produce spatter stains generally between 3 mm - 25 mm in width. The shearing of the spindle itself often produces smaller satellite stains (e.g., accompanying stains) that appear in and around the primary stains. The accompanying satellite stains produced by the spindle breakup are much smaller than the primary stains, but are easily recognized as satellites associated with the drip stains. They do not appear independent of the drip stains. Similar to the circumstances of cast-off, the size of the resulting stain is a function of many variables, which include the surface area on which the drop forms, the flow rate at the source, the distance the drop falls, the viscosity of the blood, and the surface to which the drop falls.

Drips as a taxonomic classification present themselves as a single stain or a random distribution of individual stains.

Drip Trail

(OSAC/ASB definition): A bloodstain pattern resulting from the movement of a source of drip stains between two points.

Figure 3.9 A drip trail. Passive drops of blood falling as a function of gravity to a surface can produce drips (randomly oriented stains), a drip trail (a series of stains that lead from one point to another), or a drip pattern as observed in Figure 3.8.

Drip trails are another form of pattern associated with a gravity-induced mechanism. As drops form on an object and that object is moved, a linear series of drip stains are deposited on the target(s). See Figure 3.9. Based on intermittent movement, drip trails often present as complex patterns, involving combinations of drip trails, drips stains, and drip patterns.

Edge Characteristic

(OSAC/ASB definition): A physical feature of the periphery of a bloodstain.

The ASB definition is not explicit, but an obvious association in the definition is the edge of any spatter stain (e.g., circular/elliptical shaped stains). Spatter edges can demonstrate smooth, scalloped, or spiny margins or highly deformed margins (the latter most often associated with substrate interactions).

This "edge characteristic" is also applicable when discussing non-spatter stains. As discussed in Chapter 4, non-spatter stains are initially differentiated based on the margin of the stain. One class will demonstrate regular margins (e.g., consistent with fluid flow) and a second with irregular margins.

Expiration Pattern

(OSAC/ASB definition): A bloodstain pattern resulting from blood forced by airflow out of the nose, mouth, or a wound.

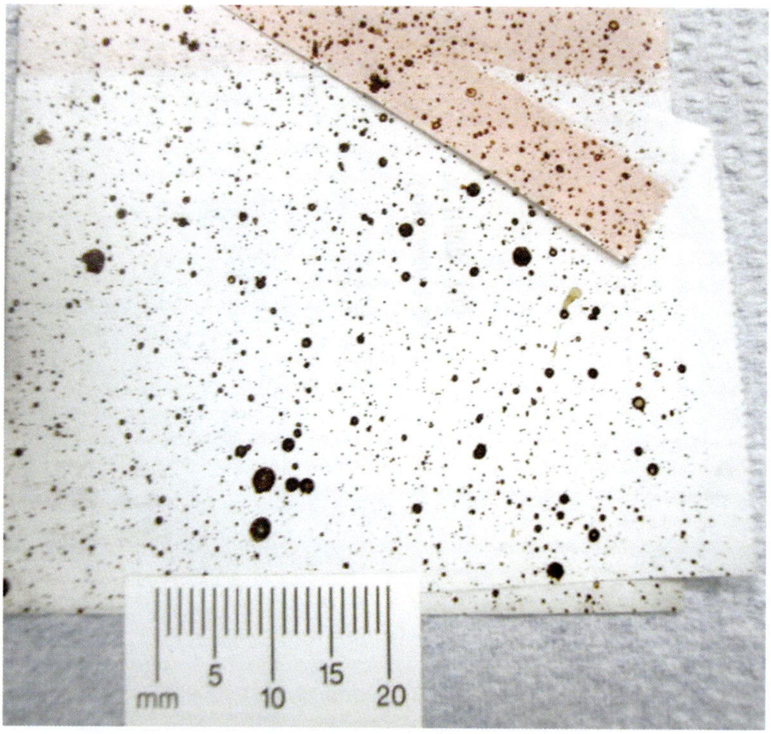

Figure 3.10 Expiration stains are blood dispersed by some breathing mechanism (e.g., breathing, gasping, or coughing). These stains were produced as a beating victim was face up in the scene and various articles including the receipt were adjacent to the head. Note the small vacuoles in many of the stains.

Synonyms: Expectorate, Expiration Stain

Expiration patterns are produced when blood is ejected by the force of air from the mouth, nose, or a wound associated with the lungs (e.g., a "sucking chest wound"). As blood is interjected into an airway, breathing results in ejections of a mix of air and blood. See Figure 3.10.

The disruption mechanism for the blood is the movement of air in the passageways. The blood breaks up into relatively small droplets that radiate outward from the source. The nature of the force behind the air movement affects the size of spatter produced, thus coughing, gasping, or breathing forcefully can produce various sized spatter. The range of size of individual expiration stains varies greatly. Additional characteristics such as mucous strands and small vacuoles allow specific verification of this pattern. If these characteristics are absent, expiration patterns may mimic impact patterns.

In some limited situations, such patterns can be produced by significant compression of the chest (e.g., when CPR actions are undertaken), even when the victim is no longer capable of breathing.

Flow

(OSAC/ASB definition): A bloodstain resulting from the movement of a volume of blood on a surface due to gravity or movement of the target.

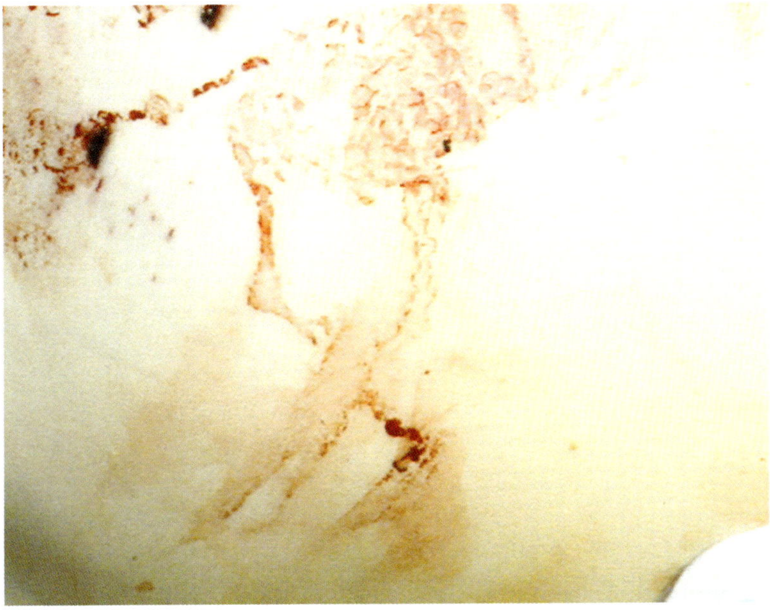

Figure 3.11 A flow pattern on the abdomen of a victim. Flows must follow gravity and they can be quite telling. In this instance, the flow with the distinct dogleg (the top left side of the X) demanded the victim was positioned contrary to the suspect's claims.

Flows are a specific classification of patterns described in Chapter 4. Flows are produced whenever a volume accumulation is subjected to gravity in a manner that causes the accumulation to move or when a volume is present on a static object and that object is moved. See Figure 3.11.

Forward Spatter Pattern

(OSAC/ASB definition): A bloodstain pattern resulting from blood drops which can be produced when a projectile creates an exit wound.

Synonyms: Forward Spatter

In instances of perforating gunshot wounds, the passage of a bullet exiting the body produces an exit wound. Ejection of the forward spatter from the exit is believed to be both a function of the bullet exiting the body and the collapse of the temporary wound cavity.

Gush

The escape of a volume of blood in a streaming ejection, often under pressure from a breach in an artery, or the heart.

Synonyms: Projected Stain, Arterial Gush

A "gush/splash" is a specific classification of non-spatter stain in the taxonomic classification system. Gushes are produced when streaming ejections are directed into the same general area. The interaction of the blood volume striking a surface produces a highly irregular margin on the primary stain(s) and significant satellite stains and spines radiating away from the primary stain. The underlying mechanism for all projected patterns (e.g., spurt, splash and gush) is this streaming ejection.

The associated pattern, splash, is only differentiated by the level of irregularity in the margin and the number of spines produced. As there is no objective approach to identifying this characteristic (e.g., how many spines define a gush from a splash), the taxonomic approach combines the gush/splash as a single pattern type.

Insect Stain

(OSAC/ASB definition) A bloodstain resulting from insect activity.

Synonyms: Fly Spot

Activity by flies and other insects interacting with blood in the scene may produce post-incident artifacts that on first appearance appear as dynamic spatter stains. The insects may either regurgitate or track small amounts of blood onto items within the scene. See Figure 3.12. They may appear as a series of small spatter in an unusual location (e.g., high on a wall, ceiling, or window disassociated from other patterns) or have small linear features protruding from them, giving them a sperm-like shape.

Impact Pattern

(OSAC/ASB definition) A bloodstain pattern resulting from an object striking liquid blood.

Synonyms: Impact Spatter

An impact pattern is a specific classification of patterns. When a force is directed at a blood mass at some point source (e.g., blunt force to a wound, a gunshot), the blood must displace

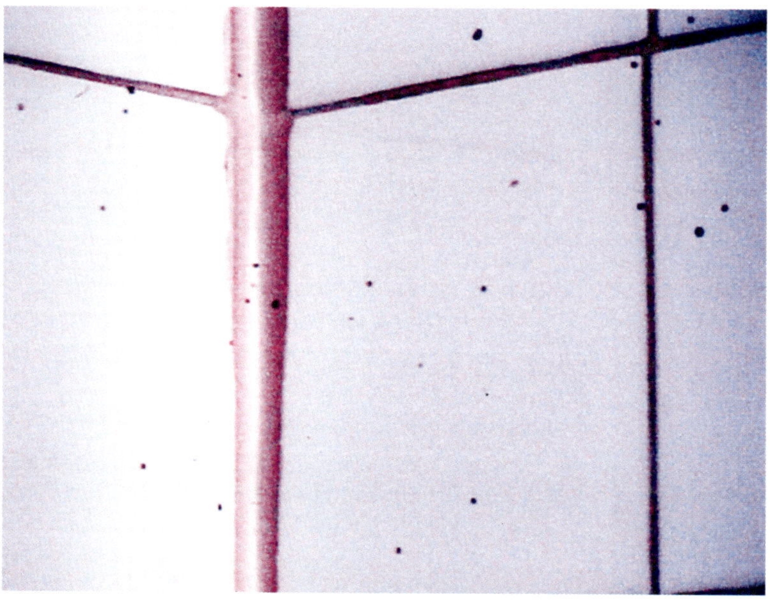

Figure 3.12 A fly spot pattern. Fly spot stains found on the bathroom tile at the scene of a natural death. The decedent lay in the scene undiscovered for some time and fly activity was found throughout the scene. (Photograph courtesy of Detective Mike McGuffey, Covington Police Department, Covington, KY.)

and ultimately does so as a series of small droplets that radiate in a three-dimensional fashion outward from the source. When these small droplets encounter surfaces, they collapse and produce a series of radiating spatter (small circles and ellipses, the majority of which are 4 mm or less in width). The radiating aspect of the pattern, generally referred to as the pattern's dispersion cone, may be quite wide (e.g., radiating in a 360 degree fashion) or this dispersion cone may be relatively small and directed (e.g., radiating in a 15 degree arc). This radiating distribution will be evident based on consideration of the varying directional and impact angles present (taking into consideration different target orientations) and/or by the level of distribution.

Linear or Curvilinear

A characteristic associated with spatter patterns, in which spatter are deposited in a linear or curvilinear distribution. This is best recognized by the parallel orientations of the individual spatter directional angles.

Linear or curvilinear distributions of spatter include drip trail, spurt, and cast-off. The stains may be oriented in a single line or multiple lines with parallel directional angles. The parallel nature of the directional angles is not meant to suggest the directional angles are always in line with the line produced by the pattern, but rather the directional angles of stains lying next to one another are so oriented. These directional angles often change across the length of the pattern, where the directional angle orientations slowly change with the pattern.

Mist

Bloodstain patterns characterized by a mist-like appearance. Mist stains are identified as having widths 0.1 mm and smaller, although larger stains may be present in the overall pattern.

Synonyms: Atomized Blood, Misting Stain, Misting Effect

When a blood mass is exposed to a significant force or impulse, it may be reduced to a fine spray of tiny droplets, resulting in a mist. These stains are most typical of situations in which the source disruption is an explosive force. The most encountered scene source for mist (but not exclusively) is gunshot.

The individual stains may be macroscopically indistinguishable, meaning one cannot visually observe individual circles and ellipses without magnification. The mist often appears as a reddish hue on the target. Individual stains may appear in large numbers (exceeding hundreds) or may be very limited. It is a generally recognized principle in BPA that as the force applied to a blood mass is increased, this results in a decrease in the size of the associated spatter. Mist stains reflect this principle. McDonell initially identified mist as being stains of no more than 0.1 mm in size, but recognized that mist routinely appear intermixed with larger spatter stains. The value of recognizing mist is in limiting the potential source event(s). Explosive force (often associated with, but not limited to gunshot) is necessary to produce such small stains. Additional examples that may produce mist-sized stains are high-speed machinery and explosions.

The term mist was introduced to replace the term high-velocity impact spatter (HVIS), but with the clear recognition that the width determination must coincide with the original size associated with HVIS (0.1 mm or smaller). The HVIS definition later evolved to include stains as large as 1 mm. Such stains are not mist.

Non-Spatter Stains

Any stain or pattern other than those defined by the spatter group (e.g., patterns not composed of small circular or elliptical shaped stains).

Non-spatter stains are a broad classification of stain/patterns described in Chapter 4. They entail a broad group of patterns that involve contact, volume accumulations, and certain ejections (e.g., gush/splash). Any pattern in which the primary stain(s) are not small circular- or elliptical-shaped stains is a non-spatter stain/pattern.

There are two primary sub-categories of non-spatter stains: Regular Margin and Irregular Margin stains. A regular margin is cleanly demarcated (e.g., that associated with fluid flow). Irregular margin stains represent anything other than a regular margin, and often involve scalloped edges or associated spines.

Parent Stain

(OSAC/ASB definition): A bloodstain from which a satellite stain(s) originated.

Synonyms: Primary Stain

As a droplet collapses on a surface, several variables act on the droplet (e.g., target surface, surface tension, inertia, and velocity), all acting in their own way to either hold the mass of the droplet together or break it apart. Very often, during this collapse, smaller droplets detach during the collapse of the droplet forming satellite stains. The breakup of the spindle of blood connecting a drop to the mass it is dripping from also produces these smaller satellite stains. The larger stain is the "parent," the smaller stains are the satellites. A parent may spawn multiple satellite stains, but a satellite has only one parent stain (see Figure 3.13).

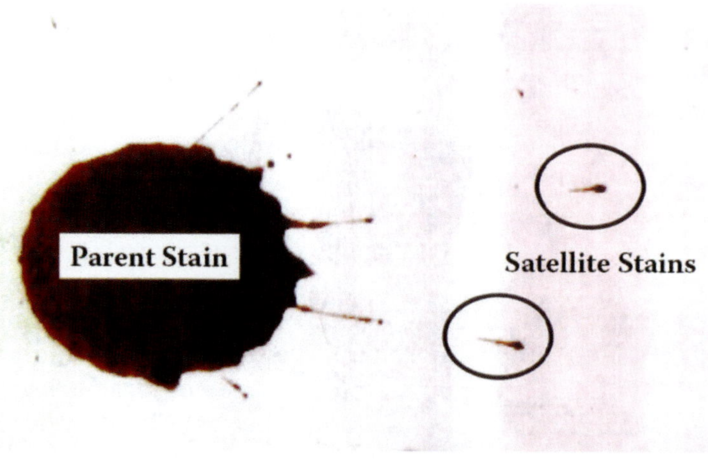

Figure 3.13 A parent stain with several attached and semi-attached satellite stains, as well as two detached satellite stains. Note that the tail of a detached satellite points in the opposite direction of tails and spines associated with the parent stain.

Particularly in impact events, the primary reason for making a distinction between the parent and satellite is directionality. Parent stain tails point away from the origin, whereas satellite tails point toward the origin.

Pattern Transfer

A pattern created by the transfer of blood from one object to another in which a recognizable characteristic or image is present in the pattern.

Synonyms: Transfer Stain, Transfer Pattern

Pattern transfers are a specific classification of patterns produced by interactions of a bloodied object with other objects and surfaces. Pattern transfers are a very common stain present in bloodstained scenes. Due to blood's adhesive qualities, objects that become bloodied tend to stain any other object they contact. Figure 3.14 shows the pattern caused by a butcher knife on carpet. Figure 3.15 is the pattern created by a lug sole boot on the back of a victim's T-shirt.

At times, pattern transfers may demonstrate specific characteristics that help the analyst identify the object producing the pattern. In many instances, the analyst will recognize a stain as some form of pattern transfer, but will be unable to associate it with a specific object. Pattern transfers may be patent (e.g., visible) or latent (invisible to the naked eye). Various enhancement techniques (e.g., luminol, LCV) are utilized to visualize latent pattern transfers.

Perimeter Stain

(OSAC/ASB definition): An altered stain consisting of its edge characteristics, the central area having been partially or entirely removed.

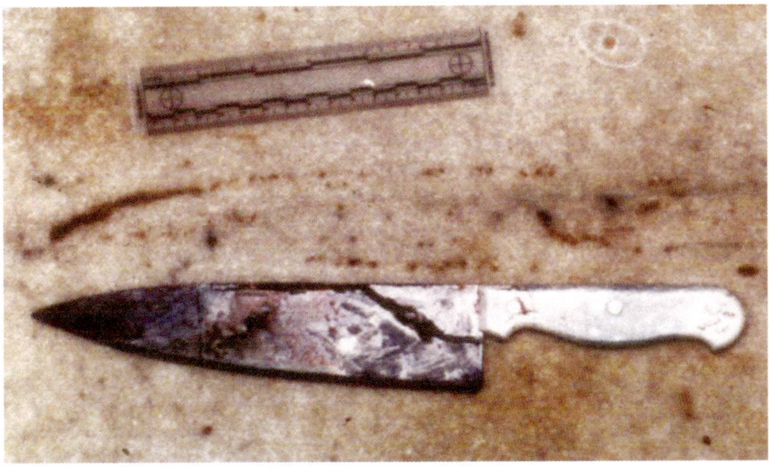

Figure 3.14 A pattern transfer on carpet associated with a bloody butcher knife. The knife was placed adjacent to the pattern transfer only after collection of samples and forensic evaluation of both items of evidence in order to show consistency between the pattern and the weapon. Note the heavy flow along the top edge of the blade toward the tip and the matching saturated area of the pattern transfer.

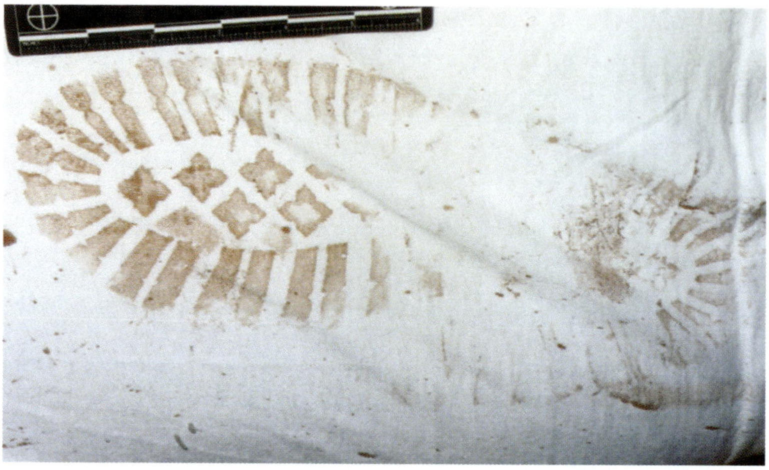

Figure 3.15 A pattern transfer of a lug-soled boot on the T-shirt of a murder victim. The most common pattern transfers involve hands, feet, shoes, and weapons.

Synonyms: Skeletonized Stain

A perimeter stain involves a bloodstain that although disturbed, still reflects some aspect of its original shape and size (see Figure 3.16). In certain instances where a preexisting stain is disrupted (e.g., a wipe through a spatter stain), the result is a stain in which the original outline or some portion thereof (its perimeter) is evident while other aspects of the stain may be absent or smeared. Once deposited, blood dries from the outer perimeter inward toward the center of the stain. Should the stain be disturbed prior to completion of the drying process, the dried portion often remains. A perimeter stain can occur within any type of stain (e.g., spatter, pool, flow). The width of the remaining perimeter is very much a factor of the length of time between deposition of the original stain and its subsequent disruption.

Pool

(OSAC/ASB definition): A bloodstain resulting from an accumulation of liquid blood on a surface.

A pool is a specific classification of patterns. It consists of a volume that accumulates on a given surface. It has no specific shape, conforming to surface contours on which it forms.

Progressive Change in Shape

A characteristic observed in spatter stains, in which there is a gradual consistent alteration of the shape of the stains within the pattern. This progression may manifest itself by the stains becoming more or less elliptical. An example is a cast-off pattern that may begin as near 90-degree impacts and slowly change to elliptical stains as the swing continues.

Projected Pattern

(OSAC/ASB definition): A bloodstain pattern resulting from the ejection of blood under hydraulic pressure, typically from a breach in the circulatory system.

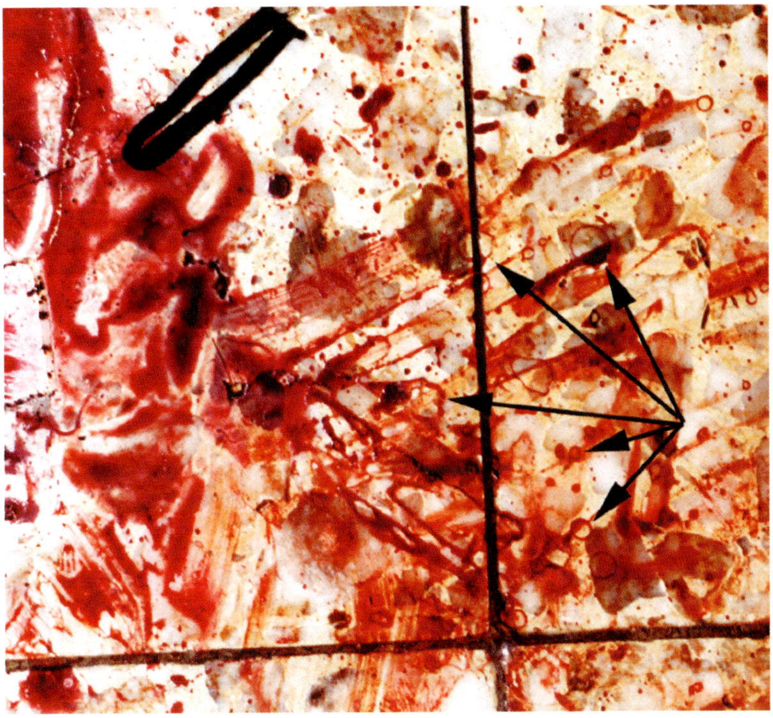

Figure 3.16 Multiple stains showing a perimeter stain (skeletonization) intermixed in a wipe and partial palm print. The movement of fingers and a hand through the stains disturbed the spatter. The individual claiming to have found the bodies some hours after their deaths stated he became bloody while checking the bodies and moving his hands in this area. A partial palm print in the pattern was identified to this person, but the limited level of perimeter stain disproves his claim as to when this palm print was created. The stains were clearly disturbed soon after they were deposited. (Photograph courtesy of Claire Dawson-Brown, Travis County District Attorney's Office, Austin, TX.)

The authors do take issue with the ASB inclusion of sources only involving hydraulic pressure. The most effective definition is patterns created from streaming ejections, often pressurized, but not exclusively pressurized. See Figure 3.17.

Synonyms: Spurt, Arterial Spurt, Gush, Arterial Gush, Splash

There are several associated issues with both the ASB term "projected" as well as the ASB definition. Regarding the latter, the association of the projected pattern to a mechanism of "hydraulic pressure" is too limiting. Hydraulic pressure (e.g., static and dynamic pressure) is generally associated with discussions of fluid flowing within and out of a contained area (e.g., pipes, arteries or veins). But projected patterns are produced just as easily and almost as often by streaming ejections flowing external to the body, without pressure and simply as a function of gravity-induced velocity. This stream from either pressurized or non-pressurized ejections breaks up in the same way; gravity and air resistance shear the stream, producing undulating masses and/or stable droplets. The resulting patterns are effectively the same.

Projected as a term of art creates significant and unnecessary ambiguity. The term projected has a standard definition in modern English, which includes the idea to throw forward or cast out into space some thing. By that definition, most dynamically produced

Figure 3.17 A projected pattern resulting from a cut carotid artery. In this instance, the blood was projected generally perpendicularly onto the target, resulting in the "trail" appearance. Still evident in the stain are spines (indicating the force of projection) and several small indications of pressure fluctuations. (Photograph courtesy of Jeff Svoboda, Fort Collins Police Department, Fort Collins, CO.)

spatter are ultimately "projected." Impact spatter are "projected" outward by the application of a force to the blood mass. Acceleration of blood adhering to an object that is swung results in the blood masses being "projected" away from the object. A failure to consider such ambiguities has plagued BPA (including your authors) for some time and continues even now in the ASB terms.

It is also rather important to recognize "how" this term originally came to be included in "suggested" language. At SWGSTAIN, the Terminology and Taxonomy (T^2) subgroup presented what they felt were effective terms to the entire group. These terms were offered after detailed consideration over several years. Specifically, when the term(s) associated with streaming ejections were presented to the group, there was pushback. Members from the Canadian representation did not like the terms offered. This contingent preferred their term of art "projected." The Chair of SWGSTAIN proposed a resolution to this objection (ignoring T^2 detailed considerations over several years): asking for alternative terms offered up from the floor by anyone and then using a subsequent majority rule vote to decide upon which term to adopt. Projected was adopted by a show of hands after only a short discussion of the "likes" and "dislikes" of the group in attendance. This process was the method of creation and adoption for several of the SWGSTAIN adopted terms. One can easily trace the ASB terminology clearly back to the SWGSTAIN terminology. As will be evident in Chapter 4, the authors continue to use the terms gush/splash and spurt when discussing specific classification of "projected" patterns.

Projected patterns are often the result of a breach or artery (e.g., pressurized system). Pressures associated with arteries are greater than veins; although depending upon which aspects of the two systems we compare, this difference may not be great.[6] Typical patterns created by arterial ejections often exhibit both the increase and decrease in the artery. Blood flowing from a wound in volume and with velocity, even as a function of the force of gravity, will also produce projected patterns. Such flowing volumes don't simply drip off an object, a streaming ejection will flow from the source and break up as any other streaming ejection based upon shearing effect of gravity and air resistance. Brodbeck also discussed that in some instances veins can produce somewhat pressurized streaming ejections as well. This gives obvious validity to not using the term arterial as an adjective.[7]

Radiating

A characteristic primarily associated with certain spatter patterns. The directional angles of the individual spatter in the pattern demonstrate a clear angular deviation in their outward flight paths.

A radiating aspect is a significant characteristic in differentiating spatter-type patterns. The various forces creating spatter will lead to patterns in which there is an obvious linear aspect (e.g., spurts, drip trails, and cast-off), a random aspect (e.g., drips with no absolute consistent directional angles), and the radiating aspect (e.g., impact, expiration) where there is evidence of an angular deviation in the outward flight paths of the droplets. This radiating nature assists in isolating both the area of convergence and the area of origin for such patterns.

Ricochet Stain

Blood, which impacts an object and bounces off or interacts with another droplet producing an anomalous directional angle compared with other associated spatter.

Bloodshed in a violent confrontation is a very dynamic situation. The highly dynamic manner in which parent stains and satellite stains interact with themselves and/or a target can produce stains that demonstrate a directional component that is counter to the obvious pattern producing it. Ricochets are known to occur in situations involving large-volume ejections such as Gush and Splash, where there is significant interaction associated with the satellites produced. Although not common, two droplets may collide, resulting in a redirection. It is also possible for a droplet to strike a surface and due to surface characteristics not collapse and instead bounce off that surface. The ricochet stain, although uncommon and an anomaly, may lead to some confusion when considering the concept of "related spatter" (discussed in Chapter 4) if not recognized.

Satellite Stain

(OSAC/ASB definition): A smaller bloodstain that originated during the formation of the parent stain as a result of blood impacting a surface.

Synonyms: Satellite Spatter, Secondary Spatter, Wave Cast-off

As a droplet impacts a surface and collapses, the mass of droplet is subjected to significant shifts. As this occurs, surface tension acts to contain the liquid as an integral entity. But the various disruptive forces acting on the mass often makes this impossible. During collapse, smaller volumes may form which are initially connected to the parent drop by a thin spine-like structure of blood.

If sufficient velocity exists during the collapse of these structures, smaller droplets form that are thrown from the parent stain and deposited on surrounding surfaces as smaller individual stains. These newly formed droplets may be attached, detached, one, or many and are known as satellite stains or satellite spatter. Refer back to Figure 3.13.

Satellites are also produced when volumes of blood interact (e.g., blood dripping into blood or streaming ejections interacting with a surface). In such instances, these satellites may involve spatter-like stains, spiny features, or small irregular-shaped stains.

Saturation Stain

(OSAC/ASB definition): A bloodstain resulting from the accumulation of liquid blood in an absorbent material.

Saturation stains are a specific classification of patterns. They occur when blood is deposited or absorbed into a porous surface. As a classification, saturations are not particularly illuminating, as compared with dynamically produced patterns (e.g., impact, cast-off). Very often saturation stains simply mar other patterns of interest.

Scallop

Scallops are an edge characteristic found in various spatter stains.

In droplets that impact a surface generally between 35 degrees and 70 degrees, an edge characteristic often encountered is a series of small wavy protrusions that occur primarily on the side opposite contact. These protrusions give the appearance similar to a seashell and are generally referred to as scallops or scalloping.

Serum Stain

(OSAC/ASB definition): The stain resulting from the liquid portion of blood (serum) that separates during coagulation.

The fibrin chemical cascade associated with blood coagulation generally leads to the solids (cellular components of blood) being drawn together in a non-soluble fibrin mesh. As this occurs, the liquid component of blood, the serum, flows outward surrounding this clotted mass. This liquid, which is primarily water, may appear as a pinkish or straw-colored fluid.

Spatter Stains

(OSAC/ASB definition): A bloodstain resulting from an airborne blood drop created when external force is applied to liquid blood.

Spatter is a broad classification of stains resulting from droplets that have been put in free flight and subsequently impact some surface. Small masses of blood that have been in free flight by whatever means subsequently collapse on a surface creating generally circular or elliptical shaped stains.

Spatter stains are produced by any number of actions, including dispersion at a point source (impact), dispersion from an object in motion (cast-off), breakup of a stream of blood ejected into the air (spurts), and by drops that form as a function of gravity (drips). For a full discussion of these distinct pattern types, refer to Chapter 4.

The size of the spatter in any pattern is dependent upon a complex interaction of the force associated with the breakup mechanism and the volume available for breakup.

Splash Pattern

(OSAC/ASB definition): A bloodstain pattern created from a large volume of liquid blood falling onto a surface.

Splash is a classification of non-spatter pattern, gush/splash. A splash volume generally involves low-pressure streaming ejections, but also involves circumstance when the volume is ejected without pressure, but en masse.

As explained in the definition of gush, there is no clear objective manner to differentiate a gush from a splash. Splashes tend to have fewer spines and satellites based on a lesser force behind disruption. Lacking a quantitative expression (e.g., a gush has more than ten spines, a splash has ten or less), in the taxonomic approach gush/splash is a singular category of classification. The recognition of this force factor may allow the analyst to better understand whether the pattern was ejected with force (e.g., gush) or with less force (e.g., splash), thereby assisting in isolating an opinion of the possible scene source.

Spines

Linear characteristics evident in both single drop stains and volume stains.

Surface tension in a liquid is a viable cohesive force, which attempts to hold the liquid mass together as a single entity. In the transition from a mass to droplets or smaller masses, long narrow columnar fluid structures form between the mass and the forming droplet. Technically, these are known as fluid jets. Droplets often detach from the ends of these structures or the columns devolve into droplets themselves. If the latter does not occur, these small fluid columns or their remnants fall to the target creating linear stains referred to as spines. See Figure 3.18.

Spines, particularly lengthy spines, are a good indication of force. An excellent example is when a quantity of blood present on a surface is then stomped in by a shoe or foot.

Swipe

(OSAC/ASB definition): A bloodstain resulting from the transfer of blood from a blood-bearing surface onto another surface, with characteristics that indicate relative motion between the two surfaces.

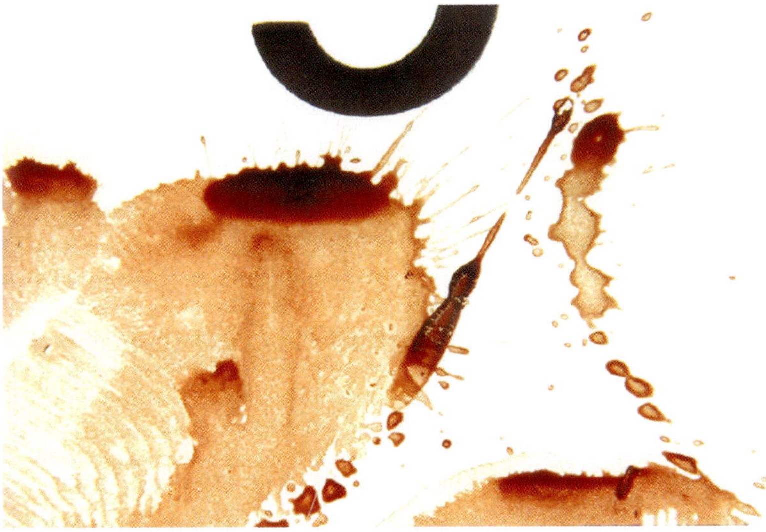

Figure 3.18 Spines are linear features that are found in a variety of patterns. They occur in large and small volumes of blood. Note the linear features at the tip of the finger pattern transfer.

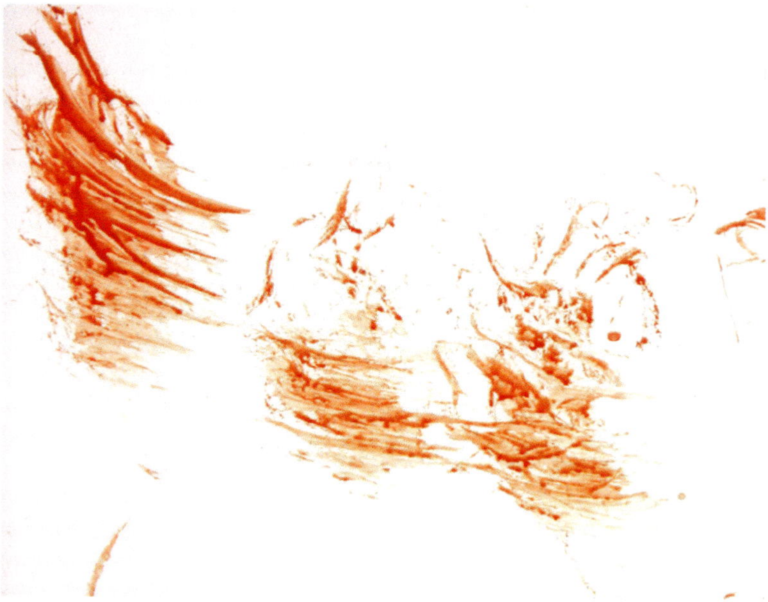

Figure 3.19 A classic swipe pattern created by bloody hair. The lateral motion of the swipe (i.e., the hair is moving left to right) is often evident by examining characteristics such as feathering of the pattern. (Photograph courtesy of Jouni Kiviranta, National Bureau of Investigation, Helsinki, Finland.)

A swipe is a specific classification of transfer stain. It represents one of several forms of stains produced through contact. In a swipe, the blood is introduced to a second surface by a bloodied object. This contact involves relative motion between the two surfaces and this motion is demonstrated through characteristics such as striations, feathering, and diminishing of blood across the pattern. An example is bloody hair contacting a surface during movement. See Figure 3.19.

Target

(OSAC/ASB definition): A surface onto which blood has been deposited.

Synonyms: Target Surface, Substrate

The surface on which a bloodstain is deposited is the target or substrate.

Transfer Stain

(OSAC/ASB definition): A bloodstain resulting from contact between a blood-bearing surface and another surface.

Synonyms: Contact Stain, Smear

In the taxonomy approach, the contact is generally synonymous with the transfer stain and is a broad classification of pattern types produced by contact. These include any contact/transfer stain, not further definable (e.g., wipe, swipe, or pattern transfer). Many smudges and light deposits that are not further definable represent this broad category.

Void

(OSAC/ASB definition): An absence of blood in an otherwise continuous bloodstain or bloodstain pattern.

Synonyms: Shadowing, Ghosting

If a target is exposed to some bloodstain-producing event and a secondary object is either present or introduced between the blood source and the target, this secondary object is likely to prevent stains from reaching the primary target. As a result, the main target is likely to exhibit some aspect of a void, an area within an otherwise continuous pattern that lacks staining. Voids occur across many different patterns, and are the *absence* of staining; therefore, voids are not described as basic "pattern" type. They are simply an adjective associated with other patterns. Voids can be probative and immensely helpful in the analysis. This is particularly true when the secondary target has been moved from its original position. A void often exhibits characteristics of the size and shape of the secondary target.

Verification that a void exists demands a clear absence of staining within the pattern involved. Spatter patterns eventually end. It is not uncommon for inexperienced analysts to claim the presence of a void, when the pattern simply ended. Clearly demarcated edges of the void within the pattern generally assist the analyst in recognizing it is a void.

Wipe

(OSAC/ASB definition): An altered stain resulting from an object moving through a preexisting wet bloodstain.

Wipe is a specific classification of a contact pattern. A wipe represents a disruption of some previously deposited blood. The primary characteristic of a wipe is a smear where a preexisting stain is evident based upon a perimeter stain. The function of distinguishing a wipe is primarily one of sequencing. The wipe must occur after the initial deposit. Not all wipes are probative, but knowing the sequence of bloodstain events may add clarity to the overall events.

Summary

Language is fluid and discipline-specific terms obviously evolve. Some authors/analysts may disagree with the inclusion or exclusion of certain terms presented in this text and certainly the desire to use only the OSAC/ASB terms may resonate with others. As discussed, the latter are at times ambiguous and incomplete; therefore, the authors feel that certain terms added are important, certainly when considering the taxonomic approach to classification.

The taxonomy chapter will introduce a specific classification system that better reflects "why" certain terms presented here are included. Combining the various terms with an understanding of the taxonomic classification system will provide any analyst a functional and objective approach to utilize and apply to their analysis.

It should be noted that technical terms are a necessary evil when evaluating and reporting a bloodstain analysis or when defending some analytical conclusion. It should be stated that in court an analyst may favor a more generic/lay method of phrasing his or her responses, but technical terms are still necessary for the actual analysis. The author is reminded of a comment made by the legendary Texas Ranger John Aycock some years ago when discussing terminology.[8] Paraphrasing Mr. Aycock, he stated in part "My jurors don't want to hear about radiating spatter, they want to know how the pattern got there." Every analyst is responsible for ensuring the judge and jury understand the conclusions associated with the analysis. *How* that is ultimately accomplished is best left to the individual analyst to decide.

Notes

1. Personal conversations – R.M. Gardner and Patricio Leyenda, 10 Dec 2020.
2. Guralnik, David B., Ed., *Webster's New World Dictionary*, William Collins, Cleveland, OH, 1979, p. 49.
3. All current OSAC adopted terminology presented is listed in "Terms and Definitions in Bloodstain Pattern Analysis" AAFS Standards Board, ASB Technical Report 033, 1st Ed, Nov, 2017, WA DC.
4. MIT Lecture Notes, "Lecture 5: Fluid Jets" http://web.mit.edu/1.63/www/Lec-notes/Surfacetension/Lecture5.pdf, 15 Feb 2021.
5. MacDonell, Herbert L. and Brooks, Brian A., Detection and Significance of Blood in Firearms, *Legal Medicine Annual*, 1977, pp. 183–199.
6. Brodbeck, Sikle, M.C., Reflections upon Arteries and Veins – A Plea for 'Spurt Patterns', *IABPA News*, Volume 23, Number 2, June 2007, pp. 5–14.
7. Brodbeck, Sikle, M.C., Reflections upon Arteries and Veins – A Plea for 'Spurt Patterns', *IABPA News*, Volume 23, Number 2, June 2007, pp. 5–14.
8. Personal conversation between Johnnie Aycock and Ross Gardner.

The Pattern Diversity Principle
Classifying Bloodstains

4

Powell, combining descriptions by the National Academy of Sciences and the American Association for the Advancement of Science, offered a simple but effective description of what entails a scientific theory:

> A scientific theory is a well-substantiated explanation of some aspect of the natural world, based on a body of facts that have been repeatedly confirmed through observation and experiment. It refers to a comprehensive explanation of some aspect of nature that is supported by a vast body of evidence. One of the most useful properties of scientific theories is that they can be used to make predictions about natural events or phenomena that have not yet been observed.[1]

Predictability lies at the heart of science; the ability to recognize certain cause and effect relationships within some accumulated body of knowledge and find practical value in that recognition. As described in Chapter 2, the foundational theory of BPA states:

> *Blood as a fluid responds to forces in a predictable manner.*

That predictability is recognized to exist in the three sub-theories or principles of BPA. The most critical of these is the Pattern Diversity Principle, the classification of patterns. Pattern Diversity can be articulated as:

> *The variations in combinations of blood volumes and forces acting on those volumes lead to recognizable classes of patterns.*[2]

The Pattern Diversity Principle starts with ***qualitative*** observations of the effects resulting from the basic forces acting on blood masses; observations that have been recognized by the forensic and scientific community for over 170 years. It is not concerned with secondary, tertiary, or deeper affects or effects. It certainly is not attempting to describe quantitative physics equations of the underlying fluid dynamics of these highly dynamic interactions. The "hidden structure" of such relationships obviously exists and at some point, perhaps future research will reveal some aspect of this hidden structure that will assist BPA in a practical fashion. But even assuming such research is forthcoming, the true practical value of BPA lies in that which has already been observed with respect to patterns.

Pattern Diversity lies in the comparison of physical characteristics observed in an unknown stain with the known physical characteristics of basic pattern types. The value of that comparison is recognition of the underlying mechanism behind an unknown stain's creation, the ability to use the characteristics to include or exclude the unknown stain to a given classification. Without regard to future research, this basic idea will not change, other than potentially refining the means of comparison.

DOI: 10.4324/9781003408734-4

Comparison of clear, articulable characteristics lies at the heart of the Pattern Diversity Principle. There must be clearly defined criteria (physical characteristics) for the comparison. Such comparisons cannot be based singularly on an "I'll know it when I see it" or "I've seen it a hundred times before" mindset. Granted, with sufficient knowledge and experience, nearly instant visual recognition of some pattern types is often a fact of life for experienced analysts. Even if unable to specifically articulate what those characteristics are immediately, they do recognize them. But such visual recognition must be followed by a functional scientific validation. There must be a directed attempt to refute that initial visual recognition. Otherwise, visual recognition is simply an oversimplification of comparative science. True comparison in BPA seeks to understand and functionally identify differences and similarities between the different classes of pattern types. This is only possible when articulable criteria are established and utilized, criteria defined through the qualitative observations of known mechanisms.

For example, the criteria associated with a cast-off pattern should include the following:

- A series of related circular- or elliptical-shaped stains.
- In a linear or curvilinear orientation.
- The stain shapes within the pattern change, becoming more or less elliptical.

When Figure 4.1 (the area outlined in red) is considered against these criteria and the criteria are met, the pattern can then be objectively considered as a potential cast-off pattern. But

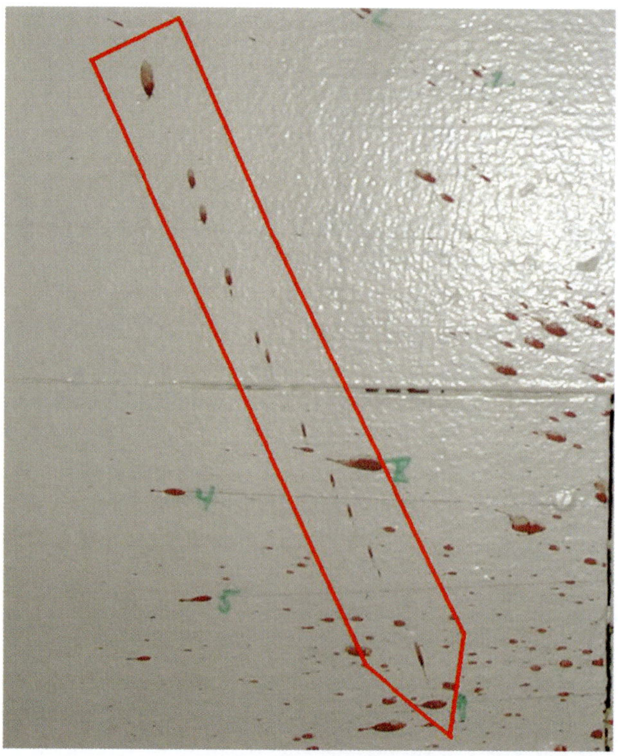

Figure 4.1 Bloodstain Pattern Analysis considers standard criteria and compares them against a questioned stain. The pattern outlined in red has small circular and elliptical stains deposited in a linear orientation. These criteria help support a classification of cast-off, blood flung from an object in motion.

that is only part of the question. Characteristics of other pattern types must also be considered. There are many pattern types that involve spatter (circular or elliptical stains). The fact that the pattern does not radiate allows the pattern to be excluded from any pattern type that involves radiating spatter. The lack of evidence of flows emanating from individual spatter allows the pattern to be excluded from the spurt grouping, another form of linear spatter. The recognition that the gamma angles are altering (becoming oriented left to right on the vertical surface and not falling consistent with gravity) excludes the pattern from the drip trail group. Only with confidence of an exclusion of other associated spatter types can the pattern be classified as cast-off alone. It goes without saying that an experienced analyst may well recognize these differences and similarities in a near instant, but until they have considered refutation of other types, the classification is not complete. And if challenged, the analyst must be capable of articulating these similarities and differences clearly. Classification of bloodstains always involves a comparison of physical characteristics of the unknown pattern to criteria found in known pattern types. This evaluation is a deductive process; ultimately either the stain meets certain criteria, or it does not.

Classification versus Overall Opinion

Classification is the most important step in the overall Bloodstain Pattern Analysis process. Done properly, it sets the stage for any subsequent analytical conclusions. Ultimately using the classification along with associated evidence and the scene context, the analyst may present a more refined conclusion or opinion regarding the specific unique scene source event. This distinction between the classification (what kind of pattern is it based on physical characteristics) and the analyst's opinion regarding the source event (based on all known information how did this pattern come to be in this scene) is a nuance often lost on the novice analyst. For example, consider Figure 4.2. At the scene, the analyst may be presented with large-volume drops that have rained down across a surface. The analyst will likely end up with a very general classification, as the pattern is merely spatter-type stains with no other characteristics. That general classification, however, when combined with information such as additional spurt patterns in the scene and evidence of a single arterial wounding (stab to the heart), may lead the analyst to provide a *concluding opinion* that the stains are the result of a particular arterial source. Given the limited physical characteristics, the classification decision might be tenuous or certainly limited. That may not prevent an analyst from presenting a more defined opinion, after applying the full methodology of BPA to the data available. It should be evident, however, that classification (a deductive process) must remain true throughout the inductive argument. For instance, one cannot classify the stain as a flow and then inductively decide it is a smear. The ultimate opinion expressed by the analyst can always become more distinct, but it cannot defy the initial deductive classification.

Classification versus Definition

Classification of patterns is quite distinct from the mere definitions of patterns. Valid classification systems set out some form of physical criteria to judge a questioned stain against distinct characteristics that can be compared. Definitions, on the other hand, are

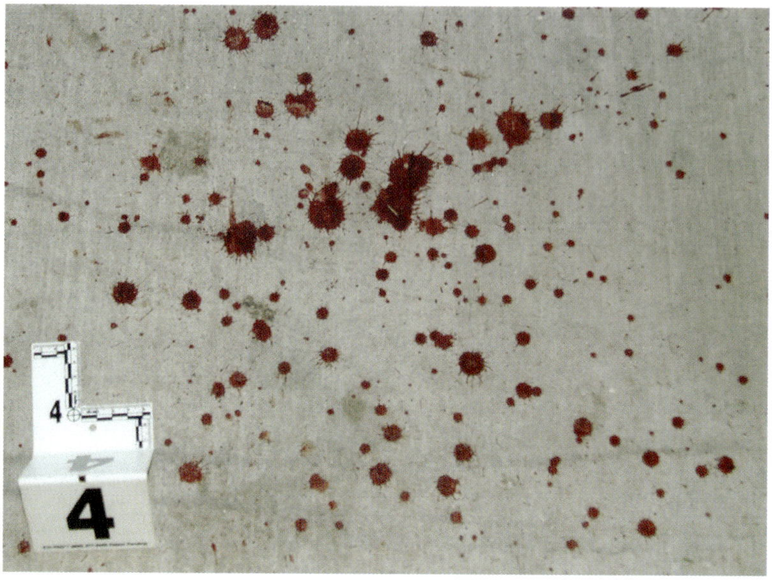

Figure 4.2 It is not uncommon to encounter patterns that have few physical characteristics. This pattern consists of spatter (drops that have been in free flight). Some of the individual stains appear large. Are they arterial stains that have rained down or simply passive drips? A deductive classification would be to claim the pattern as "spatter stains" and nothing more. This would not prevent the analyst from making a more refined conclusion based on scene context.

simply a statement of meaning. The definition of cast-off is "a pattern created when blood is flung or detached from a moving object." This definition is certainly effective for understanding the meaning of the word cast-off, yet it in no way explains what characteristics the analyst used to arrive at a conclusion of "cast-off." Classification goes beyond mere definition.

Numerous classification systems currently coexist in Bloodstain Pattern Analysis, which often confuse those new to the discipline, but these differences are simply a matter of perspective. Each classification system starts with some basic issue or perspective (e.g., size of spatter, general mechanisms of creation, shape of stains) which then leads to the various classifications. These different systems, no matter what their initial perspective, ultimately define the same basic pattern types. The names assigned to these patterns may vary, but the patterns themselves are visually and generally agreed upon. These will include:

- Blood dispersed from a point source by a force or impulse (e.g., impact pattern, mist, expectorate)
- Blood dispersed over time from an object in motion (e.g., cast-off pattern)
- Blood dispersed in streaming ejections (e.g., spurt and gush/splash pattern)
- Blood dispersed as a function of gravity (e.g., drip pattern, drip trail, drip stain)
- Blood accumulating and/or flowing on a surface (e.g., pool, saturation, flow)
- Blood deposited through transfer/contact (e.g., smear, wipe, swipe, pattern transfer)

There are differences within these basic pattern types that allow further discrimination (e.g., recognition between expectorate and impact spatter). Such discrimination doesn't change our understanding of the underlying basic mechanism; it simply refines it. These variations (the classes of patterns) and their associated criteria are *reproducible phenomena*. These classes/groupings are the functional limits of predictability associated with Pattern Diversity. As certain as there may be other distinct mechanisms that produce patterns, such mechanisms do not present specific criteria that allow discrimination. For example, cessation cast-off, defined as the creation of a spatter pattern produced when a bloody object suddenly stops its motion, creates no absolute criteria that would allow its discrimination. Variations occur in cessation patterns, such that no specific physical criteria are present that an analyst can rely on. Spatter produced by cessation cast-off are deduced through context (e.g., as a BPA opinion).

Given the dynamic nature of events producing bloodstains, not every instance of bloodstaining can be defined absolutely; there may be circumstances in play that might make classification between even the basic pattern types difficult. Ambiguous patterns do exist, patterns where the differentiating criteria are not evident. For example, there may be a pattern where spatter are oriented in a linear fashion, but additional criteria are absent. In such circumstances, it may be impossible to differentiate between the three types of linear spatter (cast-off, drip trail, or spurt). This situation forces the need for a system of classification that has clear parent–child relationships. In the example described, the criteria present demand a classification of some form of linear spatter. Based on classification alone, all three linear forms of spatter are still potentially valid explanations. Despite this ambiguity, all other spatter types are excluded, as well as the non-spatter groups. Typical of all class characteristic evidence, such exclusions have probative value.

If there is confusion in the many systems of classification, it revolves around how one objectively decides what a particular pattern is. What defined criteria exists in each system to judge the various types of patterns? This is not always apparent. Many systems of classification fail to articulate these criteria clearly. The best system for resolving this failure is a taxonomic approach, as described in this book.

Why a Taxonomic Classification System?

Keeping the concept of criteria versus definition issue in mind, most of the classification systems have failed in one sense. Many offer a hierarchal flowchart of some nature along with effective definitions of the various classifications; but the specific details which the analyst examines are buried deep in narratives or illustrated through pictures. This situation begs the question: How is the practicing analyst able to examine a questioned stain objectively and what criteria will they apply when deciding on a classification?

The presented taxonomy is a set of rules previously defined in *Bloodstain Pattern Analysis With an Introduction to Crime Scene Reconstruction 3rd Edition.*[3] The taxonomy mindset is derived from biology where organisms are classified by shared characteristics. In biology, the various levels of classification, phylum, order, family, genus, and species are defined by shared taxon. They also effectively create a parent–child relationship, top to bottom. When broad taxon are shared (e.g., for phylum or order), the broader the ultimate classification. As one moves down the groups (e.g., family, genus, or species), more specific characteristics/taxon are utilized to differentiate the various groups. This allows a more refined classification as to the nature of the organism. Specific criteria reduce ambiguity in the process and provide a foundation for any conclusion offered. Specific criteria provide for an objective analysis.

The hierarchical nature (clear parent–child relationships) in the taxonomy become significant in practical application. They reduce subjectivity. Each level of the hierarchy represents criteria to consider, a decision point. The analyst examines the item for the specific criteria listed. The criteria for the classification either exist or they do not. When a level is reached and ambiguity exists, rather than forcing a decision the analyst returns to the preceding parent level where the requisite criteria were met. As in the prior example, presented with some form of linear spatter, but insufficient additional characteristics that would allow one to conclude spurt from cast-off, objectively the pattern is classified as linear spatter. While "linear spatter" is not a term currently found in the AAFS Standards Board, ASB Technical Report 033, 1st Ed., published in 2017, it certainly, and clearly, describes a distinct bloodstain pattern. This is another benefit to a hierarchical approach to classification, wholly consistent with the general approach to class characteristic evidence. The use of the parent-level "linear spatter" that share certain class characteristics ensures that children of this parent level are not inappropriately excluded (e.g., spurt, cast-off, drip trail).

As described in Chapter 2, BPA and specifically the three predictable areas of BPA, including the Pattern Diversity Principle, can be described using Woodward's causation relationship ideas. Under the Invariance Under Intervention mindset, these causation relationships do not have to be "lawful." Within a given range of variables, if the relationships remain constant (invariant) and meet the refutation approach, they are scientific. These relationships are ultimately presented as counterfactual statements (e.g., if this, then that). For example, *if* the pattern is linear spatter, *then* the characteristics present will include related circles and ellipses, oriented in linear or curvilinear orientations.

This overall idea can be generally described in the following example:

X – is the force mechanism, where
- X1 is a point source dispersion
- X2 is accelerated motion
- X3 is a streaming ejection
- X4 is gravity
- X5 is contact
- X6 is an accumulation based on gravity

Y – is a blood mass acted on

Z – is the physical characteristics in the resulting pattern (*significantly synopsized for purposes of this example*), where
- Z1 – circles and ellipses, radiating distribution, changing in shapes further out in the pattern
- Z2 – circles and ellipses, linear distribution, changing shape on a single surface
- Z3 – circles and ellipses, linear distribution, with flow *or* irregular margins, with associated spines and satellite stains
- Z4 – circles and ellipses, generally 3–25 mm in size *or* irregular margins with random shaped and oriented spatter around the margin
- Z5 – irregular margins with striations, feathering OR demarcated patterns
- Z6 – regular margins, accumulating or extending across a surface

For simplicity of the example, we will limit the associated variables as well:

Q – is acceleration of the force disrupting the blood mass

R is the volume of the blood mass

If X1 (a point source dispersion) is the cause of Z1, then as we keep that force mechanism unchanged (invariant), while altering either Q or R (force and volume variables), the Z1 characteristics will be present and remain invariant as well. If we change the X variable to X2, while keeping Q and R stable, Z will become variant, changing and becoming consistent with the Z2 characteristics.

Invariance Under Intervention recognizes that this X1–Z1 relationship is valid within a given range or circumstances. Pattern Diversity recognizes that if associated variables significantly change, this relationship may break down. For example: adjusting the disrupted volume (R) significantly for impact, say to a large pool, the effect Z1 will become variant manifesting additional characteristics such as those found in gush patterns. If we alter the substrate (Q) to a highly porous substrate, the Z1 characteristics may become so muted that they cannot be discerned.

This example although highly synopsized (e.g., the description of the characteristics, as well as the listing of associated variables), demonstrates the application of Woodward's causation relationship description to Pattern Diversity. It can be applied to any predictable claim within the principle. For each force mechanism (the cause), there are associated characteristics (the effects). For each subcategory of pattern within any force mechanism (e.g., for point source dispersion: impact, mist, or expiration), similar descriptions can be defined. Additionally, for any of these relationships, the range in which they remain predictable (invariant) is generally understood. The foundation for all Pattern Diversity cause and effect relationships described lie in qualitative observations of known source events. Within the taxonomy, they are presented as simple counter-factual statements associating the cause with the effect: *If* the criteria set "Z" is present, *then* the classification (and its associated force mechanism) is X.

A Taxonomic Classification System for Bloodstains

Prior to the 3rd Edition of *Bloodstain Pattern Analysis, With an Introduction to Crime Scene Reconstruction,* no true taxonomic classification existed in Bloodstain Pattern Analysis. James et al. represented a "taxonomy" in their book, but lacking were specific criteria for the various patterns. When present, these criteria were buried deep in the general narrative.[4] James simply presented a new hierarchical format. Wonder provided an initial criterion-based approach, yet her efforts ultimately returned to using mechanistic explanations.[5] Without question hierarchies are important, but it is criteria that define a taxonomic approach. Without specific physical criteria that can be related back to a hierarchical parent–child mindset, no truly objective conclusion is possible. At each decision point in the hierarchy, the analyst must have characteristics to evaluate.

It should be stated clearly that in terms of the history of Bloodstain Pattern Analysis, the question is not whether criteria exist in other systems, they simply have not been articulated or functionally isolated to the most objective criteria.

While a member of the FBI's Scientific Working Group on Bloodstain Pattern Analysis (SWGSTAIN), author Gardner first suggested the need to define these criteria more clearly. SWGSTAIN adopted the idea and what was the Terminology subcommittee became the Taxonomy and Terminology (T^2) subcommittee. Yet, despite these efforts over the period of 2002 through 2011, no consensus could be agreed upon for several reasons. None of this related to the *lack of specific characteristics* in patterns. Subsequently, the OSAC Bloodstain

Pattern Analysis subcommittee has pursued a similar attempt, yet to date no published criteria have been forthcoming from said groups. In 2008, Bevel and Gardner initially published the taxonomy. The need was clear and the potential for criteria from other groups was not even on the horizon. Over the last 17 years, that taxonomy has been revised slightly to better articulate the manner the original criteria were described, but the criteria themselves are still the same. It is not that the description of the criteria cannot be better articulated in some manner in the future. Recent scientific reviews of various classification systems indicate that this approach, using specific unambiguous articulable criteria, is critical component moving forward for the discipline.[6] Bevel, Gardner, and Associates (BGA) have taught this approach to analysts for the last 15 years and have always been clear that there may be better ways to describe the characteristics and/or quantify some of them. But lacking current consensus from established groups responsible for the future of BPA should not prevent any organization or analyst from adopting a more clearly articulated classification system. The taxonomy presented here is a functional and a valid starting point. As with all science, it may not be the end all approach. The authors offer this simple analogy as to the value of the taxonomy presented:

> If asked, most people would love to own a sleek sports car; with clean, exacting lines, all the bugs removed. But before one can have a sleek sports car there had to be its precursor, the Model T. A product that was clunky but held potential for future refinement.

The BGA taxonomy is that precursor, although refined over the past 15 years through observation and refinement of verbiage based on those using it. Are further refinements possible? Assume that to be true. A prime example is the cast-off characteristic of parallel directional angles. This concept always requires effort in classrooms to explain fully. There may be a more effective way to articulate that particular characteristic. Yet that characteristic exists, as well as the others described here. Regardless of how they may ultimately be articulated, they will still be present.

Figure 4.3 shows the first three levels of the taxonomy hierarchy. The initial category is simply "bloodstain"; the initial decision point is whether the stain in question is blood. A classification of "bloodstain" ensures first that only stains associated with blood are considered. There may be instances where no further differentiation is possible, ambiguous pattern situations that prevent further classification. The decision point then divides patterns into two broad groups, the spatter/non-spatter groups. Anita Wonder set these broad

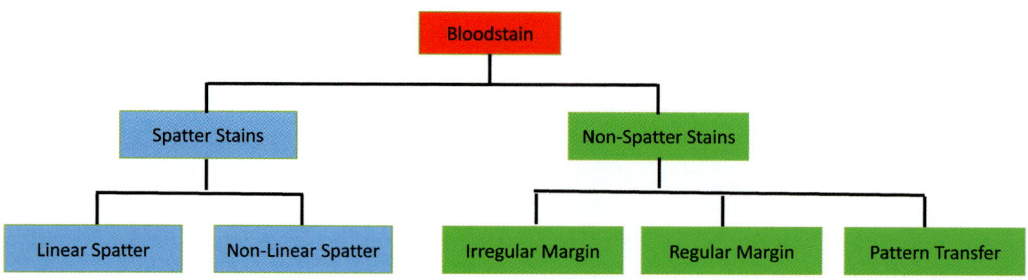

Figure 4.3 The first three levels of the bloodstain taxonomy generate two primary categories (spatter and non-spatter) and these are further divided into five lower subcategories (linear spatter, nonlinear spatter, non-spatter regular margin stains, non-spatter irregular margin stains, and non-spatter pattern transfers).

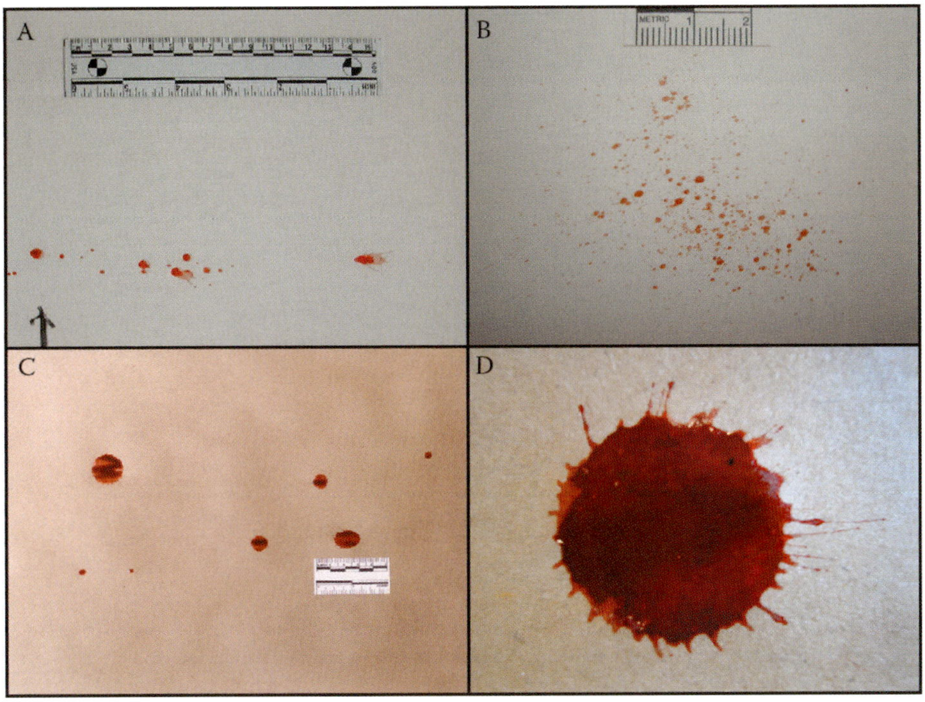

Figure 4.4 Various spatter stains. Pattern A is cast-off. Pattern B is impact spatter. Pattern C is a simple drip trail. Pattern D is single drip. The common characteristic that spatter share is generally elliptical or circular-shaped stains, which result from the impact of a free-flight droplet with a surface.

groups based on the most evident initial characteristic found in patterns, whether the stains are primarily small, circular, or elliptical shaped stains. This characteristic defines if the primary stains have been in free flight. Once disrupted and put in free flight, blood masses (except for certain large-volume ejections) generally break up into smaller masses. The resulting free flight droplets upon collapse on a target surface create generally circular or elliptical shaped stains. See Figure 4.4.

The Spatter Family

The initial decision point that the stains are made up of only small circular and elliptical stains (with or without associated flow) criterion defines the spatter group. General morphology of spatter stains includes this obvious elliptical or circular shape with or without associated flow. The parent stain in spatter may demonstrate an edge characteristic known as scalloping, and/or tails produced by the collapse mechanics of the droplet. Some of these scallops or tails may throw small satellite stains that may be connected, partially connected, or completely disassociated from the parent spatter. See Figure 4.5.

Other critical criteria/morphology concepts are explained as one moves down the spatter stain hierarchy; these include:

- Related spatter
- Linear or curvilinear distribution
- Radiating distribution

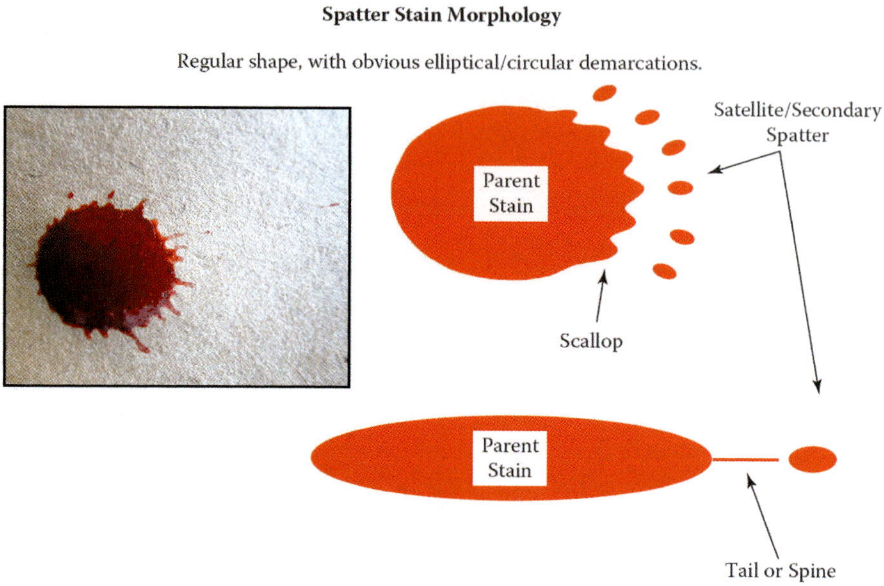

Figure 4.5 Spatter stains demonstrate specific traits (morphology). This includes the circular or elliptical shape, scallops, spines, or a tail, and secondary/satellite spatter.

- Random distribution
- Volume in individual stains
- Progressive change in shape
- Parallel directional angles that change relative to the orientation of the swing

The spatter group hierarchy is depicted in Figure 4.6 and each classification is described with any preceding family, a general definition, the specific classification criteria, and a discussion of these criteria.

Category: Spatter

Family: Bloodstain

Definition: Stains resulting from blood that has been put in free flight.

Criteria:

- One or more distinct parent stain(s) with generally circular- or elliptical-shaped demarcations. With or without flows associated with the stains

Discussion: The single criterion of a *circular- or elliptical-shaped* stain is a function of a small mass of blood that has been in free flight collapsing on a surface to produce the resulting stain. There may or may not be small flow stains originating from the individual spatter. The shapes may be symmetrical (clearly demarcated) or highly asymmetrical. Spatter are produced through a variety of dynamic mechanisms, as well as from simple gravitational force.

There are two primary subcategories of spatter: linear spatter and nonlinear spatter. See Figure 4.6. Before proceeding to these aspects, it is important to understand the concept of "related" spatter, which is one of the characteristics that must be evaluated.

"Related" spatter is understood through consideration of three basic aspects: proximity, directional angle, and impact angle. Stains must have sufficient

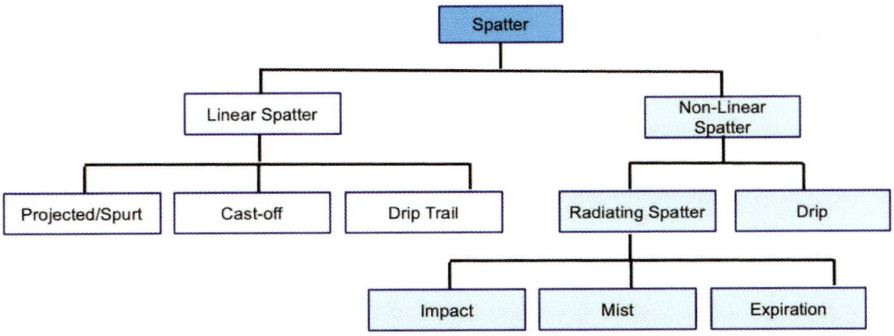

Figure 4.6 The spatter family taxonomy. Note that the lowest level of the taxonomy is divided into the common terms associated with Bloodstain Pattern Analysis. The taxonomy does not change the nature of the basic patterns; it simply identifies the shared class characteristics between patterns.

proximity to one another, taking into consideration the dynamic nature of the various spatter events. Stain shape (a function of the individual stain's impact angle) must be consistent. One should not see circular stains intermixed in the same position as the elliptical stains. The stains' directional/gamma angles must be consistent. If associated, there cannot be stains collocated that were moving in opposing directions at the time of deposition, excluding the rarely observed anomalous ricochet. With the exception of a classification of drip, the gamma angles will show either some aspect of parallel deposition, or a radiating deposition.

Dynamics: When a blood volume is disrupted by some force or impulse smaller masses are put into free flight (e.g., flung from an object, dripping, disrupted at a point source). The disrupted blood mass generally proceeds from a sheet to ligaments to individual droplets. The cohesive forces of the fluid lead to these smaller masses attaining spheroidal shapes. Then as the spheroidal droplet impacts a surface, it collapses in a relatively consistent fashion. The nature of this collapse is described in detail in Chapter 5. The collapse results in stains that have either a circular or elliptical shape.

Probative Value: Any given spatter may provide indications as to its spatial origin (e.g., its impact or directional angle). Beyond spatial origin, a single spatter has limited probative value as the range of mechanisms that produce spatter vary widely.

Category: Linear Spatter

Family: Bloodstain: Spatter

Definition: A series of related spatter stains dispersed over a surface in a linear or curvilinear orientation.

Criteria:

- Related circular and/or elliptical stains with or without associated flow
- In a linear or curvilinear orientation
- Demonstrating an evident interrelationship in the shape/impact angles
- Demonstrating an evident interrelationship of the gamma/directional angles

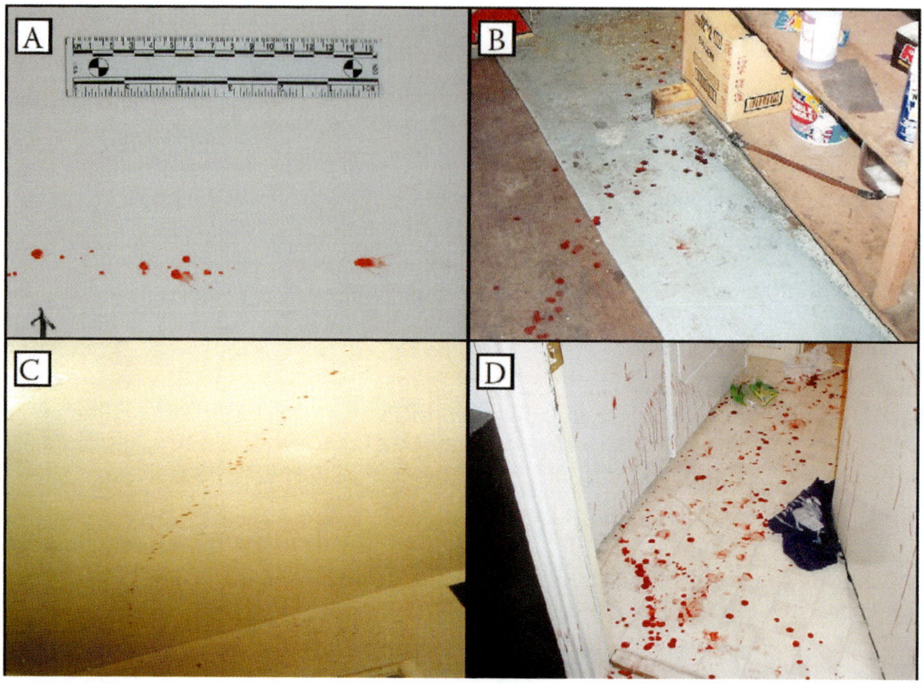

Figure 4.7 Spatter deposited in linear or curvilinear orientations. Pattern A is a cast-off. Pattern B is a drip trail. Pattern C is another cast-off. Pattern D is a projected/streaming ejection event.

Discussion: In considering linear spatter criteria, the stains will be oriented in a linear or curvilinear orientation, but not always as a single line of stains. There may be a wide pattern of linear stains, not simply one after the other. This line of stains can be straight or oriented in an arcing or serpentine pattern. At any point in the pattern, side-to-side the directional/gamma angles of the stains will lie generally parallel with one another. The directional angles may well change over the course of the pattern. See Figure 4.7.

Dynamics: The disruption mechanisms producing linear spatter include streaming ejections, blood flung from bloody objects, and bloody objects moving while dripping. These mechanisms generally involve small jets (long fluid columnar structures or ligaments) that devolve into individual free flight masses of blood (droplets). The similar vectors of the associated droplets lead to deposits of stains in lines.

Probative Value: A classification of linear spatter define the three associated classifications (drip trails, spurt, cast-off). Additional characteristics to further refine the classification may be muted or absent. If sufficient scene context exists, a more refined opinion may still be possible.

Example: Given a series of linear spatter directed to a wall where the spatter have directional angles that are not oriented downward, drip trail is clearly excluded.

Even as a general classification linear spatter is still valuable in exclusion of certain scene mechanisms.

Category: Projected/Spurt

Family: Bloodstain: Spatter: Linear Spatter

Definition: Stain patterns created when blood is ejected in a stream, often under pressure and typically encountered when an artery or the heart is breached

Criteria:

- Related circles and ellipses
- In linear or curvilinear orientation including Vs, arcs, or a serpentine pattern
- With evident volume which is best defined by flow from individual spatter stains
- Large elliptical stains may accompany the pattern

Discussion: The spurt pattern is a form of what OSAC/ASB define as projected. Spurts are produced through a streaming ejection, a jet. The stream is often ejected under pressure (e.g., an arterial source), but a stream produced by blood flowing with velocity (e.g., a wound bleeding freely and coming off a finger, elbow or object) can produce a classic spurt pattern as well. The stream source itself may be in motion within the scene or the stream may be ejected in varying directions. Unlike other forms of spatter breakup, the ejected stream may or may not produce only stable drops/spheroids. If allowed sufficient time in free flight, even the slightly larger and unstable volumes will collapse producing circular and elliptical shapes. The larger volume present in the resulting droplets often produces a flow from the individual spatter. Such flows are uncommon in other types of spatter deposition. It is not uncommon to observe a change in the pattern's linear aspect produced by changes in the systolic and diastolic pressure of the circulatory system. This can result in a wave-like, arcing or serpentine patterns. See Figure 4.8. If the source of the stream changes orientation to the target, this leads to diversity in appearance of the linear aspect of the pattern, producing a completely different pattern, the gush/splash.

Dynamics: The stream originates from some point (e.g., a breached artery, blood flowing off a finger). This stream, known as a jet in fluid dynamics, shears apart during flight from both gravity and air resistance acting on the streaming mass. This differs from breakup at a point source (e.g., impact) or ejection from different points in space (e.g., cast-off). The fluid column is directed in some orientation and as the column breaks up the individual masses or drops follow similar flight paths, resulting in linear distributions. It is not uncommon to have several of these droplets strike beside or on top of one another. Together their combined volumes coalesce flowing across the surface like a single droplet's collapse, creating associated large elliptical stain(s). See Figure 4.9 The excess volume present in the droplet, when deposited on a vertical surface, often leads to a small flow that emanates from an individual spatter stain.

Probative Value: A spurt classification limits the potential mechanisms to a streaming ejection. Potential scene mechanisms are limited to arterial events or situations where a volume is flowing off a surface. The spurt's directional indicators provide information as to the orientation of the stream to the surface.

When a spurt pattern ejects across both a vertical and horizontal surface, it is not uncommon that a linear pattern will appear on the horizontal surface, leading to or coming from the vertical aspect of the pattern. This is observed

Figure 4.8 A projected/streaming ejection pattern, often called a spurt. An arc and linear deposition can be observed on the wall to the left. Flows from individual spatter are present in these stains as well. Large volume spatter that fell to the floor can be seen leading to and extending away from the pattern on the wall. (Photograph courtesy of Steven Chancellor, United States Army Criminal Investigation Command, and Donald R. Schuessler, Department of Public Safety, Eugene, OR.)

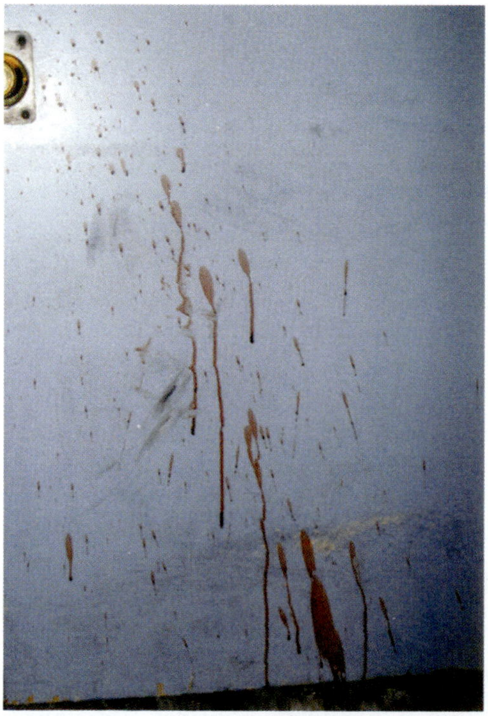

Figure 4.9 Large elliptical stains are found in spurts as well. These occur when a large mass of blood impacts a surface, or multiple individual spatter impact at the same time. (Photograph courtesy of Steven Chancellor, United States Army Criminal Investigation Command, and Donald R. Schuessler, Department of Public Safety, Eugene, OR.)

in the lower left of Figure 4.8. Such patterns are very indicative of the general location of the source of the stream at the time of ejection.

One significant issue regarding a spurt pattern is when it occurs solely on a horizontal surface. In such instances, the volume indicator, flow from individual stains, is completely absent. In such instances, other indicators such as shapes that suggest other than perpendicular impact angles (e.g., as in dripping actions) are often present, but such patterns are often misidentified as drip trails. Patterns of this nature, when found solely on a horizontal surface, are likely to be classified as linear spatter.

Category: Cast-off

Family: Bloodstain: Spatter: Linear Spatter

Definition: Stains created when blood is flung or projected from an object in motion. Cessation cast-off, produced when that motion stops, may exhibit singularly cast-off characteristics, but often demonstrate additional characteristics.

Criteria:

- Related circles and ellipses
- In a linear or curvilinear orientation
- With parallel direction angles that change relative to the orientation of the swing
- On a single surface the shapes of the stains will change, becoming more or less elliptical[7]

Discussion: Blood adhering to an object that is subsequently accelerated (both in speed and direction) produce linear patterns called cast-off. In terms of the acceleration, it's not enough that the object simply increase speed, there must be a slight change in direction of the object as well. This movement causes droplets from the object or a small stream of blood that flows off the object and subsequently breaks into droplets to continue to move in its initial direction. In either case the droplets are released at different points in time, at differing angles to the surrounding targets, but in a vector or path oriented with the swing of the object. As a result, the stains are deposited in lines that may be straight, curved, or any combination. As previously described, this linear aspect may manifest itself as a series of single stains in line with one another or they may be deposited in a wider pattern. See Figure 4.10. As the droplets are ejected in the same direction as the object, generally on a single surface the directional angles of the individual stains are consistent and follow the general direction defined by the actual arc of the swing. This idea, however, is a bit more complex depending upon the orientation of the swing to the target and certainly when the cast-off is deposited across multiple surfaces. In the right orientation to a wall (a given height of the fulcrum point of the swing) and assuming a downward swing, some of the stains in the upper aspect of the pattern might well strike the surface with an upward movement. These stains may reflect a slight upward directional angle, while the stains that struck more perpendicular to the wall and those that were deposited as the swing continued downward on the wall will reflect downward directional angles. The directional component of the resulting stains is always relative to the target's orientation to the release point of the droplet during the swing. If a single cast-off crosses

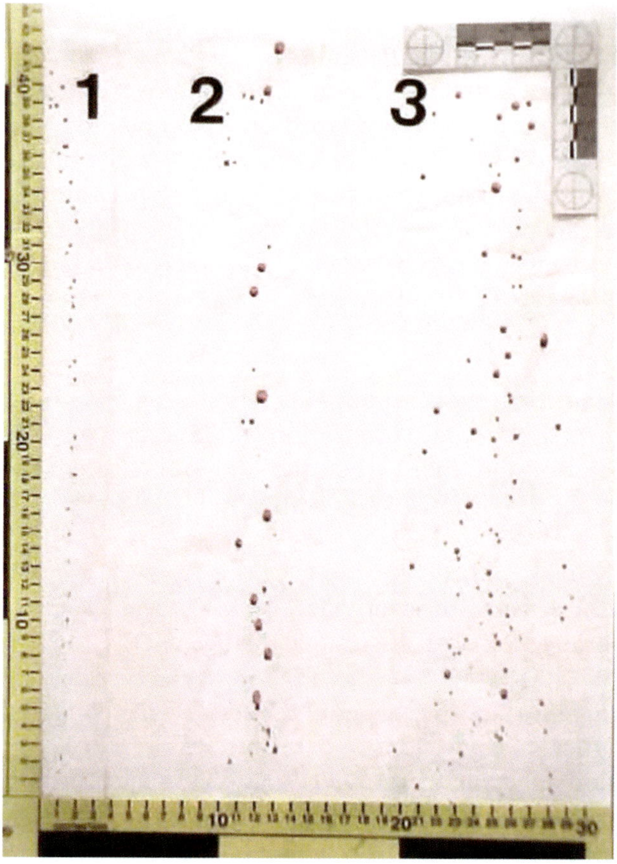

Figure 4.10 Three different cast-off patterns produced by three events. The first was created by swinging an edged weapon, the second with a bat, and the third with a piece of lumber.

multiple targets oriented 90 degrees to one another, the two components of the cast-off created can actually point to one another. Whenever evaluating cast-off of this nature, one must effectively consider how the individual drops are striking the surface.

The concept of parallel directional angles does not mean the directional angles are always "in line" with the overall path, as this interplay is quite dynamic. Rather, it indicates a consistency of the directional angles at any point along the pattern. Looking at stains deposited adjacent to one another, the directional angles will lie generally parallel. They are not radiating out from one another.

Cessation cast-off is produced when the bloody object ceases some motion, and is often present along with the linear cast-off pattern. Cessation cast-off may demonstrate a slight cast-off appearance, but more often display additional characteristics (e.g., a slightly radiating distribution). To understand the creation of cessation, it is important to realize the cast-off droplets projected during the swing do so generally at a tangent to their release point, typically 60–90 degrees, depending upon whether droplets are produced at the bloodied object, or from a small column of fluid which then breaks up into droplets. When the motion of the object ceases, any remaining blood continues to move

Figure 4.11 A distinct characteristic of a cast-off pattern is that because the drops are ejected at different times from the object in motion, their directional angles (gamma angle) will change over time to follow the change in direction of the swing of the object. Note the change in directional angle between A1 and A2.

in the direction the object was moving at the moment it stopped. This blood is projected forward of the object. These cessation droplets are likely to appear as radiating spatter ejected from a point source. If directed forward during a blow (e.g., toward a victim), cessation cast-off may be completely lost in any associated impact spatter. In rearward swings, they may appear on any surface (to include the backs of individuals swinging items). There are no absolute characteristics associated with cessation. Patterns of this nature, when recognized, would likely be described and classified as nonlinear spatter. Recognition that they are in fact cessation and related to swinging a bloody object is very much a function of the context in which they are found.

Dynamics: As a bloody object is accelerated, the inertial energy sets the blood in motion. The adhesive and cohesive properties of the blood are overcome and blood can detach directly from the object or as a small column (what Phua referred to as a ligament) which flows off the object. In the latter, gravity, air resistance, and Plateau–Rayleigh Instability lead to the column devolving into individual droplets, but all following similar paths leading to the inline distribution. The droplets are released along their individual paths at different points in time and space. So, each droplet has a distinct impact angle to the associated surfaces. Stains associated with a cast-off pattern on a given surface will always manifest either an increasing or decreasing impact angle across the pattern. See Figure 4.11. In some instances, this change in shape may be slight, but it will be present.

Probative Value: Cast-off patterns represent a bloody object being swung. The nature of that object is not singularly an offensive weapon. Any bloody object can produce a cast-off (e.g., bloody hands or appendages in motion).

The volume present in the individual droplets produced by any cast-off action, including those produced by different swings of the same object, can vary dramatically. As volume of the droplet affects the size of the resulting spatter, no specific size of spatter can be delineated. Generally, the smaller the surface area bearing the blood, the smaller the resulting stains. But volume factors can eliminate this correlation. Effectively, a cast-off is a cast-off and specific identification of what was being swung is difficult based solely on the pattern. Scene context (e.g., what items may have been bloody and potentially were in motion) may allow a more refined conclusion as to the scene source.

Category: Drip Trail

Family: Bloodstain: Spatter: Linear Spatter

Definition: A pattern of individual spatter deposited in a linear orientation on a surface, demonstrating movement of the dripping blood source from one point to another.

Criteria:

- Related circles and ellipses
- In a linear or curvilinear orientation
- Primary stains demonstrate a generally consistent size range (typically within a width range of 3–25 mm)
- Directional angles will correlate to gravitational force
- The stains lead from one point to another

Discussion: Spatter produced by dynamic actions (e.g., streaming ejections, accelerated motion) typically create smaller volume droplets. This in turn results in smaller diameter stains. The only force acting to disrupt a blood mass in drips and drip trails is gravity, so the droplets tend to be larger.

The typical "stable" drop is on the order of about 0.06 ml. producing a free flight droplet with a diameter of about 5–5.5 mm. Depending upon the substrate, a drop of this size collapsing on a surface has the potential to create stains with widths of easily 14 mm in size. Should the drop form on an object with a large surface area, the volume present in the drop may be greater, resulting in larger stains. It is not uncommon to see drips that measure 20–25 mm in size. Objects with smaller surface area typically produce much smaller drops, resulting in smaller stains generally 3 mm in size. Prior studies place the smallest parent stain produced by drip as no smaller than 3 mm.[8] This figure is a function of the minimum volume required for gravity to overcome the adhesive and cohesive forces. Taking into consideration this volume, the substrate the stain falls to, and the speed of the droplet as it falls, any stain smaller than 3 mm can be considered to have formed from some additional force beyond simple gravity (e.g., a slight acceleration of the object as the drop forms). Thus, a pattern consisting of stains smaller than 3 mm is excluded as a drip or drip trail. Although no absolute maximum size of a drip stain has been defined, patterns with stains exceeding 25 mm in diameter should be considered cautiously.

The blood trail is simply a deposit of drip stains deposited over time as the dripping blood source moves from one point to another. Given a single source event (e.g., blood dripping from an individual), the range of stain sizes evident in the pattern will remain generally consistent relative to surface characteristics. If the object dripping blood moves across varying targets (e.g., moving from tile onto carpet), the size of the stains may change as a function of this substrate change. See Figure 4.12.

Drip trails can be produced from either replenishing or nonreplenishing blood sources. The size of stains associated with a nonreplenishing source will

Figure 4.12 A drip across multiple targets. A drip event deposited stains onto the concrete and the door mat. Note the dramatic size change between the stains deposited on the concrete and the carpet. The primary difference in size is not associated with the volume of the drip, but rather with the nature of the target surface.

Figure 4.13 To be a drip trail, the stains must be deposited in a linear fashion that leads from one point to another.

ultimately decrease over time. The size of stains present in the drip trail may vary due to falling height. If a significant change in the dropping height occurs, the pattern will reflect a change in spatter size. Drip trails lead from one point to another as in Figure 4.13.

The shapes of the stains in a drip trail may alter depending upon the speed of the dripping source. If the bloody object is moving slowly, many of the stains will be generally circular (70–90 degrees), but the impact angles may randomly increase or decrease as a function of the speed at which the dripping object is moving. Assuming a dripping bloody object in motion with no other associated movement (e.g., some swing associated with the object), the stains can manifest impact angles as low as 40 degrees.

Drip trails can be deposited on both horizontal and vertical surfaces. Whatever the surface, the stain's directional angle must correlate to gravitational effect. For instance, in Figure 4.14, the stains all appear round having fallen generally perpendicular to the surface. For drips falling onto a vertically oriented surface, the shapes will be elliptical and pointing generally downward. In the latter case, there are often associated horizontal lines of drips leading to or away from the vertical surface.

Dynamics: As blood collects on a surface, gravity acts on the mass pulling it away from the source. A small connecting column of blood (referred to as necking in fluid dynamics) stretches between the source and the forming drop. Gravity's downward effect will ultimately shear this column and the parent drop from the blood source. The parent drop is released when gravity overcomes the cohesive and adhesive forces in the fluid column. Some part of the column itself also shears from the source. This column devolves, producing small droplets, primarily from the action of surface tension (Plateau–Rayleigh Instability) within the column. These smaller droplets fall along with the parent, often producing much smaller accompanying stains around the parent stain. The series of parent drips from the object in motion produce the resulting inline distribution.

Probative Value: Drip trails indicate a bloody source is in motion while dripping. The source of the drip may be replenishing (e.g., an individual bleeding) or a nonreplenishing source (e.g., a bloodied object). Even in the latter, extensive drip trails have been observed. Assuming that the object dripping blood

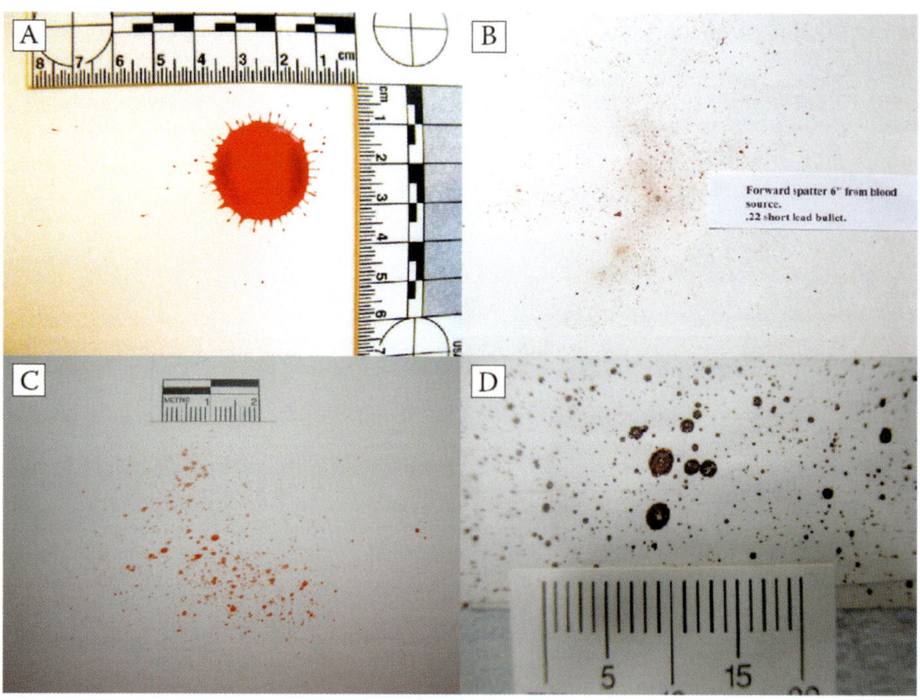

Figure 4.14 Various forms of nonlinear spatter. A is a single drip. B is a mist pattern produced by gunshot. C is an impact pattern. D is an expectorate pattern.

is moving at some speed, directional aspects of the more elliptical stains in the drip trail may provide an understanding of that motion. Drip trails that lead away from a victim may be highly probative based on their DNA. Granted, the blood of the victim may be adhering to some object as it is removed from scene or area, but it is also possible that the attacker was injured in some fashion and they are the dripping source.

Consider that post-incident artifacts do occur. This is particularly true of drip trails. Consideration must be given to the thought that first responders created stains at the scene through movement or when transporting the body from the scene. A good practice is to study the first photographs taken by the scene photographer and subsequent pictures that may show differences in bloodstaining between the two time frames.

Category: Nonlinear Spatter

Family: Bloodstain: Spatter

Definition: A series of related spatter stains dispersed over a surface other than in a linear orientation, either radiating or deposited in a random fashion.

Criteria:

- Related circles and ellipses
- Demonstrating no linear aspect

Discussion: Nonlinear spatter are spatter stains that demonstrate no linear deposition. See Figure 4.14. Directional angles of nonlinear spatter will either demonstrate a radiating nature or a random deposition. This random aspect is a statement that there is neither a radiating, nor linear dispersion.

Dynamics: The disruption mechanisms producing nonlinear spatter exclude streaming ejections, blood flung from bloody objects, and objects moving while dripping. The blood may be dispersed from a point source and thus radiate, or it may result from a gravitational drip from an object that is not in motion, producing a random distribution.

Probative Value: Nonlinear spatter is a general classification, which does exclude mechanisms associated with linear spatter and non-spatter type stains. Thus, it is still valuable in exclusion of certain scene mechanisms.

Category: Radiating Spatter

Family: Bloodstain: Spatter: Nonlinear Spatter

Definition: A series of related spatter stains dispersed over a surface in a radiating distribution.

Criteria:

- Related circles and ellipses
- Demonstrating a radiating dispersion

The radiating spatter pattern results from some volume of blood broken up at a point/area by some force or impulse. As a result, the blood is dispersed outward, ultimately breaking up into droplets that generally are no greater than 2 mm in diameter.[9,10] This dispersion process generally creates a sheet of blood, which devolves into ligaments, which devolve into individual droplets. The resulting stains produced by the collapse of the droplet on a surface are circular to elliptical, and based primarily on their directional angles, demonstrate they are radiating away from one another. The pattern can demonstrate this radiating aspect either through considering the individual stain directional angles or by evaluating the nature of the distribution (e.g., the stains are distributed closely together and further out in the pattern they become more distributed).

Dynamics: As the blood source at some area comes under a shearing stress, the blood must displace away from this area. The transition from a blood mass to individual droplets is highly complex, with thin fluid sheets and columnar structures ejecting out from the source along whatever avenues of escape exist. These structures are ultimately stretched beyond the ability of the cohesive forces to keep the fluid in a single mass. When their cohesive forces are overcome, these structures devolve into individual drops. The droplets disperse in a three-dimensional fashion, producing a conical cloud of droplets (referred to as a dispersion cone) moving out and away from the source. This dispersion cone may radiate outward in a very narrow fashion (e.g., radiating outward in a tight 15-degree orientation) or it can disperse in a 360-degree circle around the source. As each droplet is following an individual vector, the resulting stains have varying impact and directional angles that correlate back to the source.

Probative Value: A classification of radiating spatter indicates an inability to further define the three associated classifications (impact, mist, expectorate)

Additional characteristics to further refine the classification may be muted or absent. If sufficient scene context exists, a more refined conclusion may still be possible.

Example: Given a series of radiating spatter deposited on fabric, which would tend to mute or prevent the creation of vacuoles in the associated stains, the lack of blood in an airway would exclude expectorate.

Even as a general classification, radiating spatter is still valuable in exclusion of certain scene mechanisms.

Category: Impact Pattern

Family: Bloodstain: Spatter: Nonlinear Spatter; Radiating Spatter

Definition: A radiating pattern of small individual stains created when a blood source is broken up at a point source by some force or impulse.

Criteria:

- Related circles and ellipses
- In a radiating distribution
- Radiating from a central point
- With a progressive change in shape, the further out in the pattern
- Various sized stains, the majority of stains measuring 4mm or less in width

Discussion: Typical droplets associated with impact events are generally no greater than 2 mm in diameter.[11,12] These droplets are ejected outward from the source onto various surrounding surfaces. When the droplets collapse, the result is small circular and elliptical stains. The majority of these will be 4 mm or smaller in width. Larger stains may be present in the pattern, but the preponderant spatter widths are these smaller stains.

In impacts directed at an acute angle to the target, the radiating distribution is quite dramatic. The various directional angles will radiate out almost like spokes on a wheel. In an impact oriented perpendicular to a target (e.g., droplets striking at 70–90 degrees), the radiating distribution is more subtle and difficult to visualize. In such circumstances, detailed evaluation of the directional angles of the individuals spatter stains should indicate they are radiating. The dispersion of the stains on the target will also demonstrate this radiating nature. In this latter aspect, the stains will be tightly grouped in a central core of the pattern, whereas as one moves toward the outer edge(s) of the pattern, the stains will become more dispersed.

"Impact" is one term that demonstrates the inferences associated with the mechanistic terminology used in BPA. In common use, this term might suggest to the layperson a blow (e.g., some blunt force). Impact patterns are not limited to only blunt force circumstances; they include weapons' strikes and projectiles (e.g., bullets) striking a blood source. Impact can also include events where some force disperses a volume of blood already outside the body. A *potential* future resolution to the use of any mechanistic term and its associated inference is offered at the end of this chapter.

Dynamics: A blood mass, as a liquid volume, is generally incompressible. When an impulse is directed to the blood mass, the blood must displace and will seek a path of ejection along whatever avenues of escape exist. See Figure 4.15. The ejection of droplets from this blood mass may produce a dispersion cone that radiates out from the source in a very broad fashion (e.g., ejected in a 360-degree circle around the source) or in a particularly focused fashion (e.g., in a tight 15-degree dispersion cone away from the source). Many variables affect this dispersion cone, including the available avenues of escape, the size or shape of the wound, the presence of any intermediate targets (e.g., clothing or hair), and the orientation of the object as it contacts the source. As all the droplets radiate out from the same source, the resulting stains will have

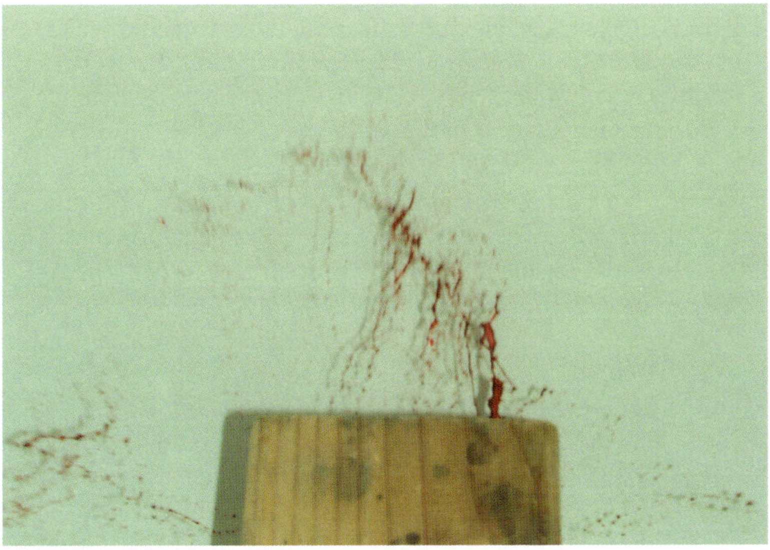

Figure 4.15 Impact spatter radiates out from a central area. In this impact, the drops can be seen ejecting forward along the leading edge (350° to 20°), to the left (270° to 290°), and to the right (80° to 90°).

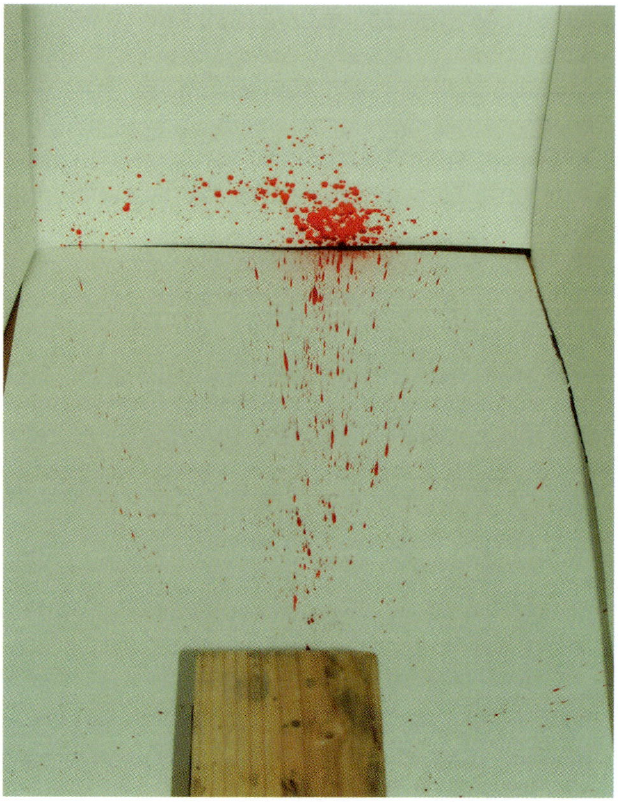

Figure 4.16 The resulting pattern created by the impact observed in Figure 4.15. Note the horizontal pattern clearly radiates outward from the displacement point. Not all of the stains on the right and left are captured in the photo (they carried to the side walls), but some stains are visible showing the full cone of ejection.

directional angles pointing back to the source. See Figure 4.16. The further out the droplets are ejected from the source alters their individual impact angles. Thus, in the pattern, on a given surface the further out in the pattern, the more acute the angles of impact.

Probative Value: Impact patterns define some impulse directed at a blood source. In a scene context, this generally involves blows to an individual or bullet strikes (either forward or back spatter ejected from exit or entry wounds, respectively). In instances where an area of origin can be isolated, this places the event in a relatively specific location. Thus, subsequent claims of where these actions occurred can be corroborated or refuted. Additionally, the number of specific impact patterns present may corroborate or refute claims as to how many such actions occurred.

The limiting concern of impact patterns is in understanding specifically what action created it. In a given scene, there may be several scene unique mechanisms (e.g., multiple blows by different objects) that could produce an impact pattern.

Category: Mist

Family: Bloodstain: Spatter: Nonlinear Spatter: Radiating Spatter: Impact Pattern

Definition: Extremely small spatter (<0.1 mm) created when blood is exposed to an explosive force.

Criteria:

- Related circles and ellipses
- In a radiating distribution
- A majority of the stains will all be generally circular and less than 0.1 mm in width
- Routinely intermixed with larger impact spatter

Discussion: When a significant force is involved in the dispersion of a point source blood mass, the resulting spatter are very small. This category was historically termed "High Velocity" (HVIS); however, there were varying and contradicting size descriptions as to what define true HVIS (e.g., 0.1 mm up to and including 1 mm). MacDonell's initial definition of size (0.1 and smaller) is appropriate. Stains with a greater width (0.15 up to 1 mm) routinely occur in impact type events where explosive force is not present. Nevertheless, as McDonell reported, it is not uncommon that larger stains will be present intermixed within the overall mist pattern.[13]

Dynamics: As with all impact events, in the presence of a shearing force or impulse, the blood must displace. It does so as a conical cloud (a dispersion cone). One distinction of these 0.1 mm-sized spatter, different from larger impact spatter, is they will rarely present elliptical shapes. The Weber number of such small droplets (relating to the droplet's surface tension and ability to flow upon collapse across a surface) results in generally circular stains.[14] The droplets land and simply dry, rather than flowing across the surface. As such, recognition of the radiating characteristic in mist is based solely on dispersion.

A radiating pattern that demonstrates no 0.1 mm and smaller spatter is not specific evidence of a lack of explosive force. In other words, one cannot exclude explosive force under those circumstances. Particularly in gunshot events, the motion of the small mist stains produced as back spatter (e.g., exiting the entry wound) can easily be redirected and dispersed by the pressure wave produced by the weapon's discharge. The presence of such tiny 0.1 mm spatter, however, effectively points to some form of explosive force. Recognition of the Mist from typical Impact allows greater differentiation of potential source events.

Claims of the presence of the 0.1 mm spatter characteristic are based on actual measurement. Small stain sizes (>0.1–0.5 mm) are difficult to validate visually. See Figure 4.17. One effective consideration is that most individuals cannot visualize the actual circles or ellipses of the mist with their eyes; magnification is required. If one can visualize the spatter, it is likely larger than 0.1 mm. Lacking confidence that a majority of the stains are 0.1 mm and smaller, the pattern is appropriately classified as Impact. Recognizing that a majority or the preponderant stain sizes are >0.1 mm is not difficult. There may be hundreds of larger stains (0.2–1 mm) present in some aspect of the mist pattern, but intermixed in the same area will be a pink-red discoloration, which upon magnification will reveal the presence of the 0.1-mm stains. Often, these smaller stains will be so numerous, it would be difficult to count them accurately. When presented with this pink hue discoloration, there may be literally

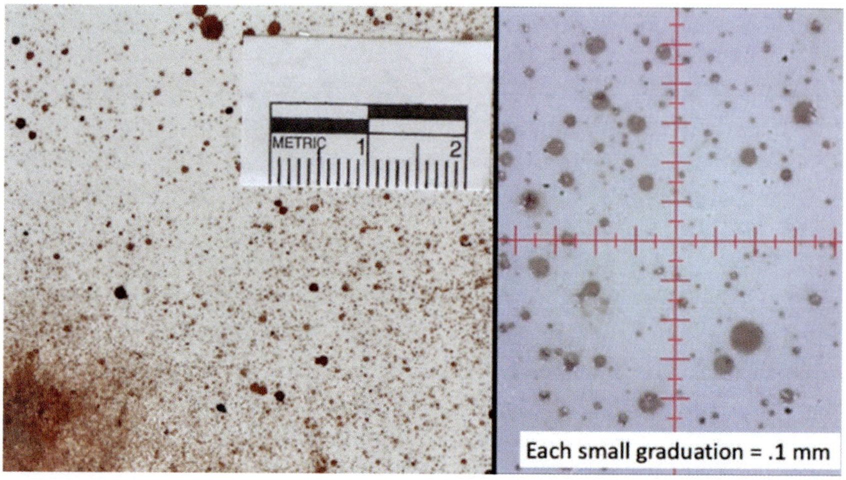

Figure 4.17 On the left is a typical mist stain. Larger impact stains are present, but for the actual mist (0.1 mm and smaller), one cannot visualize the actual stains. They appear as reddish hue on the target. On the right is the mist observed under magnification. Note there are no elliptical stains, the Weber number of such small droplets (relating to the droplet's surface tension and ability to flow during collapse) results in only generally circular stains.

thousands of stains in a single square inch. There are instances when the mist is less pronounced. Under such circumstances, the analyst may have to be more precise in deciding what is the preponderant stain size present before drawing a conclusion.

Probative Value: In crime scenes, mist is produced by gunshot, explosive force, and even circumstances involving high-speed machinery. Recent efforts also require consideration of certain but very limited expectorate events, as these research efforts experimentally produced 0.1 mm stains with air pressure, although not in great numbers.[15] The value of the presence of mist is limiting potential scene source mechanisms. The nature of the mist (specifically the lack of elliptical stains) prevents the ability to conduct any formal Area of Origin analysis, yet the dispersion cone may provide indicators as to the orientation of the source. Interestingly, the dispersion cone ejects generally perpendicular to the defect (e.g., a gunshot entry or exit wound). As the droplets producing mist are very small, their individual sizes prevent them from travelling any significant distance from the source. True mist rarely travels more than 4–6 feet from the source unless some air current is involved.

Category: Expiration Pattern

Family: Bloodstain: Spatter: Nonlinear Spatter: Radiating Spatter

Definition: Spatter created when blood is forced from the mouth, nose, or wound to the respiratory system, ejected by air pressure.

Criteria:

- Related circles and ellipses
- In a radiating distribution
- Various sized stains
- With vacuoles (bubbles or bubble rings present in the stain) and/or mucous strands

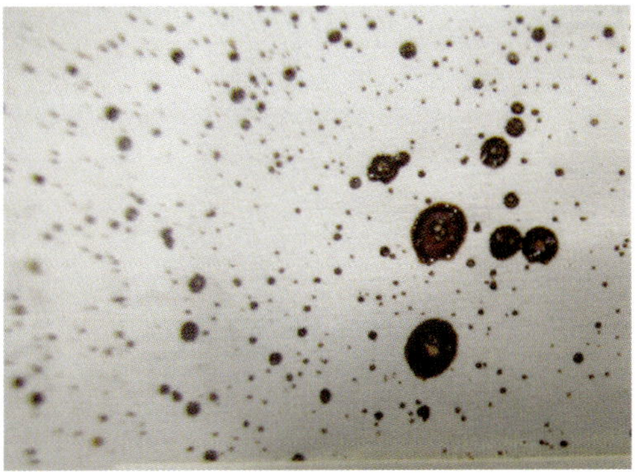

Figure 4.18 Expectorate stains on paper. Note the small circular vacuoles in many of the individual spatter stains. These form after air bubbles mix with the blood in the airway, the droplets are dispersed, and the bubbles burst as the droplets collapse.

Additional considerations include:

- The presence of blood in an airway
- The presence of epithelial cells or other chemical properties supporting a respiratory source (e.g., amylase)

Discussion: In airway injuries (e.g., the nose, mouth, trachea, lungs) or in instances where blood is drawn into an airway, or in certain instances when CPR is administered, air and blood intermix. If the victim is capable of breathing, the movement of expelling air while breathing can produce spatter. These spatter eject in a conical fashion, but one less symmetrical as those observed in impact and mist. These stains are not aspirated; they are produced by expiration.

Dynamics: The mixture of air and blood is ejected as a function of breathing out. Air movement is the dispersion force. Unlike impact where the blood source is more isolated (e.g., a gunshot wound or laceration), the air–blood mixture is in the airway. Once expelled, as the droplets collapse any small air bubbles adhering to the droplet burst, leaving behind small vacuoles (small circular voids in the spatter). These vacuoles are the most specific criteria used to support an expectorate event. Such vacuoles may appear in a few or many of the resulting spatter stains. There may be one or many vacuoles in any given spatter stain. See Figure 4.18. The analyst should seek multiple vacuoles, not just one or two examples in the pattern. Vacuoles are rarely present in any other impact event, with the exception of an extended beating directed at a particular bloody area. In that instance, it is possible that over time air and blood may mix and a few vacuoles may be present in the resulting impact spatter. In terms of size of stains, varying sizes may be present in expectorate.[16] As breathing changes moment to moment, the size of the stains found in any expectorate event can vary dramatically.

When spatter are ejected from the mouth, the blood is often diluted by saliva. This produces a mixed fluid (blood and saliva) that has a higher viscosity than simply blood. Droplets may be ejected that have interconnecting strands of fluid. These more viscous strands can survive to the substrate, creating lines connecting two or more spatter. These linear features are called

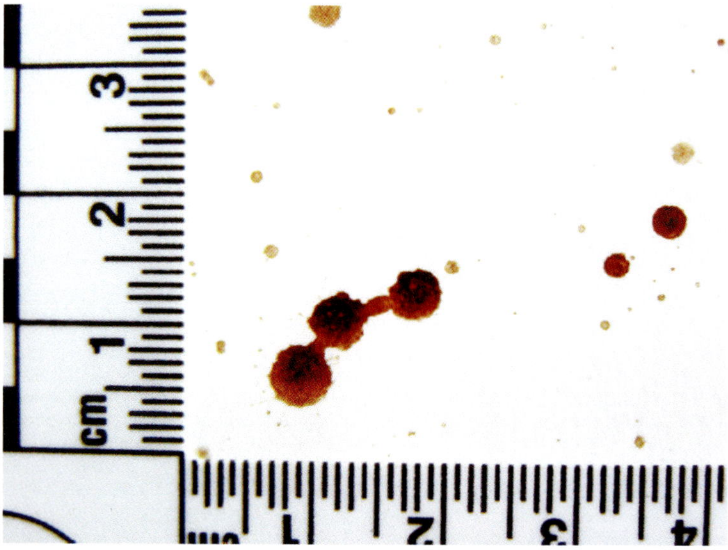

Figure 4.19 Mucous strands that connect various spatter stains are another common characteristic found in expiratory spatter. Note that dilution of the spatter also occurs on occasion.

mucous strands. These mucous strands connecting the various spatter may not be significant in number, but when present they are quite evident. See Figure 4.19.

If chemical or microscopic analysis is possible, expectorate may demonstrate the presence of epithelial cells and/or various chemical compounds associated with the mouth or respiratory system. The most common of which is amylase. These criteria however are identified only through subsequent detailed examinations, which is not always available or timely. Care must be taken in how items of evidence are handled as breathing directly on the evidence can deposit amylase. This would likely occur if the analyst were examining the evidence with a magnifying loop where their eye contacts the eyepiece while breathing next to the evidence.

Probative Value: Expectorate patterns may provide indicators as to the orientation of the victim at the time they were injured and breathing. Expectorate is often problematic in terms of classification. The vacuoles may not appear on absorbent targets and mucous strands are not always present. Lacking these characteristics, the pattern will appear as some form of radiating spatter. A conclusion (not a classification) of an expiration pattern is often a function of scene context. A necessary consideration at the scene or at autopsy is the presence of blood in the victim's mouth, airway, or lungs. Lacking evidence of blood in the airways would generally preclude expectorate.

Category: Drip(s)

Family: Bloodstain: Spatter: Nonlinear Spatter

Definition: Spatter resulting from blood dripping from an individual or otherwise bloodied object in a random orientation.

Criteria:

- One or more circles or ellipses
- In a random distribution

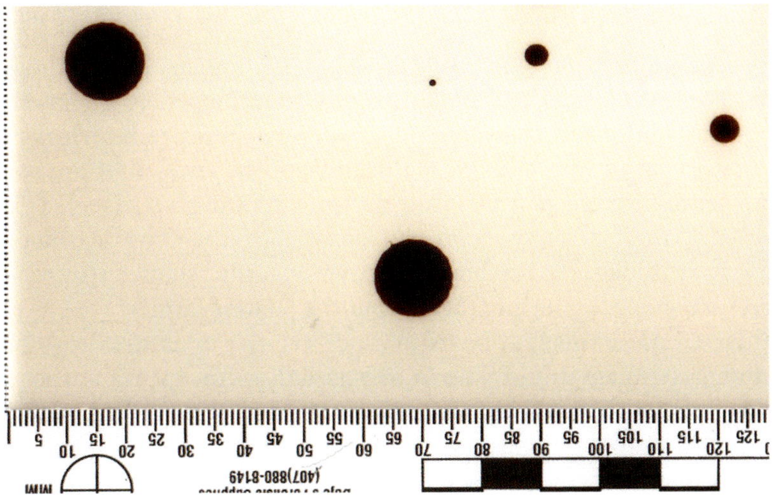

Figure 4.20 Stain size of passive drips can vary greatly. Parent stains will typically range from 3 mm to approximately 25 mm. Smaller satellite droplets, the follow-on drops associated with the breakup of the ligament between the parent and the source may appear as well.

- Primary stains demonstrate a generally consistent size range (typically 3–25 mm)
- Directional angles will correlate to gravitational force

Discussion: Drips are a product of gravity acting on the blood source. As with all drip actions (drip and drip trail), the stains generally range in size between 3 mm and 25 mm in diameter. See Figure 4.20. Drip stains may occur as a single stain or a number of individual stains. Smaller satellite stains may also be present, the follow-on stains produced by the breakup of the ligament of blood connecting the parent drop to the source. Their number, if present, are few and they will be in proximity to a parent stain. These smaller stains are satellites of the breakup and as with most satellite stains do not conform to the size range described.

The drip entails a situation where drops are deposited one or in number, creating a series of individual stains in a random orientation on the target. The associated impact and directional angles will reflect downward paths wherever they are deposited (vertical or horizontal surfaces). This random deposition is a function of minor variations due to movement of the dripping bloody object. Drip patterns (blood into blood) are often present as well, if the dripping bloody object is static for any period, producing a complex pattern.

Dynamics: Drip stains are produced as previously described for all gravitational spatter. The parent stains are a larger volume than that typically observed in impact events. The small satellite stains (follow on stains) created by the breakup of the connecting column of fluid fall in similar paths as the parent. They may land within the parent, coalescing and therefore be absent or they may be slightly offset and in proximity to the parent.

Probative Value: Drip stains have a limited value. Like the drip pattern (blood into blood), they indicate a general lack of motion of the dripping bloody object.

The Non-Spatter Family

Any pattern consisting of other than only circles and ellipse with or without flow belong to the non-spatter group. These will include many patterns where no associated satellite spatter form. But this family also includes stains where there are numerous satellite/spatter stains present. In this group, satellite stains and secondary spatter are commonly produced by various mechanisms (e.g., volume ejections striking a surface). This might seem a contradiction (e.g., there are spatter present, so why are they not classified as "spatter"), but also present in such patterns are obvious larger irregular shaped parent stains with no elliptical or circular shape, excluding them from the Spatter family.

The margins of these stains may be regular or irregular in nature. Regular margins are best defined as consistent with fluid flow. When fluid flows on a surface, the resulting margin is generally clean and uninterrupted (e.g., as found in many pools or flows). Irregular margins are margins that appear otherwise. See Figure 4.21. Such irregular margins may include spines, feathered edges, or a scalloped appearance. Remember spatter with scallops (an edge characteristic of an individual ellipse with no larger irregular margin stain) alone precludes a *spatter* from ever being considered in the non-spatter family. In contrast, a series of drips deposited on top of one another (e.g., as found when blood drips into blood) will present a larger stain with an irregular margin, along with associated spatter. See Figure 4.22.

The irregular margin group is composed of a wide range of stains that can be difficult to functionally describe in detail, nevertheless the margin characteristics will be evident. The entire non-spatter hierarchy is depicted in Figure 4.23.

Other critical criteria/morphology concepts are explained as one moves down the non-spatter stain hierarchy, these include:

- Feathering
- Striations
- Random shaped and oriented spatter on the margin.

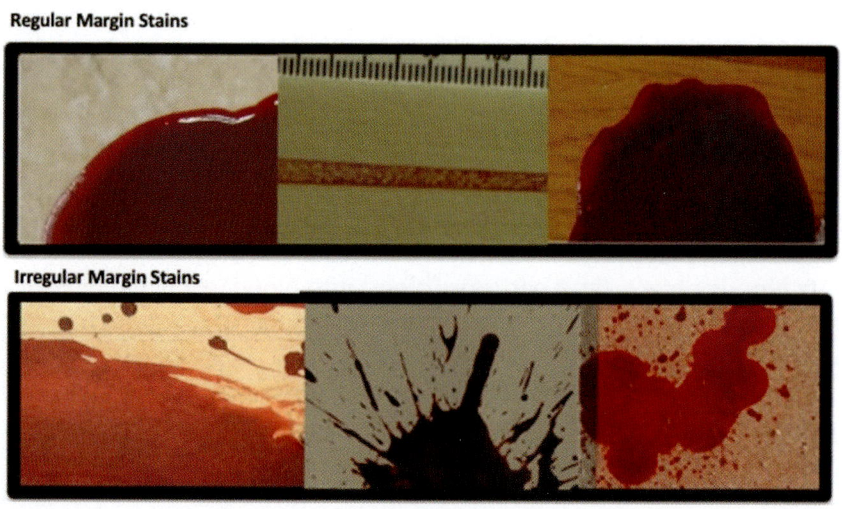

Figure 4.21 The primary observational difference in non-spatter stains are the margins (an idea first proposed by Philippe Esperanca). Regular margins are consistent with fluid simply flowing. Irregular margins are anything else.

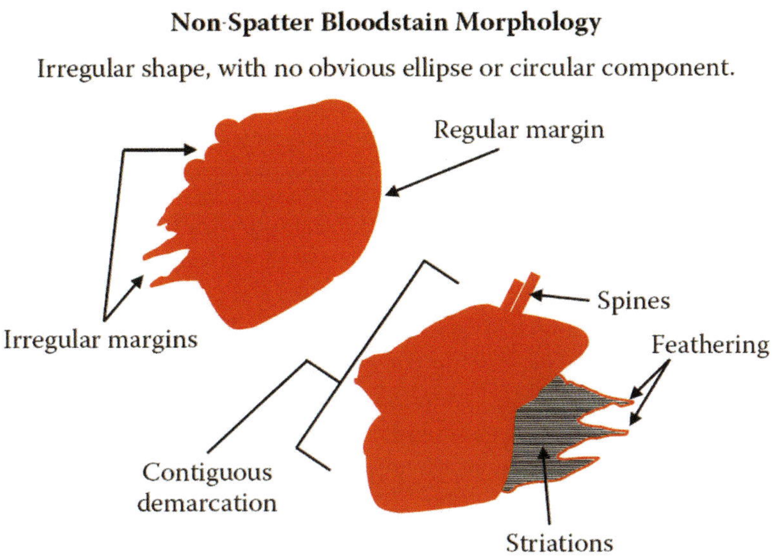

Figure 4.22 Non-spatter stains demonstrate specific traits (morphology). This includes parent stains that have a shape other than an ellipse or circle, feathering of the edges of the stain, striations in the body of the stain, spines, and even secondary spatter.

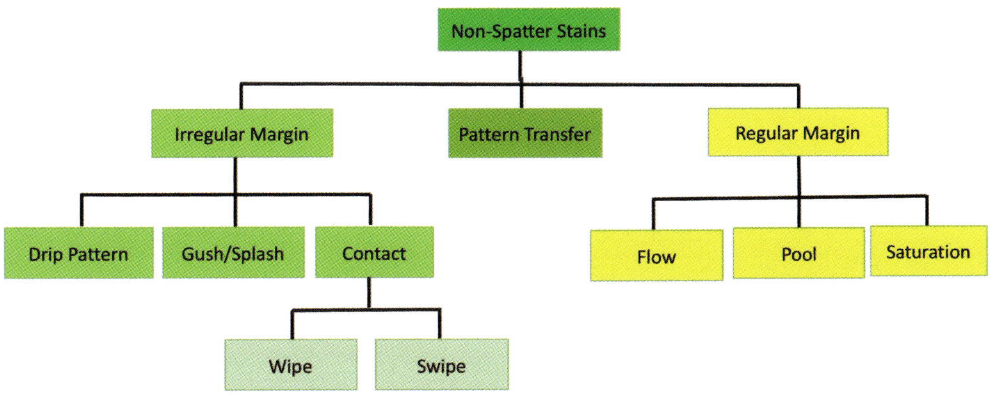

Figure 4.23 The non-spatter taxonomy relationships. Once again, if we examine the lower levels of the taxonomy, we find the typical pattern types (e.g., wipes, swipes, pattern transfers, gush, and drip pattern).

Category: Non-spatter

Family: Bloodstain

Definition: Any stain or pattern other than those defined by the spatter group.

Criteria:

- Presence of primary stain(s) with no evident elliptical or circular component

Discussion: Mechanisms associated with this group include contact, accumulations of blood, and ejections of large volumes. Blood accumulations such as pools may have a circular component, but that characteristic is clearly an obvious distinction from the small circular stains associated with spatter. Unlike the spatter family, non-spatter stains have highly diverse appearances.

Category: Irregular Margin Stain

Family: Bloodstain: Non-spatter

Definition: A stain with an irregular margin.

Criteria:

- A non-spatter primary stain
- With an irregular margin

Discussion: Mechanisms associated with this category include streaming ejections, fluids dripping into fluids (e.g., blood into blood) and most stains resulting from contact.

Dynamics: Volume ejections of blood lead to significant interactions of the volume with the surface, producing the irregular aspect. Dripping action into the accumulation produces the irregular aspect in the main stain (a large irregular stain), but also produces irregularly shaped and oriented satellite stains. In contacts, the interaction of the objects (when one or more of the objects is bloodied) produces the irregular characteristics (e.g., displacement of blood, feathering, striations). Refer back to Figure 4.21.

Probative Value: An irregular margin stains excludes a variety of scene mechanisms. The diverse nature of the irregular margin stains will usually allow further differentiation.

Category: Projected – Gush/Splash

Family: Bloodstain: Non-spatter: Irregular Margin Stain

Definition: An irregular margin pattern created when blood is released in volume as a streaming ejection or a volume ejection as a mass. The stream may be, but is not necessarily, under pressure.

Criteria:

- A non-spatter stain
- With an irregular margin, exhibiting numerous spines and/or secondary satellites stains
- Often accompanied by large elliptical stains

Discussion: Although produced by the same mechanism as spurt (a streaming ejection), the gush differs dramatically in appearance. The spurt is a discrete pattern of individual spatter in a linear orientation. These occur primarily when the stream is ejected at an acute angle to the target or when the distance from the target allows breakup of the stream into individual drops. A gush/splash stream in contrast is directed at the same general location on the target, resulting in significant interaction within the blood volume as it strikes the surface. The gush/splash produce a very distinct irregular margin stain.

This streaming mass is unstable in flight. As in the case of the spurt pattern, once ejected the stream is acted on by the external forces of air resistance and gravity. The volume may be directed as a stream into some aspect of a target, or the volume may separate into smaller unstable volumes all ultimately striking the surfaces. Whether the ejection breaks up or not, the stream or small volumes and droplets strike in the same general area. As these blood volumes land on the target, they interact producing the highly irregular margin of the gush/splash. These interactions also produce the associated satellite spatter and detached spines radiating away from the primary stain. Smaller volumes

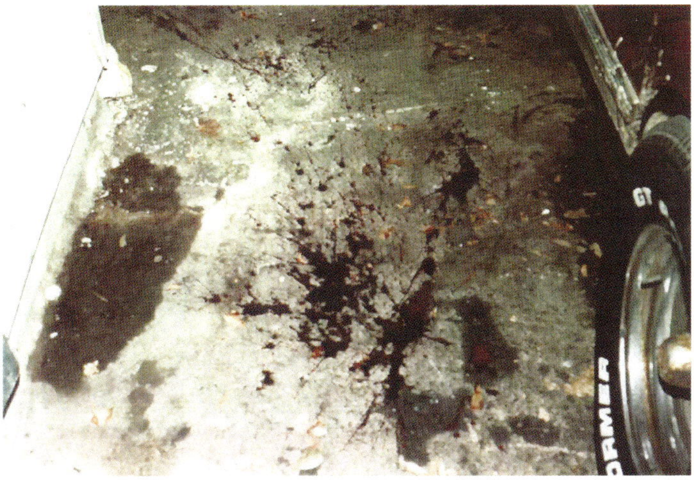

Figure 4.24 A gush pattern. A gush is a large-volume deposit with spines and secondary spatter in and around the periphery creating a very irregular or spiny margin. This gush was created by a gunshot injury to the femoral artery.

may separate sufficiently in flight to produce large elliptical stains around the periphery of the primary stain(s). See Figure 4.24.

A point of interest with the gush/splash category is the inclusion of less forceful ejections (e.g., unpressurized) in the category. Blood can flow in volume (e.g., from a large venous source) with velocity. For example, blood in volume streaming down an arm does not simply "drip" off the elbow or fingers, it comes off in a streaming ejection. No pressure is involved in the stream, merely velocity produced as gravity affects the volume. The patterns produced are similar to the patterns produced by pressurized ejections.

If the pressure behind the ejection is minimal or absent, the resulting collapse on the surface produces fewer spines and secondary spatter. Patterns of this nature are often called a splash by other classification systems. The primary difference between the two is a function of force, something the analyst can never measure. This presents an interesting classification problem (e.g., Is it gush or splash?). This is a question with no objective resolution. Both patterns exhibit irregular margins and spines. The only observable difference may be the number and length of the spines present. The question then becomes: Can a quantitative criterion be established? For example, counting the number of spines? Ten spines or less is a splash, 11 spines or more are gush? And if so, where or how is that number defined? The only valid resolution is to keep the two together as the gush/splash.

This is not to say the force distinction will not be evident to the analyst, as when presented with a pattern with significant spines and satellites. When recognized and the pattern is considered within the unique scene context (applying the full BPA methodology), this may aid the analyst in providing some ultimate opinion/conclusion as to the potential source event.

Regarding low force circumstances, analysts may encounter instances of extremely low force ejections that are effectively ambiguous.. The result may be a primary stain that appears almost poured onto the surface. Presented with such an ambiguous pattern, the only valid classification may be "non-spatter."

This is also true when one considers the drip pattern mechanism in relation to streaming ejections, both involve blood interacting with blood. The resulting patterns are quite similar, in the sense that they will both have irregular margins, and often significant satellites. The critical random shaped satellites observed in the drip pattern can also occur in the gush (depending on the orientation of the ejection). What is clearly different between the two, however, are the spines produced in the gush. These do not occur when gravity is the only force. In a low-force ejection on a horizontal target that lacks spines or has very few, the resulting ambiguity may prevent distinguishing between the gush and the drip pattern.

Dynamics: The volume from a streaming ejection is directed at an area on the target. This precludes greater breakup of the volume's mass from air resistance and gravity. Whether this is a stream directed over time (e.g., a series of streams coming from a breached artery) or a volume falling en masse (e.g., as can occur in significant heart injuries), the blood volume lands in the same general area. The interaction of the blood and the target, as well as the blood interacting with itself as it displaces on the target, leads to the creation of spines and satellite stains. These characteristics produce the irregular margin aspect.

Probative Value: A gush/splash identifies the source mechanism as a streaming ejection, eliminating other source mechanisms. The production of the spines and satellites may provide indicators as to the general direction of the ejection.

Category: Drip Pattern – Blood into Blood

Family: Bloodstain: Non-spatter: Irregular Margin Stain

Definition: A pattern resulting from two liquids, at least one of which is blood, one dripping into the other, resulting in an irregular shaped accumulation with satellite spatter that are randomly shaped and oriented around the primary stain.

Criteria:

- Non-spatter stain
- With an irregular margin
- Exhibiting satellites with random shapes and orientations around the immediate margin
- Few if any spines

Discussion: A drip pattern, also referred to previously as blood into blood, is produced when liquid drops (typically blood) fall onto the same surface. As repetitive drips occur, a volume accumulates on the target. The impact of the individual drops into the liquid results in a significant increase of secondary spatter from the accumulation. The margin of the accumulation becomes irregular as a function of drips falling into and overlapping each other. Because of the relative motion of the liquid on the target (the volume may be rolling or oscillating), secondary spatter are ejected in random directions out from the primary stain. Around the margin of the primary stain, many of these satellites demonstrate highly irregular directional angles. As satellites are ejected, instances occur near the margin, in which one satellite lands on or in contact with another. The two small volumes coalesce, leading to highly irregular shapes other than just circles and ellipses. These stains will differ distinctly from the spatter observed in radiating patterns. Further out in the pattern however, this random shape characteristic will disappear and the satellite spatter will appear as a more typical radiating series of circles and ellipses. Drip patterns rarely demonstrate spine features. See Figure 4.25.

Satellite spatter produced in these patterns may become disassociated from the primary stain. If the accumulation occurs on an object that is later moved (e.g., as in the case of a shoe), the satellite spatter would be found on the surrounding surfaces. When the wearer walks away, the primary irregular shaped pattern leaves as well. All that is left are the random distribution of satellites along with a possible void where the shoe was positioned. See Figure 4.26. Although not common, in such an instance the spatter themselves might initially be classified as some form of radiating spatter. The random shape and orientation of the satellites, if present, will resolve this issue.

This is another pattern term that presents certain potential ambiguities based on the prescribed term when presented in court. In general conversation on the stand, drip stains, drip trails, and blood into blood patterns are routinely referred to as "Drip" patterns. But it is also true, the prior term blood into blood represents a problem, in that the pattern may involve blood and some other liquid. A potential future resolution for pattern naming/labeling is provided at the end of this chapter to resolve such ambiguities.

Dynamics: As individual drips fall to the same general area, they produce a small accumulation. Additional drips then interact with that accumulation. The four phases of collapse occur but with slight differences. No major differences were noted for the contact/collapse phase. The droplet still collapses from the bottom up. In the displacement phase, however, there appears to be more blood forced into the expanding rim. Blood present on the target is also

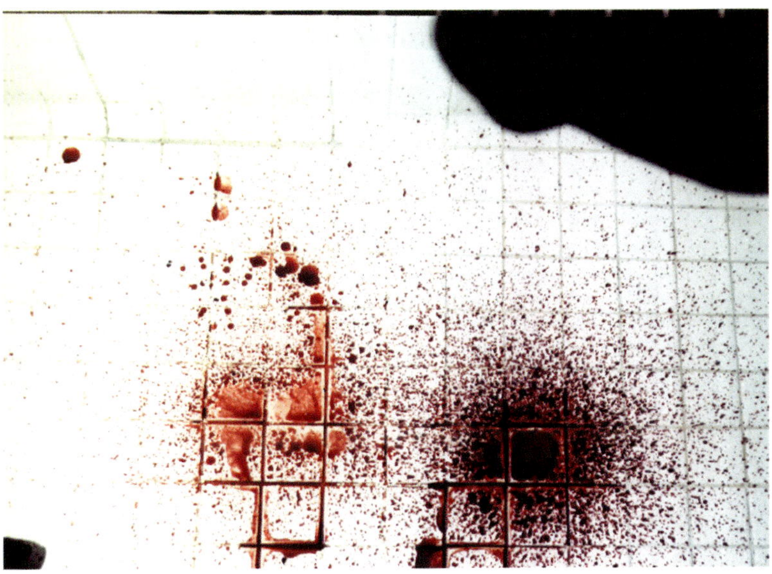

Figure 4.25 Another non-spatter pattern is the drip pattern, also known as blood into blood. It too has an irregular margin. An associated characteristic of the drip pattern is the random distribution of irregular-shaped spatter around the margin of primary stain and the lack of spine features.

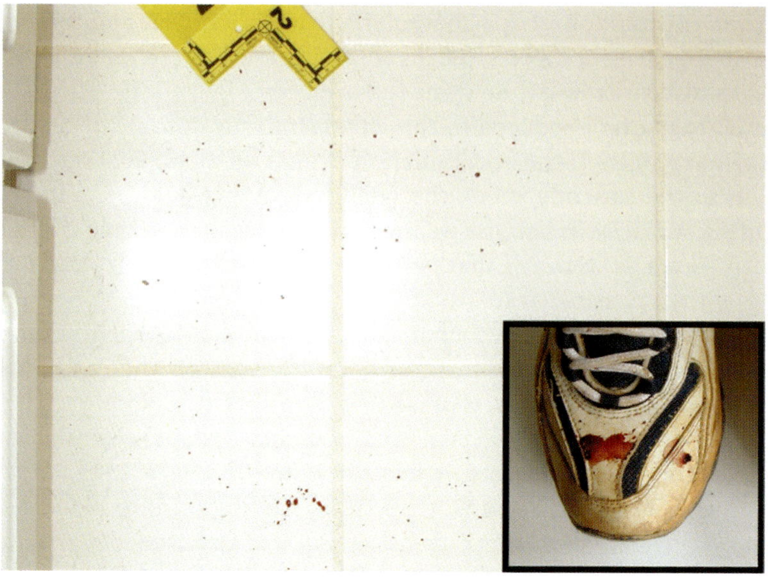

Figure 4.26 An interesting feature of a drip pattern is that in some instances the primary stain can be disassociated from the spatter produced. In this instance, the parent stain (the accumulation) was on the shoe. When the subject walked away, only the random irregularly shaped spatter was left in the scene.

drawn into the rim, in addition to the blood present in the droplet. This is true in instances where the impact occurs into a pool of blood or even another drop of blood.

As a result of the increased volume and lack of a solid target, we see a less-defined displacement phase and an early development of the dispersion phase. The blossom is far more pronounced and has been said to resemble a sea anemone.[17] The protuberances extending from the blossom are more pronounced and almost always result in satellite spatter. These spatter are thrown off in nearly every direction. This result is quite evident in the classic drip pattern, where satellite spatter surrounding the pool extend out multiple inches from the central pattern. In retraction, the only added result of a liquid-to-liquid impact is a rolling motion, which develops in the pool. If other droplets impact the target while this motion is evident, the dispersion of the satellite spatter is even more random. As a result, the satellites are deposited in very random orientations, distinct from impact events. Drip patterns are produced over time, so satellites are deposited adjacent to as well as in contact or on top of other satellites previously deposited. These satellites may coalesce or cause additional displacement after their initial deposit which produces the irregular shapes observed in the satellite spatter around the margin.

The farther a satellite spatter is ejected from the accumulation, the more dispersed they become. The stains are less likely to interact with other satellite stains. Thus, the irregular shapes and orientations are only found around the margin of the primary stain. Further out, the satellite stains will appear much as typical impact spatter do, simply as radiating spatter.

Probative Value: The presence of a drip pattern demonstrates that a dripping blood source was static for some period. Within any blood trail, there may be instances of drip patterns.

Category: - Contact or Transfer[18]

Family: Bloodstain: Non-spatter: Irregular Margin

Definition: Any stain or pattern created by the transfer of blood from one object to another through some form of contact. This will effectively include any form of contact, whether lateral motion is involved in the contact or not.

Criteria:

- Non-spatter stain
- With an irregular margin.
- Demonstrating any of the following:
  - A contiguous boundary.
  - A feathered boundary.
  - Striations in the body of the stain.
  - Evident displacement of blood, demonstrated primarily by diminished volume of blood across the body of the stain.

Discussion: These stains run the gamut of contact stains with their varied irregular shapes. In the original taxonomy, the classification smear was synonymous with contact, but due to inferences associated with the term smear, we have chosen to use contact. These contact stains will appear in a wide range of shapes, literally impossible to fully describe. The margin of the stain may be intermittent, some parts showing a regular margin, others irregular or fully intermittent with no regular margin present at all. Some aspect of the margin may appear feathered. The stain will reflect various combinations of contiguous and noncontiguous feathered edges along its margin or irregular staining anywhere within the primary stain itself.

The contiguous boundary aspect consists of a side(s) of the stain that are effectively marked by a continuous or nearly continuous edge of blood, presenting a distinct boundary. Refer back to Figure 4.22. The feathered aspect presents itself as a side(s) with intermittent demarcations that are staggered like barbs of a feather. The stain may demonstrate striations (lines of blood), that add to the feathering appearance. See Figure 4.26

Dynamics: The contact interaction of two objects with blood is highly diverse. The nature of the contact may involve lateral motion between the objects, displacing blood as that contact occurs. Contact may displace blood within the stain into the edges producing the contiguous boundary aspect. Lateral motion may result in a diminishing of volume of blood within the stain. Blood is deposited heavily at first, but as the contact continues less blood is present to be deposited. These features reflect displacement of the blood. Depending upon the additional characteristics present, a contact/transfer is often further classified as either a wipe or swipe.

Probative Value: The presence of a contact/transfer limits the mechanism to some form of contact. All contact involves some relative motion between the surfaces coming into contact, however the stain may or may not demonstrate indications of that motion. Figure 4.27

Figure 4.27 An example of the feathering and striation characteristics often found in a contact stain created by some lateral motion.

Category: Wipe

Family: Bloodstain: Non-spatter: Irregular Margin: Contact

Definition: Any stain or pattern created when an object moves through a preexisting bloodstain.

Criteria:

- Non-spatter stain
- With an irregular margin
- Demonstrating any of the following:
  - A contiguous boundary
  - A feathered boundary
  - Striations in the body of the stain
  - Evident displacement of blood, demonstrated by diminished volume of blood across the body of the stain or removal of blood within the perimeter of the stain
- Presence of a preexisting stain/volume

Discussion: The wipe is a contact stain involving disruption of a preexisting stain. The wipe can demonstrate some or all of the parent contact characteristics, but additionally some form of preexisting stain must be evident. One characteristic for recognizing the preexisting stain is the presence of a perimeter stain (an outline in dried blood of the original stain). This original stain boundary may

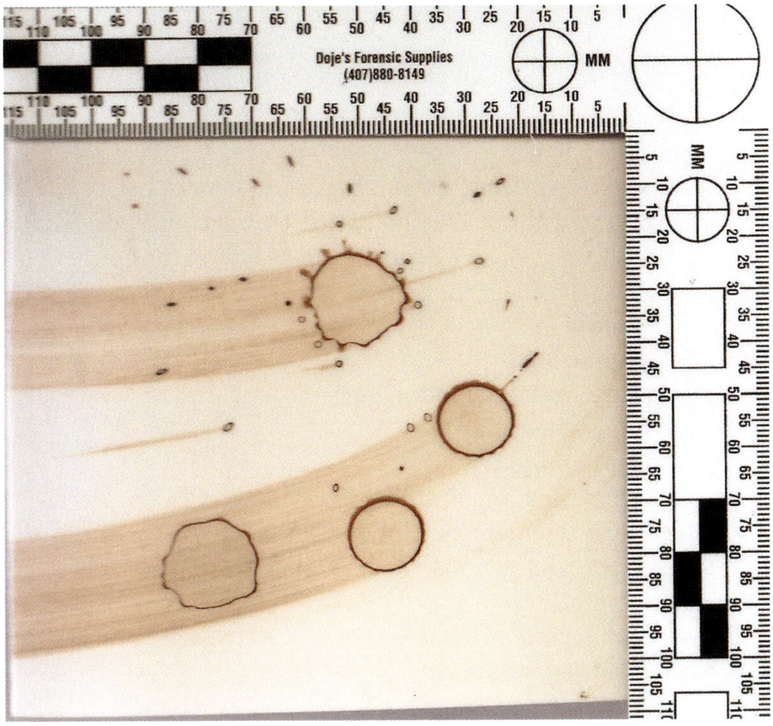

Figure 4.28 The presence of the perimeter stain associated with the original deposition of blood, whatever its nature, is the defining characteristic of the wipe.

or may not demonstrate a completely dried outer margin depending upon the amount of time that passed after the original stain's deposition or the volume involved in the preexisting stain. See Figure 4.28. Even when a perimeter stain is not clearly present, disruption or displacement of the blood may validate the existence of the preexisting stain.

The wipe pattern represents one of the few potential confusing aspects in the Taxonomy system. As will be discussed, alterations of patterns are considered adjectives when discussing patterns, they are not specifically a classification; however, by definition, the wipe is an *alteration* of another pattern.

Thus, most wipes can effectively be classified a complex pattern, which are described and discussed later in this chapter. In any instance where a sufficient perimeter stain exists to recognize the nature of the preexisting stain, the "wipe" is an integral part of a complex pattern. Yet the wipe has historically held a "classification" status. The individual analyst may choose to classify the pattern specifically as a wipe or as a complex pattern. Neither decision is incorrect.

Dynamics: Stains dry from the outer perimeter inward. So even when a disruption occurs and blood from the preexisting stain is fully displaced, this outer perimeter may be evident. If a particular motion is involved in the disruption, the blood will be displaced in a similar orientation and this may be immediately evident.

Probative Value: The value of a wipe is recognizing that some subsequent action was ongoing after an initial deposit of blood. This may allow a general sequencing of the associated events. The condition of the perimeter stain may also provide indications as to how long after initial deposition, the wipe occurred. The wipe remains a distinct "classification" primarily due to the historical mindset and its probative value.

Category: Swipe

Family: Bloodstain: Non-spatter: Irregular Margin: Contact

Definition: Any stain or pattern created by the transfer of blood from a bloodied surface on to another surface by some form of lateral motion.

Criteria:

- Non-spatter stain
- With an irregular margin
- Demonstrating any of the following:
  - A contiguous boundary
  - A feathered boundary
  - Striations in the body of the stain
  - Evident displacement of blood, demonstrated by diminished volume of blood across the body of the stain or removal of blood within the perimeter of the stain.
- Often demonstrating a contiguous stain boundary opposite the feathered and striated side of the stain.

Discussion: The swipe is an irregular margin stain distinct from the wipe. It has no preexisting stain associated with it. In the swipe, an object introduces blood onto a target. The nature of the stain will reflect obvious lateral movement as the contact occurred, in the form of feathered edges, striations in the stain, and a diminishing of blood across the stain. The difference between the contact and the swipe is the former does not effectively reflect associated lateral movement between the surfaces.

The initial side of contact in a swipe is often defined by a contiguous boundary, while the side opposite will reflect any of the characteristics described: a feathered appearance, striations, accumulation of blood on the other boundaries or diminished blood across the body of the stain.[19] Just the same, not all swipes will produce this effect, eliminating any statement as to the direction of the swiping motion. See Figure 4.28. If this condition is present, this provides indicators as to the direction of the swipe.

Dynamics: In the swipe, blood is brought to a surface by contact with a bloodied object. At the point of contact, this often (but not always) produces a clear contiguous boundary. As contact diminishes and is lost, this results in striations, feathering or a diminishing of the blood being transferred.

Probative Value: The value of the swipe is recognizing that blood was brought to the surface that is then stained, distinguishing it from the wipe where the blood preexisted on the surface. This is not always a particularly probative consideration, but it can be. Depending on the associated margins (e.g., when there is one contiguous and one feathered), it may be possible to determine the general direction in which the blood was deposited. See Figure 4.29.

Category: Regular Margin

Family: Bloodstain: Non-spatter

Definition: A stain with a regular demarcated margin.

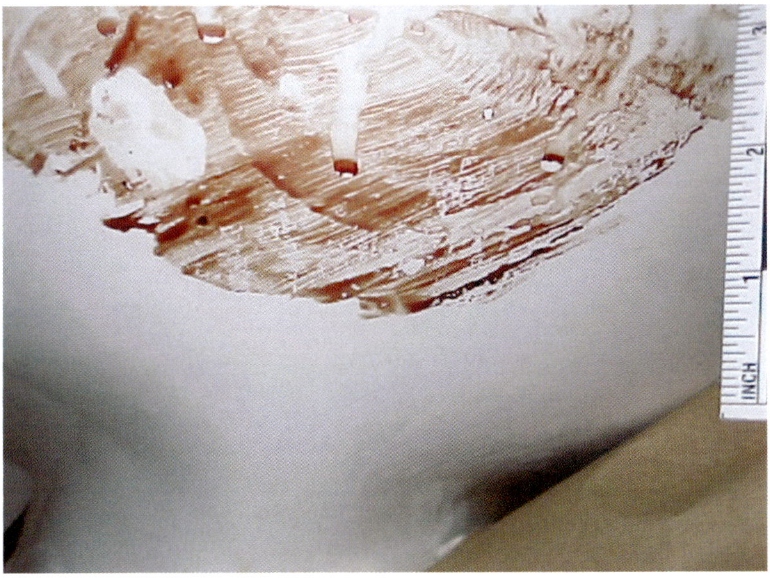

Figure 4.29 A swipe pattern. This blood was deposited when a bloody object made contact with the ceramic surface. Note the characteristics of striations in the body of the stains and some feathering in the right boundary. The left boundary is essentially contiguous, thus the general direction of motion in this instance is left to right.

Criteria:
- Non-spatter stain
- With generally regular margins

Discussion: This group represents volume accumulations, where liquid blood is pooling or flowing in some form. Refer back to Figure 4.21.

Category: Pool

Family: Bloodstain: Non-spatter: Regular Margin

Definition: An accumulation of liquid blood responding to gravity and conforming to container/surface characteristics of the pooling area.

Criteria:
- Non-spatter stain
- With generally regular margins
- In an obvious accumulation.
- Without any specific shape, but conforming to the surface contours.
- May demonstrate serum separation or coagulation

Discussion: Accumulations of blood form in a variety of fashions and on any number of surfaces, thus they have no specific shape. The accumulation will conform to surface contours. They demonstrate a distinct liquid boundary (a regular margin). Pools may show coagulation or separation of the blood cells from the plasma (serum separation). See Figure 4.30.

OSAC/ASB provided a quantitative volume of 20 ml in their definition of a pool; however, the basis of this specific volume and "how" it would be clearly and objectively defined in any unknown stain is unclear. Effectively, defining a "volume accumulation" will always be a somewhat subjective determination by the analyst. But this is not a difficult decision; an accumulation will be more

Figure 4.30 A pool with evident serum separation. The straw-colored fluid at the lower right side of the pool is the serum.

than the mere coalescing of a few drops on a surface. Should another analyst or counsel argue that some small "pool" is not a pooling, this is one decision most juries should have no trouble resolving. More to the point, such an argument is unlikely to be significantly probative in any investigation.

Dynamics: When a volume of blood accumulates on any surface it will obey the effect of gravity, settling into in the lower aspects of the surface. Depending on that surface, it may attain any shape (e.g., blood settling into grout on a flat surface). Surface tension of the fluid produces smooth generally regular margins. Pools often coagulate and this coagulation may in certain instances provide an indication as to when the accumulation occurred. Coagulation is discussed in greater detail in Chapter 8.

Probative Value: Pools are not particularly probative, but obviously must be explained as to the source of the blood that has pooled. This may assist in understanding the position of the source when the pooling occurred.

Category: Saturation

Family: Bloodstain: Non-spatter: Regular Margin

Definition: An accumulation of liquid blood absorbed into a permeable surface.

Criteria:

- Non-spatter stain
- In an obvious accumulation
- With generally regular margins
- Without any specific shape, but conforming to the surface contours
- Absorbed or wicked into a permeable surface

Discussion: Volumes of blood accumulating on permeable surface produce saturations. The blood may wick into the surface (e.g., fabric) or simply flow through the surface (e.g., sand). Ultimately the liquid blood is drawn into the surface in some fashion. Saturations may have associated clotted material or

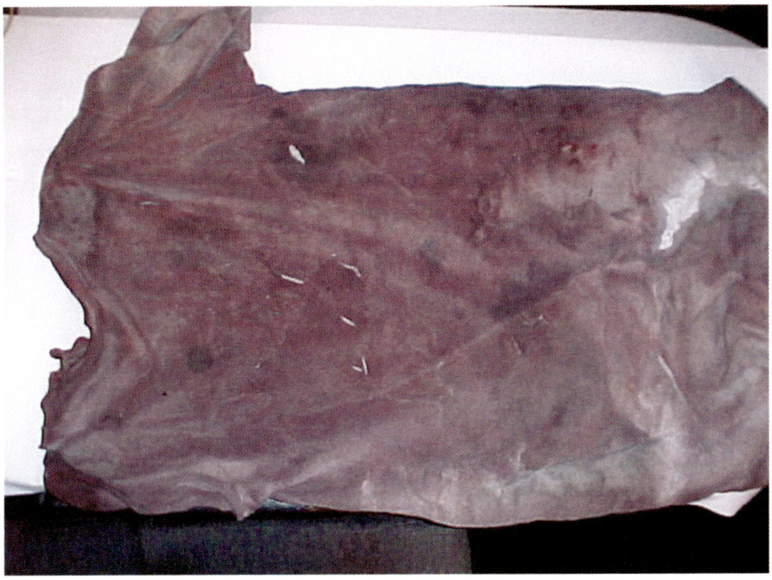

Figure 4.31 When a volume accumulates on a permeable surface, it is absorbed into the surface producing a saturation stain. Saturation stains tend to mar other patterns of interest. In this instance, pattern transfers on the back of the stabbing victim might have been of interest, but the saturation destroyed anything that may have been present.

an overlying pooling when the volume is greater than the surface can absorb. See Figure 4.31.

Dynamics: Blood in contact with a permeable surface may be drawn into the surface through capillary action, or when accumulating on permeable surfaces simply flow into the surface or any combination of both.

Probative Value: Saturations generally mar other patterns of interest and provide little probative knowledge. They do indicate the presence of a volume of blood, which like the pool may assist in understanding the orientation of the source of blood to the target. In some saturation stains, a gravitational effect (a flowing aspect) may be present. Such stains are primarily classified as saturation (e.g., blood saturating a shirt at the point of a wound). This gravitational effect, if present, may assist in understanding orientation of the target as the saturation occurred.

Category: Flow

Family: Bloodstain: Non-spatter: Regular Margin

Definition: The movement of liquid blood as a mass under the effect of gravity.

Criteria:

- Non-spatter stain
- With generally regular margins
- Leading from one point on a surface to another

Discussion: A flow has demarcated edges (regular margins). The flow may be narrow, wide, straight, or curved. A flow may move across a surface freely following gravity or follow surface irregularities. A flow associated with a nonreplenishing source (e.g., some volume accumulation that was deposited) is normally

Figure 4.32 A minor flow on the interior of a shirt. This flow helps establish that the individual was in a torso-erect position following the wound that produced the flow.

referred to as a passive flow. Flows produced by an actively bleeding injury, which continues to feed the accumulation, are referred to as active flows.

Dynamics: Blood must obey the effect of gravity. When a volume is present or introduced on a surface, the surface tension seeks to keep the volume together. Whether active or passive, gravity as well as any orientations or irregularities of the surface will cause the accumulation to move accordingly.

Probative Value: The orientation of a flow on a surface may provide indications as to how the surface was positioned in the scene (e.g., on both objects and individuals). See Figure 4.32. Flows that defy expectations based on the object's position when found will indicate some alternative orientation as the flow occurred.

Category: Pattern Transfer

Family: Bloodstain: Non-spatter

Definition: Any stain created by the transfer of blood from one object to another in which a recognizable characteristic(s) or image is present in the pattern.

Criteria:

- Non-spatter stain
- Exhibiting demarcations, curves, or some other characteristic and/or a two-dimensional image of an object.

Discussion: In the original published taxonomy, the Pattern Transfer was listed under the Regular Margin stains. This decision was honestly a rationalization on the author's part; the "pattern" created the regular" aspect. In subsequent use and instruction, numerous students commented on this approach as it created certain contradictions and confusion. Pattern transfers routinely demonstrate what could be considered both regular and irregular margins. An obvious solution was offered by a group of students attending a class presented at the Oregon State Police Academy; place the Pattern Transfer as a direct child of Non-Spatter, in line with both Regular and Irregular margin stains. This

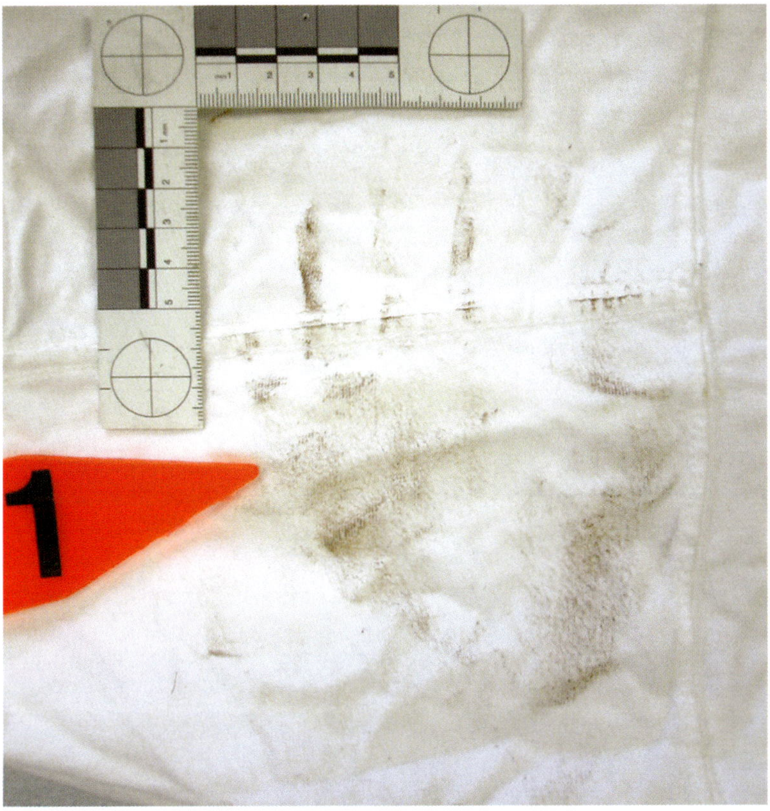

Figure 4.33 A pattern transfer on a T-shirt. A bloody hand-made contact with the victim's shoulder, leaving behind a light pattern transfer.

adjustment allows it to be differentiated immediately, without demanding consideration of the margin itself.

Pattern transfers are contact stains where the contact leaves characteristics associated with the object's surfaces or a two-dimensional image of the bloodied object. A pattern transfer may be distinct, as in the case of a pattern created by a knife blade, footwear, and bloody friction ridge surfaces. See Figure 4.33. Just as common are patterns that are vague in terms of recognizing the bloodied object. These may simply demonstrate angular and other sharp demarcations indicating something was in contact, but not providing sufficient characteristics to allow recognition of the object.

Due to the adhesive quality of blood, pattern transfers routinely occur as a series of similar pattern transfers, deposited over time on a surface or surfaces. "Repetitive" is not a classification, it is simply an adjective used to describe a number of similar pattern transfers.

Pattern transfers may be visible (patent), partially visible, or invisible (latent), the latter two requiring chemical enhancement to view. The individual patterns within a series of patterns may not be exact matches of one another. This is simply the result of the object contacting the surface in different orientations and the diminishing of blood with each contact. See Figure 4.34.

Figure 4.34 A repetitive pattern transfer. Multiple finger marks on a cup from a crime scene. Repetitive marks encountered at a crime scene may include shoes, feet, hands, or weapons.

Dynamics: As blood has an adhesive quality, any contact between a bloodied object and another surface leads to a transfer of blood. Most of these transfers (smears and swipes) have highly ambiguous shapes. Given a circumstance that enough blood is present on any edges or high points of the bloody object, characteristics of these edges or high points may transfer during contact. On occasion, the presence of blood in low points on the object can create a negative pattern transfer, such as in the tread of a shoe. In such instances the pattern is still evident, but the high points are not the area bloodied.

Probative Value: Pattern transfers can be highly probative. They may indicate the specific nature of an object in contact (e.g., footwear, feet, hands, knives) or at a minimum provide indicators as to the nature of the bloody object in contact. They help analysts understand what objects have been bloodied and where and how those objects were oriented at the time of contact. The presence of repetitive pattern transfers may assist in understanding the sequence of the contact and even the direction of contact (e.g., circumstances involving footwear, feet or hands).

The Complex Pattern

The classifications presented above create, in effect, boxes (specific classes) that unknown stains are compared against and placed into when the appropriate characteristics are present. Many stains encountered at scenes will fit distinctly and neatly into these classifications. But bloodletting events represent highly dynamic events and not every stain produced and encountered by an analyst will fit into these classifications cleanly.

The complex pattern is a necessary tool in the taxonomy approach. The idea presumes that patterns will be encountered that demonstrate characteristics of multiple categories. A complex pattern is one that incorporates characteristics of several categories and cannot be assigned to a specific category. Complex patterns are recognized as a problematic area within bloodstain classification, simply because they routinely involve highly dynamic circumstances with multiple characteristics not easily defined.[20]

Complex patterns occur in two primary situations; complex patterns created by highly dynamic events, and when multiple distinct actions deposit blood onto the same surface any of which may or may not occur contemporaneously. In the latter, the stain on top of stain situation may make it impossible to distinguish the associated patterns from one another. To be clear, it is not that such situations (e.g., pattern deposited on pattern) always create patterns indistinguishable from one another. But in some situations, this intermix of the patterns presents itself in a very confusing fashion. When such confusing conglomerate patterns appear, they are treated and classified as complex patterns.

The other form of complex patterns result from highly dynamic events that are also quite common and encountered with some level of regularity. There are several examples of common complex patterns. Drip trails are an excellent example. As the blood source is moving, the classic drip trail is produced, but any hesitation in that movement will produce drip patterns or associated random drips. Additionally, if the bloody dripping object (e.g., a hand or knife) is suddenly swung or even flicked, a small cast-off pattern may suddenly present itself, emanating from the drip trail.

Another example of the complex pattern includes combinations of spurt and gush. If a single streaming ejection is changing orientation to the target, both linear spurt characteristics and the irregular gush/splash characteristics will appear within the pattern.

A final example of a common complex pattern encountered involves stomps or slaps into large volumes of blood (e.g., pools). The resulting patterns often include gush, impact, and contact characteristics. Such patterns rarely fit into a single classification.

Additionally, highly dynamic situations may occur in bloodletting that produce patterns that simply do not fit into a classification. In such instances, even considered within the scene context, the analyst may never fully understand the nature of "how" the pattern was produced.

The complex pattern as a tool of taxonomy allows the analyst to account for the extremely dynamic conditions found in bloodstained scenes, without assuming every bloodstain must fit the classic patterns. Presented with a pattern that does not fit neatly within a specific classification, the analyst simply identifies it as "complex," articulating those characteristics that are present. Recognition of the various characteristics present may allow the analyst to understand some aspect of the underlying mechanism(s) producing the pattern. But it is entirely possible when presented with a complex pattern, no conclusion will be forthcoming.

The Ultimate Value of Taxonomy

A taxonomic approach to classification brings objectivity to the analyst's conclusions. Only after specific, predefined physical characteristics are used to validate or refute a belief, can a classification be considered scientific. The taxonomy articulates those criteria. The presence of the parent–child relationships is also important. When presented with ambiguous patterns, rather than forcing a conclusion beyond the evident characteristics, intermediate classifications (e.g., linear spatter, radiating spatter, irregular margin stains) provide an objective classification that does not inappropriately exclude some potential child classification. On the other end of the spectrum, the complex pattern also allows the analyst to not force a classification in the face of patterns produced by highly dynamic circumstances. It is not that other systems of classification are invalid; it is simply that taxonomy with its defined criteria and decision points provides the analyst with a more objective way (ergo more scientific) to achieve and support conclusions.

Bloodstain Pattern Analysis Decision Map (D/Map)

The basic premise of Bloodstain Pattern Analysis is that based on physical characteristics (e.g., size, shape, volume, orientation), the analyst can differentiate patterns created by different forces involved in bloodshed events. To support any claim about a pattern, the analyst must be able to articulate what characteristics led them to their decision. A BPA decision can't be "because I said so." As the taxonomy demonstrates, characteristics are shared within groups. Consideration of certain characteristics in a proper order allow an orderly elimination of certain patterns (the proper use of all class characteristic type evidence), which leads to a refinement of the basic pattern type. Experienced analysts recognize most patterns quickly, even if they cannot specifically articulate the characteristics they observe. This is a function of experience. If experience matters, how do we assist analysts without years of experience or better yet the analyst in training? What guides their classification efforts in order to make defensible objective decisions?

A solution is the BGA Decision Map. Philippe Esperança first suggested the idea of a decision map concept for Bloodstain Pattern Analysis based on his training and experience in entomology.[21] His initial ideas and input significantly aided the development of the initial BGA decision map. Through years of practical usage and input by those employing the document, multiple iterations have occurred to refine the questions posed. The enclosed decision map (D/Map) is a tool that leads any analyst, with any level of experience, through a series of decision points about the questioned pattern. Remember always that basic scientific method demands that questions be posed and answered. The D/Map, at least for classification, provides those questions.

In effect the D/Map is an "expert" system. Bart Epstein first proposed the idea that some expert system for BPA should be possible, 20 or more years ago. Something that would capture the knowledge of an expert and extend this knowledge in a functional tool for those new to the discipline.[22] The D/Map provides that knowledge. Questions are presented using the most critical and significant characteristics in a defined order, that effectively eliminates other classes of patterns. This will lead the analyst to an objective result.

It is imperative that one recognizes the D/Map is merely a tool of the analyst. It does not replace the trained analyst. It is necessary to have a full understanding of the nuances associated with patterns and the various substrates and situations found in any scene. The D/Map simply considers specific critical characteristics that should allow for differentiation. Upon arriving at some result (some classification), the analyst must challenge that decision by considering all of the criteria associated with said result and any opposing hypothesis. Only then can the classification be considered complete. The greatest value of the D/Map is in defining for the new or less experienced analyst the questions and order of questions that best leads to an objective classification, and articulating the foundation for their conclusion. After using the D/Map for a period, analysts will find they will follow the questions without needing to specifically refer to the actual document. The questions and order presented become second nature; however, when presented with an ambiguous pattern, even the experienced analyst may find it useful to refer back to the D/Map.

There are two sides to the D/Map: the spatter side and the non-spatter side. See Figures 4.35 and 4.36. The organization of the BGA D/map consists of blue diamonds, colored rectangles, ellipses, and the associated lines interconnecting them. The diamonds represent the question to be posed. The lines associated with the Y(es) or N(o) responses to that question, direct the analyst to the next question. Rectangles represent the associated criteria already considered up to that point. The ellipses represent the most defensible classification, given the questions posed and answered. The D/Map is based primarily on excluding pattern types, and pushing the classification to the remaining potential explanations.

The rules for applying the D/Map are simple. The answer to any question posed is either Yes, No, or I'm unsure. If a yes or no answer is forthcoming, the associated Y and N line directs the analyst to the next question. If the question cannot be answered, the analyst stops at the last point of agreement (the ellipse leading to the question being considered). Whatever level this may be, it is considered to be the most objective classification possible. Questions presented cannot be skipped, nor can one randomly choose what questions, or in what order the questions will be answered. One must begin at the top (Q: Are the stain(s) only small circles and ellipses w/without flow) and follow the appropriate Y and N lines of questions. Ignoring the order of the questions will effectively invalidate the entire document and process. One obvious question not included on the D/Map: "Is this blood?" The D/Map assumes the analyst has already considered this question.

It should be obvious that complex patterns cannot be evaluated with the D/Map. The D/Map assumes a situation where the pattern is one of the distinct classes/groupings. Any attempt to classify a complex pattern is likely to immediately make it obvious that the pattern is indeed complex, as bifurcated answers are not possible.

Initially, the taxonomy and D/Map may be slightly foreign. In practice, its ease of use becomes evident. Rarely are students, who are introduced to the concept and examples of the characteristics, unable to immediately apply the D/Map tool practically. Once understood and applied, the questions and order of questions becomes second nature. The D/Map appendix in this book provides several practical examples of applying the D/Map and Taxonomy. It would be extremely useful for the reader to move to that appendix and follow through the examples to better understand the application of the D/Map. The D/Map is now available as an App for both iPhone and Android users (search Bloodstain ID Decision Map).

A Final Consideration: Ambiguity and Contradictions in Word Terminology: Labeling Patterns

Throughout the discussion of terminology and pattern classification, the authors commented on the fact that many of the existing terms in use today are mechanistic (e.g., impact, projected, expectorate, drip). Terms utilized can also be contradictory (e.g., projected) or ambiguous (e.g., Drip Pattern) and certainly suggestive based on common English usage (e.g., impact). If two different analysts use different terms from different classification systems when describing the same pattern to the court, or during supporting

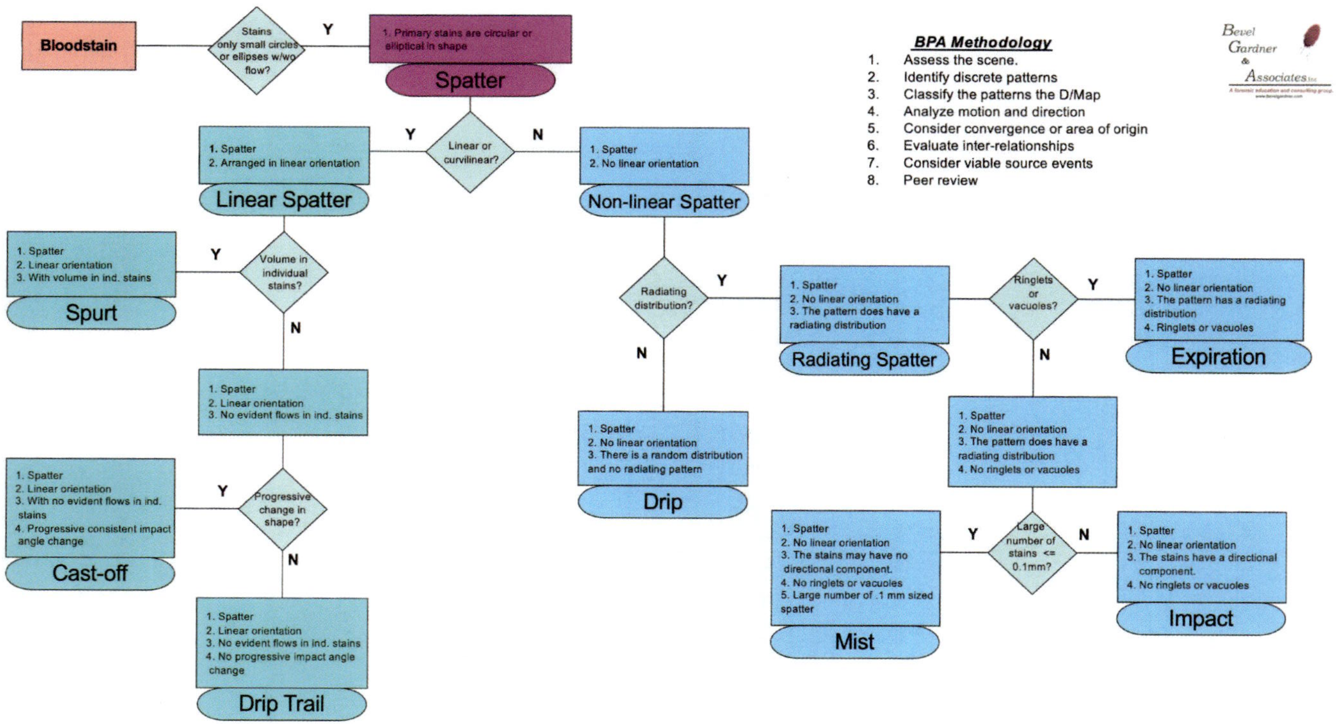

Figure 4.35 The spatter side of the decision map. By answering the first question: Are the stains only circles and ellipses with or without flow with a "Yes"; this leads the analyst to additional questions to further isolate a potential spatter classification.

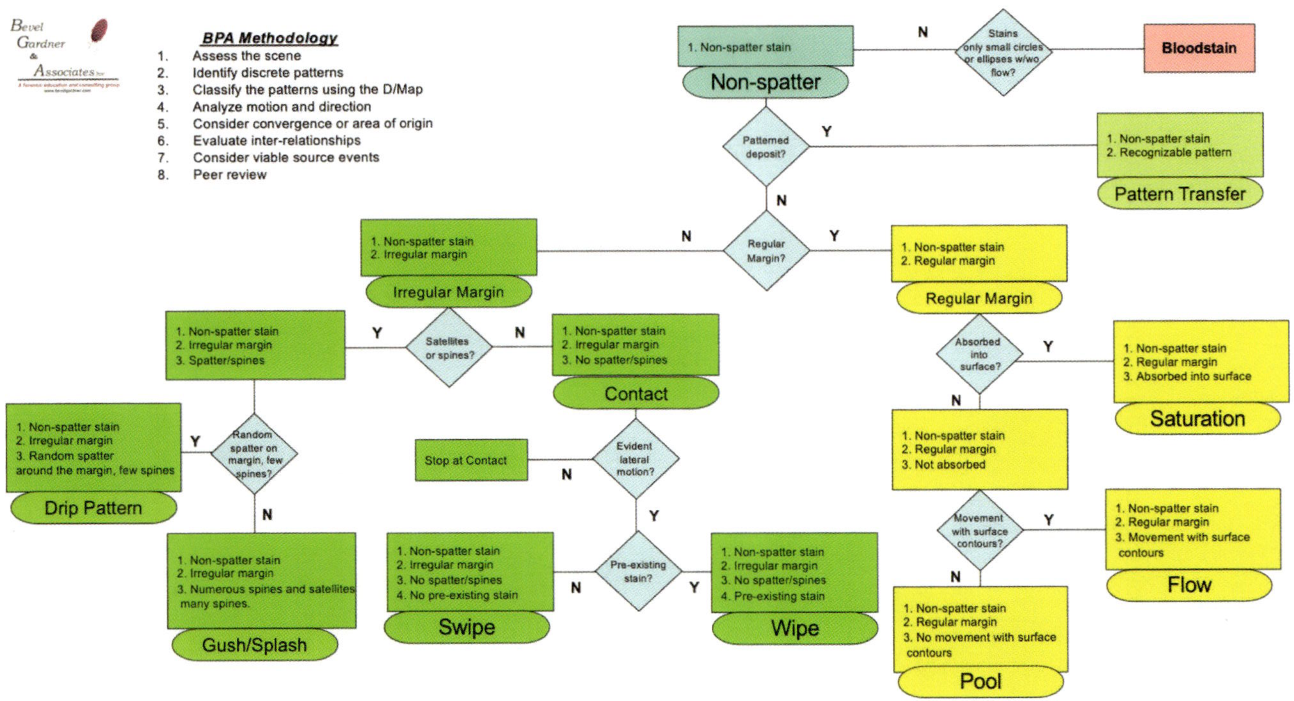

Figure 4.36 The non-spatter side of the decision map. By answering the first question: Are the stains only circles and ellipses with or without flow with a "No"; this leads the analyst to additional questions to further isolate a potential non-spatter classification.

testimony one uses similar terms (e.g., describing a specific "drip pattern," then referring to a drip trail or a pattern of random drips as a drip pattern), this may lead to confusion by a juror or imply to the jurors that BPA is not consistent (e.g., two analysts describing the same pattern in completely different verbiage). Finally, the use of English terms without recognition that literal translations in other languages do not always exist in a discipline practiced throughout the world is problematic. Given the concern of OSAC in establishing a singular vocabulary for patterns (which only makes sense as a concept), the following idea offers a potential resolution to those issues. Adoption of terms/words, and certainly mechanistic words, regardless of who defines what words may be used, is likely to continue to plague the discipline. Carolyn Gannet's list of over 500 words associated with BPA effectively proves that point.[23]

This concern over mechanistic labels remains an issue for the future of BPA.[24] Taking an idea practiced in other sciences, there is a very simple solution to these issues that would resolve most of the problems effectively. This idea is presented here merely as food for thought to those charged with furthering the discipline of BPA in the future. Adoption, however, would be exceedingly simple.

Rather than using words as labels for patterns, such word labels could easily be replaced by an alphanumeric label. The adoption of an alphanumeric label prevents two analysts describing a given pattern to the court, from using two different terms and causing potential confusion to those listening. Each analyst could identify the pattern by this alphanumeric label (for example "NSIR2"). The analyst is then left to describe in whatever vocabulary or language they feel comfortable with (e.g., this is a projected pattern, or this is a spurt pattern) to the jury, as well as explaining "what" that means. Alphanumeric labels would prevent inappropriate inferences based on some listener's prior definition (e.g., drip pattern, impact) that contradicts a BPA technical definition. Any analyst, regardless of where trained or in which language they were trained, could use this label when discussing patterns with any other analyst, anywhere.

To be clear, an alphanumeric label approach *forces* expectations that some form of criteria-based approach is present. There must be established parent–child relationships, with shared class characteristics that define that relationship and specific articulable criteria to produce any effective alphanumeric label system. There must be a way to clearly define what entails a pattern with any given label. If this suggests a bias by the authors, then yes, guilty as charged. But just the same, it should be clear by now that some form of taxonomy *is necessary*, regardless of how that taxonomy is ultimately articulated. A potential alphanumeric labeling approach is shown in Figures 4.37 and 4.38. The current associated word labels for the stains are shown in the figures for clarity, but would obviously be omitted. To be clear, the ideas presented in the figures certainly do not represent the *only* way an alphanumeric system might be defined.

OSAC Decision Process Chart and the D/Map

In 2019, OSAC released a document intended to offer analysts decision points to utilize when evaluating patterns. This document can be viewed at the OSAC website.[25]

In the author's opinion, there are significant issues with this document. This includes the use of highly ambiguous characteristics (some never discussed in the history of Bloodstain Pattern Analysis), as well as introducing context into the actual classification process. Following published methodology, patterns are evaluated first based on then stain

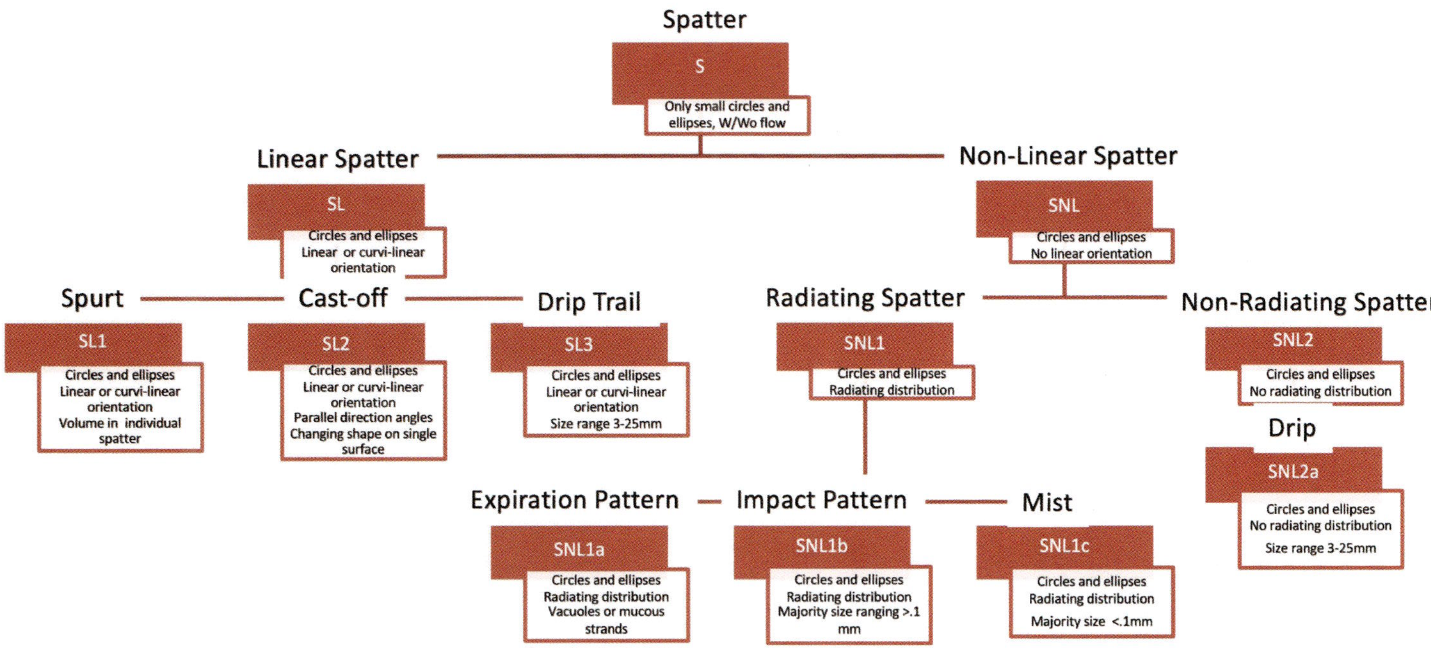

Figure 4.37 A potential solution to the ambiguity associated with word labels for patterns is an alphanumeric label. Rather than using a word label (shown here simply for clarity), an alphanumeric label is assigned to classifications. This is an example of how the spatter group might be configured.

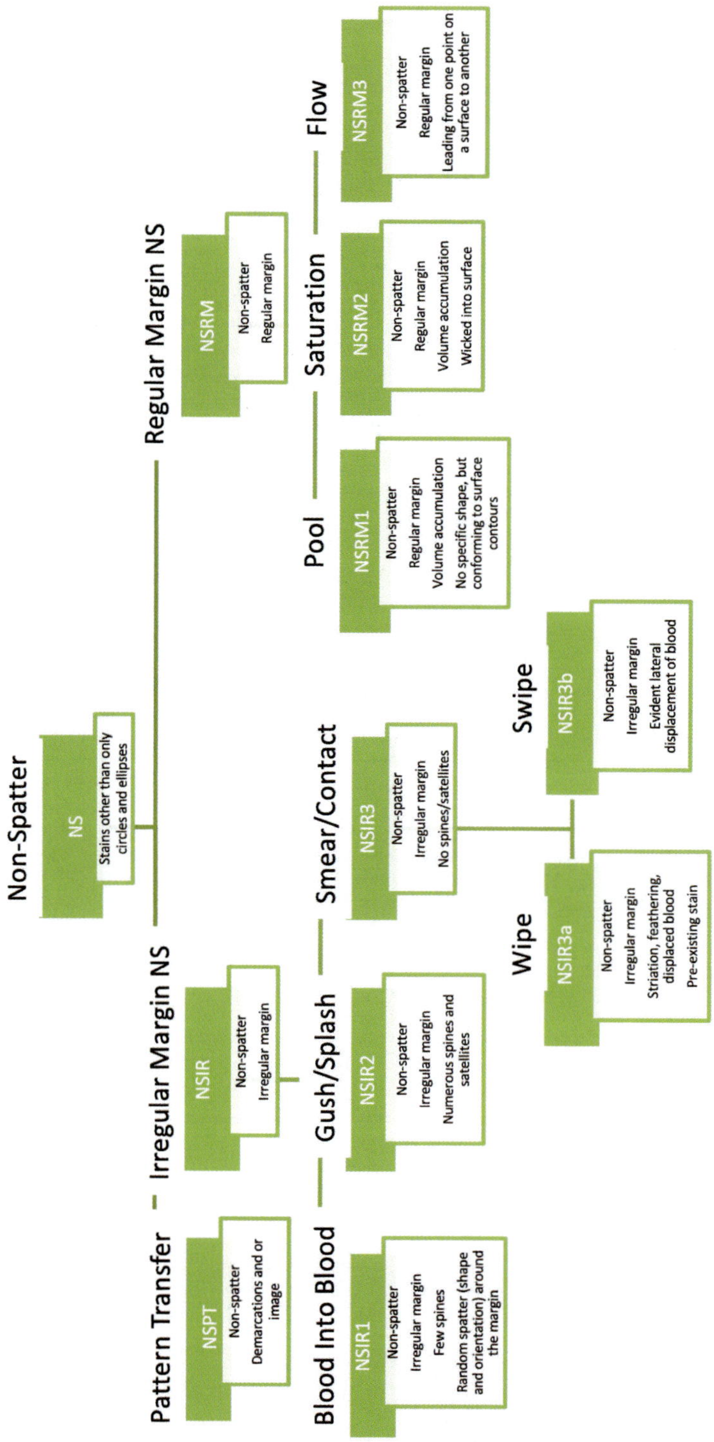

Figure 4.38 A potential solution to the ambiguity associated with word labels for patterns is an alphanumeric label. Rather than using a word label (shown here simply for clarity), an alphanumeric label is assigned to classifications. This is an example of how the non-spatter group might be configured.

itself and only then is context considered. Another concern with this document is the introduction of questionable decision points that can result in inappropriate inclusions or exclusions. As presented, there are no true parent–child relationships; different paths on the chart can lead to similar results demanding no viable solution when presented with ambiguous patterns. The OSAC Appendix details various concerns with this currently published OSAC "Process Map."

The BGA D/Map was developed over years. Published in 2008, it was put into practical application by varying organizations, in BPA training, and validated through an academic research effort. It came under immediate practical scrutiny. Many lessons were learned in terms of verbiage used in the D/Map questions, including the order of questions posed and even whether a question was best answered "Yes" or "No" for effective elimination of patterns. Through practical application, the BGA D/Map has been revised many times to resolve issues associated with building a true "expert system." In the authors' humble opinion, although the OSAC attempt was certainly a step in the right direction, the issues described ignore the obvious class characteristics the Taxonomy approach provides, and require significant reconsideration. Specific issues relating to the published OSAC Decision Process are discussed in a later appendix.

The D/Map remains a practical tool, tried and tested in the field. The authors are unable to explain OSAC's decision to publish a proposed "expert system" that clearly has not been practically tested or pushed to its limits, all the while ignoring a published taxonomy that has been in use for over 15 years. The authors do however wish them well as they experience the lessons the authors experienced through practical application and development of the D/Map.

Summary

The Pattern Diversity Principle is the heart of Bloodstain Pattern Analysis. Rarely is any conclusion offered in BPA without first considering "what kind of pattern is this." Every conclusion offered in BPA starts with consideration of the pattern type. Without classification, BPA is effectively moot.

But how one goes about classification is significant. On any given week, in classrooms across the world, students new to BPA produce patterns from known mechanisms in order to familiarize themselves with their visual appearance and prove that this predictability exists. Once in the field, this knowledge is obviously helpful. "I'll recognize it when I see it" and "I've seen that before" ruled the discipline of BPA for literally years. As true as this may be, such statements ultimately lack the foundation necessary of science in a court setting. The authors consistently see reports from experienced BPA analysts who operate under the "I'll recognize it" premise who are often correct in their classification conclusions. Simply because it "typically" works is not reason to allow the discipline to continue down that path. The authors have made a clear and consistent argument that if BPA is to continue as a scientific discipline, there must be specific articulable characteristics that any analyst can point to in achieving their conclusion. If not, then how is it we "*know it*" when we see it"?

Scientific rigor demands an unknown be compared to the known by some specific means; only then is a conclusion possible. That comparison must be based on specific criteria. The Taxonomy approach provides the analyst with those criteria. The taxonomy as presented may not be an "end all" product; refinement in the future is possible. Perhaps additional quantitative criteria can be defined for different patterns; perhaps someone more versed in language will find a simpler way to describe the specific qualitative criteria offered. But these published criteria are real, they exist and the basic idea behind them, no matter how they may ultimately be articulated, isn't going anywhere. It can't, it is basic fluid dynamics. This taxonomy is your basic Model T! Make it into a Ferrari and the authors will applaud such effort. But the basic concepts that made the model T work will remain. Science always seeks to refine its understanding, and this is certainly true of the taxonomic approach.

What a criteria-based classification, the taxonomy, provides science is twofold. First, it returns the discipline back to the expectations of science. Claims of subjectivity in forensics abound. Any approach in any discipline that removes subjectivity is good. Remember that BPA is a class characteristic evidence, thus like all class characteristic evidence it is best for eliminating possibilities (e.g., this pattern is not part of that group). Elimination must be based on articulated criteria. Exclusion from some group/classification significantly limits the potential explanations for creation. This is a critical component of BPA's application in the courtroom, where invalid claims as to the source of any given bloodstain do occur. Regarding exclusion, the recognition of the characteristics found in a parent classification held by the children of that parent is critical as well. Presented with a linear pattern of spatter on a wall, but lacking the additional characteristics that would allow exclusion of either spurt or cast-off, an objective classification decision of "linear spatter" ensures that some potential source explanation is not inappropriately excluded.

Second, the taxonomy provides for defensible conclusions. No matter how well the "I'll know it when I see it" approach might work, even for experienced analysts, it will ultimately fail when presented with ambiguous stains. In the instance of poorly trained analysts and those with limited experience, this approach routinely fails. Armed with the taxonomy and presented with a stain and the ability to point to and articulate specific physical characteristics lends confidence to the analyst for any subsequent conclusion. If challenged in court with "What led you to that conclusion?," the response of the analyst is immediate. The analyst points to and describes the criteria present. When alternative claims are made "Well couldn't that stain be the result of…" the analyst can define specifically why that possibility is or is not eliminated. Recognizing the specific criteria and the ability to articulate those criteria moves the conclusion far beyond the visual "I'll know it when I see it." Remember also that a duty of every scientist is to attempt to refute some belief, no matter how strongly they may feel about it. Without criteria how is that possible?

Analysts must understand that classification is merely recognition of the basic underlying mechanism of creation. It alone does not define a specific scene mechanism. If that is possible, it occurs only after evaluating the entire scene and context using the established BPA methodology. Certainly, not every possible mechanism can be defined. What the Pattern Diversity Principle of BPA provides is recognition of several basic mechanisms and as described some limited differentiation within these mechanism groups. This currently includes the following:

- Blood dispersed from a point source by a force or impulse (e.g., impact patterns, mist, expectorate patterns)
- Blood dispersed over time from an object in motion (e.g., cast-off patterns)
- Blood dispersed in a stream (e.g., spurt and gush/splash patterns).
- Blood dispersed through the air as a function of gravity (e.g., drip patterns, drip trails, and drip stains)

- Blood that accumulates and/or flows on a surface (e.g., pools, saturations, and flows)
- Blood deposited through contact (e.g., smears, wipes, swipes, and pattern transfers)

Classification is the first and most important step of BPA. Only when the basic mechanism is understood, can the analyst look at a unique scene and begin to isolate potential explanations from the scene for a given pattern. This is of course the goal of BPA. As John Aycock so aptly made the point, jurors want to know how did this stain come to be in this scene. Any such conclusion (one based on the entire methodology and predictability of Bloodstain Pattern Analysis) is only possible based on an objective classification conclusion. One cannot hope to explain a forest, until they understand the trees.

Notes

1. Premature rejection in science: The case of the Younger Dryas Impact Hypothesis James Lawrence Powell University of Southern California, Los Angeles, CA, USA, Science Progress 2022, Volume 105, Number 1, p. 37.
2. Gardner, R. and Griffin, T., Foundations of the Discipline of Bloodstain Pattern Analysis, A Response to the Report of the National Academy of Sciences, *Journal of Forensic Identification*, Volume 60, Number 4, 2010, p. 480.
3. Bevel, T. and Gardner, R.M., *Bloodstain Pattern Analysis; Bloodstain Pattern Analysis With an Introduction To Crime Scene Reconstruction*, 3rd Edition, CRC Press Inc., Boca Raton, FL, 2008, Chapter 4.
4. James, S. et al., *Principles of Bloodstain Pattern Analysis Theory and Practice*, CRC Press, Boca Raton, FL, 2005, p. 68.
5. Wonder, Anita, *Blood Dynamics*, Academic Press, San Diego, CA, 2001.
6. Hook, E., Fieldhouse, S., Flatman-Fairs, D., and Willimas, G., Bloodstain Classification Methods: A Critical Review and Look to the Future, *Science and Justice*, Volume 64, Number 4, July 2024.
7. Phua, Zai Rong and Hauw, Jiw Boon, Probing the Formation and Characteristics of Downward Cast-off and Cessation Patterns, *Journal of Forensic Identification,* Volume 72, Number 4, (1), Sept 2022, p. 4.
8. Gardner, R.M., Defining the Diameter of the Smallest Parent Stain Produced by a Drip, *Journal of Forensic Identification*, Volume 56, Number 2, Apr 2006, p. 210.
9. Gardner, R.M., Deformation Levels of Droplets Created by Impact, presentation to the International Association of Bloodstain Pattern Analysts, Albuquerque, NM, Nov 9, 1996.
10. Van Netten, A.A. and Dewey, J.M., Blood Spatter 2 Final Report, DMA Report 97/2, Dewey, McMillin and Associates, LTD., Victoria, BC Canada, Mar 97.
11. Gardner, R.M., Deformation Levels of Droplets Created by Impact, presentation to the International Association of Bloodstain Pattern Analysts, Albuquerque, NM, Nov 9, 1996.
12. Van Netten, A.A. and Dewey, J.M., Blood Spatter 2 Final Report, DMA Report 97/2, Dewey, McMillin and Associates, LTD., Victoria, BC Canada, Mar 1997.
13. MacDonell, H.L. and Bialousz, L.F., Flight Characteristics and Stain Patterns of Human Blood, National Institute of Law Enforcement and Criminal Justice, LEAA Report PR-71-4, U.S. Government Printing Office, Washington, D.C., 1971.
14. Carter, Fred, *Physics of Bloodstain Analysis* – Lecture Notes, 1999, p. 17.
15. Forensic implications of respiratory derived blood spatter distributions
    David Denison a, Alice Porter b, Matthew Mills c, Robert C. Schroter, *Forensic Science International,* Volume 204, 2011, pp. 144–155.
16. Forensic implications of respiratory derived blood spatter distributions

David Denison a, Alice Porter b, Matthew Mills c, Robert C. Schroter, *Forensic Science International,* Volume 204, 2011, pp. 144–155.
17. Metropolitan Police Laboratory, Blood in Slow Motion, Droplet Dynamic Video, London, England, 1991.
18. The OSAC/ASB term "transfer" includes a broader category of stains, beyond that intended by "smear," but it is a functional term.
19. Gardner, Ross M., Directionality in Swipe Patterns, *Journal of Forensic Identification*, Volume 52, Number 5, Oct 2002, pp. 579–593.
20. Hook, E., Fieldhouse, S., Flatman-Fairs, D., and Willimas, G., Bloodstain Classification Methods: A Critical Review and Look to the Future, *Science and Justice*, Volume 64, Number 4, July 2024, p. 418.
21. Philippe Esperanca, verbal communications while attending SWGSTAIN meetings, 2005–2006.
22. The authors apologize for any ambiguity, as this was a verbal discussion by Epstein many years ago. The specific date and location has been long forgotten.
23. As a participant in SWGSTAIN, Carolyn Gannet provided a database she created of terms used throughout the history of Bloodstain Pattern Analysis. This database consisted of over 500 terms.
24. Hook, E., Fieldhouse, S., Flatman-Fairs, D., and Willimas, G., Bloodstain Classification Methods: A Critical Review and Look to the Future, *Science and Justice*, Volume 64, Number 4, July 2024, p. 419.
25. OSAC website, https://www.nist.gov/osac/bloodstain-pattern-analysis-subcommittee, February 20, 2020.

Stain Shape and Vector Correlation

Issues of Motion and Spatial Origin

5

The second significant aspect of predictability in BPA lies in the principle of Stain Shape and Vector Correlation (SSVC). This principle involves understanding motion associated with blood deposition and the origin/position of bloodstain actions occurring within the scene. The most recognized aspects associated with SSVC is consideration of impact and directional angle, but these are not the only indicators of motion found in bloodstains. One way of articulating the general principle is to say:

> *The shape of certain bloodstains provides indicators as to the direction of deposition as well as to the spatial origin of the blood.*[1]

Establishing the motion or spatial origin involved in bloodshed events allows the analyst to better understand the actions that occurred during an incident. The nature of this motion includes the following:

- Droplet impact angles – identifying the angle a given spatter droplet intercepted a substrate, as it collapsed on that substrate.
- Droplet directional angles – identifying the direction a given spatter droplet was moving relative to the substrate, as it collapsed.
- Seeking to isolate or limit areas of convergence and origin for impact events. (Detailed in the next chapter)
- Understanding the directional component of blood trails.
- Understanding relative motion associated with disruptions of a stain, as well as depositions of contact stains.
- Motion associated with deposition of repetitive pattern transfers.
- Motion indicated by flows.
- General sequence of actions.

Except for Area of Origin (discussed in Chapter 6), this chapter explains these varying ideas associated with defining motion and spatial orientation associated with bloodstains.

General Sequence of Actions

This first issue is not as distinct or as absolute as other considerations of motion, such as direction and impact angle. Nevertheless, it does provide indicators that may be helpful to the overall analysis. A general rule can be applied in Bloodstain Pattern Analysis that is reciprocal to a rule found in a suspected arson investigation. In any fire investigation, the investigator looks for the point where the fire caused the greatest damage. Typically,

DOI: 10.4324/9781003408734-5

the point of greatest damage is the point where the fire burned longest, which may be the potential origin. In bloodstained scenes, analysts apply this approach in reverse. The location where one finds the greatest bloodshed is likely to represent latter aspects of the bloodshed event. Those areas where we find the least blood (e.g., spatter and drips) is likely at or near the point where the bloodshed started.

This rule for bloodstained scenes exists for two reasons. As the attacker inflicts injuries, the level of bleeding is likely to increase as greater damage and breaching of the circulatory system occurs. The resulting damage decreases the victim's ability to escape and retain mobility. The end effect is an increase in bleeding at any given location in the latter stages of the event as compared to the earlier stages. The authors do not suggest this is an absolute rule. But considered in many scenes, it will guide the analyst. Those locations with only limited staining likely lead to areas of greater staining. As the victim sustains different wounds, the patterns may change, but the increase in bleeding should be apparent.

For example, if the first blows of an attack create a small laceration, one may see evidence of drips. As the victim retreats to another area, this bleeding from the laceration continues. At the second location, the attacker strikes the victim again. Blood is present on the body from the first wound, thus the blow itself creates more dynamic spatter stains. These blows might create a widening of the wound or additional wounds. This in turn increases blood loss and additional dynamic staining. Blood may be accumulating in the clothing to the point it reaches a saturation level and begins dripping or producing significant contact patterns. Now attempting to defend against the blows rather than escape them, or perhaps dazed as a result of the blows, the progress of the victim may begin to slow. The volume of blood lost at any point increases as the victim remains in that location longer. Throughout the events, the victim's hands and clothing become stained. These bloodied surfaces in turn stain walls and surrounding items. The more blood, the more soaked the items; the more soaked the items, with additional movement the more staining present. It is by viewing the stains in this simple fashion that analysts may discern a general flow of the overall incident and the actions that define it.

The obvious exception to this rule is a situation in which the body is moved to a secondary scene or location. In these instances, the absence of blood is significant and will become an alternative indication of sequencing actions.

Droplet Collapse Dynamics

In order to understand why both impact angle and directional angle are definable requires an understanding of the collapse of a stable droplet onto a target. This collapse leads to a flow of the droplet's volume across the target surface. This flow may be symmetrical or asymmetrical due to several variables. Characteristics of the substrate itself and/or foreign material on the substrate can lead to asymmetrical outflow. Symmetrical outflow leads to symmetrical or well-formed stains; asymmetrical outflow leads to asymmetric or irregular-shaped stains.

Before one can consider collapse, one must consider the droplet's behavior in free flight. In flight, what characteristics does a blood drop have? It should be noted that previous authors have suggested that the information gleaned from raindrop research (water droplets in free flight) is not applicable to blood drops, as the former are a Newtonian fluid, and the latter are a non-Newtonian fluid. The primary difference between Newtonian and non-Newtonian fluids is their behavior under pressure (a volume of fluid flowing in a tube).

Despite these differences of flow behavior, the nature of blood drop behavior in free flight has through empirical experiment and research proven quite similar to that described for water drops, so we feel this research is still applicable to our understanding.

Blood drops in flight ultimately achieve a spherical shape. James E. McDonald, in discussing raindrop shapes and answering why a fluid droplet is spherical, commented:

> Surface tension always tends to reduce the surface of a free mass of liquid to the smallest area it can achieve; an isolated drop of liquid not distorted by external forces is pulled by its surface tension into a spherical shape.[2]

Dr. Alfred Carter also commented on the issue of shape and surface tension with regard to blood drops stating: "With regard to the spherical shape of liquid droplets, it can be explained entirely by the surface tension of the liquid."[3]

Figures 5.1 and 5.2 demonstrate liquid blood forming on the tip of a screwdriver and the subsequent detachment of a single drop. As evident in Figure 5.1, the forming droplet, while still attached to the blood volume on the screwdriver, takes the shape of the teardrop. Once fully detached, this shape becomes generally spherical. The droplet is not a perfect sphere, as the droplet is oscillating while in flight. See Figure 5.2.

McDonald in considering how the force of surface tension affects fluid droplets found that small droplets (1 mm in diameter and less) were almost perfect spheres. Larger drops,

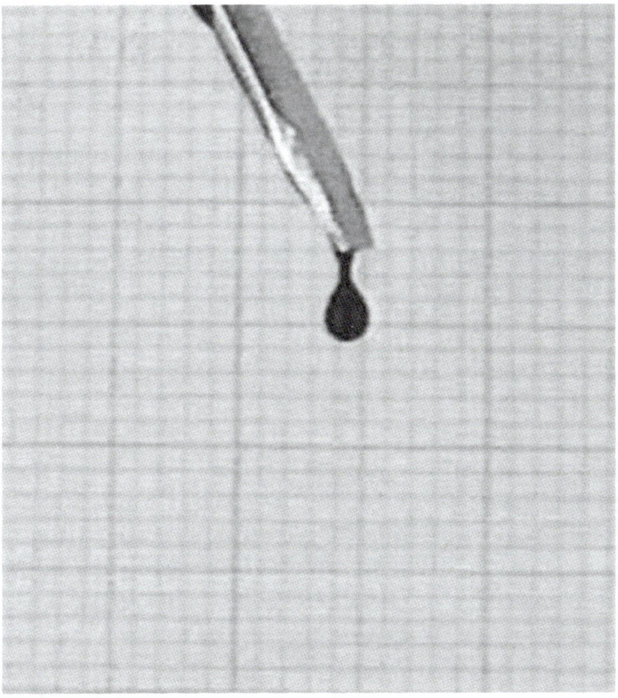

Figure 5.1 A drop forming on the tip of a screwdriver. At this point, the weight of the liquid droplet at the tip has almost overcome the ability of surface tension and adhesion to the surface to hold it intact with the fluid running down the screwdriver. The forming droplet takes on a "tear-drop" shape as gravity pulls it down and away from the tip. The structure connecting the forming droplet and the fluid on the screwdriver is nothing more than a thin ligament of fluid blood.

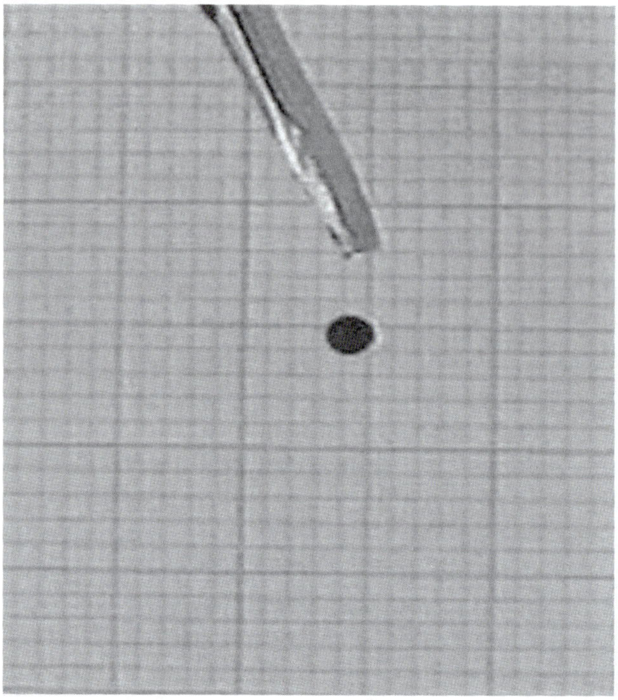

Figure 5.2 Gravity acting on the mass of the droplet has overcome the force of surface tension and in the connecting ligament structure and the droplet is now detached. Once in free flight the droplet is generally spherical, although it is oscillating.

however, were unable to maintain a true sphere shape in flight and often resembled nothing less than a "hamburger bun."[4] These oscillations in the droplet can cause significant alterations in the shape of the drop. Surface tension acts to shape a droplet into a sphere, but surface tension is not the only factor in reducing these oscillations. As many authors have noted, water droplets oscillate to a great extent when falling in air. The surface tension of water is considered to be 70 dyne/cm, whereas blood has a surface tension of only 50 dyne/cm.[5] If surface tension alone were the only force damping these oscillations, we might well expect blood droplets to oscillate considerably. In fact, they do not, based on their viscosity.

To understand these oscillations, imagine for a moment that we could sit a droplet on a table (like a water balloon) and that it would retain its shape. If pressure were applied to either the side or top, the shape of the droplet would change to accommodate the liquid volume moved by that pressure. The volume would not change, only the shape. While in flight, oscillations in droplets have the same effect on the shape, but the force for change comes from within the droplet. Initial oscillations of the blood in a droplet are created when the drop is dispersed at the blood source.

The closest analogy we can offer is that, like a wave, once produced, the oscillation moves through the droplet rebounding within it. The resulting change is cyclic, that is, it causes a range of recurring deformations from the perfect sphere. This results in a somewhat evenly distributed shape deformation. See Figure 5.3. Once initiated, this oscillation rebounds in the droplet, but is constantly damped by the force of surface tension and viscosity. As a result, the level of this deformation will decrease over time. Just like water

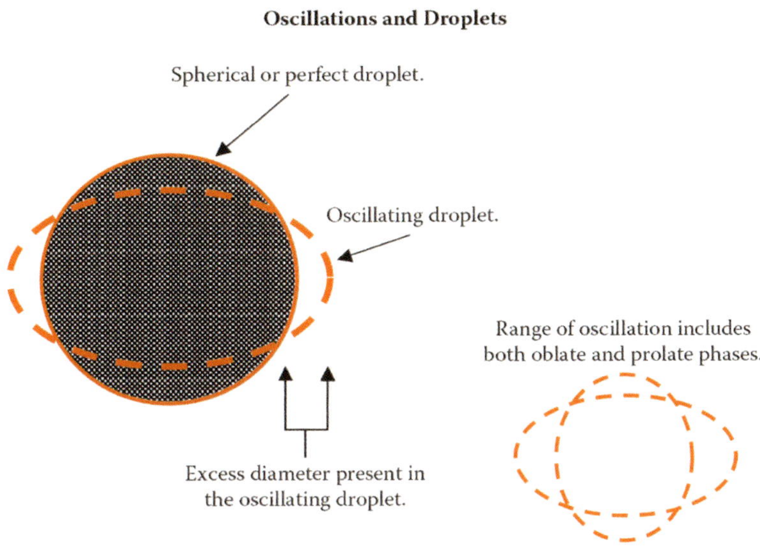

Figure 5.3 Oscillations in droplets. Because of their formation, droplets oscillate. This causes a shift of the droplet's mass from a perfect sphere to an oblong spheroid. The oscillation is dynamic; that is, it shifts until the oscillation decays completely. As such, the droplet moves through a wide range of shapes as the oscillation shifts back and forth between the oblate and prolate phase.

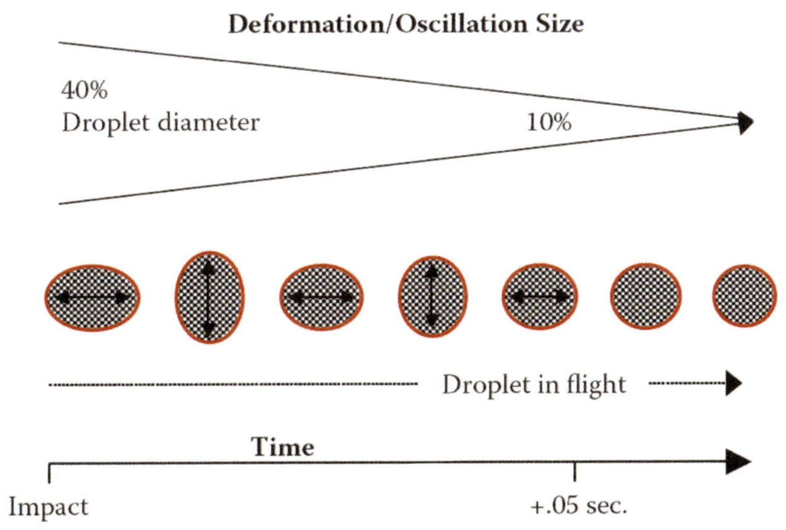

Figure 5.4 Oscillations in droplets move through an entire range of deformations. The most important aspect of the oscillations is that they decay with time. The longer the droplet is in flight, the smaller the remaining oscillation.

droplets, the smaller the blood drop, the faster the oscillations will damp and disappear. See Figure 5.4. This allows the droplet to return to a nearly spherical shape in short order.

We understand that our blood droplet's behavior is to seek a spherical shape as it follows a given flight path to its destination, but what happens when that spherical-shaped droplet impacts a given surface?

As with all fluids, on impact these small volumes of fluid have no ability to withstand a shearing stress. When the droplet contacts an unyielding surface, the volume of the droplet must displace. The volume will flow across that surface and the behavior of the flow defines the ultimate shape of the resulting stain. This leads to the classic characteristics of a spatter stain, a circular or elliptical shaped stain.

Over a two-year period, Gardner studied impacts using both stroboscopic photography methods and stop-motion video. The droplet impacts involved varying targets, including coarse sandpaper, rough weave fabrics, poster board, and Plexiglas. Impact angles were also varied. As a result of these observations, four distinct phases of this flow were identified.[6]

These phases appear independent of surface texture or impact angle, although all these issues may have some affect on the subsequent outflow. The phases, although slightly different on different surfaces, occur in nearly all impacts. The only caveat we make about these observations is that there appears to be a point of diminished volume for very small droplets at which the later phases are altered. The Weber number in physics is a dimensionless number proportional to the amount of kinetic energy in a fluid drop, divided by its surface tension energy. Extremely small blood drops have very small Weber numbers. As Carter reported, studies suggest that if a Weber number were low enough, a droplet could theoretically retain its shape (not collapse or flow across the surface) or even rebound off a surface, much like a drop of mercury.[7] This may explain why in very small droplet impacts (e.g., those associated with mist stains) directionality is rarely noted in individual stain shapes. The Weber number is too low. The drops simply impact without collapsing as do larger drops, appearing generally circular with no ellipses noted.

In considering the observations, accept them as just that, empirical observations, which help us understand the flow of the droplet's volume across the surface to create a stain.

Your authors choose to call these four phases of impact:

Contact/Collapse, Displacement, Dispersion, and Retraction. For purposes of the discussion, the following assumes a solid substrate or target. Keep in mind that the extent a target is hydrophobic (a non-wetting surface) or hydrophilic (a wetting surface) also affects the outflow, but only in terms of the extent of certain aspects of the outflow observed in the four phases. Absorbency of the target can also play a prominent role in affecting the symmetry of the outflow. Additionally, when a droplet encounters a highly textured surface such as carpet fibers, the outflow in the displacement phase is significantly affected.

Contact/Collapse

This phase begins when the droplet (a sphere) contacts the substrate or target. The droplet begins to collapse at the initial point of contact of the sphere. Those sections of the sphere not in direct contact with the target remain intact and generally unaffected until they too collide. As a result, halfway through the collapse, one would see half a sphere or droplet centered about the point of impact.

As this collapse occurs, the blood present at the collapse point is forced outward creating a rim around the droplet. This rim or boundary grows as more of the droplet's volume is forced into it. The flow to the rim has been referred to as an "involution"; however, it does not appear to fit that definition since the blood flows outward and not inward.[8]

Figure 5.5 A droplet in late contact/collapse phase. A small dimple is evident in the center of the developing stain. This is what is left of the still collapsing droplet.

The angle of impact affects the droplet's collapse by defining the nature of the rim and blood flow into it. For instance, in a stain impacting at 90°, the blood flows into the rim equally on all sides. See Figure 5.5. As a result, we see a circular stain with a generally uniform diameter.

In stains impacting at more acute angles, the droplet's momentum forces the blood from the collapse into a more directional outflow. Consider an impact at 20°. Although the rim is still evident across the entire periphery of the expanding stain boundary, the blood is flowing primarily into the area of the rim opposite the droplet's initial point of contact. See Figures 5.6 and 5.7. This directional outflow in combination with the droplet's skimming movement across the surface during collapse helps explain the elliptical shape of the stain.

Surface characteristics have an influence on the contact/collapse phase. Surface irregularities found on rough targets can initiate an irregular outflow of blood in the collapse; however, this irregular outflow is far more evident in the displacement phase.

Displacement

In this phase, the sphere has collapsed onto the target and the majority of the blood volume from the droplet has displaced to the boundary rim of the resulting stain. Dr. Balthazard, in his early efforts, also noted this particular phenomenon.[9] To this point, there is no apparent disruption of the surface tension. The droplet's mass, although exposed to considerable shifts in its shape, is still one entity and has not broken apart.

Often evident on the boundary rim of the collapsing droplet are smaller developing droplets or short spines. See Figure 5.8. These appear to form as a result of the blood displacing from the central portion of the stain into the rim. These formations may eventually lead to the creation of satellite spatter during the next phase of the impact. The central area of the stain retains some blood, resulting in a sheet of blood interconnecting the entire structure. Nevertheless, nearly all of the liquid volume of the original droplet is now within

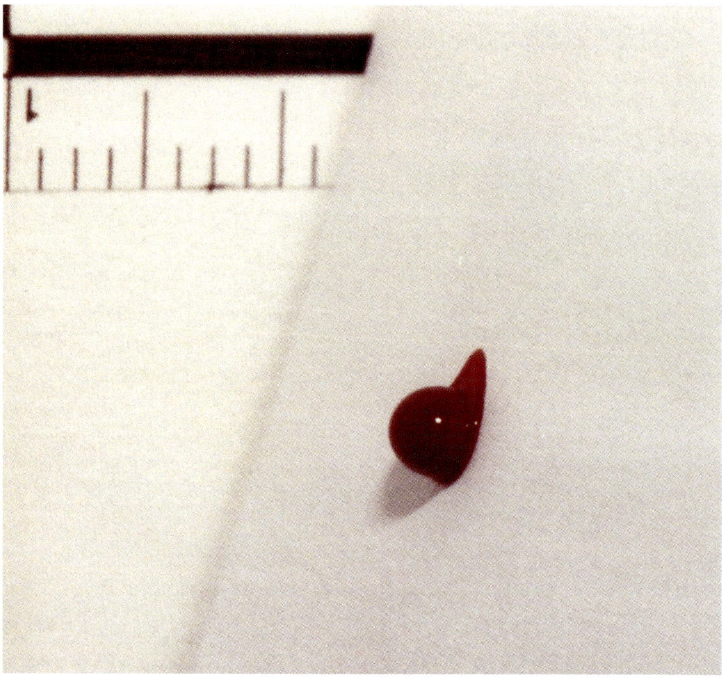

Figure 5.6 Early contact collapse in an acute angle impact. Note that the droplet retains its shape as it collapses from the bottom-up.

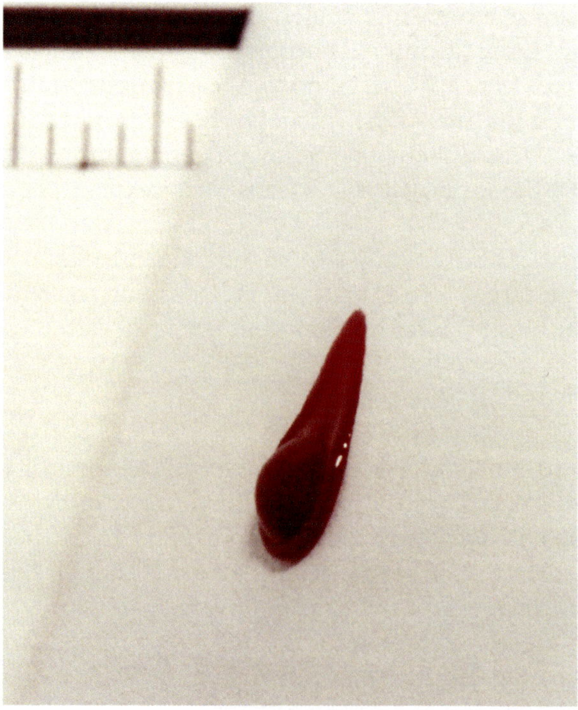

Figure 5.7 Late contact collapse in a droplet impacting at an acute angle. In this photograph, about half of the droplet is still intact and evident in the center of the expanding stain. Note that the majority of the blood is being forced into the rim forward of the direction of its movement across the surface.

Figure 5.8 Displacement phase in a 90° impact. Note the displacement of the liquid to the rim of the developing stain. All motion up to this point in the collapse is generally lateral (outward).

the structure of the rim. The outflow movement in displacement is generally laterally to the target, the blood flows out and away from the initial point of contact.

Although displacement is only the second phase of the impact, it is interesting to note that the area defined in the displacement phase (its contact line) typically defines the overall dimensions excluding any spines of the resulting stain. If the liquid laterally displaces 4 mm in this phase, the resulting stain will measure 4 mm. It is important to note that displacement does not occur on a 1:1 ratio. That is, a 4-mm-diameter droplet does not create a 4-mm stain. We have yet to define the true nature of this displacement ratio, but it has been empirically noted to be as much as 270% (e.g., a 4-mm droplet may create an 11-mm stain). Given the variations of viscosity in different drops, this is to be expected. It is also believed the displacement ratio drops as the volume drops, but this area requires further study.

Impact angle affects the displacement phase only in the nature of the developing protuberances that will ultimately spawn satellite drops. As was evident in the contact collapse phase, 90° impacts result in development of a symmetrical rim, blood flows symmetrically outward in all directions. Thus, in displacement, the small developing droplets in the rim may also appear symmetrically, surrounding the entire rim. Acute-angle impacts lead to these protuberances only in the forward edge of the developing rim. A single developing droplet, as found in a 10°–20° impact, results in the classic wave cast-off stain. For example, Figure 5.9 exhibits displacement, but prior to the development of any protuberance(s).

Any of these small protuberances in the rim can result in an individual satellite spatter, scallops, or spines. For instance, in 50° impacts, one often observes a bear claw effect in the resulting stain. Each claw (the scalloped aspect) is the result of a protuberance that developed in the rim during displacement.

The surface texture of the target plays a prominent role in the displacement phase. It is not that a rough target surface "bursts" the droplet, resulting in the irregularly shaped stains we often find. In fact, the old "bursting balloon" analogy is severely lacking. In the

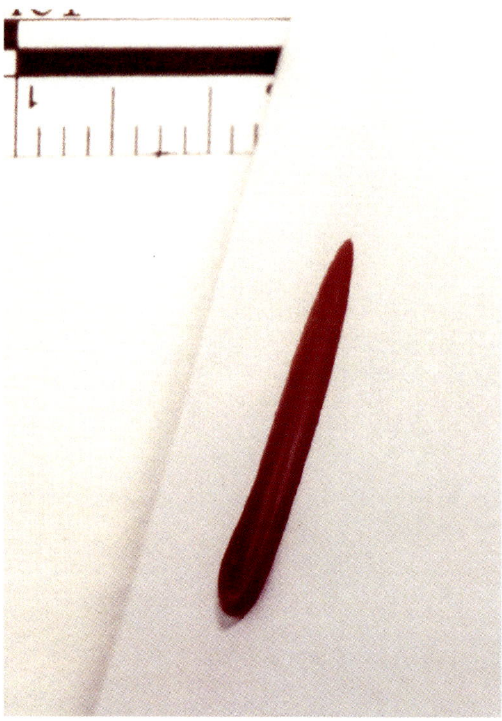

Figure 5.9 Early displacement of a droplet impacting at an acute angle. Note the distinct rim in the forward edge.

balloon analogy, the skin is not an integral part of the fluid. In the blood droplet, however, the "skin" created by the surface tension is the fluid. It doesn't burst; it simply shifts its shape.

On a rough surface, the volume doesn't flow evenly across the surface, minor irregularities in the surface often cause asymmetrical outflow. The volume in one part of the rim may be large, while in another it may be small. The protuberances and spines that form will also be irregular based on this difference in volume. This irregular outflow will usually result in an asymmetrical collapse of the entire structure later in the retraction phase.

Irregular outflow produces what analysts would likely call a distorted, irregular, or excessively spattered stain. Figure 5.10 shows stains created by similar droplets falling the same distance, but onto different surfaces. The droplets falling onto the rougher surface display distinct shape distortion. We can never totally ignore the effect of the target surface on the collapsing droplet, as it may limit subsequent analysis.

As previously mentioned, the displacement phase is effectively muted when a droplet impacts a surface such as carpet. The volume flows into and down the fibers and fabric loops rather than displacing laterally. This is particularly evident in pile-type carpets, but age of the carpet and condition of the carpet fibers and loops all play a role in this aspect.

Droplet velocity also plays a role in the displacement phase. As other authors have indicated, terminal velocity is reached for a given droplet when air resistance counteracts the effect of gravity.[10] This results in a constant velocity.

If we drop similar volume drops from different heights, the resulting stains will exhibit different overall sizes, resulting from different levels of displacement. All things being

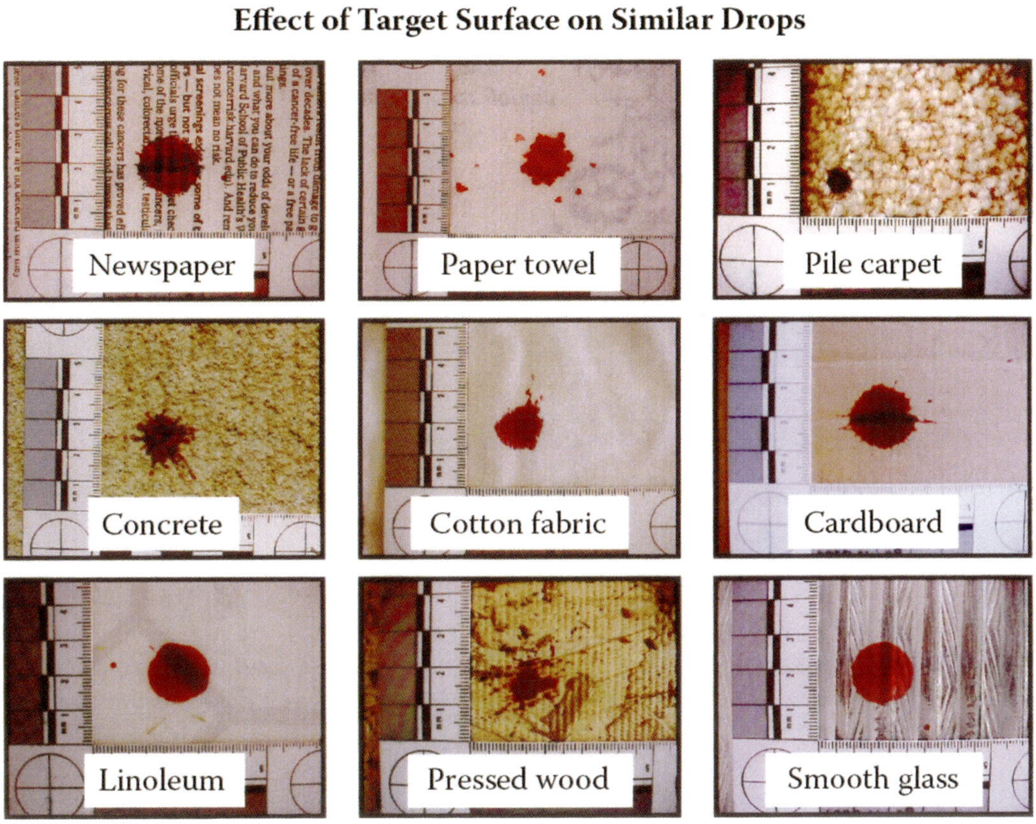

Figure 5.10 Target surface and stain shape. Target surface characteristics affect the shape of the resulting stain. The smoother the surface, the more likely the stain will be symmetrical with fewer satellite spatter. The rougher the surface, the more likely the stain will be asymmetrical with a greater number of satellite spatter produced.

equal, a stain caused by a drop falling 3″ will be smaller than one caused by a drop falling 3′. This is due to the increase in speed of the droplet as it falls. The farther it falls, the faster it falls, up to its terminal velocity.[11] However, dropping distance experiments also indicate that before the droplet reaches this terminal velocity, no further lateral spread (displacement) will be evident in the resulting stain. At a height of approximately 90–100 in., any additional displacement occurring in the droplet above this height will be so minor as to impossible to measure effectively. See Figure 5.11.

In the 1970s, analysts often made a correlation between the size of a stain and the distance that it fell. These conclusions were based on the effect of velocity on the droplet (e.g., the effect of its falling speed to its lateral spread in displacement), but this presumed a constant volume for a drop. Subsequent efforts show that the volume of a drop varies greatly depending upon the nature of the surface on which it forms. The greater the surface area available will allow more blood volume to collect in the resulting drop (realizing there is a maximum volume possible for a stable droplet). Unfortunately, the many combinations of droplet volume and droplet speed leave too many variables for this knowledge to be of any precise use in analysis. The only caveat we offer is if a droplet falls very near the surface (e.g., generally at 3″ or less) the volume will displace minimally. The blood will not displace significantly, appearing more as a convex mass.

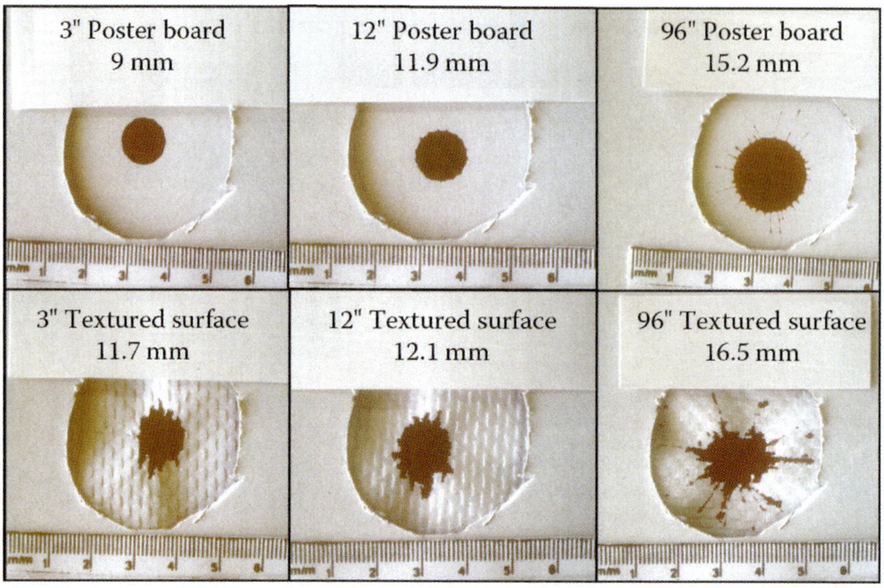

Figure 5.11 The further a gravity-induced droplet falls, the faster it accelerates up to its terminal velocity. The result of this increase in velocity is an increase in the resulting stain size, up to a point. These drops (all of equal volume) fell 3″, 12″, and 96″, respectively, onto the two target surfaces. The size of the stains increases with the increase in dropping height for nearly all surfaces, with the exception of significantly absorbent targets (e.g., carpet).

The inertial force, whatever put the droplet in motion, is the primary force affecting the displacement phase. Nevertheless, the cohesive forces are still active within the fluid and keep the drop as a single entity.

Dispersion

The dispersion phase offers a glimpse at perhaps the most elegant behavior of the droplet during impact. Results include the crown or blossom effect in near 90° impacts. See Figures 5.12 and 5.13. The extent that the substrate is or is not water repelling (e.g., is the surface hydrophilic or hydrophobic) defines to what extent the crown or blossoming affect occurs. For example, the silicon dioxide in most glass leaves it with a slightly polar charge, making it somewhat hydrophilic. The crowning affect observed in such instances is significantly smaller than those observed on paper targets.

Figure 5.14 shows the dispersion phase in an acute angle impact. The development of the droplet at the end of the emerging spine is often referred to as a wave cast-off. Although commonly reported, there is no whip or backlash effect evident in the wave cast-off.[12] It is simply flowing forward.

This phase is easily categorized by the action of the rim. Blood volume flows into the rim and any developing protuberances. Here inertial forces are still forcing the blood to flow outward, but cohesive forces become more significant. Rather than displacing only laterally, surface tension acts within the liquid causing an upward outflow off the contact line with the surface, resulting in the blossoming effect. As the volume of liquid increases in the small protuberances around the rim, these structures become unstable. If sufficient

Figure 5.12 Early dispersion in a droplet impacting at 90°. Note the surface tension is still a factor, although the droplet's mass has shifted dramatically. Movement of the droplet's mass is upward and outward from the contact line.

Figure 5.13 Late dispersion. Note the small dimples present on the rim of the blossom. In some cases, these may detach and cause satellite spatter. Also, note that the movement of the liquid in the dispersion phase is upward from the contact line and slightly outward.

Figure 5.14 A classic wave cast-off developing during the dispersion phase in an acute angle impact.

remaining inertial force is present, the structures continue in this now outward and upward movement. These structures may break from the main volume, resulting in the creation of satellite spatter.

The angle of impact plays an important role in defining the dispersion phase. In droplets impacting at or near 80°–90°, the result is the classic blossom effect. In this instance, the rim and protuberances rise upward and slightly outward from the contact line around the entire periphery of the stain. Refer back to Figure 5.12.

In stains impacting at the more acute angles, the blossom builds more like a wave on one side of the rim. This wave in turn creates one or more ligament structures. As the wave builds, surface tension is still a force affecting the fluid. The leading aspect of the wave rises slightly up off the target; this causes the contact line with the substrate to narrow. Surface tension pulls the liquid that is away from the target in the forward aspects of this wave back together. This action results in the forward edges of the volume pulling toward one another, which effectively closes the contact line of the droplet resulting in the classic elliptical shape. Any spine(s) present in these instances extend out beyond the rim of the wave and often spawn a satellite droplet. At the end of each spine or ligament, one often observes a small droplet forming.

Retraction

The retraction phase is the final phase in the development of the stain. It appears to result from the effect of surface tension attempting to pull the fluid back into a single entity. The

Figure 5.15 Retraction in a droplet impacting at 90°. The blossom retracts to its contact line with the surface, the area defined in the displacement phase.

inertial forces that caused the shifts in the droplet's volume on the surface are now either overcome by surface tension resulting in a complete retraction of the liquid back into the stain, or the remaining inertial force overcomes the surface tension present in the spines resulting in the formation of a tail, scalloped protuberances on the boundary of the main stain and/or satellite spatter.

Particularly in acute angle impacts, if the remaining inertial force is sufficient, each spine/ligament will be stretched beyond the ability of surface tension to hold the fluid intact. Given these opposing forces, satellite droplets routinely detach. It is only at this point in the overall impact sequence that the surface tension of the droplet is overcome, resulting in the creation of satellite spatter. The cohesive forces are unable to draw these ligaments back to the main stain, and they simply fall to the target, producing tails and satellite stains.

Any blood retracted to the parent stain coalesces and generally levels itself across the stain to some degree. See Figure 5.15. Particularly in less acute impacts (70°–90°), one often observes a thick outer rim in both the wet or dried stain.

The angle of impact has some effect on the resulting stain in the retraction phase. For instance, in the 80°–90° impacts, no part of the blossom's structure, excluding any spines, adds to the stain's ultimate length or width. The structure of the blossom completely retracts to the original contact line boundary evident in displacement.

In the more acute angle impacts, some of blood present in the ligament retracts to the parent stain, but in many instances remnants of the ligament simply fall to the target. The ligaments create the "tadpole tails" and scallops we so often see in elliptical stains. In acute angle impacts, the mass of the droplet in the ligament cannot be retracted and it falls to the surface producing a tail and/or a satellite stain. See Figure 5.16.

Figure 5.16 Retraction in an acute angle impact. Inertia of the satellite droplet overcomes the surface tension in the connecting ligament resulting in the complete detachment of the satellite. The ligament falls to the target, creating a tail.

In summing up our observations on the phases of impact, Figure 5.17 graphically depicts all four phases for varying impact angles. These observations allow the analyst a better understanding of why the spatter stain has the characteristic circular or elliptical shape and why both directional angle and impact angle are definable.

Understanding the Impact Angle

One of the truly significant aspects of Stain Shape and Vector Correlation is impact angle. This angle (also known as the alpha angle) is viewed from a side view (the XZ plane) on any plane/target. It is created by the intercept of the target by the droplet's vector. The subprinciple of impact angle can be articulated as follows:

> *The collapse of a free flight droplet on a surface produces a stain with a circular or elliptical shape. The ratio between the lengths of the short and long axes of the resulting stain has an empirical relationship to the angle at which the droplet struck the target.*

This concept is easily described utilizing Woodward's causation definition. Using the simplified example from Chapter 2, where

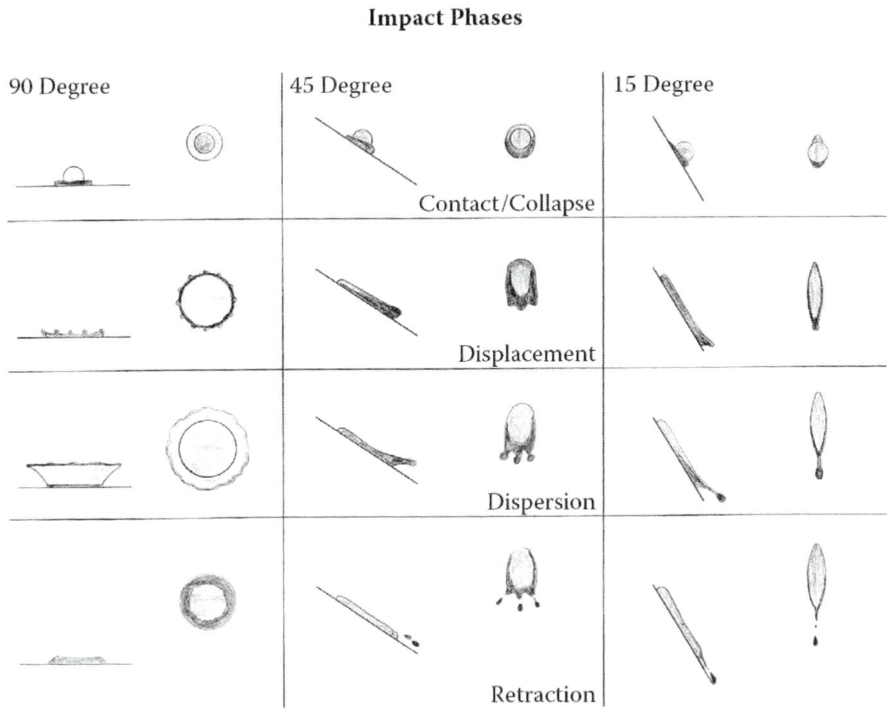

Figure 5.17 A graphic showing the four phases of impact as they occur in varying angles of impact. The drawing on the left side of the box depicts the view from the side; the drawing on the right of each box depicts the view from overhead. What is evident from this figure is that the collapse of a droplet is a complex process.

X – is the angle of impact, where:
- X1 is 10°
- X2 is 45°
- X3 is 90°

Y – collapse of a stable droplet upon impact
Z – is the resulting width to length ratio of the stain produced.
- Z1 is 0.174
- Z2 is 0.707
- Z3 is 1

Q – is the substrate on which the collapse occurs.
R – is the mass of the stable droplet involved.
S – is the distance the droplet was projected before collapse.

Setting X to 10° (keeping it invariant throughout each iteration of the experiment), a series of experimental iterations are pursued in which Q is varied using a variety of relatively smooth surfaces (e.g., tile, gypsum, glass). R is varied in another series by changing the volume of the droplets produced. Finally, in a third series S is varied, by moving the target surface varying distances from the source. Under those circumstances, the resulting width to length ratio (Z) will remain consistent (invariant) at approximately 0.174.

In a new series of experimental iterations, Q, R, and S are left invariant throughout, while changing X to 45°. Z will then change to approximately 0.707. Under these circumstances, any change to X (any given degree for the impact angle) will result in a predictable change in the Z value (the width to length ratio).

Recall, causation relationships are not exceptionless. In this relationship, the substrate (Variable Q) is a critical concern. Should Q be changed to an absorbent or highly textured surface, this will affect the symmetrical collapse (the outflow of blood) of the droplet. This in turn will affect the symmetry of the shape in the resulting stain. The lack of a well-formed (symmetrical) ellipse will make it difficult, if not impossible, to isolate and accurately measure the two axes. Certain substrates (e.g., some fabrics) will produce stains where the two axes do not follow this relationship at all. When presented with an asymmetrical stain, predictability of the impact angle relationship cannot be relied on, simply because one cannot measure accurately the length and width of the resulting stain.

Dr. Victor Balthazard et al. are credited with having recognized this empirical relationship exists between the length and width of the resulting stain and the angle at which the droplet impacts a surface. This relationship can be described in simple terms. The more circular the shape of the stain, the more perpendicular it struck the target surface (e.g., closer to 90°). Whereas the more elliptical the shape of the stain, the more acute the angle of impact (e.g., closer to 10°). See Figure 5.18. With practice and experience, the analyst by observation alone will recognize the general angle of impact, based solely on the shape. This angle can also be computed based on this relationship. Given a well-formed stain, where we can accurately measure the width and length of the two axes, we can easily establish the impact angle using Dr. Balthazard's concept.[13] See Figure 5.19.

MacDonell, using Balthazard's recognition of this empirical relationship (e.g., the use of the length/width ratio), refined the approach for defining impact angle. MacDonell chose instead to use a width/length ratio in conjunction with the trigonometric Sine function. This approach is the standard in use today.[14]

To apply this approach to defining impact angle, it is important to understand that in right triangles certain relationships exist between the internal angles of the triangle and the length of a right triangle's sides. The Pythagorean theorem establishes that the sum of the sides (the side opposite and the side adjacent the acute angle) of a right triangle when squared is equal to the hypotenuse (the side opposite the right angle) squared ($SO^2 + SA^2 = Hyp^2$). See Figure 5.20. To maintain the equality inherent in this relationship, a change to any one of the lengths of a side in the triangle demand a corresponding change to the others. For any given angle (1°–90°), there is a unique empirical ratio between the various sides of the triangle. From this understanding, the trigonometric relationships arose. These relationships are the trigonometric functions of sine, cosine, and tangent. These relationships are in no way dependent upon the factors found at the crime scene; they are mathematical in nature. As with all mathematics, these relationships allow application of abstract ideas to explain real world objects. MacDonell realized the intercept of the droplet and target (its impact angle) is analogous to these right triangle relationships. (Although one might infer from the drawing that a 1:1 relationship exists between the size of the droplet and the size of the resulting stain (e.g., AB = ab). This is not the case. As the droplet collapses, there is a lateral spread of the liquid. We do not yet fully understand this lateral spread and resulting ratio; however, it has no apparent effect on the actual application of the process.)

To understand this correlation, imagine a right triangle formed between the droplet and the target surface as the droplet strikes. Figure 5.21 outlines how this triangle might look. A blood droplet in flight is in the shape of a spheroid. Therefore, in viewing Figure 5.21, Line DE of the sphere, its width can be considered equal to Line AB its height. This

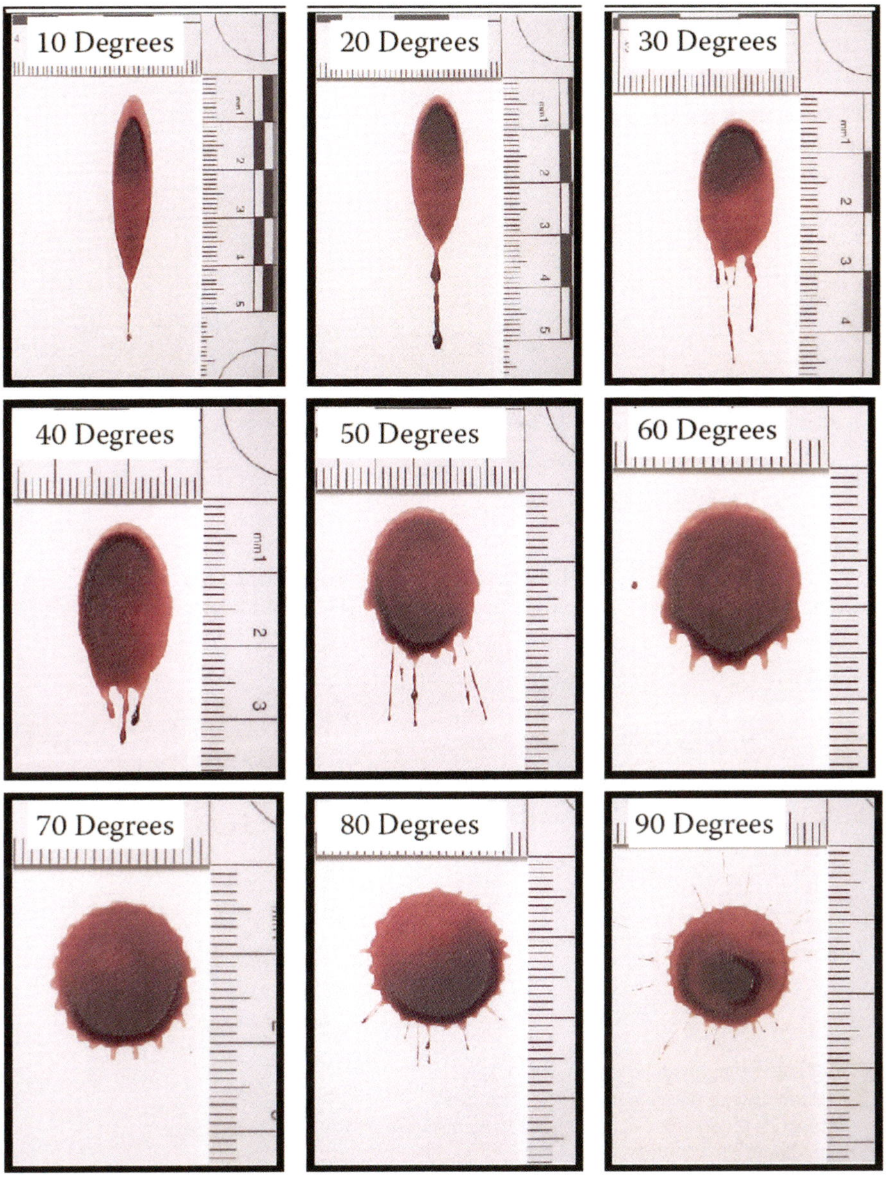

Figure 5.18 Impact angle and stain shape. There is an empirical relationship between the shape of a stain and the angle at which it struck a target surface. The more elliptical the stain's shape, the more acute the angle of impact. The more circular the stain, the less acute the angle and closer it fell to 90° on the target.

correlates to the line ab in the triangle, the Side Opposite (SO) of the triangle. The line bc represents the Hypotenuse (Hyp) of the triangle.

An analogy can then be drawn between Line ab and bc and the width and length of the resulting stain. See Figure 5.22. Based on the analogy, Line ab is represented by Line LM of the stain and Line bc is represented by Line JK of the stain. The Sine (the empirical ratio) of an acute angle formed *i* in the figure is equal to the length of the SO divided by the length of the Hyp.

As a result of this analogy, two known quantities of the stain can be applied to the Sine formula. A circle or ellipse has two axes, its width and length. By measuring the stain's long

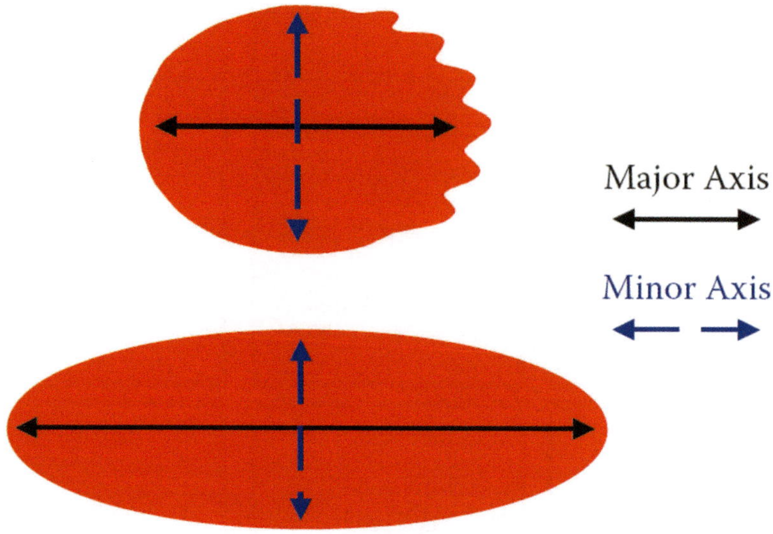

Figure 5.19 Our concept of a "well-formed" stain means simply that if we divide the stain along the major and minor axes of its ellipse, the opposite halves would be generally equal to each other.

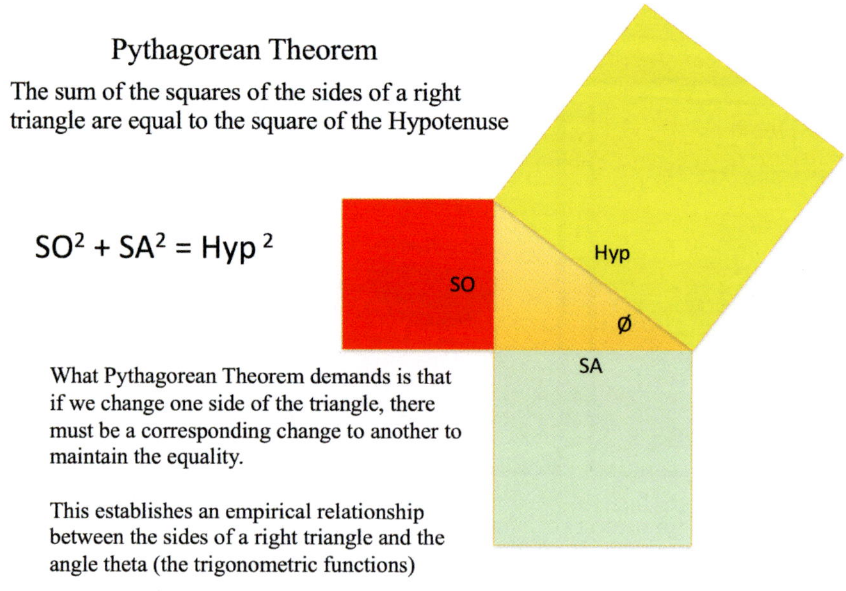

Figure 5.20 Pythagorean theorem established that there is a relationship between the squared sides of a right triangle and the hypotenuse squared. Any change to the length of one side requires a change to the others to maintain the equality of the theorem. This established the trigonometric functions.

axis (line JK) and its short axis (line LM) and applying the measurements to the following formula using a scientific calculator, the droplet's impact angle becomes evident:

Sine *i* = Width (ab)/Length (bc)
Inverse Sine (ASN) *i* = Impact Angle

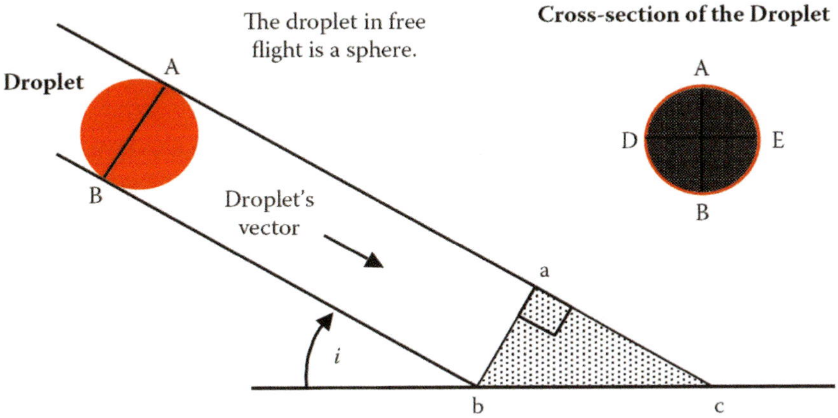

Figure 5.21 The relationship of the droplet to an imagined right triangle. Using the sine function and this relationship, the analyst can establish the angle *i*. This is the droplet's impact angle.

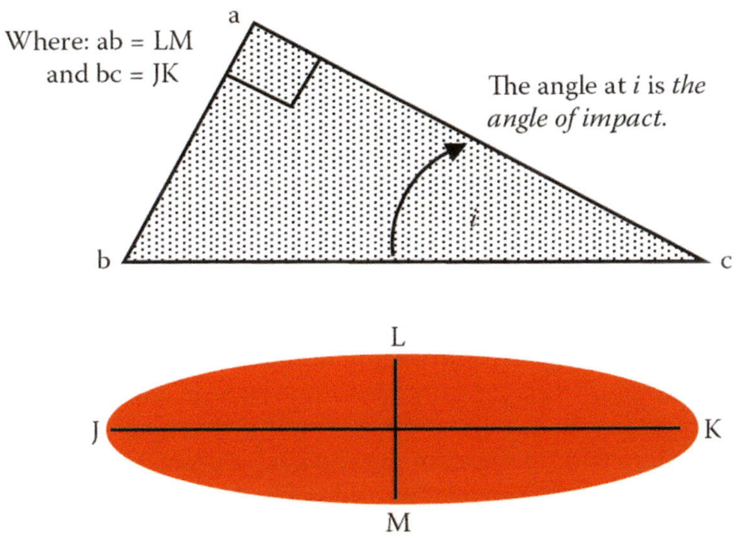

Figure 5.22 We can draw an analogy between the right triangle formed in Figure 5.21 and the ellipse of the bloodstain. Line ab is analogous to line LM, as is bc to JK. Thus, the length and width of the stain are quantities we can measure and apply using the sine function to determine the impact angle *i*.

The Sine is simply the ratio between these two axes, the side opposite and the hypotenuse (the width and length of the stain). This ratio will always be 1 or less. Using a scientific calculator, the Sine of the angle is converted to the angle (using the Inverse Sine function). The Inverse Sine function isn't a mathematical calculation. The scientific calculator simply looks up that unique Sine number and its corresponding angle in the calculator's stored trigonometric table.

Example:

Width = 3 mm
Length = 5 mm

0.6 (Sine *i*) = 3/5 (Width/Length)
Inverse Sine *i* (0.6) = 36.8
Impact angle = approximately 37°

It is important to recognize that the formula provides the analyst with an estimate of the impact angle. The precision of the resulting math for the angle (e.g., the significant digits) should not be construed to mean a similar precision in the definition of the angle. Issues related to the ballistic path of the droplet and the precision of the stain measurements precludes one from accepting this angle as absolute. The more elliptical the stain (up to 10°) the smaller the error. It has always been recognized that circular-shaped stains presented a greater error level. Both Rowe and McGuire's studies at George Washington University demonstrate that when dealing with droplets impacting between 10° and 45°, the error rate is approximately plus or minus 2°–3°. This error rises to 6°–7° for droplets impacting at 45°–60° angles. After 60°, the error rate rises dramatically and for circular stains, it may be as high as plus or minus 15°. As will be discussed, a similar error concern exists for defining directionality. Error rate concerns mandate the analyst carefully consider what stains they utilize for area of origin determinations.

Lacking a scientific calculator, one need to only seek a physical trigonometric table. Such tables are found in a variety of references. The width and length of the stain are measured and divided in the same fashion as McDonell suggested (W/L). This number is then located on a physical Sine function table to find the correlating impact angle. A Sine table simply eliminates the need for a scientific calculator.

Stain Measurement

Measuring the stain is obviously critical, as this provides the analyst with the values of the stain's axes measurements. In considering measurement, the analyst measures only the main body of the stain. This measurement must exclude any portion of the satellites, scallops, or spines present in the stain. To accomplish this, one need simply envision a perfect ellipse superimposed on the stain. By choosing a point on the stain that naturally completes the ellipse, the remaining scallop and tail portions are not subjectively included in the calculation. See Figure 5.23.

It is important to understand that the inclusion of any excess portion of the spine will change the overall width to length ratio. For example, consider a stain with an actual 5 mm length, a 4 mm width, and a 0.5 mm scalloped tail. Given such a stain, the following is possible:

Correct Measurement and Evaluation

Length:	5 mm
Width:	4 mm
Impact Angle:	53°

Incorrect Measurement and Skewed Evaluation

Length:	5 mm plus 0.5 mm scallop
Width:	4 mm
Impact Angle:	46°

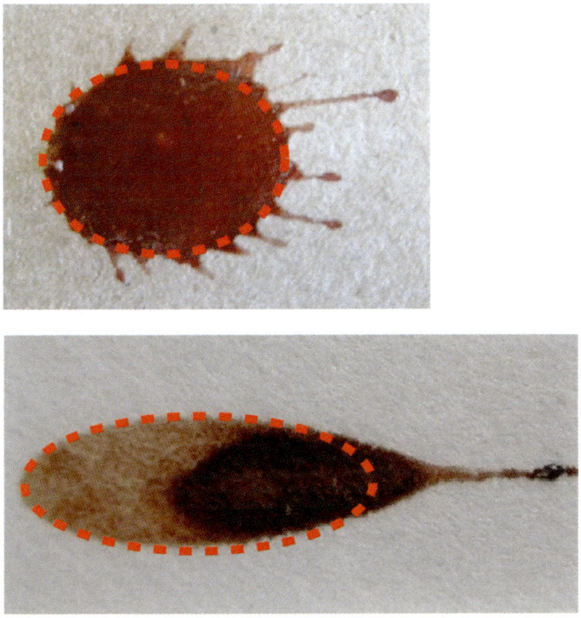

Figure 5.23 By comparing the ellipse formed by the body of the stain to an imagined ellipse, one can generally distinguish those portions of the stain to disregard during the measurement process. Any inclusion of scallops, spines, and satellite spatter in the measurement will result in a skewed estimate of the impact angle.

In this example, the small excess scallop adds an error of 7° to what is already accepted as an *estimation* of the impact angle. Obviously, by including these excess portions, the analyst can change the calculated impact angle significantly. Unfortunately, visually there are no absolute rules for closing out the ellipse. The analyst must judge each stain individually and attempt to eliminate these scalloped edges and tail portions from the measurement. Figures 5.24 and 5.25 show examples of what portions of the stain the analyst should include and exclude. Although the tail structures will only cloud the length measurement, scallops may appear on both the sides and end of a stain, making it more difficult to identify the true edges of the stain both in the width and length aspects.

Closing out the ellipse and locating the end of the stain is an important consideration. It is only through practice that one learns to do this well. The use of an ellipse template can help in identifying the end of a bloodstain ellipse. See Figure 5.26. This end point of the stain (the point farthest from the source and usually scalloped or attached to a tail) is often hidden within the tail and satellite structures. This makes it difficult for a new analyst to differentiate bloodstain lengths, particularly those in the 10°–35° range. By simply comparing the bloodstain to a manufactured ellipse template, it becomes easier to recognize the natural closing of the ellipse. For example, Figure 5.27 shows a 20° ellipse overlaying a 10° stain. The outline of the template obviously does not align with the outside edges of the bloodstain. Figure 5.28 shows a 10° template overlaying a 10° bloodstain and the two are nearly a perfect fit. Current virtual techniques for defining Area of Origin (discussed in the next chapter) utilize an ellipse matching technique to reduce associated error.

As bloodstains vary in volume, they will also vary in size, even if they impact at the same angle. For the ellipse template to be a functional means to evaluate bloodstains at the crime scene, the analyst must have a set of ellipses not only for the varying degrees they

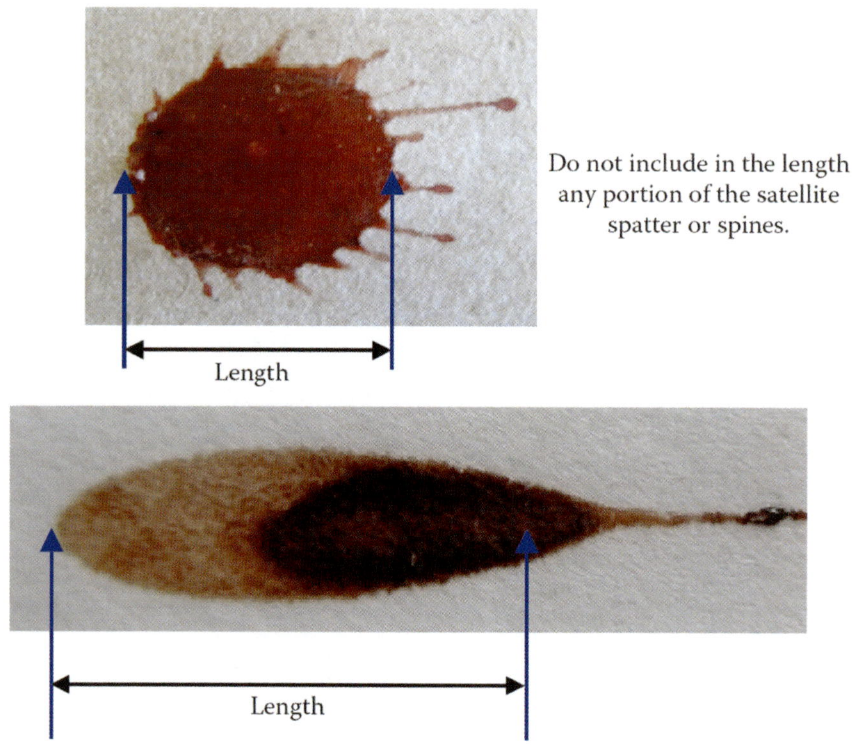

Figure 5.24 The length of the stain is measured along the major axis of its ellipse. The analyst must exclude from the measurement any spines, scallops, tail, or satellite stains that may be present, using only the true ellipse.

Figure 5.25 The width is measured along the minor axis of the stain's ellipse. Although satellite spatter are less a concern when measuring this axis, scallops along the outer edges of the stain are often encountered. Do not include scallops in the measurement.

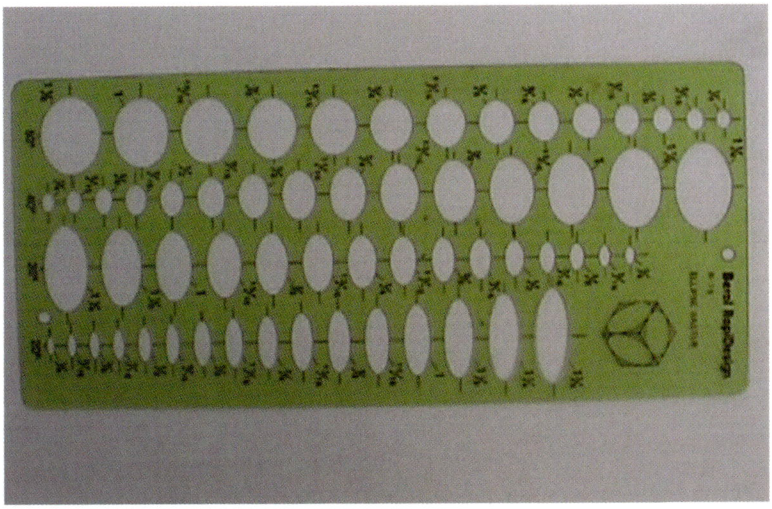

Figure 5.26 An ellipse template. Although not convenient for on-scene work, the comparison of the stain to an ellipse template in training and for practice assists the analyst in recognizing what portions of the stain's ellipse to measure.

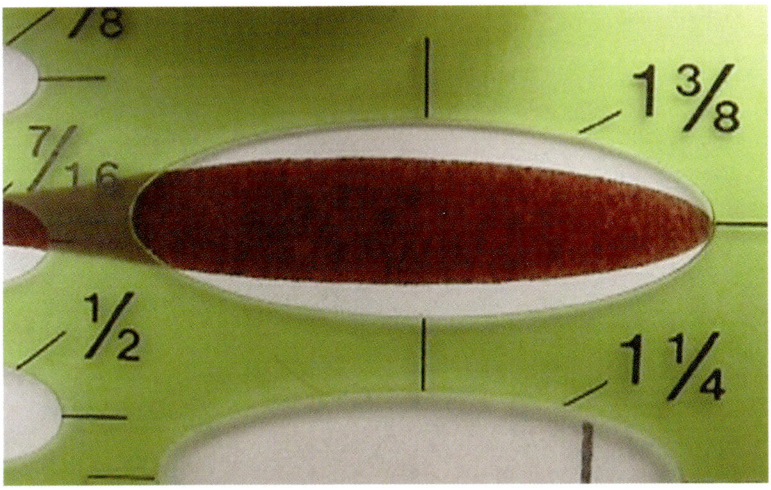

Figure 5.27 In this example, an ellipse of 20° overlays a stain with a 10° impact angle. The comparison makes it evident the two do not match.

require, but also to match a wide size range of bloodstains they may encounter. As a result, templates as a measuring mechanism are not particularly feasible as a scene technique, due to the problem of carrying the templates around. The template method is excellent in training. Students find it gives them a greater level of confidence in identifying the natural closing point of the stain.

The specific methods the analyst uses to measure the stain are unimportant, so long as the analyst is comfortable with both *what* they are measuring and their measuring device. Options include the following:

- Measuring the stain with a ruler, or micrometer
- Measuring the stain with a drafting divider
- Measuring the stain with a photographic loupe

Figure 5.28 In this example, an ellipse of 10° overlays the same stain as shown in Figure 5.27. The fit is almost perfect. Note at the left end of the ellipse, the stain's tail extends well past the perfect ellipse. The area of the stain extending beyond the ellipse should be ignored in the measurement of the stain.

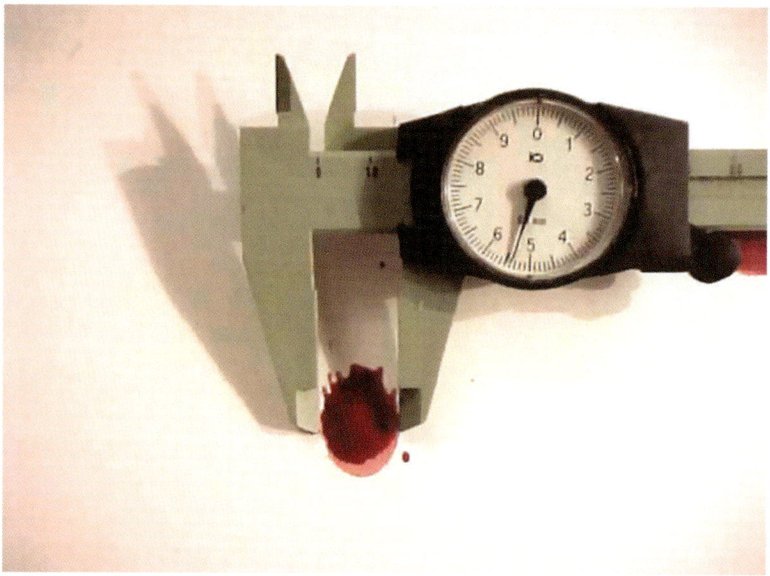

Figure 5.29 Measuring a stain with the micrometer. The analyst matches the inside edges of the caliper to the stain edges and then reads the stain measurement using both the scaled bar (10 mm) and the needle gauge (5.4 mm). The short axis of this stain's ellipse measures 15.4 mm.

The use of a ruler is certainly functional; however, the smallest scale present on the ruler limits the measurement. For example, if the ruler is scaled to 1 mm, the analyst can only estimate the measurement to the closest 0.5 mm. Any estimate beyond the 0.5 mm would be subjective. The micrometer offers an advantage, as most are scaled to the level of 0.1 mm. When one considers that the typical stains being measured are not the large 4–10 mm stains, but rather the smaller 1–4 mm stains found in impact events. With such small stains, the use of the micrometer is much more functional. See Figure 5.29.

Figure 5.30 A photographer's loupe in use to measure a stain.

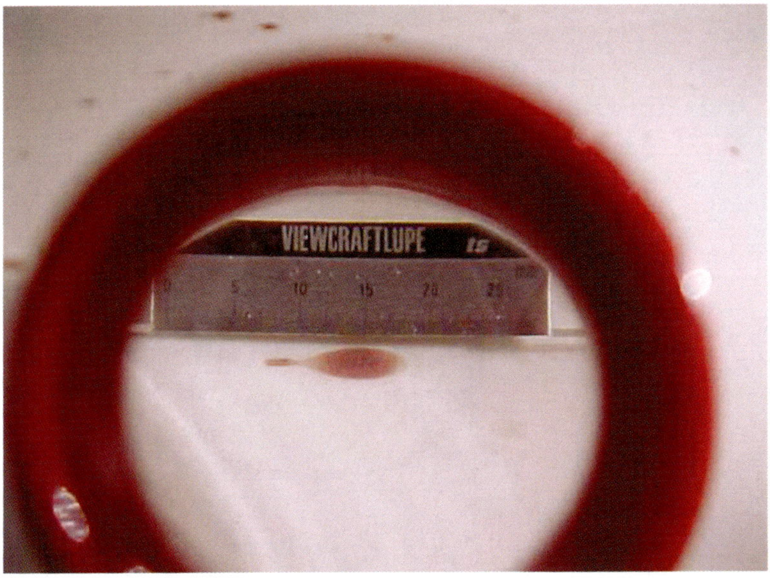

Figure 5.31 The view through the loupe. The loupe should include a functional scale situated in the viewfinder in a manner that will allow easy comparison of the stain to the scale.

The use of a photographic loupe is effective, but presents the previously discussed limitation of the included scale. Loupes come in a variety of styles and configurations and not all have a reticule, scale, or functional measuring mechanism. Before acquiring one, the analyst should verify that there is a scale and that it is functional for their intended purpose. See Figures 5.30 and 5.31.

A third method for measuring a bloodstain is with a drafting divider. Charlie Green first suggested this approach, involving measuring the back half of the ellipse and doubling

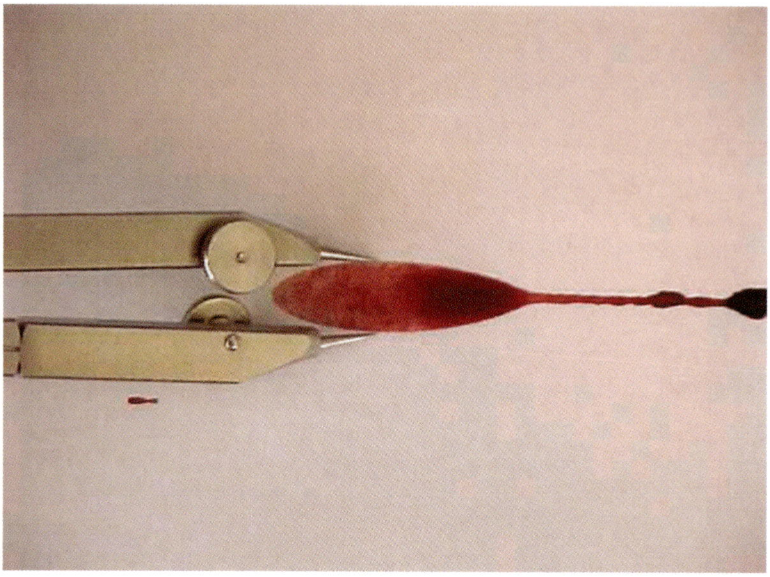

Figure 5.32 In the first step, match the divider points to the widest part of the outside edges.

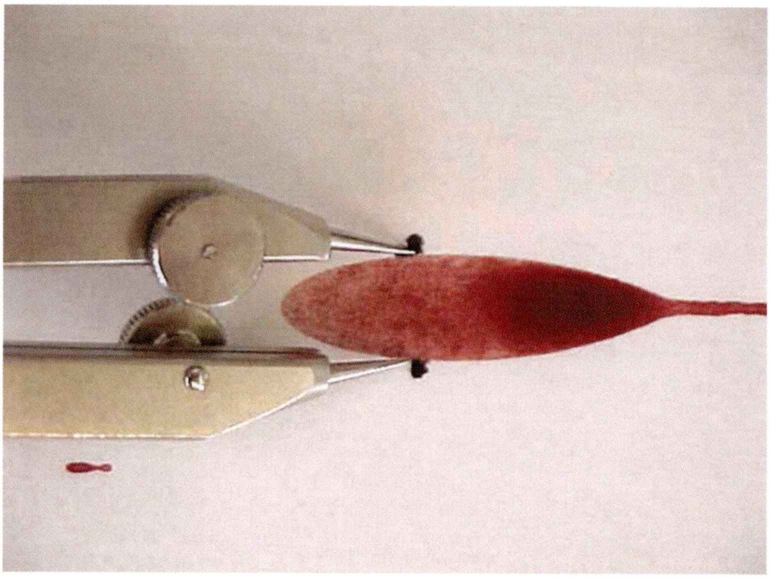

Figure 5.33 Without adjusting the arms, place them at the top (the leading edge of the stain) and move them down the body until the arms appear to touch the edge of the stain. Mark both points.

this number to identify the length.[15] The analyst adjusts the dividers to match what appears to be the widest part of the short axis. See Figure 5.32. Maintaining this adjustment, the analyst runs the divider arms down from the top (the leading edge of the stain), along the length of the stain. The analyst then marks the point where the divider arms both touch the bloodstain. See Figure 5.33. With the same adjustment, the procedure is followed from the opposite end of the stain. Once again, the point where the arms touch is marked. See

Figure 5.34 Without adjusting the arms, move them in the same fashion from the bottom of the stain (the edge with the tail and scallops) back toward the middle of the stain until the arms appear to touch the edge of the stain again. Mark these points.

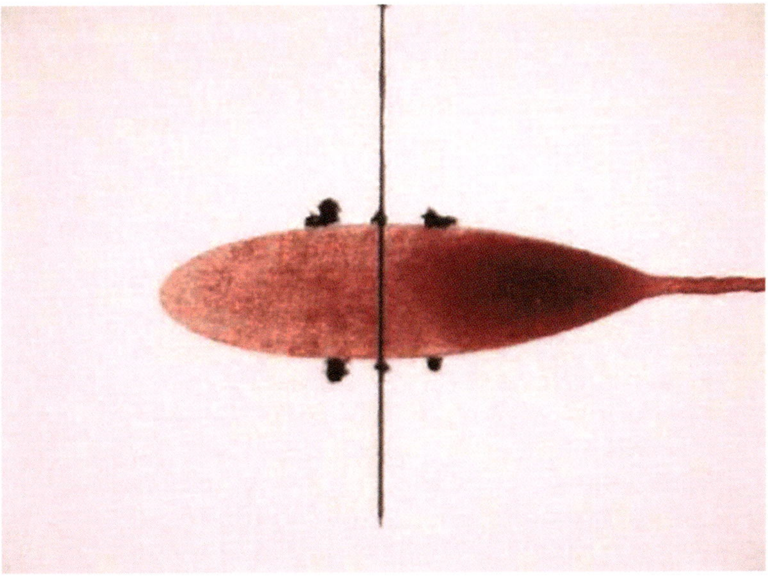

Figure 5.35 By splitting the difference of the measurement between the two marks, we identify the widest part of the stain. This often appears as an area, when in fact it is really a point.

Figure 5.34. It is actually a point, but our eyes tend to see it as an area. The comparison and marking from both the top and the bottom of the stain ensures finding the widest point of the short axis. This process should identify the widest area of the short axis (its width) of the stain.

The analyst splits the difference between the two previously established marks. See Figure 5.35. This third mark is the middle of the long axis. The divider points are placed

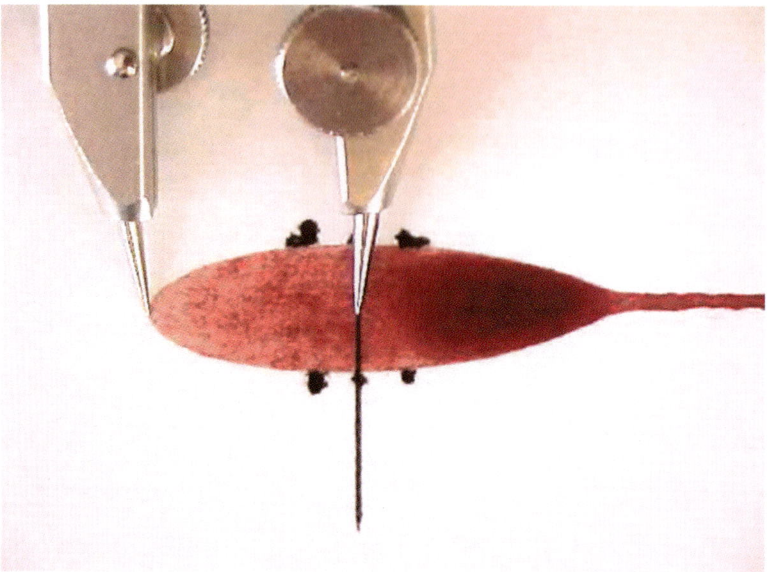

Figure 5.36 To locate the length of the ellipse's long axis, place one arm on the center line and the other on the top (leading edge) of the stain. This measurement doubled identifies the length of the long axis.

at the central mark and at the top of the stain and determining this measurement. See Figure 5.36. This identifies a true half of the ellipse. By doubling this measurement, the analyst finds the length of the long axis.

Understanding Directional Angles

Directionality defines a spatter droplet's motion relative to the target's XY plane (its directional orientation as viewed from above the target). The concept of directionality can be articulated as:

> *The collapse of a free flight droplet onto a surface produces a stain with a circular or elliptical shape. These stains may have spines, scallops, or satellite stains. These characteristics, if present, in conjunction with the long axis of the stain define direction of travel for the droplet at the moment of impact.*

The directional angle causation relationship can be described in the following simplified example, where:

X – is the orientation the droplet intercepts a target in the XY plane:
  X1 is oriented toward 0°.
  X2 is oriented toward 90°.
  X3 is oriented toward 180°.
  X4 is oriented toward 270°.
Y – is the collapse of a stable droplet upon impact.

Z – is the orientation of long axis in the resulting stain and general orientations of the primary associated scallops, tails, or satellites.
- Z1 – the long axis is oriented 0°–180° (relative to some reference point, e.g., up, or North) and the primary scallops, tails or spines are generally oriented at 0°.
- Z2 – the long axis is oriented 90°–270° and the primary scallops, tails, or spines are generally oriented at 90°.
- Z3 – the long axis is oriented 0°–180° and the primary scallops, tails, or spines are generally oriented at 180°.
- Z4 – the long axis is oriented at 90°–270° and the primary scallops, tails, or spines are generally oriented at 270°.

Q – is the substrate on which the collapse occurs.
R – is the angle the stable droplet intercepts the substrate.

Assume X is set to 0° (X1) and kept invariant. In one series of experimental iterations, Q is varied, to include a variety of relatively smooth surfaces (e.g., tile, gypsum, glass). R is varied in a second series by changing the impact angle at which the droplets are directed onto the target.

Under those circumstances, the resulting long axis of the stains will be oriented 0°–180° relative to the XY plane and the location of the primary satellites will be oriented at 0° (Z1).

As previously explained in Chapter 2, that is only half of the effort for establishing this causation relationship. In a new series of iterations, Q and S are kept constant throughout, all the while changing X to X2, X3, or X4. Under those circumstances, Z will become variant in a consistent fashion (e.g., X2 producing Z2).

Once again, causation relationships are not exceptionless. The substrate (variable Q) is the most critical concern. Should Q be changed to an absorbent or highly textured surface, this will affect the symmetrical collapse of the droplet. This in turn will affect the symmetry of the ellipse in the resulting stain. The lack of a well-formed (symmetrical) ellipse will make it difficult, if not impossible, to isolate the major axis of the stain. Additionally, as the angle of impact (variable R) increases, stain shape will become more circular. As the stains become more circular, it will be difficult to isolate the major axis itself or understand where the most satellites are located. Thus, there are practical limits to applying the directional angle causation relationship.

As described, the collapse of a droplet onto a substrate does so in a relatively defined fashion. Within a wide range of circumstances, this collapse produces a symmetrical stain, either circular or elliptical.

As the liquid droplet contacts a surface, the droplet begins this collapse. Inertia keeps the volume of the droplet moving along the same path it was traveling before its encounter with the target. The blood present in the droplet flows outward to the edges of the stain during the collapse, creating either an elliptical or a circular stain depending upon the angle of impact. Figure 5.37 details how this motion occurs. Whatever the shape, the resulting stain, unless impacting at 90°, will have both a major and a minor axis. The major axis (the long axis of the ellipse) is always aligned with the path of the droplet.

It is the long axis of this circle or ellipse that begins the definition of the droplet's directionality. This axis defines a line with two possible droplet directions. See Figure 5.38. In viewing the stain, the analyst must seek additional information to eliminate one of these two directions. Satellite stains, scallops, or spines associated with the stain provide the analyst a means for making this elimination and identifying the direction of travel.

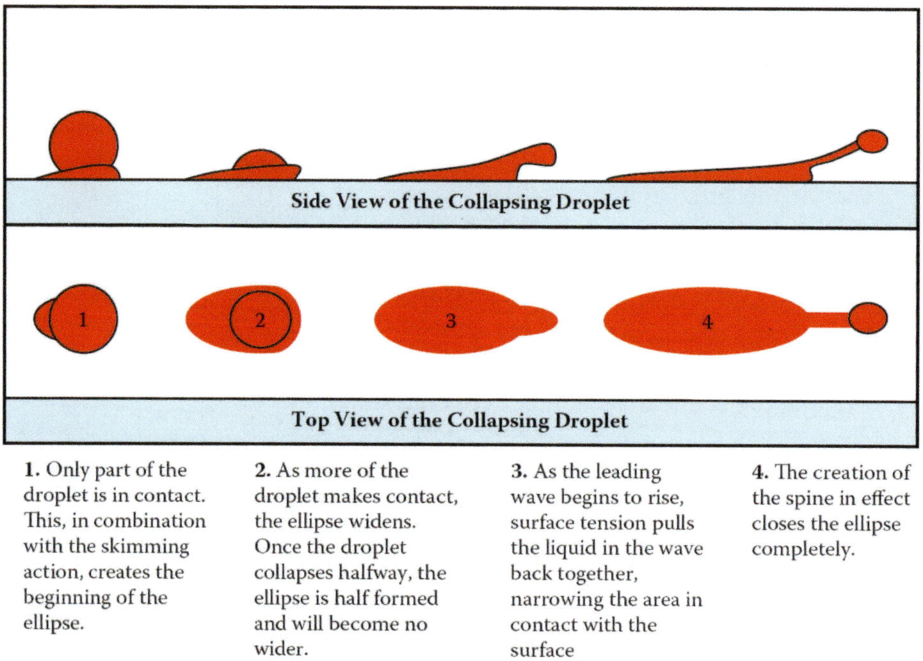

Figure 5.37 A graphic of a collapsing droplet impacting at an acute angle. This demonstrates how the long axis of the ellipse develops and is oriented along its direction of travel.

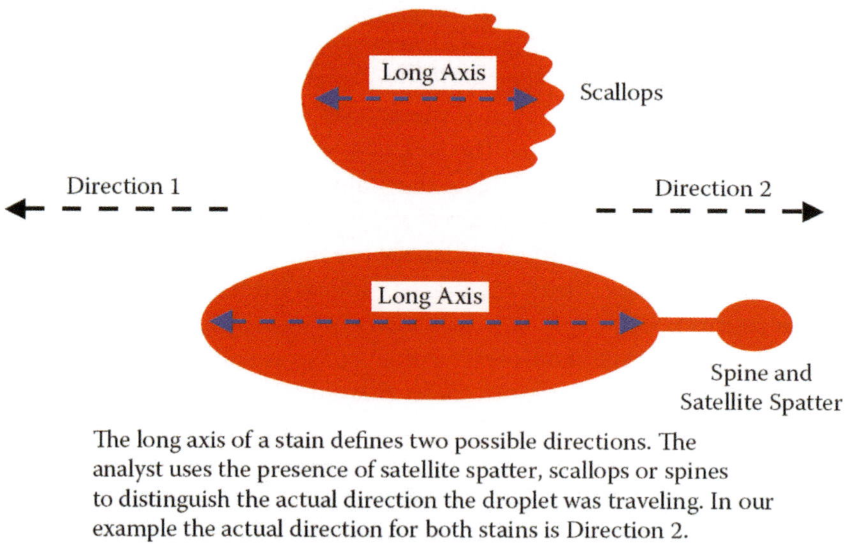

Figure 5.38 The presence of spines, scallops, satellite spatter, or tails help the analyst identify the direction the droplet was traveling at impact. These structures are found opposite (or at least concentrated opposite) the point where the droplet made first contact with the substrate, its leading edge.

The blossoming effect produced by the outflow of the volume in the droplet begins as the droplet collapses. The exact nature of this outflow is defined by several variables, including the angle of impact and the nature of the target surface (e.g., the surface hydrophilic or hydrophobic). As this blossom develops, surface tension of the liquid pulls the blood into distinct protuberances in the expanding rim of the blossom structure. These protruding formations in many instances break from the blossom structure, creating satellite stains. Whether these protuberances break completely free or simply form spines or scallops, their presence in relationship to the parent stain defines the vector motion of the droplet at impact. The majority or all will be found on the side opposite the initial point of contact.

The path these spines and satellite droplets follow is defined by the outflow of the blood as it impacts. In instances of 70° to 90° impacts (the more circular stains), scallops, small spines, and satellite stains may be evident all around the parent stain. The side associated with the directional motion generally reflects a greater number or more pronounced indications of the scallops and spines. In distinctly circular stains, it may be impossible to isolate this characteristic. See Figure 5.39. As the stain becomes more elliptical, this characteristic becomes more evident. See Figure 5.40.

Stain shape is critical in understanding directionality, as there is associated error in finding the long axis. When considering 70° through 90° impacts, the resulting stain shapes are more circular. This makes it difficult, if not impossible, to define a specific long axis. Surface disturbances or slight irregularities present in the target can lead to deviations in the outflow of blood. This can affect the location of resulting spines or satellite stains. Such deviations, combined with our ability to identify only a general direction in these instances, make the trustworthiness of the directionality in circular stains much lower than those found in elliptically shaped stains.

Fortunately, clarity of this direction increases as the impact angle decreases (e.g., moves toward a 10° impact). The more elliptical the stain, the more clearly the long axis is defined. Regarding the location of the satellites, in droplet impacts between 40° and 60°, the nature of the outflow of blood is likely to create a number of spines and/or satellite spatter oriented

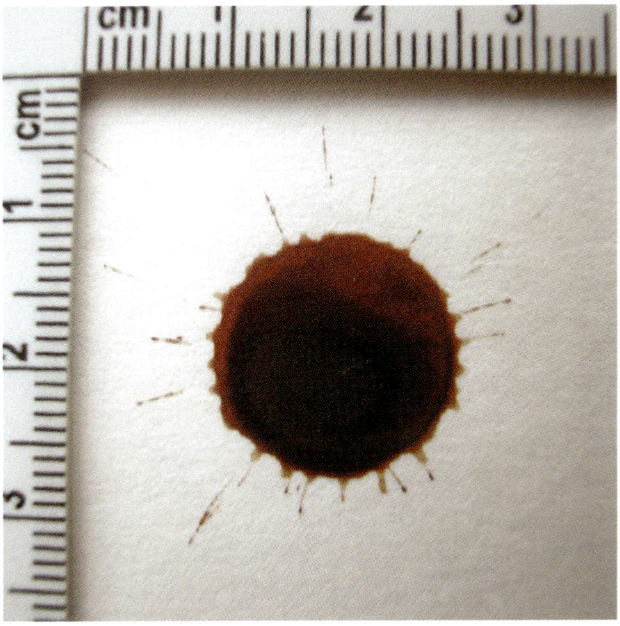

Figure 5.39 A 90° impact. Evident are small scallops, which appear around the entire periphery of the stain. There is no true long or short axis.

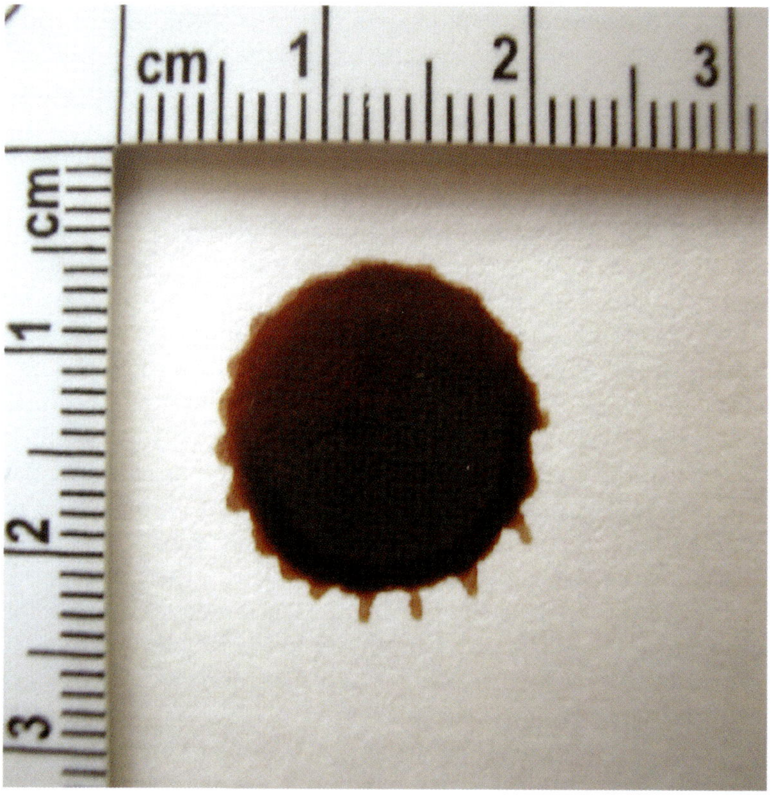

Figure 5.40 A 70° impact. Note the greater concentration of scallops on one side of the stain. They are opposite the droplet's origin and point in the general direction the droplet was traveling. Directionality in generally circular-shaped stains is not as clear as in the more elliptical stains because the scallops are not concentrated along a single vector.

to one side of the stain. The position of such spines and satellite stains allows the analyst to more accurately define the path of travel. See Figure 5.41.

In impacts below 40°, the resulting stains are highly elliptical. In such instances, the outflow of blood travels more tightly along the expanding forward edge of the collapsing droplet. Such an impact is likely to create a single tail. See Figure 5.42. Consideration of where this tail exits the parent stain in combination with the major axis of the parent makes it very clear as to the directional aspect of the droplet.

The long axis of the stain is the true indicator of directionality. In these elliptical stains, the direction of travel for the satellite spatter often matches that of the parent stain. But minor redirection of the satellite or tail can occur, resulting in a spine or tail pointing slightly offset to the actual direction of travel. Using the full length of the tail in this alignment may throw off the directional angle. See Figure 5.43. Directional alignment is always oriented to the long axis of the stain.

Recalling the discussion of parent and satellite stains, it is important to differentiate between the two. Both the parent and satellite may have a tail. While the parent tail points in the direction of the droplet's travel, satellite tails point in the opposite direction. They are oriented back toward where the satellite came from. If the analyst were to confuse a satellite for a parent, this would obviously confuse the investigative findings. In practice, it is quite easy to distinguish between the two. See Figure 5.44.

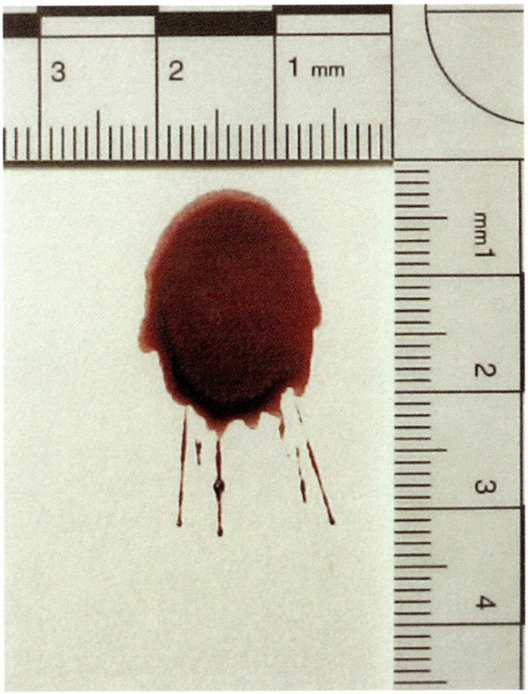

Figure 5.41 A 50° impact. In this instance, the scallops have spawned significant tails. Once again, these tails point generally in the direction the droplet was traveling. Although directionality is not as specific as when we find a highly elliptical stain, the analyst can still effectively define the droplet's directionality.

Figure 5.42 A 10° impact. This stain has a single tail and satellite. In combination with the long axis of the ellipse of the stain, it provides a very specific indication of the droplet's direction of travel.

Figure 5.43 The spine associated with elliptical stains will not always align exactly with the directional angle. The red line divides the major axis of the stain's ellipse and together with the tail indicates its directionality. Note that the spine shows slight redirection to the left.

Figure 5.44 A disassociated satellite stain. These stains have a very distinct tadpole look to them, which is different from the elliptical shape seen in parent stains.

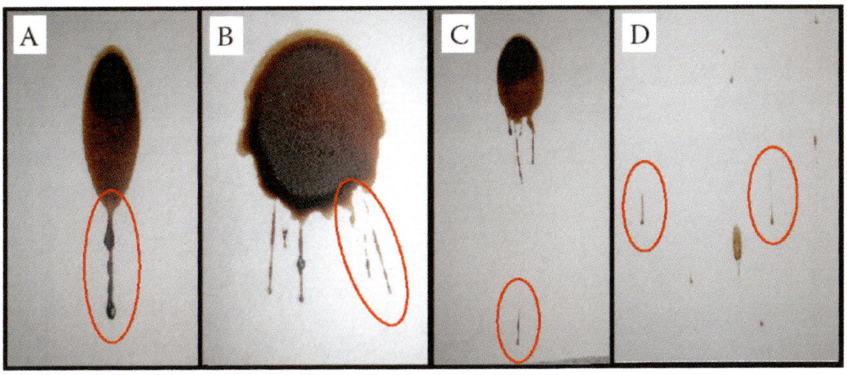

Satellite spatter can be found:
A – Attached to the parent stain
B – Semi-detached from the parent stain
C – Detached and in proximity to the parent stain
D – Detached and disassociated from the parent stain

Figure 5.45 Satellite spatter can appear in many forms. They can be attached, semi-detached, detached and in proximity, or totally disassociated from the parent. In the latter case, it is imperative that the analyst recognize the difference between satellites and parent stains.

Satellite stains will occur in the scene in various fashions, several of which may be detached from their parent. See Figure 5.45. These include:

- Attached to the parent stain
- Semi-detached to the parent stain
- Detached but in proximity to the parent stain
- Detached and significantly disassociated from the parent stain

In circumstances where the satellite is attached, semi-attached, or detached and in proximity, the correlation of parent to satellite is relatively easy. But in instances of acute angle impacts, detached satellites can end up some distance away from the parent stains. The shape of these satellites is somewhat dissimilar to the shape of a typical parent stain. Satellite stains have a tadpole look to them, which is different from the elliptical nature of a parent stain. Refer back to Figure 5.44. If the analyst is properly trained and has spent any amount of time looking at spatter patterns, it is difficult to mistake satellite stains as parent stains.

In practice, it is simple for the analyst to visualize the directionality of any given droplet. The analyst simply draws a line down the center of the long axis of the stain (no matter what its shape), looking for and then using the presence of any tails, scallops, or satellites to understand the direction. The more elliptical the stain, the more defined the directionality. The variations of interpretation of where this long axis is represents the error associated with directional angle, which becomes a consideration when pursuing a convergence or area of origin. See Figure 5.46.

As previously stated, target surface is an important variable. A rough substrate can impair the ability to read a stain. Yet a rough or irregular surface alone, particularly in instances of the more elliptical stains, doesn't mean directionality can't be defined at all. Consider a worst case, such as a stain found on asphalt. The resulting stain may well be

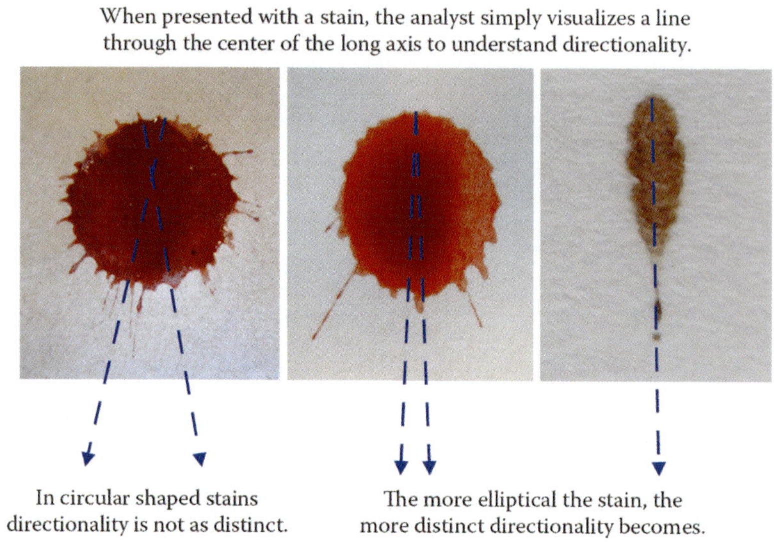

Figure 5.46 To visualize or demonstrate directionality in a droplet, the analyst simply draws a line down the center (the long axis) of the stain's ellipse, splitting it into two equal parts. The problem becomes defining where this line belongs in the more circular stains. Finding the "center" of the stain in these instances presents a range of possibility. The more elliptical the stain, the more defined the directionality.

Figure 5.47 Even in cases of significantly rough surfaces such as asphalt, directionality is not always a lost cause. This photograph shows small satellite spatter deposited to the right side of the parent stain, indicating a general directionality from left to right.

irregular when compared to one impacting a smooth surface, but the general directionality may still be evident. Irregular outflow of blood in the parent stain does not always mask directionality entirely. See Figure 5.47.

Given the preceding discussion, the following is true regarding directionality:

- The more elliptical the stain, the more accurate the definition of directionality.
- Less elliptical stains mandate a very distinct cautionary approach. Stains above 70° are extremely difficult to evaluate.

- The smoother the target, the less likely irregularities will occur which might mask the directionality.
- General directionality may be evident even on rough surfaces, but the lack of a distinct long axis in the stain demands a cautionary approach.

A final word of concern regarding directionality of a droplet is that directionality is in relation to the *source* of the blood. Although directionality is generally evident, it is not a certain conclusion for two reasons. First, when speaking of directionality, we are discussing only the path a droplet was traveling at the time it impacted a given target. Although uncommon, there are instances when ricochet situations are encountered. Ricochets are possible and must be considered.

A second concern is if the target itself is capable of motion at the time of bloodshed; consider both the directionality and the angle of impact cautiously. Motion of a target makes it difficult to establish directionality in any fashion. If one is not confident the target was generally stationary, a single stain by itself may tell the analyst little. However, when examining such items, the presence of radiating patterns of spatter should make this concern moot. Although multiple spatter could fall onto a moving target, their resulting directional angles would likely be consistent and parallel. In such an instance, it is the motion of the target that produces the directional effect. Each directional angle would be generally parallel to the next. If the individual spatter on the object are radiating outward, as typical impact patterns do, the directional angles of the spatter must be a function of their flight paths and not from motion of the target.

Recently, a term was introduced that is associated with directionality. "Reverse directionality" has been used by some authors to explain the presence of random directionality of satellites seen in splash patterns.[16] Reverse directionality refers specifically to what we have described as ricochet stains. Your authors find this term inappropriate for one simple reason. Directionality is the path a particular drop was moving at the time of impact relative to a substrate, defined by the resulting shape of the stain. Physical laws, which do not change, ordain this shape. The collapse of the droplet when it strikes a target will produce the same general effect every time. The shape of a parent stain tells the analyst one thing; the shape of satellite stain tells the analyst another. These are the rules, and they are always in play. The term "reverse directionality" suggests to laymen that the droplet was going in a direction opposite that described by its shape. The resulting inference has the potential to be used to suggest that the most basic issue in Bloodstain Pattern Analysis is not valid. The authors have already encountered unqualified analysts who made exactly such a claim in court. In one instance, an analyst attempted to explain that the directionality of impact spatter was opposite that indicated by the shape because of "reverse directionality." The implication of the analyst either defied a basic understanding of droplet behavior or resulted from their inability to recognize a parent from a satellite. There is no instance we are aware of in which a stain's shape (applying the rules as described) reflects a direction of motion opposite that as described. There simply is no "reverse directionality."

As intended by the original authors, "reverse directionality" describes the odd and random way spines and satellites eject out from a primary stain associated with streaming ejection type events. It was discussed in relation to a "splash pattern" but the effect can be observed in any large volume ejection that is directed relatively perpendicular to a target. Such behavior may well entail the ricochet type stains, stains caused when some aspect of a droplet literally bounces off a surface and is redirected to another surface or

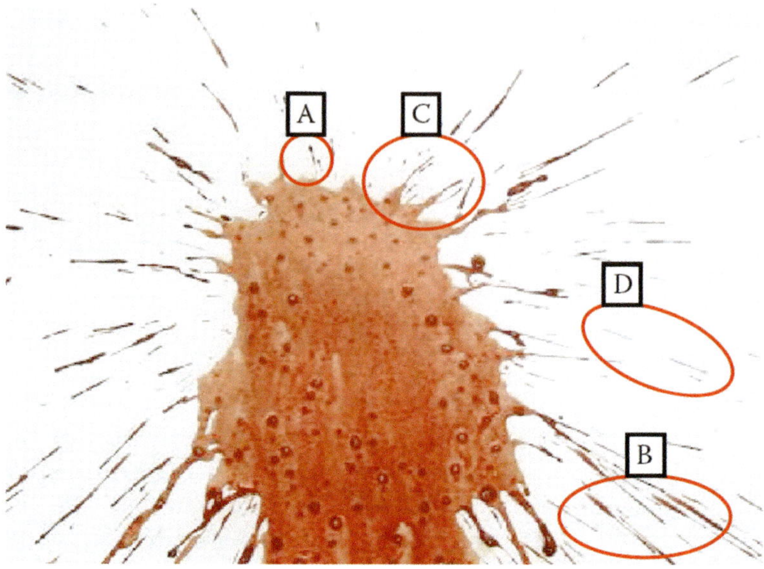

Figure 5.48 When a large volume of blood impacts a surface, the interaction creates several interesting characteristics related to directionality. These include: (A) detached satellite stains ejected from the primary stain, (B) spatter ejected from the primary stain, (C) spines emanating from the primary stain, and (D) long intermittent linear stains. There is no doubt that some collisions and deflections occur in such a complex event, but this in no way alters our general understanding of the direction of deposition.

while in mid-flight a droplet collides with another droplet, leading to a redirection. This area has never been studied in detail, but it is a highly complex event and such actions have been observed. The resulting stains surrounding the primary stain include satellites ejected from the primary stain itself (marked as A in Figure 5.48), spatter and their satellites ejected out from the primary stain (marked as B in Figure 5.48), spines ejected out from the primary stain (marked as C in Figure 5.48), and the odd linear patterns (marked as D in Figure 5.48). It is believed the latter characteristic results from small droplets that are traveling at extremely acute angles to the target surface. All these actions are occurring near simultaneously to the impact of the volume. In some instances, they may well collide producing deflections (ricochets) of some sort. The resulting stains will however reflect the direction the individual droplet was traveling at the time of impact.

Understanding the Motion Component in Blood Trails

Previous consideration of stain directionality leads to a major consideration of blood trails. As individuals become injured and move, or bloody items are moved within a scene, blood trails are likely to occur. Can the directional motion associated with the blood trail be confidently defined?

The drops resulting from these dripping actions strike the surrounding floors and surfaces. As they break free, they are moving generally with the same momentum and in the same direction as the item from which they fell. The combination of gravity and the momentum imparted by the motion of the object from which it fell causes the drops to

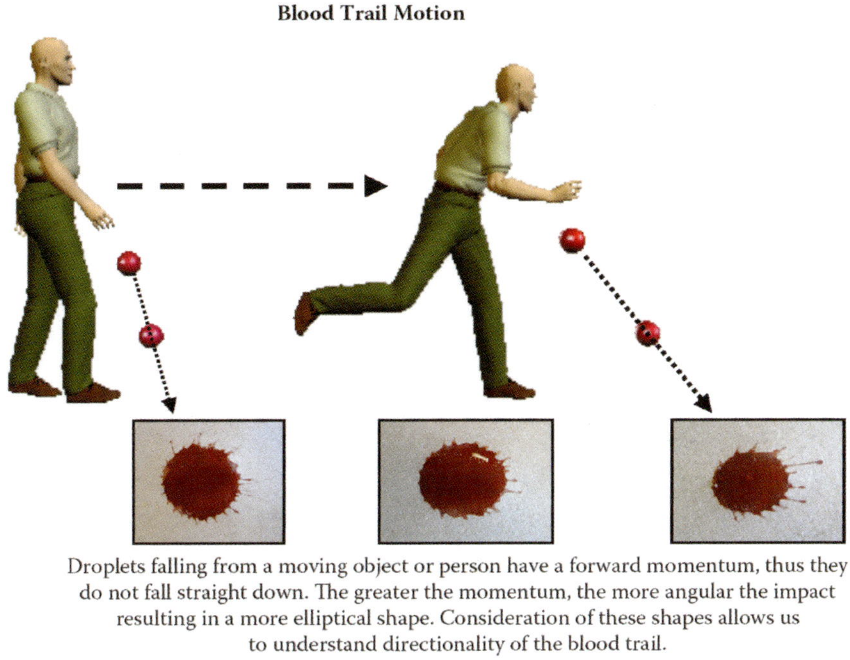

Figure 5.49 Blood trail motion is defined by considering the directionality of the individual drops across the entire pattern.

impact the ground at varying angles, but all generally greater than 40°. The resulting stains in the blood trail show evidence of this angle, as well as the droplet's directionality. See Figure 5.49.

Directionality of a blood trail is not based on a single stain's directional component. It must be considered across the pattern. It is important to recognize that the shapes of the drip stains produced in a drip trail will routinely alternate. Stain shape may change based on the speed the dripping object is moving. Slower speeds will produce less elliptical stains, faster speeds will produce more elliptical satin. If the object the drops fall from has some secondary movement (e.g., an arm swinging back and forth), the drops will not fall at a consistent angle to the surface. Thus, stain shapes may alternate between circular to elliptical and back again, any number of times within the pattern. It also possible to locate a few stains whose directional component is opposite of the actual direction the object was moving in the scene based on swinging of the object. Within the overall pattern, stains that fall opposite the object's direction of travel are few in number. The droplet's path to the surface is a function of vectors and vectors are always summed. A significant backward swing of the object (e.g., swinging the arm) has to be considered with the vector associated with the forward movement and can produce a slightly rearward vector. The ultimate vector followed by any given drip is a function of combining the associated vectors. Forward movement imparted to the droplet by movement of the object through the scene combines in either an additive or subtractive manner with that caused by any backward swing of the arm. Thus, the backward aspect is often canceled or reduced. The majority of stains in the pattern should demonstrate the primary movement of the object dripping the blood.

Figure 5.50 depicts a blood trail on a coarse irregular surface; even here directionality is evident in the resulting bloodstains. Figures 5.51 and 5.52 depict individual stains

Figure 5.50 A blood trail pattern on concrete. The placards allow us to orient the location of specific stains and their shapes to the pattern as a whole.

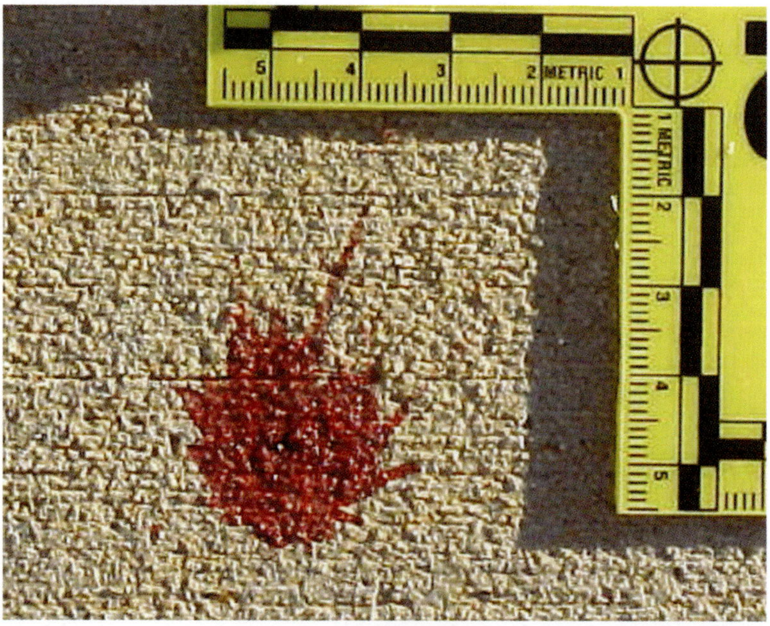

Figure 5.51 Stain #2 from the previous pattern, an elliptical stain with evident directionality. The directionality evident in the pattern is away from the camera as shown in Figure 5.50.

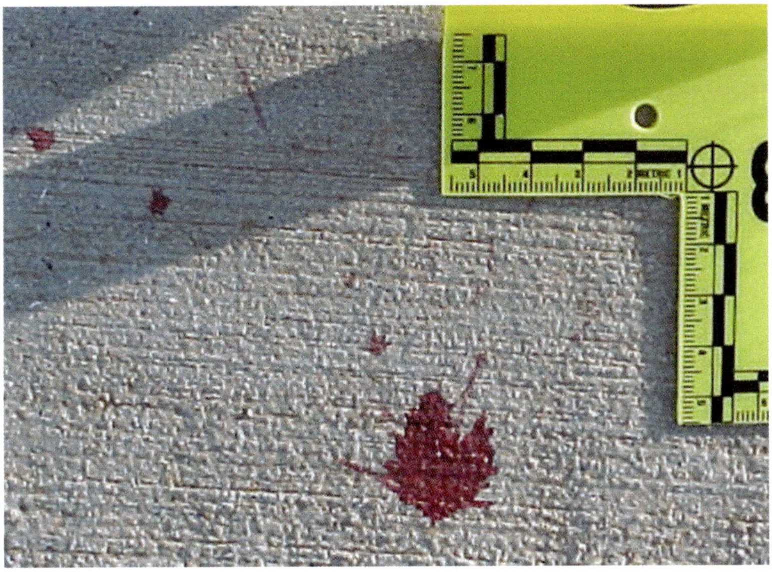

Figure 5.52 Stain #3 from the pattern in Figure 5.50. This stain is significantly irregular in its shape. Satellite spatter are still evident to identify directionality. Note the satellite spine approximately 4 cm to the left of the placard.

Figure 5.53 A generally circular stain from the pattern in Figure 5.50. Taken as a single stain, the shape is not helpful, but considering the shapes evident along the whole pattern, we can still be confident of the overall direction of travel.

with elliptical shapes. Although anything but symmetrical, it is very apparent the general direction the droplet was moving at impact. Figure 5.53 depicts a stain further along the blood trail. This stain is far more circular; given this stain by itself no directionality would be evident.

The pace or speed the object is moving in the scene while the blood is dripping is important from a practical perspective. As discussed in droplets falling perpendicular to a surface, the resulting stains are generally circular, and directionality is difficult to define. As the pace or speed of the dripping object increases, that forward momentum translates into a more acute angle of impact on the surface below. When more elliptical stains are found across some aspect of the overall pattern, this most likely indicates a greater forward pace of the object moving in the scene. The elliptical shapes of Figures 5.51 and 5.52 suggest a faster forward movement. A series of circular shapes like those in Figure 5.53 would indicate a slower forward pace. An extended series of different shaped stains at different points along the trail may well indicate a change in the speed at which the item is moving through the scene. Unless the speed of the associated movement is sufficient to produce elliptical stains, it may be impossible to identify an overall direction of movement.

The presence of a drip pattern (blood falling into blood) at any point along the blood trail may also come into play, indicating that the item dripping the blood had become stationary for some period of time. Drip trails commonly demonstrate the idea of a "complex" pattern as described in Chapter 4. This is true where both a drip trail and drip pattern appear together. Another complex occurrence often encountered in a blood trail is the presence of cast-off patterns emanating from the trail. Assuming an individual is moving as blood is dripping from an arm or from an object being carried, it takes nothing more than a significant flick of the wrist to create a cast-off event in any given direction. Where typical drips fall with correlating shapes generally 45° and greater, if a cast-off action is imparted into the overall movement, a series of cast-off spatter (more elliptical) may appear somewhere within the overall pattern. This effectively creates a slight or partial cast-off pattern extending out of the drip trail.

Understanding Relative Motion of Disruptions and Depositions of Contact Stains

Another consideration for determining motion is a stain produced from contact. Wipes and swipes were previously defined; both can play a prominent role in defining the motion of the contact.

As items become bloodied and stains are created, actions at the scene may still be proceeding. These stains often become disturbed by other actions (resulting in wipes), detailing the associated motion of the disturbance. This motion is almost always obvious in instances of wipes by the displacement of the blood. Figure 5.54 shows a large volume wipe present on a kitchen floor. Another wipe example encountered is when a blow creates a spatter pattern and individual spatter are subsequently disturbed. The spatter had specific boundaries to begin with. As the blood within the stain boundary is wiped through, it is moved in the same direction as the item disturbing it. See Figure 5.55.

In a situation involving a swipe where a bloody object deposits blood onto a second surface, once again direction may be evident, but not always as clearly as in the case of a wipe. Depending upon how the bloody object encounters the target, the contact and exit boundaries for the stain may be similar or radically different. If the edges are similar, there may be no functional method to establish the specific direction in which the swipe was deposited.

Generally, in a swipe, we seek to locate evidence of a thinning of the stain's appearance or color to help define the motion. As the blood volume is deposited against the target, the

Figure 5.54 A large volume wipe pattern on a floor. The displacement of blood allows us to know in which direction the motion of the wipe occurred (for the largest stain, the movement is toward the bottom of the picture).

volume available to be smeared against the target decreases. The exit/departure portion of the swipe often demonstrates this "thinning" characteristic as a result.

In addition to the thinning of the color and consistency of the stain, there may be linear features leading away from the departure side of the stain. These linear features result when the bloody object loses contact with the target surface. See Figure 5.56. Such features are commonly referred to as "feathering" of the stain. Unfortunately, if the initial contact and exit boundaries demonstrate similar characteristics, defining direction is difficult. This may result from both the movement and the characteristics of the contacting object. When feathering is present in both boundaries of the stain, one is unlikely to have confidence in what is an entry or exit boundary. The presence of feathering on a single side of the stain is usually a significant characteristic that allows for easy identification of the departure side of the stain.

In the past, analysts have claimed that "feathering" is not a functional means of evaluating directionality of the wipe or swipe. This belief may well be based on the possible presence of feathering in both boundaries. We would agree in such instances. These authors

Figure 5.55 A spatter pattern that was wiped through after deposit. The wiping and displacement of the liquid blood leaves evidence of right to left motion by whatever caused the wipe.

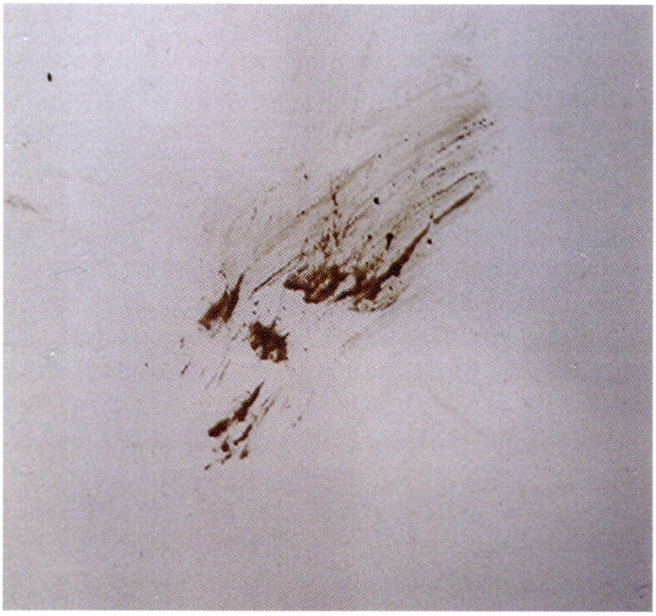

Figure 5.56 A swipe pattern on a plate glass window. The linear features in the stain that run diagonally up and to the right, as well as the thinning of the blood volume in this direction, indicate the motion of the swipe as up and to the right.

offer two alternative methods for evaluating direction of the swipe and wipe: displacement of the volume of blood and the evaluation of the "gritty" nature of the blood based on drying characteristics.[17]

As previously discussed regarding wipes, when present displacement of the blood on a single aspect of the stain is a functional method of determining direction. Displacement in a swipe can suffer the same limitation as feathering regarding defining directionality. Depending upon the means of application of blood by the swiping object, displacement of the blood can occur in both directions, although this is far less likely. The "gritty" nature of blood is not a characteristic your authors feel is a functional means for defining motion. There is no known research that effectively defines what "gritty" entails.

In 2002, author Gardner conducted a study of directionality in swipe patterns. A primary conclusion of the study was that three characteristics routinely appeared on the departure side of the swipe mark. These included feathered edges opposite a contiguous boundary, striations in the body of the stain, and diminished volume.[18] Based on this effort and our experience, we feel comfortable stating that feathering must be considered in relation to all other characteristics of the stain. When the stain is considered in a holistic fashion, feathering is a significant characteristic for defining direction of movement of the swiping object. It is not absolute, thus caution is always advised.

Motion Indicated by Repetitive Pattern Transfers

Repetitive transfer patterns can also define motion in the scene. A repetitive transfer pattern occurs when an item becomes bloodied and then encounters a given target a number of times.

Repetitive transfer patterns occur as a result of any number of actions, feet, and footwear are frequently the major cause of such a series of patterns. The item encounters a blood source, as in the instance of a shod foot stepping into a puddle of blood. See Figure 5.57. As the individual walks, blood is deposited each time the shoe contacts the target until the blood is depleted.

Defining this motion is based upon evaluating the dimming or diminishing of the pattern as the number of contacts increase. If all things are equal, the first contact leaves a greater volume than the next. Each subsequent contact depletes the available blood on the object, leaving less to make the next pattern transfer. If patent (visible) patterns are present, invariably latent (invisible) patterns are also produced and present. The latent patterns will require chemical enhancement to locate, but upon enhancement are likely to demonstrate this diminishing characteristic as well.

Motion Indicated by Flows

Blood flows also provide critical information about motion and orientation. Obviously, blood flows will obey physical laws, specifically gravity. Of concern to the analyst are flows that indicate orientation or movement on the object during or after the flow occurs. Flows on the body or a scene object that appear irregular or defy expectations regarding gravity may point to earlier positions of a victim or object.

Figure 5.57 A repetitive pattern transfer resulting from a shoe stepping into the large pool of blood at the top of the photograph. The wearer then stepped backward. The patterns continued into a hallway until they became latent.

Another irregular flow that the analyst may encounter is due in part to capillary action. When an intervening object is exposed to a flow, or in some way dams a flow, the blood will tend to follow the contour of this object. Particularly, if the object is later moved, the resulting flow pattern will likely appear abnormal or out of place.

Gravitational effect (the flowing of a fluid) is often apparent even when a distinct flow pattern is not present. Saturations on clothing worn by those involved may demonstrate this gravitational effect. The length (the long axis, whatever its final shape) of a saturation stain is oriented to a vertical axis of the item when the saturation occurred. Perhaps not as clear as a classic flow, such effects can clarify general orientations of the individual at the time of the saturation.

Summary

The principle of Stain Shape and Vector Correlation is a critical aspect of the predictability in Bloodstain Pattern Analysis. Varying bloodstains provide excellent indications to the analyst about motion and spatial orientation. Through experiments, training, and experience, impact angle correlations are easily recognized and understood. Understanding the minor nuances of directionality requires little effort because we observe liquid droplet behavior daily. Whether raindrops, bird droppings, or milk splashes, this concept is easily observed and recognized. These ideas of impact and directional angle can be applied to any individual stain(s) to understand the general spatial source of the stain or, in the case of impact patterns to multiple stains to define an area of origin, which is discussed in the next chapter.

We also come to Bloodstain Pattern Analysis with experience in swipe and wipe patterns. Whether encountered as paint, ink, or dirty mop water on a floor, the underlying motion or action is generally obvious.

General sequencing of events by considering the volume of blood at different points in the scene may require a little more experience; nevertheless, its basis is still evident.

Notes

1. Gardner, R. and Griffin, T., Foundations of the Discipline of Bloodstain Pattern Analysis, A Response to the Report of the National Academy of Sciences, *Journal of Forensic Identification,* Volume 60, Number 4, 2010, p. 484.
2. McDonald, James E., The Shape of Raindrops, *Scientific American*, Volume 190, Number 2, February 1954.
3. Carter, A.L., Personal communications, July 1992. Physics & Computer Bloodstain Analysis course, University of Central Oklahoma, May 1998.
4. McDonald, James E., The Shape of Raindrops, *Scientific American*, Volume 190, Number 2, February 1954.
5. Strand, Fleur L., *Physiology: A Regulatory Systems Approach*, Macmillan, New York, 1978, pp. 239–241.
6. Gardner, Ross M., Bloodstain Droplet Dynamics and Oscillation Damping, presentation at the 76th Annual Training Conference. Int. Assoc. of Identification, St Louis, MO, July 1991.
7. Carter, Fred, *Physics of Bloodstain Analysis* – Lecture Notes, 1999, p. 17.
8. Pizzola, P.A. et al., Blood Droplet Dynamics – I, *Journal of Forensic Sciences*, Volume 31, Number 1, 1986, p. 44.
9. Balthazard, Victor, Piedelievre, R., Desoille, Henri, and Derobert, L., Etude Des Gouttes De Sang Projecte, XXII Congres De Medicine Legale, Paris, France, June 1939, p. 17.
10. Carter, Alfred and Podworny, Edward J., Computer Modelling of Trajectories of Blood Droplets and Bloodstain Pattern Analysis with a PC Computer, IABPA Training Conf., Dallas, TX, Dec. 1989.
11. MacDonell, Herbert L., *Bloodstain Patterns*, Golas Printing, Elmira Heights, NY, 1993, pp. 9–11.
12. MacDonell, Herbert L., *Bloodstain Pattern Interpretation*, Laboratory of Forensic Science, Corning, NY, 1982, p. 7.
13. Balthazard, Victor, Piedelievre, R., Desoille, Henri, and Derobert, L., Etude Des Gouttes De Sang Projecte, XXII Congres De Medicine Legale, Paris, France, June 1939, p. 25.
14. MacDonell, Herbert L., *Bloodstain Pattern Interpretation*, Laboratory of Forensic Science, Corning, NY, 1982, pp. 44–53.
15. Charlie Green suggested this approach in the mid nineteen-eighties at a training program in Colorado Springs, CO.
16. James, Stuart H. et al., *Principles of Bloodstain Pattern Analysis, Theory and Practice*, CRC Press, Boca Raton, FL, 2005, p. 82.
17. Hoymeyer, Jane M. Ph.d, "Blood Evidence" IAI Workshop Notes, unpublished.
18. Gardner, Ross M., Directionality in Swipe Patterns, *Journal of Forensic Identification*, Volume 5, Number 52, 2002, p. 579.

Stain Shape and Vector Correlation

Area of Origin Considerations

6

As discussed in Chapter 5, the shape of a generally symmetrical spatter stain provides specific indicators as to the general origin of that droplet. The width to length ratio of the resulting circle/ellipse defines the angle of impact of the droplet relative to the plane as it collapses onto a target. The long axis of the ellipse, combined with the associated tails and satellites, defines the droplet's directionality relative to this plane. Taken together, these two indicators define a trajectory for the droplet. To be sure, this indicated trajectory is a straight-line path, droplets in flight long enough eventually follow a ballistic path so there are certain limitations inherent in this information. Those limitations are discussed later in the chapter. But given a droplet with sufficient momentum (as most impact droplets have) and a short flight path, this straight-line trajectory is an effective indicator of the droplet's movement through space.

In impact events, point source displacements, many droplets are dispersed from the same area by the impact. Having originated from a singular area, each droplet's individual trajectory should lead one back to the source of the displacement. This is the corollary developing from the ideas associated with impact and directional angle, the idea of Area of Origin (AO).

Area of Origin as a technique provides the ability to define or at least limit the common spatial source that numerous spatters may or may not have in three dimensions. This effort can help isolate the potential location of an impact event by identifying the range of what is possible and impossible based upon those individual flight paths.

Area of Origin as a technique is directed at impact patterns and only impact patterns. In cast-off, it is a given that the individual droplets originate from distinct points in space as the item creating the cast-off is swung. Thus, tracing back the trajectories does not provide an AO type convergence of the lines, but rather helps demonstrate the general arc of the swing. This can be helpful and demonstrative to the overall investigation, but it is not the same as Area of Origin. Additional research with cast-off may produce additional potential information, but this has yet to be established.

Balthazard et al. first suggested that AO was possible. MacDonell later defined a specific technique (stringing) to isolate the AO. Since that time multiple approaches have been developed, the most efficient of which is virtual stringing using software. The various approaches (physical stringing, tangent method, and virtual stringing) require slightly different data in each technique, but effectively there are five steps required in some fashion:

- Identify related well-formed spatter stains in the pattern.
- Identify the directionality of the stains.
- Identify the area of convergence for the pattern.
- Identify the impact angle of the individual stains.
- Combine this information for Area of Origin.

DOI: 10.4324/9781003408734-6

Identify Well-formed Stains in the Pattern

The concepts associated with recognizing well-formed stains was initially discussed in the previous chapter. During collapse, the mass of a stable droplet flows across a relatively smooth target surface in a symmetrical fashion. The symmetrical outflow of the liquid during the collapse results in a symmetrical shaped stain, often referred to as "well-formed." Figure 6.1 contrasts the difference between symmetrical and asymmetrical stains.

Identify Directionality of the Stains

Chapter 5 explained the process of defining directionality in spatter stains. The ability to recognize directionality is critical to locating the area of origin. As previously discussed, there are limitations imposed by the various shapes of the spatter stains. Elliptical shaped stains define directionality far more effectively than circular stains. See Figure 6.2. The more ambiguous the directional angle (e.g., the more circular the stain), the harder it is to isolate a clear and specific directional orientation. The more ambiguous this directional angle, the less likely a concise convergence (tightly intercepting lines from multiple stains) will occur.

The directional motion of a droplet collapsing on a target occurs in line with the long axis of the stain. Taken alone, the line of the long axis obviously indicates two different directions. The position of the primary tails, satellites or scallops when present, defines in which direction on that line the droplet was moving. Isolating the directional angle is as important as defining the impact angle of the droplet. A haphazard approach to identifying directional angle or the lack of functional elliptical stains will result in a wide convergence. The trajectory lines of the various spatter will simply not converge tightly.

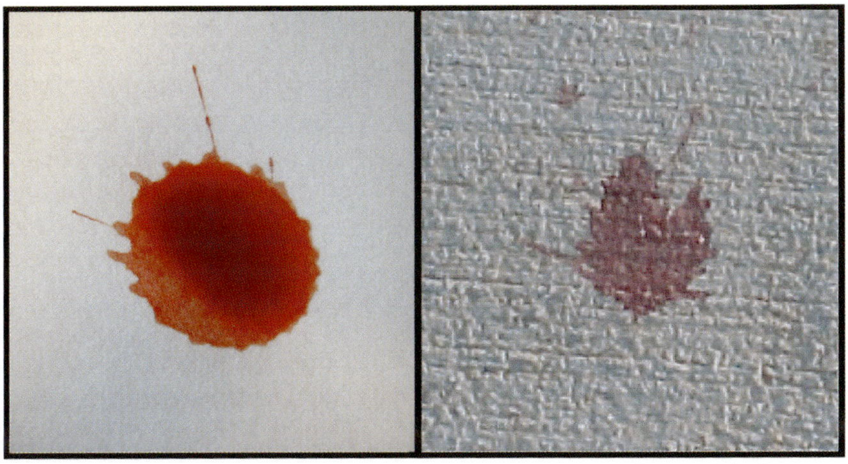

Symmetrical Shaped Stain   Asymmetrical Shaped Stain

Figure 6.1 The symmetry of the stain is important to the analyst in both defining directionality and impact angles. In pursuing an analysis of the area of origin, the analyst chooses stains that are symmetrical. The stain on the left is symmetrical with clearly defined margins. The stain on the right, having impacted onto a rough surface, is not symmetrical. Defining a clear length and width for the stain's ellipse, or identifying a long axis in such a stain is difficult.

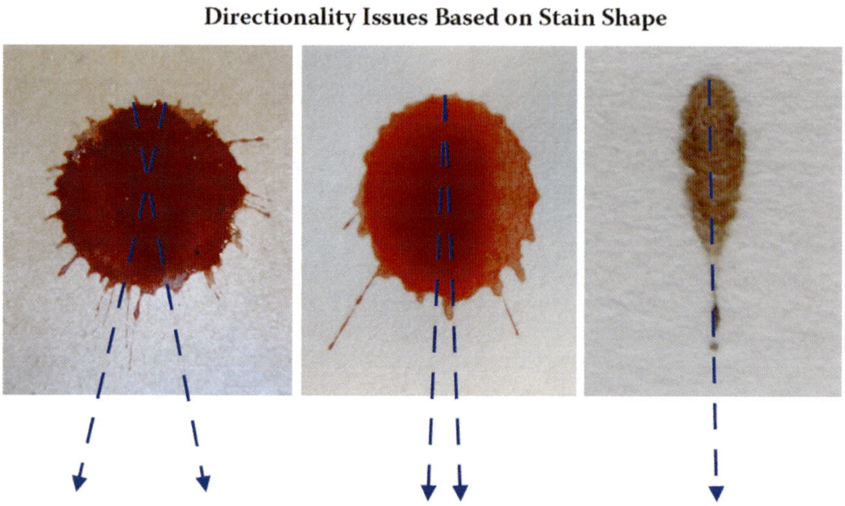

Figure 6.2 Directionality of a stain becomes more defined as the stain becomes more elliptical. In the circular stain on the left, it is difficult to know where the true long axis of the stain's ellipse is. The middle stain provides a clearer idea of this long axis, but even in this instance, different analysts might view it in slightly different ways. The elliptical stain on the right allows for a very clear understanding of where the long axis is. Different analysts viewing directionality in this stain would not vary in any significant fashion.

Directionality defines the path of the droplet as it struck a target relative to the XY plane, a top or bird's-eye view. Often this is described in general terms (e.g., left to right). This direction of travel can also be defined numerically by a specific angle. The directional angle, also known as the gamma angle, describes directionality either as a specific angle between 0 and 359 degrees, or in some instances as a positive or negative angle between 0 and 180 degrees. Generally, the reference (0 degrees) is North for patterns on horizontal surfaces and Up for vertical surfaces. The directionality is established and the long axis of the stain is then measured against the reference point. See Figure 6.3

Identify the Area of Convergence for the Pattern

Depending upon the specific question raised about a given spatter pattern, the needs of the analysis may be different. An area of convergence analysis (looking solely at the XY plane, a top view) may be sufficient to answer probative questions in some circumstances. Particularly on a horizontal surface, it is a given the three-dimensional source of the spatter (their AO) must be located somewhere above this convergence. On a vertical surface, there are additional issues for consideration. Therefore, a two-dimensional convergence on a target may, in a given context, isolate the general location of the source event sufficiently for purposes of the investigation. At other times, the analyst may require the more detailed approach, demanding the use of the full Area of Origin evaluation method.

On a horizontal surface directionality defines the direction the droplet was moving during collapse. This path is visualized by creating a reverse azimuth (a line which extends opposite along the path the droplet was following). One can be reasonably sure, the droplet originated somewhere along this line. When present on a horizontal surface, looking

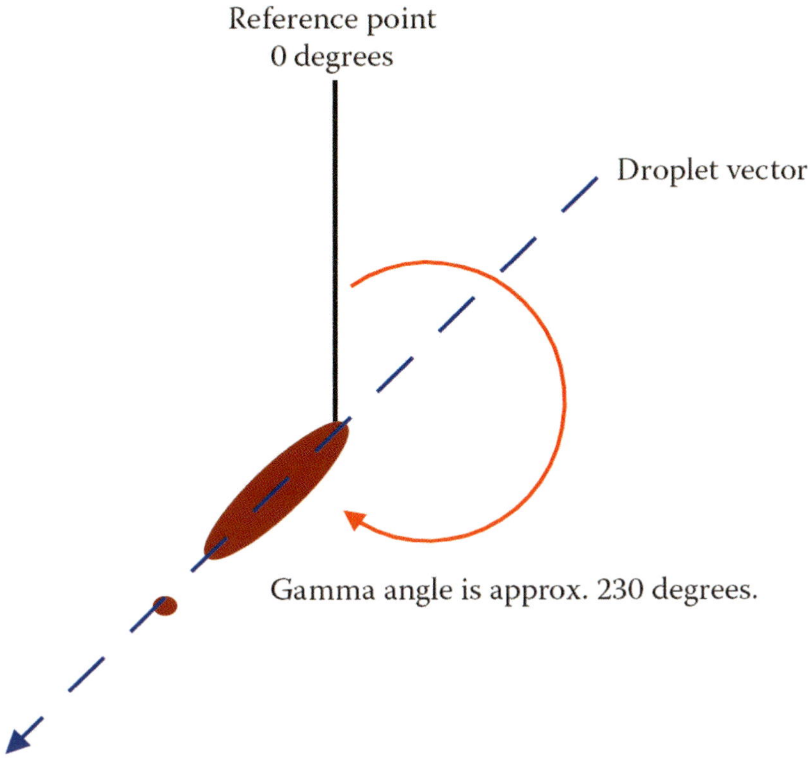

Figure 6.3 The directional or gamma angle is the angle between the long axis of the stain's ellipse and a standard reference point. When measured from this reference point, this allows it to be used by forensic software.

down on the droplet's path from this top view, there are no forces that affect the droplet's flight path. In this top view, once set in motion, a droplet will follow a straight path from its source to its destination. Gravity and air resistance affect the droplet only in the vertical plane of its flight path.[1]

For example, given Figure 6.4, Stain #1 must originate somewhere along a reverse path as indicated by the directionality of the droplet. The only limits to the origin are the room's limitations or any intermediate obstacles. If the box for Figure 6.4 represented the room boundaries and the possible reverse path for Stain #1 extends 13 ft., then the drop's origin must lie somewhere within those 13 ft. With only one stain, the resulting parameter of possible origins is broad.

Introducing a second stain and a reverse azimuth, as in Figure 6.4, a point occurs where the two paths intersect, a convergence. The process is no different from the technique known as resection in map reading. By taking compass readings from an unknown point to two known points and applying reverse azimuths from each, the unknown point is defined by where those reverse azimuths cross. In BPA the same is true, the point where the various reverse lines associated with the stains converge is the likely source for the stains.

Keep in mind that this convergence is an area. Using only two stains, the point of intersection appears as a "point." When combined for numerous stains, the individual lines created by these reverse azimuths will cross, but not at an exact location on the target (e.g., a specific XYZ point). See Figure 6.5. However, the wider the radiating pattern of

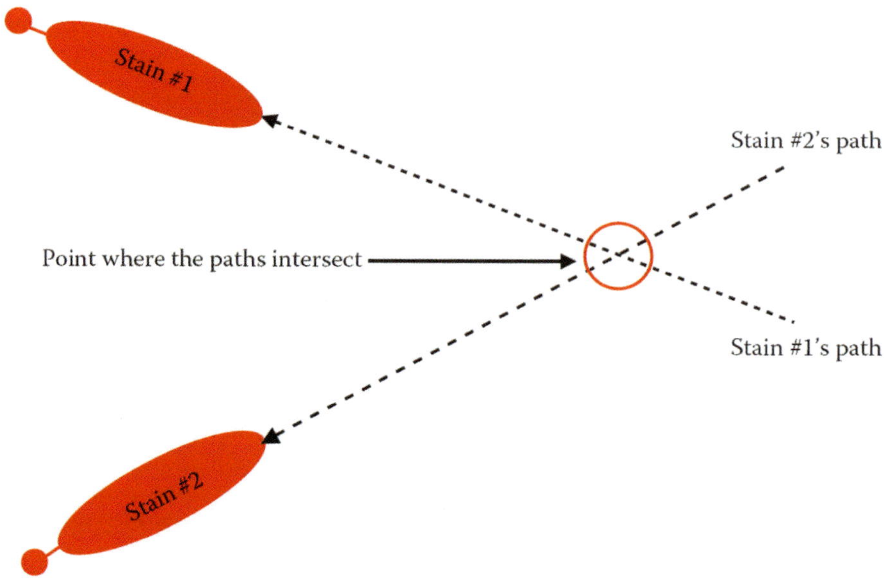

Figure 6.4 By following the reverse vector of each droplet to the point where the paths intersect, we can establish a probable point of convergence in two dimensions.

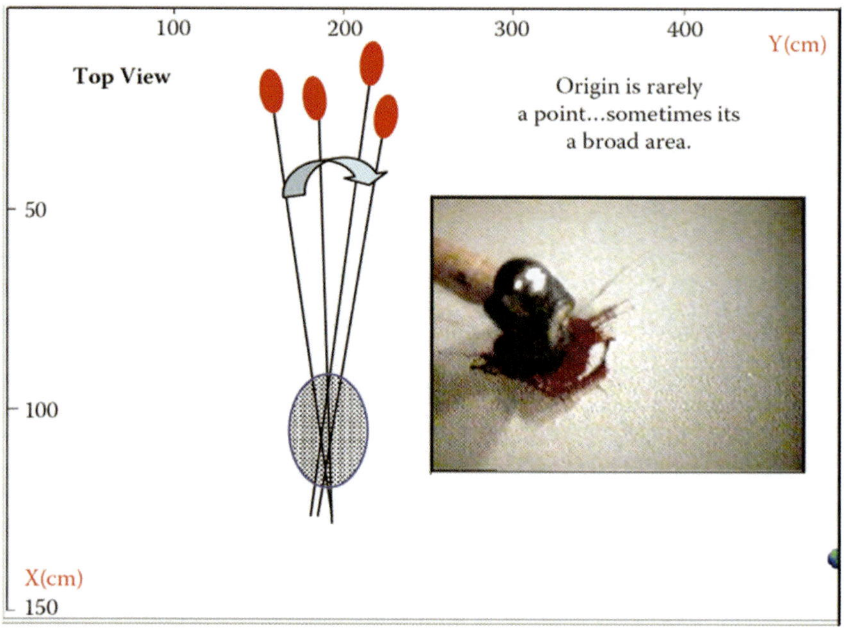

Figure 6.5 At the convergence, the lines from multiple stains will rarely cross each other at a single point. The convergence point is an "area" as well.

spatter involved, the more likely the convergence can be isolated to a more refined area. See Figure 6.6.

Convergence alone (this top view approach) certainly provides simplicity in the evaluation. There is, however, an inherent difficulty. If the circumstances limit the analysts to only a few stains, it is always possible they are the result of more than one event. That is, an

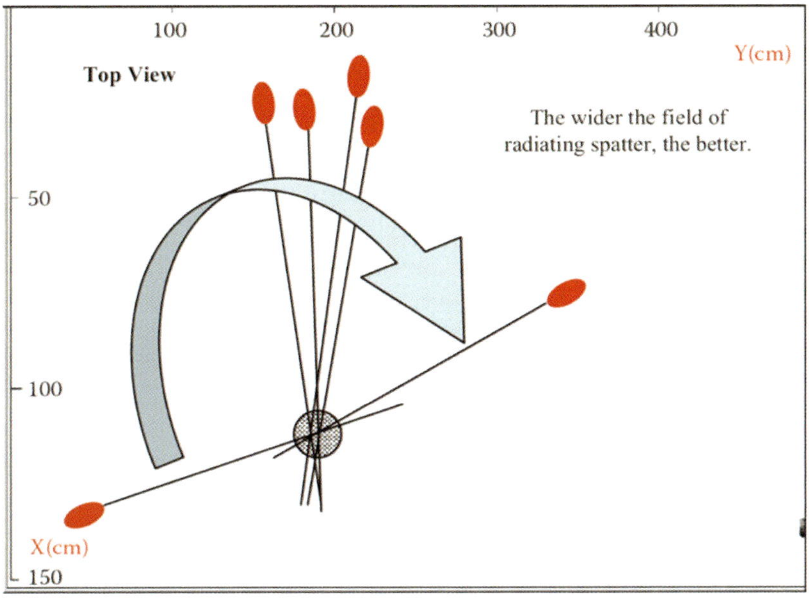

Figure 6.6 The wider the field of radiating spatter, the more defined the convergence becomes. Note that the inclusion of the far right and far left stains functionally tightens the convergence as compared to that observed in Figure 6.5.

event created Stain # 1, followed by another unrelated action that caused Stain # 2. Given this situation, the points where their reverse paths cross may simply be coincidental and have no investigative significance.

Widespread within a scene and showing no common convergence, one would easily recognize such stains as separate actions. Found in proximity to each other, the likelihood increases of a coincidental convergence of the flight paths along their reverse paths. Choosing to view this coincidental convergence as a single origin for both stains would be invalid. Presented with a situation with only a few stains to work with, it may be difficult to recognize such an error.

The concern of coincidental convergence is also true when viewing two adjacent patterns involving numerous stains. There may be a coincidental convergence for two patterns, and this might be interpreted as the source for both. Figure 6.7 illustrates the possible error when viewing patterns from two closely located impacts. This is not a particularly common issue, but it should be kept in mind.

As the number of evaluated stains increases, the level of confidence increases as well. The more paths found which intersect a given area, the more likely that a single true area of convergence exists. Even in circumstances of multiple overlying events with the spatter intermixed, the primary convergence points may still be evident. In this situation, the various paths may cross at several locations, but where we begin to find clusters of intersecting paths will establish the primary convergences of the various events.

What this top view method of analysis does not establish is the three-dimensional origin of the spatter, the area above the convergence where the spatter originated. This AO location is a position in three-dimensional space. In the top view approach, a convergence in two dimensions is defined and it is given the flight paths originate somewhere above it. Once again, the possibility of multiple events always exists. The analyst may be viewing

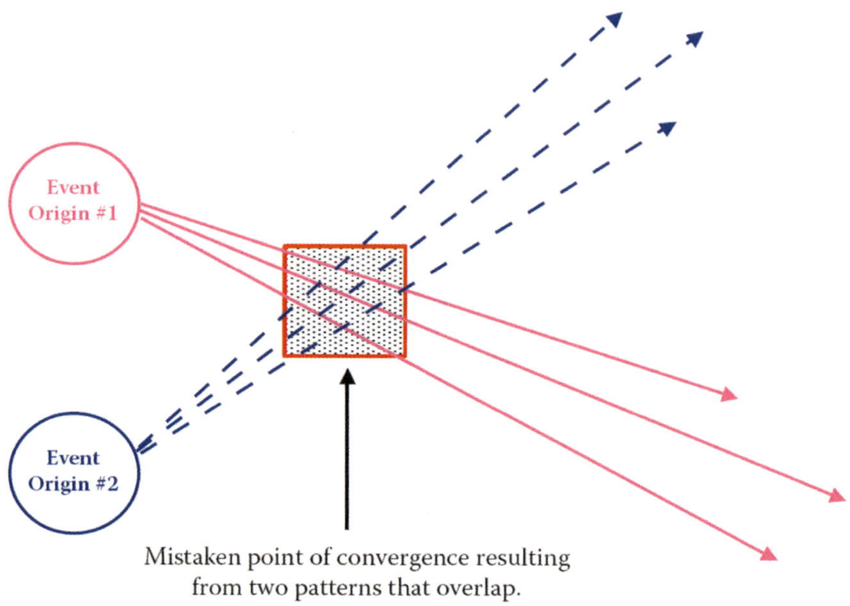

Figure 6.7 In addition to the possibility of two stains having a coincidental intersecting point, it is also possible for two patterns to overlap in this fashion. If this condition is not considered, it might well result in a mistaken area of convergence for both patterns.

spatter from two or more events that originated at the same location in a room, but from different heights during different events (e.g., one impact at 6′, another at 3′).

As described, the top view convergence technique is effective in and of itself in some instances. Used in conjunction with Toby Wolson's[2] technique of Roadmapping, which is discussed later in the book, it represents a functional method of portraying this convergence in the scene. See Figure 6.8. This information alone limits the possible area where the event could have originated. Thus, if presented with a circumstance where impact angles cannot be determined for some reason (e.g., fabric surfaces), the top view convergence defined by the long axis of the stains may still provide probative information to the analysis.

Case Example: A woman was shot once in a hotel room. Her husband claimed the shot was accidental, in that his weapon fell from a nightstand and upon striking the floor went off. Due to after market adjustments to the weapon, a ballistics examination supported the belief that the weapon could malfunction under the reported conditions.

The shot perforated the woman's right arm, and then penetrated her head, producing a significant spatter pattern, the vast majority of which was deposited across a duvet cover on the bed. The spatter pattern was obviously associated with either the arm or head entry or some combination, as there were no other associated bloodstain producing events.

The duvet cover presented a major issue in terms of Area of Origin. Any calculations of the impact angle would be unreliable given a fabric surface. But directional angles are based solely on the long axis of the resulting stains and in this instance the stains were relatively symmetrical. There were no stains on the vertical aspect of the edge of the bed, all stains radiated across the duvet and room. The pattern's deposition demanded that the wound path was not oriented vertically, but rather at the very least horizontally at the edge of the bed. There was an obvious convergence represented by the pattern and this convergence was literally feet away from the nightstand's location. Also present at the termination of the convergence was

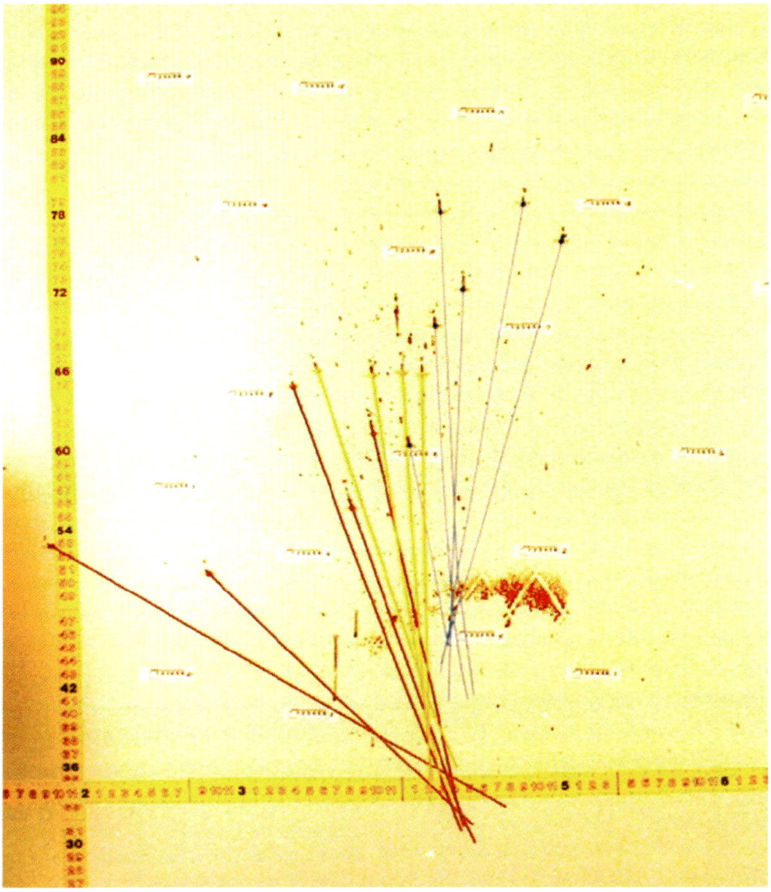

Figure 6.8 An example of using graphic tape to demonstrate a two-dimensional area of convergence in a spatter pattern.

an obvious contact stain, as well as a GSR pattern at the edge of the duvet. Neither of which correlated to the alleged location where the weapon "fell off" the nightstand. See Figure 6.9

On vertical surfaces, convergence is more problematic. In some instances, this method remains valid; but both gravity and air resistance affect the directional path of the droplet in the vertical plane. Thus, the indicated directionality is affected. This concern is discussed in detail later in the chapter.

To this point, the analysis has narrowed the scope to finding the convergence along the paths of the stains of interest. The converging lines of these reverse vectors establish an area of convergence for the pattern. To more effectively limit the source in three dimensions, flight paths in the vertical plane must be considered; demanding a side-view approach of the target. This requires including the stain's impact angle in the analysis.

Identify Impact Angles for the Stains

As described in Chapter 5, Dr. Victor Balthazard can be credited with having recognized that a relationship exists between the length and width of the resulting stain and the angle

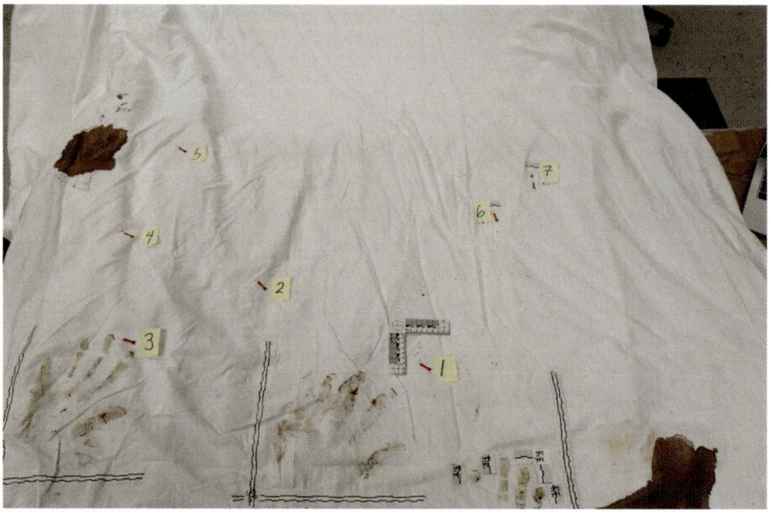

Figure 6.9 An impact pattern distributed across a duvet. The red arrows demonstrate the directionality of various stains, which generally converged at the large saturation in the bottom right corner of the photograph. This convergence demonstrated the position of the victim at the time of her wounding and was inconsistent with the claims of her husband.

at which the droplet impacts. The shape of the stain defines the angle of impact. In general terms, what this means is the more circular the stain, the more perpendicular it struck the target surface (e.g., 90 degrees), whereas the more elliptical the shape, the more acute the angle of impact (e.g., 10 degrees). MacDonell later refined Balthazard's original concept (the length to width ratio) and instead applying a width/length ratio in conjunction with specific trigonometric functions (e.g., the sine function). Given a well-formed stain, where we can accurately measure the width and length, we can easily calculate the impact angle as described in detail in Chapter 5. This process, like directional angle is integral to the AO procedure. McDonell's stringing method allows the analyst to use straight-line geometry techniques in defining the bloodstain event.[3,4]

Combine the Directional and Impact Angles to Establish an Area of Origin

Having learned to determine the impact angle, we combine the convergence point (the top view) and the impact angle (the side view) information in order to identify the area of origin for the spatter event. This is a location in three-dimensional space. For many years, this location was referred to as the Point of Origin; but a "point" suggests an actual point, a specific position in the X, Y, and Z planes. Because of the issues associated with defining directionality and impact angle, we recognize that what we are defining is an area, not a point. In some instances, this area may be quite specific. In other situations, AO may simply identify a range of what was possible and what was impossible.

There are a variety of approaches for combining this information:

- Graphing
- Tangent method

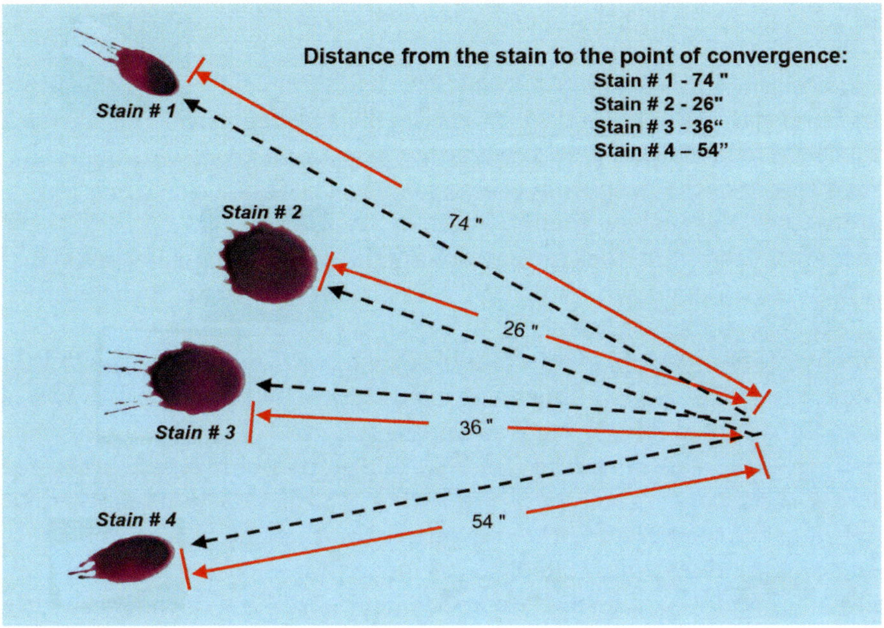

Figure 6.10 An example of using convergence evaluations. In this instance, we have four stains with a common convergence and we wish to determine if they share a common area of origin. We measure the distance from the base of each stain to the convergence. We then combine that information with the impact angle in a graph (Figure 6.11), this allows the analyst to visualize the area of origin for the spatter.

- Physical stringing
- Virtual stringing (computer-based)

For the sake of simplicity, in the following example all stains are on a single surface and only four stains are used. See Figure 6.10. The four stains' flight paths appear to have a common convergence point in the scene. The stains are "well-formed" and the analyst measures each and applies the sine formula, identifying each stain's impact angle.

Accept for the examples that based on the width/length ratio, the stains impacted as follows:

- Stain # 1 30°
- Stain # 2 60°
- Stain # 3 50°
- Stain # 4 40°

For purposes of the example, imagine that the analyst then measures the distance from the convergence to the rear (leading) edge of each stain. In this instance, the analyst found:

- Stain # 1 is 74" from the convergence
- Stain # 2 is 26" from the convergence
- Stain # 3 is 36" from the convergence
- Stain # 4 is 54" from the convergence

Graphing Points of Origin

One method of combining this information is to use a graph. To do so, the analyst draws a graph with a positive X and Z axis. Zero on the ZX axes represents the convergence; the Z axis graphs the height above that convergence. The X axis represents the target plane; it graphs the distance of the stain from the convergence. The two axes are scaled the same, with whatever scale the analyst choose.

Consider that each stain, the convergence and its area of origin create a right triangle. In the graphing technique, each individual stain's triangle is rotated (whatever its orientation) on the X axis, so they effectively all lie flat on the graph.

The analyst does this by placing a mark for each stain along the X axis at the appropriate distance (e.g., 74 ″). From this point, using a protractor, a line is drawn along the indicated angle of impact for that stain extending it to the Z axis. The analyst repeats this process for each of the stains involved. The resulting graph would look like Figure 6.11. The point where the lines converge on the Z-axis scale establishes the best representation of the area of origin for the spatter above the convergence on the target.

As we draw the lines along the angle of impact for each stain, a common Area of Origin on the vertical Z axis should become evident, assuming of course a true common origin exists. If the spatters evaluated are the result of two or more different events, which occurred at different heights, that too would become evident. Figure 6.12 is an example of what one would expect to see in a situation involving two impacts at different heights over

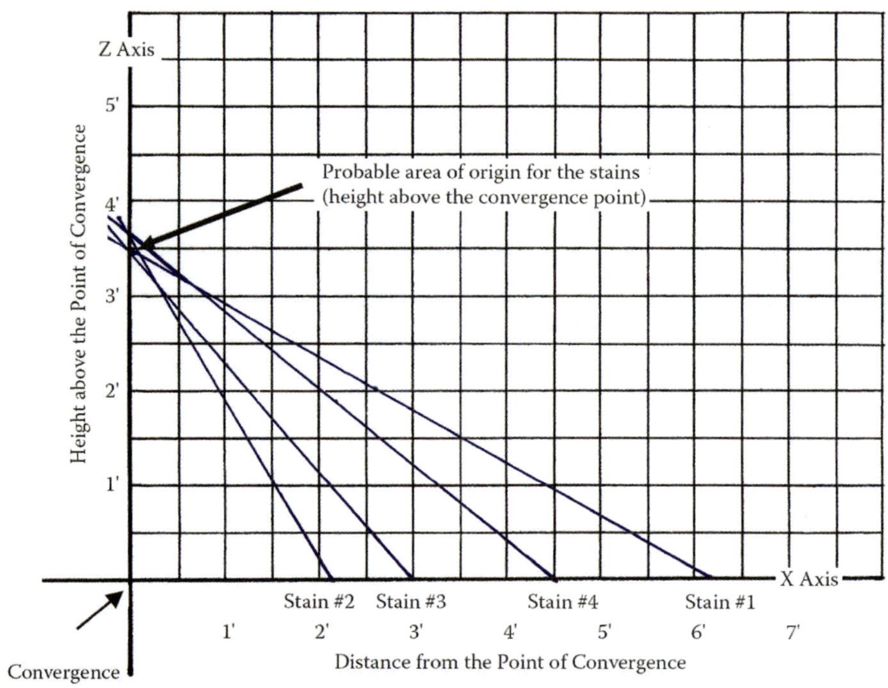

Figure 6.11 Graphing the area of origin. For each stain, we mark the distance from the convergence on the X axis and then draw a line to the Z axis at the indicated impact angle of the stain. Where the line crosses the Z axis is the probable height above the target for that particular stain. In this instance, taken together, the stains appear to have a common origin approximately 42″ above the target.

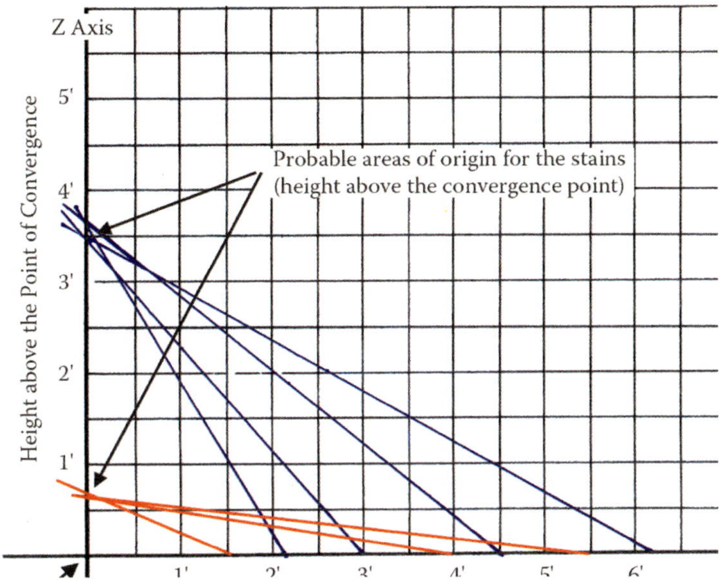

Figure 6.12 A graph indicating two areas of origin. In the scene from an overhead view, all the stains appear to share a common convergence. The inclusion of the impact angle information adds an additional dimension, making it apparent that two separate groups are present, suggesting two separate events.

the same location. Remember, the Z axis represents the area above the pattern's area of convergence on the target. When viewed from above, all the paths for the stains involved cross at this area. Viewed only from a top view perspective, one might choose to believe they originated from a single event.

By considering the impact angles in conjunction with the top view for Figure 6.12, a distinct grouping becomes evident with a common origin close to the target's surface. Another group is also evident with an origin higher than the first. This indicates two different spatter-producing impacts at that convergence point.

Defining Area of Origin with the Tangent Function

Using a scientific calculator, it is possible to simply calculate this distance above the area of convergence for each stain. The analyst does this using another relationship related to right triangles, the tangent. Figure 6.13 shows the relationships of the scene to this imagined triangle.

The first step in making this determination is to identify stains that appear to have a common convergence. As in the graphing method, the analyst measures the distance from these stains back to the point where the stains have a common intersection. The analyst also determines the impact angle of each stain.

To determine the Area of Origin, the analyst uses the following formula:

$$\text{TAN } i = \text{SO/SA}$$

where i equals the known impact angle, the Side Adjacent (SA) of the triangle is equal to the distance of the stain to the convergence, and the Side Opposite (SO) equals the unknown distance above the target surface (see Figure 6.14).

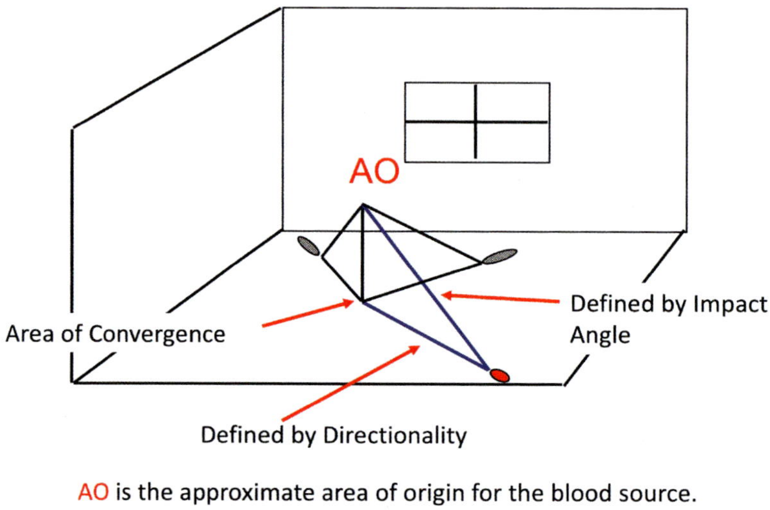

Figure 6.13 The base of each stain's position, the point in two-dimensional space where their paths converge, and their area of origin (AO) define a right triangle.

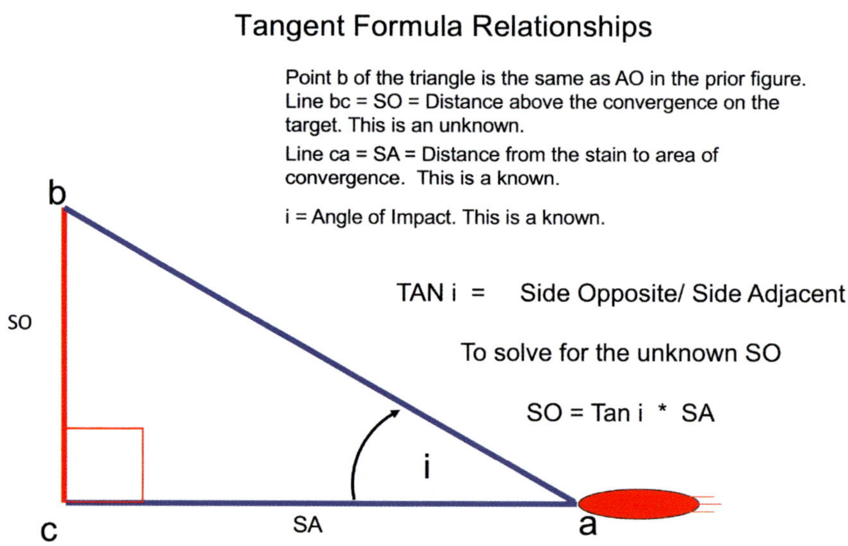

Figure 6.14 Our right triangle from Figure 6.13 is further defined. We know the distance from the stain to the convergence which is the side adjacent (SA – line ac in the triangle). We can establish the impact angle *i* using the standard Sine formula. Combining this information in the formula Tan i = SO/SA, we simply balance the equation and solve for the side opposite (SO – line bc in the triangle). Thus, the line bc or the height above the area of convergence is derived by the formula SO = Tan $i \times$ SA.

For example:

$i = 19°$

SA = 25″

TAN 19° = 0.344

0.344 x 25″ = 8″

To solve for the SO, which is our unknown height above the convergence, we simply balanced the equation by multiplying Tan *i* by the SA, giving a result of 8.66″. This procedure is convenient and provides immediate feedback at the scene. Once again, the analyst requires a calculator with trigonometric functions to use this method.

Using a number of stains from the pattern and their individual triangles this procedure is repeated. Once again, there is no expectation that each stain's information will produce an exact result with the next. The results for all included stains are averaged to provide an estimate of the height above the convergence for that pattern.

Three-dimensional Evaluations of Area of Origin

A final manner of evaluating the Area of Origin is in a graphic three-dimensional fashion. This is possible either in a physical sense, using the "stringing" technique or by using computer-aided analysis tools.

Stringing Scenes

Stringing is a basic method taught to bloodstain analysts to facilitate an understanding of the principles, using a hands-on approach. As a result of computer technology, its place in on-scene analysis is slipping into the past, but stringing is still considered useful as a visual aid to help explain "Area of Origin" to both juries and students (see Figure 6.15). Herb MacDonell is the first person to describe the process itself, yet we usually credit Balthazard et al. for setting the stage for the development of stringing.[5]

Whoever the actual originator, the process is simply a physical extension of what we've described to this point. Stains from the pattern are chosen and impact angles determined. Based on directionality and convergence points, thread or string is taped in place at the base (the initial point of contact) of the actual stain. The string is then extended back along the indicated path and angle. The most effective way to find this path is with a laser protractor. The protractor is set at the indicated impact angle and aligned with the directional angle (the long axis of the stain). The laser is then turned on and the point where the laser strikes a nearby surface is marked. A string is then attached from the stain to this point. See Figure 6.16. This process is repeated for each stain evaluated. Not only does the laser protractor reduce the subjectivity of aligning the string, it also reduces the level of labor necessary to string a scene. As more and more strings are introduced into the scene, a generalized convergence of the strings will begin to form. See Figure 6.16.

Stringing as a technique has several drawbacks. First, it requires the expenditure of immense effort. Placing the strings into the scene and finding objects to attach them to is a chore. The analyst ends up interjecting tripods or extending strings across rooms to opposite walls. This of course produces clutter, which the analyst must move around while completing the stringing process. Limitations on where the analyst can attach strings generally decreases the precision of the process.

A second and perhaps more dangerous flaw of stringing is the error evident in the process. Placing the strings at accurate angles and precisely along the indicated path can be difficult. Subjectivity can easily be introduced into the process, with analysts tweaking their strings to make them look more precise. The use of a laser protractor certainly helps eliminate some of the problems associated with this issue.

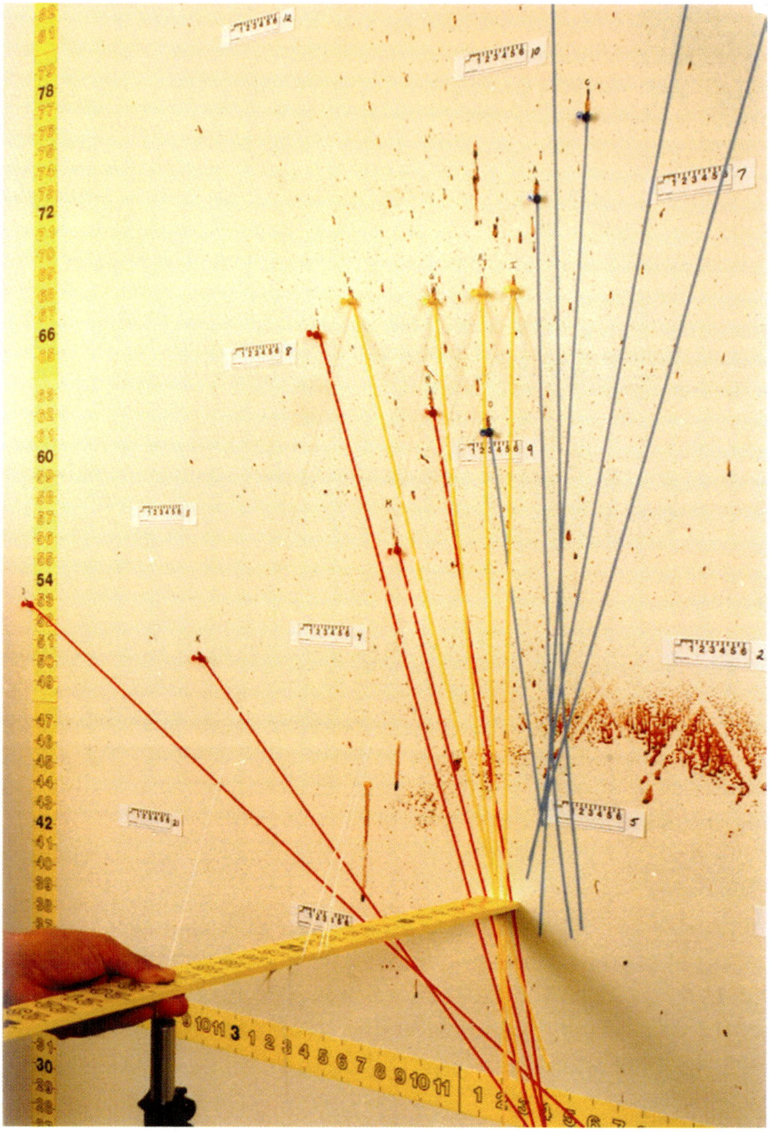

Figure 6.15 An example of on-scene stringing in combination with the use of graphic tape. The graphic tape demonstrates the convergence, while the strings demonstrate the area of origin. (Photograph courtesy of Toby Wolson, MS, Miami-Dade Police Department, Miami, FL.)

In actual casework, any stringing effort that results in an extremely tight convergence of the individual strings should be considered suspect. It is not uncommon to see physical stringing effort where the strings do exactly that, tightly converge. This is typically a function of both internal bias and the limitations of attachment points. Consciously or not, the analyst begins to "position" the strings relative to the developing convergence. It is highly improbable the resulting strings from any stringing effort would cross so precisely given the inherent limitations of the process. Physically positioning the strings in the actual scene is difficult to begin with and as the number of stains considered increases, the possibility of a precise convergence decreases dramatically (Figure 6.17).

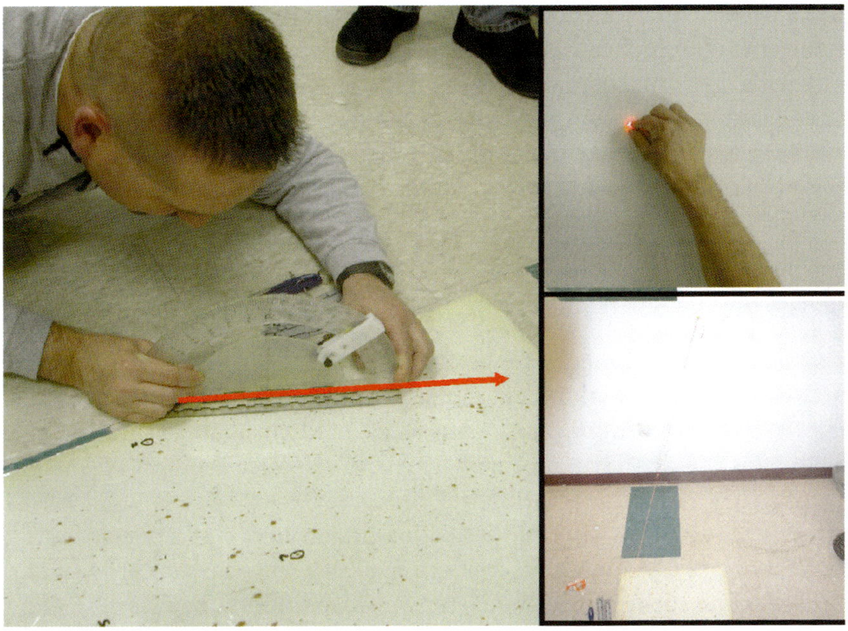

Figure 6.16 The use of a laser protractor for stringing an area of origin. The protractor is aligned to the directional angle and set at the indicated impact angle. The ends of the string are then attached to the point where the laser dot appears and the leading edge of the stain.

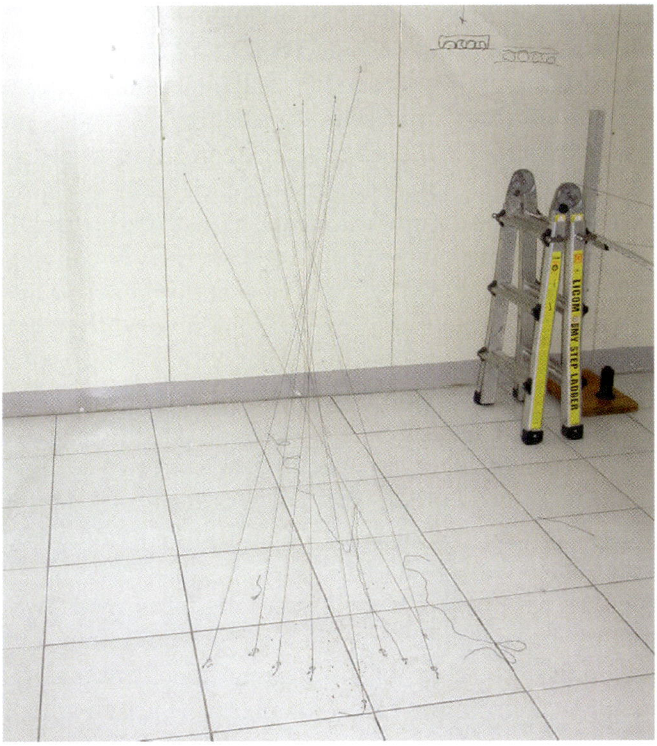

Figure 6.17 A student stringing exercise. The more strings that are added to the effort, the more effectively an area of convergence is identified.

A final drawback associated with stringing is the inherent risk associated with the process. In order to string, the analyst must get near the stained surfaces. They must physically interact with these surfaces putting them in repeated proximity to blood and all of its associated biohazards. This is particularly true when a loupe is the associated measuring device. Even with appropriate counter measures (e.g., protective clothing, masks and goggles), there is always a risk to the analyst. The use of forensic software as we will describe in this chapter effectively reduces this risk.

Virtual Stringing

Computer software advancements suggest that physical stringing should be replaced in its entirety. Numerous stand-alone virtual stringing programs were introduced over the past 30 years. With the recent developments associated with laser scanning systems, such developments offer the analyst a more efficient and effective method of analyzing spatter patterns for Area of Origin information.

One of the first virtual stringing programs introduced was *Trajectories®*. Developed by Dr. Alfred Carter of Carleton University in 1987, it allowed the analyst to evaluate Area of Origin using crime scene data. Dr. Carter improved upon *Trajectories®* and eventually released it through Forensic Computing of Ottawa (FCO) as *BackTrack®*: a full-function program that allowed for Area of Origin calculations on vertical surfaces.[6] *BackTrack®* employed actual flight path (parabola) calculations, which incorporated estimates of blood droplet volume, gravity and air resistance factors.

Shortly after *BackTrack's®* release, FCO released another program, *BackTrack/Strings®*.[7] FCO scaled down this version somewhat, changing the way it handled the Area of Origin algorithm. It had all the primary functions of *BackTrack®* but utilized a string or straight-line approach in determining flight paths in lieu of the parabola approach.

In running *BackTrack/Strings®*, the analyst measured the stains present at the crime scene. This includes identifying the stain's coordinate position, width and length, and the gamma angle (directional angle). Once entered for each stain, the program provided the analyst with a top view of the stains and associated virtual paths (the XY plane). The analyst then chose those stains that converged. The chosen lines of convergence provided the computer with an averaged CPx and CPy, a top view convergence point in the XY plane.

The analyst then moved to the side view and, again considering the paths, chose an appropriate convergence based on the position of the strings in the side view. Once chosen, the analyst saved these convergence points, and the software created a graphical representation of the probable Area of Origin.

Subsequent to *Backtrack/Strings®*, Dr. Carter produced two additional programs that worked in conjunction with each other. The first was *Backtrack Images®*.[8] *Images®* used photographs of stains, allowing the analyst to identify the gamma and impact angle for each stain on the computer.

The stains in question were photographed with a digital camera. Two pieces of information were included in the photograph. First, the analyst would draw a plumb line onto the surface, in order that *Images®* could establish a gamma angle. The analyst also placed a small readable reference scale, which *Images®* used to determine actual scale of the photograph. See Figure 6.18. Additionally, in terms of its XYZ position on the surface, each stain had to be located. Even when presented with a large pattern, where the number of stains chosen for evaluation might be 20 or more, the entire process typically took under a half-hour.

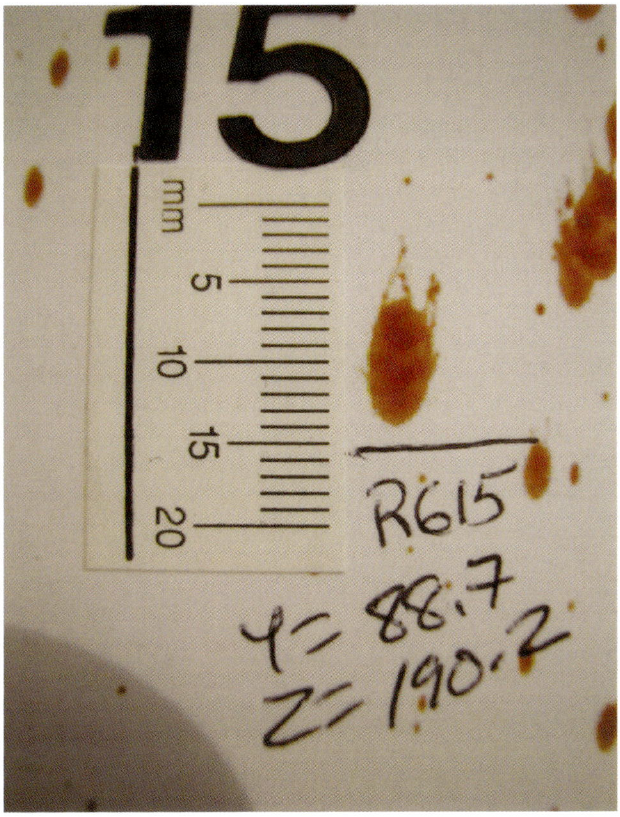

Figure 6.18 Using software such as Backtrack Images, the analyst captures certain on-scene information. For each stain, a small scale is included, as well as the location of the stain (YZ in this instance) to allow the software to accurately place the stain in the scene and establish its measurements. The analyst then draws a plumb line on the scale adjacent to the stain, which allows the software to identify the gamma (directional) angle.

The *Images*® program allowed the analyst to compare the image to a virtual ellipse on the computer screen. In terms of precision, ellipse comparison is one of the most effective methods to get precise lengths and widths of the associated stains. Once completed, the software determined the impact angle from the ellipse measurements. The information captured in *Images* was used in conjunction with a second program, *Backtrack/Win*®, which then effectively calculated the area of origin.

Backtrack/Win[9] could accept input from either Images® or information gathered directly by the analyst. Using a similar routine as found in previous Backtrack® products, it determined the various convergence points and Area of Origin. Figure 6.19 is a printout of a top view created by Backtrack/Win®. The small cross represents the established convergence point for the top view. Figure 6.20 is a side view printout of the same stain pattern. Once again, the small cross represents the side view convergence point and with both convergence calculations complete, the display lists the Area of Origin in the upper right corner. Although the Backtrack/Win® program is no longer available from FCO, it is entirely possible some of these programs may still be in use.

Another virtual stringing program, HemoSpat, followed FCO's software. Backtrack/Win® was somewhat clumsy in terms of operation, the process was not intuitive and did

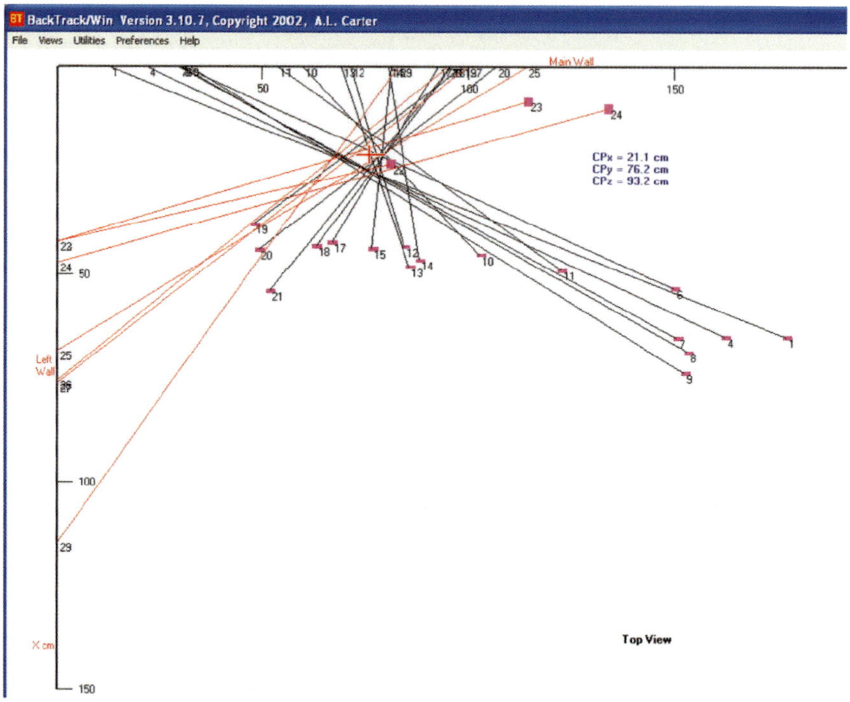

Figure 6.19 A Backtrack/Win® top view screen. This view identifies the overhead convergence in the XY plane for a group of spatter stains. The small cross is the mathematical mean for all of the points of convergence. The calculated position for the various convergences is annotated in the text at the top right.

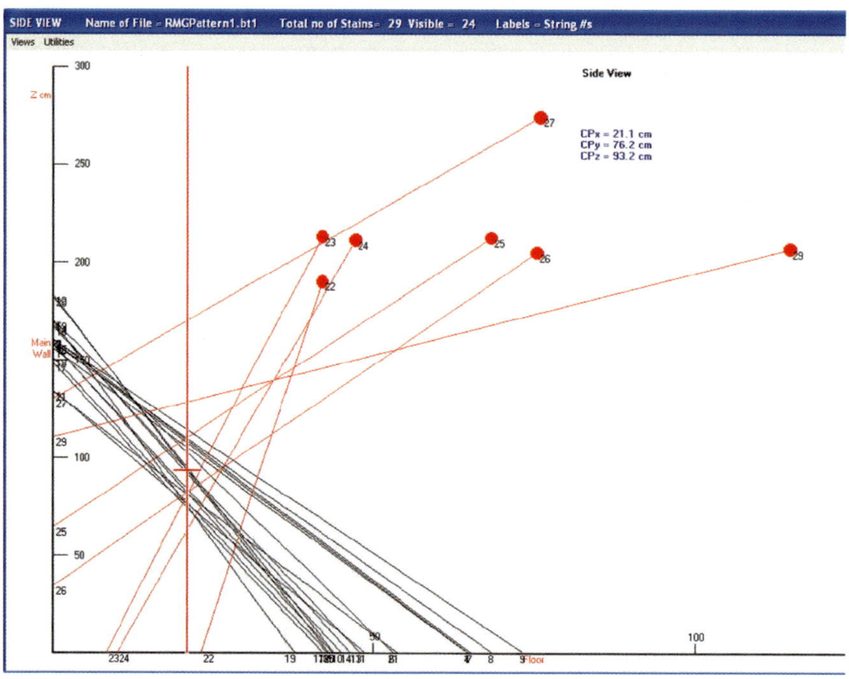

Figure 6.20 A side-view screen from Backtrack/Win®. In this view, the small line that crosses the extended line identifies the convergence of the side view (the ZX plane).

not offer a graphical user interface approach. This required a certain level of experience to move through the required steps of the operation effectively. HemoSpat developed a more intuitive approach. The various icons to conduct the analysis are presented in the order of the operation and the program offers an effective graphical user interface. As a result, the learning curve for using the software is very low. HemoSpat also provides an automatic ellipse matching process that is quite effective. In easily 70% of the stains examined, the computer produces an ellipse for comparison, using the image alone without input by the analyst. This automatic ellipse is often precise, requiring minimal subsequent adjustments by the user. Even when the program has difficulty with a particular stain photo, the ellipse adjustment controls allow the user to make such adjustments. The analyst using similar controls can also adjust the directional angle, if necessary.

Like Backtrack/Win, HemoSpat relies on photographs of stains. Small scales are introduced by the stain, a plumb line is drawn, and the photograph is taken. Photographs are moved into the program through a "drag and drop" approach. Once imported, the analyst enters the physical location of the stain into the database. Stains can be added, or removed from a group, and groups of stains can be split if needed. The program then produces a graphical representation of the convergences (top, side, and front views). See Figure 6.21.

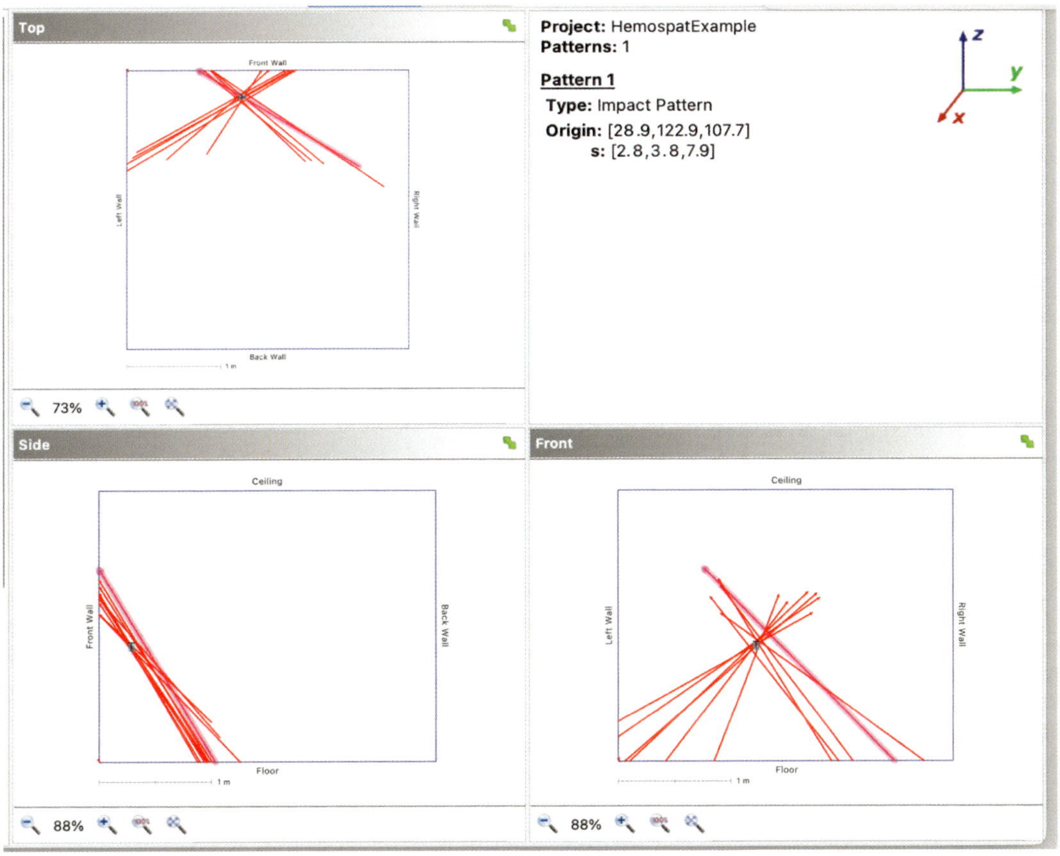

Figure 6.21 A screen capture from a HemoSpat virtual Area of Origin analysis. It provides three views of the pattern convergence, top, side and front, as well reporting the defined Area of Origin in the upper right quadrant.

Figure 6.22 A screen capture from HemoVision. The large target is placed within the area of interest. Additional smaller targets are placed on any associated surface that hold stains. Targets do not have to be fixed (e.g., position measured by the analyst). The software manages all such measurements independently. (Courtesy of Philip Joris and HemoVision.)

Another virtual AO program now available is HemoVision, the product of a Ph.D. study at KU Leven University. Unlike previous systems, this software incorporates advances in digital photogrammetry, effectively eliminating much of the physical measurements required by prior systems. The process begins with the placement of markers in the area of interest, which the software uses to integrate various input photographs. A single larger pattern marker is positioned in the area of the pattern and measured (its vertical distance from the ground). Smaller markers are then inserted in additional areas within the pattern. These require no measurements as to placement. See Figure 6.22.

The user then captures digital photographs of the entire pattern, or at least a large portion of it, from different perspectives. The software can utilize any digital image, including those produced by smart phones. The software corrects for any angular perspective created by the user when taking the photograph. Thus, hand-held photographs are fine, without the need for tripod photography, which can of course reduce the time required.

All of the overall photographs are loaded into the software, which through integration produces a reconstructed pattern. The user can include images of the pattern from any associated surface facing the camera, regardless of their relative scene orientation. From this, a complete reconstructed pattern is created. Detail images are then loaded in the software, and are automatically integrated into the reconstructed pattern. Once complete, any stain can be evaluated simply by clicking on it. The software provides the impact and gamma angles, as well as the location of the stain and of course a representation of the AO. Like HemoSpat, stains are evaluated in the software as groups, the user adding or removing stains as needed.

The software produces various outputs, including pdf reporting, augmented reality, or movie exports. An additional function is the ability to overlay the calculated area of origin over some other scene photographs, providing the viewer (e.g., the jury) with a clear scene context. See Figure 6.23.

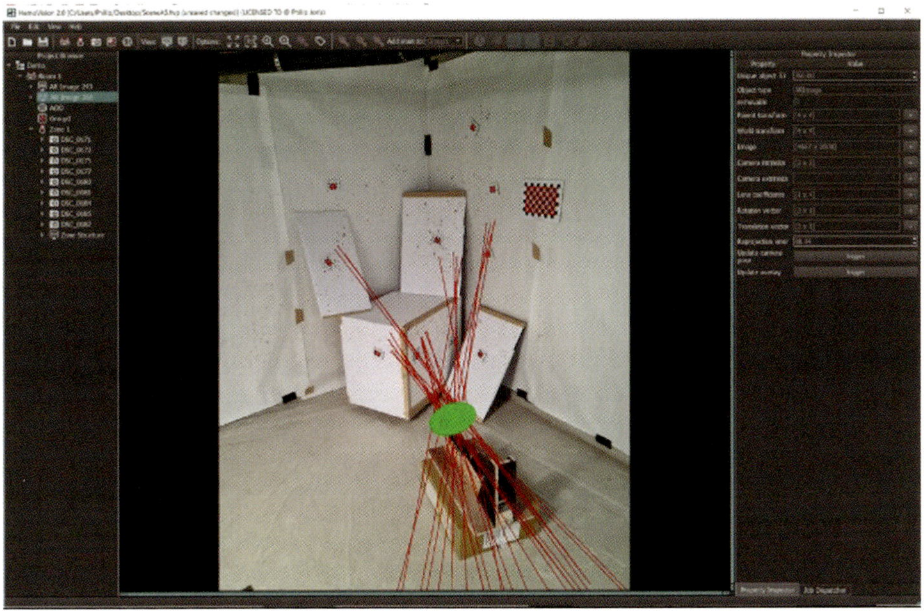

Figure 6.23 A screen capture from HemoVision. Using all of the photographs of the pattern, regardless of their orientation or surface, HemoVision calculates individual trajectories and an Area of Origin for the pattern. (Courtesy of Philip Joris and HemoVision.)

With the increased use of laser scanning systems for crime scene documentation, both Faro and Leica are developing AO routines within their software that also allow evaluation of impact patterns. Once fully validated, for users with such equipment they will functionally eliminate the need for stand-alone software.

How Many Stains Are Enough

Area of origin analysis requires the analyst to locate and choose several "well-formed" stains in the pattern, but how many stains are enough? This is a difficult question to answer outside of a specific context. The answer will depend in a great part on the nature of the pattern. Is the pattern radiating broadly or tightly? How many well-formed stains are present? Is there an indication of two impacts? All of these issues may require the analyst to conduct the AO analysis with fewer stains or suggest inclusion of a greater number of stains. It would certainly be inappropriate to evaluate two or three stains for AO, when other well-formed stains were present in the pattern. The authors would suggest that when present, at least 15–20 stains be included in any AO analysis. Presented with fewer stains (which is not an uncommon scenario), the analyst should recognize that the resulting conclusions are less certain.

Automation Efficiency or Precision, an Important Distinction

It should be evident that forensic software advances allow the analyst to more efficiently locate the Area of Origin of a given event. The computer handles the data and calculations in a more precise manner, eliminating human math errors and some of the subjectivity inherent in the stringing and stubby pencil routines. Certainly, a significant benefit of virtual stringing is the ellipse comparison techniques, which functionally eliminates error associated with manual measurement processes. What is not evident is whether a more efficient mathematical process means a more precise definition of the true Area of Origin. Other than providing more precise impact angle calculations, the authors would argue that it does not.

In a classic stringing technique, one usually limits the Area of Origin to the size of a basketball. Rarely has the need existed to define this location further, particularly by trying to define an absolute point. This limitation in precision really doesn't matter because the nature of the events that create spatter are themselves dynamic and probably cannot be pinpointed in 99% of all situations. They occur over a given area involved in the event.

Granted, in some instances the movement involved may only be a few inches, but that is distinct when compared to a precise point in space. Events rarely, if ever, occur at a precise XYZ coordinate point. Additionally, one must consider that wounds are often jagged and are certainly not defined by precise points. Thus, trying to define an absolute XYZ coordinate as an Area of Origin is really a moot issue. For the analyst to state that an event occurred at a precise point in space is almost ludicrous.

Unlike the old stringing technique, however, the computer provides us with a very precise point in three-dimensional space. Some software programs graphically depict this point and then give an absolute XYZ point in text. Having failed to read the software documentation and lacking an understanding of basic blood droplet dynamics, someone evaluating this information could easily assume the computer had narrowed the event down to this absolute location. As analysts, we recognize these points, as idealized and based on the averages contained in the data. They do not place the event in an absolute location with absolute precision.

Can one more easily compute Area of Origin using computers? Yes. Are computers more absolute in their accuracy of defining the event? No. Mere placement of a decimal point in a mathematical calculation will not eliminate the level of uncertainty that will always exist in these situations.

Limitations in Area of Origin Evaluations

Using straight-line geometry and trigonometric functions allows the analyst to identify the probable area of origin for the stains. Although generally accurate, it should be evident to the reader that all the methods make a major assumption. All assume the droplets involved travel in straight-line trajectories. Unfortunately, this isn't always true, as both gravity and air resistance can affect the droplet, creating distinct arched parabolas.

If the analyst recognizes and understands this inherent limitation, the various methods are functional and useful. They do not define the droplet's flight paths absolutely. They do, however, place specific limits on that path. Consider Figure 6.24. The impact angle for each droplet, in effect, sets a limit of possible and impossible flight paths.

A critical consideration in area of origin determinations is the difference between evaluating horizontal and vertical targets. We raised this issue earlier in the chapter and it is imperative that we fully explain these concerns. When looking at a spatter droplet's path from a top view on a horizontal surface, gravity and air resistance have no affect on its

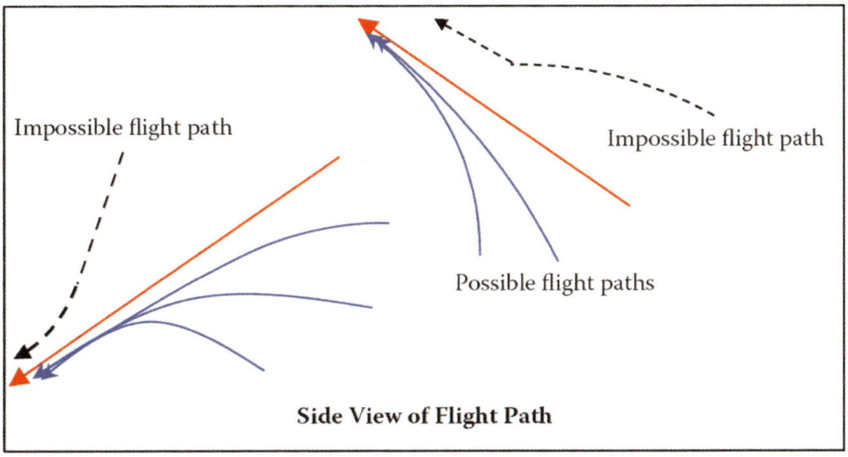

Figure 6.24 The angle of impact establishes an absolute limit for the droplet's origin. For instance, a droplet can originate at some point under the angle of impact. As it follows its trajectory, gravity alters the path making it steeper and causing the droplet to strike at the angle indicated by the stain. What cannot happen is for the droplet to alter its path on its own, defying gravity, thus striking the target from a point above the indicated angle of impact.

directionality. The top view of a pattern on a vertical surface is quite different. On a vertical surface, a top view of the pattern is the vertical plane and both gravity and air resistance do affect the flight path and the resulting directional angle.

Unfortunately, many impact patterns are found on associated vertical surfaces. There are three general situations one might encounter in a vertical surface target. The droplets may still be moving with the acceleration produced by the actual impact and thus traveling in generally straight lines. The droplets may be losing acceleration, or the droplets may have lost their initial acceleration and are falling out of their original flight vectors.

Figure 6.25 demonstrates the first circumstance. In this instance, the droplets are still moving with their initial acceleration and their flight paths are straight. When they impact the vertical surface, both the impact angle and directional angle and thus the area of origin defined by this information is a good representation of their origin.

Figure 6.26 demonstrates the second circumstance. In this instance, the droplets are losing their initial acceleration as a result of the effect of both gravity and air resistance. The flight paths are affected by this and become arced. The information defined by the resulting stains' directionality and impact angle still leads one back to an area. But because of the change in the flight path, the strings and actual flight path are beginning to diverge.

Figure 6.27 demonstrates the worst possible situation. In this instance the droplets have lost their initial acceleration from the impact. They are closing in on their terminal velocity and are now following steep downward paths. The resulting strings suggested by the stains will reflect this path alteration. The stains will have directional angles showing they were moving toward the floor and the impact angles will reflect sharp angles into the wall. The divergence between the virtual strings and the actual flight paths is significant. Any attempt at defining the area of origin with such stains will suggest an origin near the ceiling.

The more gravity and air resistance pull the droplets from their flight path, the more this effect. This alteration of the impact and gamma angle on vertical surfaces will effectively move the indicated area of origin inward and upward. It is important to note that

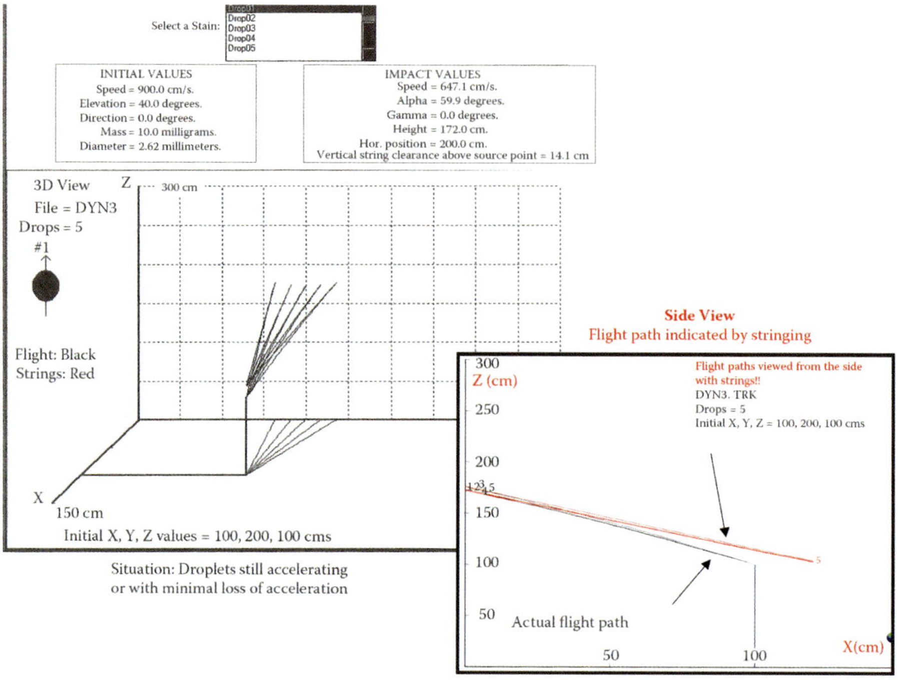

Figure 6.25 In dealing with a vertical target and the effect of gravity and air resistance, the best-case scenario is one in which the droplets have not lost acceleration. In this instance, the strings are a good representation of the actual flight paths.

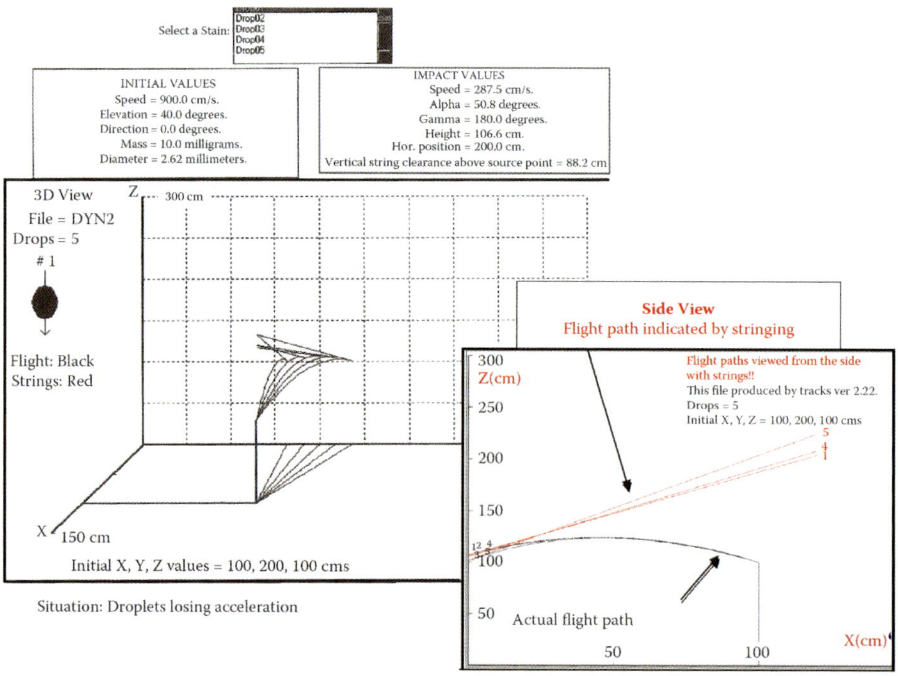

Figure 6.26 In dealing with a vertical target and the effect of gravity and air resistance, it is also possible the droplets will have lost some acceleration. As they do, the difference between the actual flight path and strings will begin to increase.

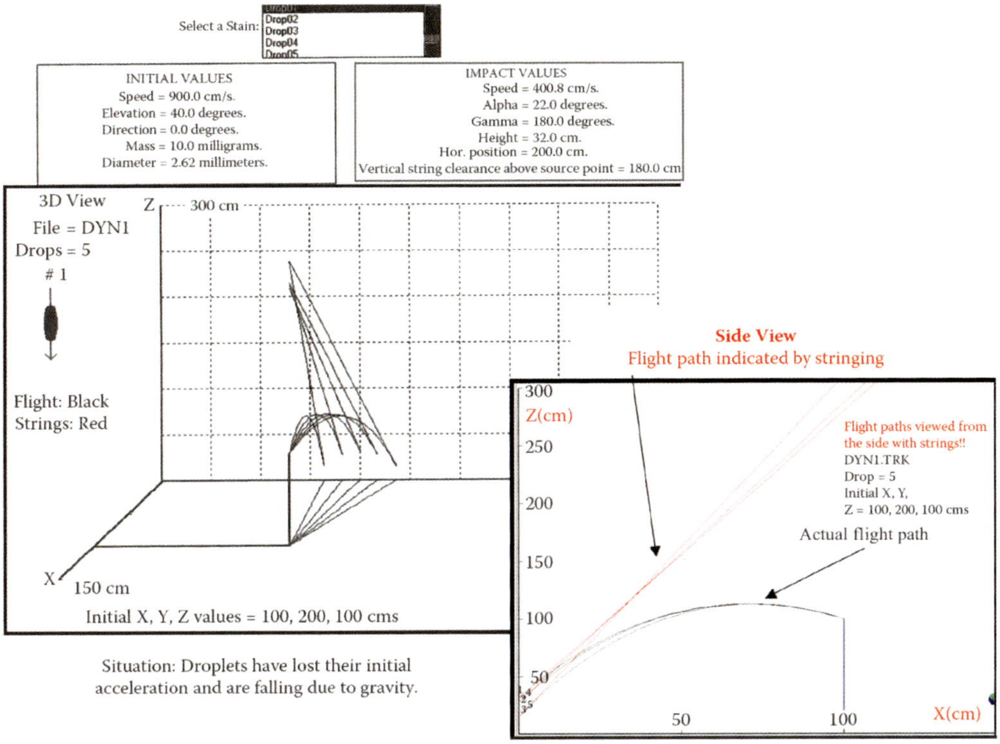

Figure 6.27 In the worst case on a vertical target, the drops have lost acceleration and are falling with gravity. The strings suggested by the impact angles will literally point to the ceiling.

not every pattern demonstrating downward moving droplets is affected by this issue. Is it possible for a pattern to be directed downward onto a vertical surface near that surface? Of course, and the droplets would still have acceleration and effectively straight-line flight paths. The stains would still provide functional directional and impact angle information. See Figure 6.28. Carefully considering the radiating effect of the pattern may allow the analyst to recognize the difference between these two distinct situations.

Even in the worst situation, the information reflected in the area of origin analysis still offers some information. If we apply the umbrella effect, any area of origin evaluation sets parameters of possible and impossible flight paths. For example, a droplet traveling downward toward the wall at an angle of 60° may have originated at some point behind the indicated angle of impact. That is, it may have been projected at 60° degrees toward the wall. As the droplet slows in flight, its flight path begins to angle downward, eventually striking the wall at the more acute angle of 40°. What cannot happen, however, is for the droplet to gain momentum on its own. Having been projected toward the target at 60°, the droplet cannot strike at a less acute angle (e.g., 70°).

The limitations involved in vertical surface AO evaluations demand caution in considering the results. The indicated Area of Origin sets an absolute upper limit for the event based on the umbrella effect (refer back to Figure 6.24), but it cannot always exclude the event from having occurred at a point beneath the indicated Area of Origin. The ability to make an exclusion is situationally driven, depending upon the number of stains available and the nature and dispersion of the stains.

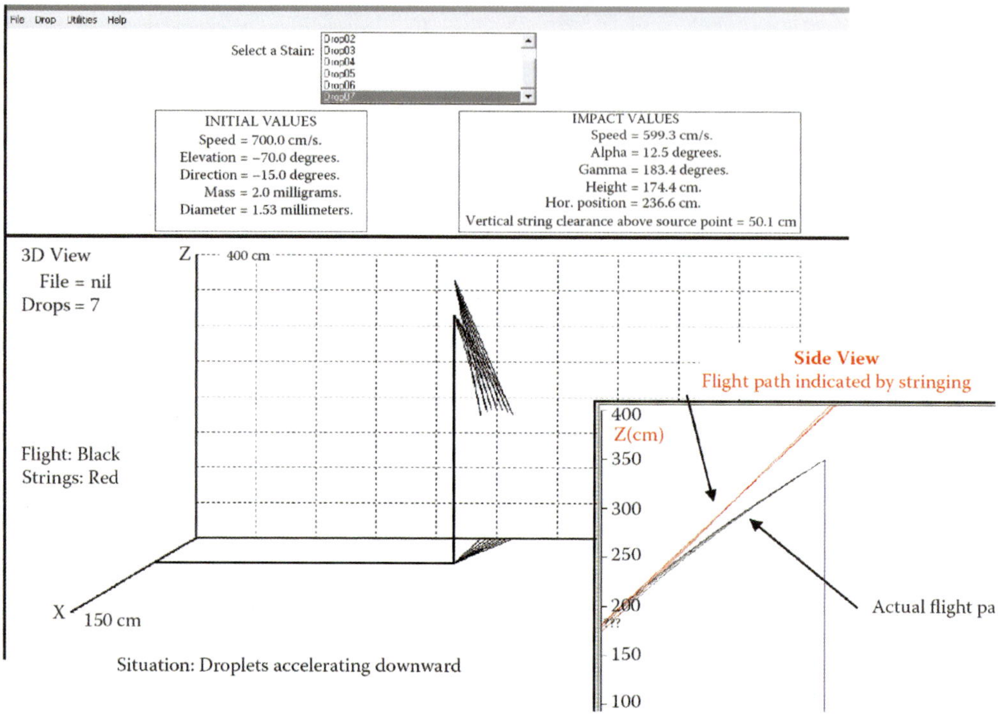

Figure 6.28 The fact that drops are moving downward on a vertical target in and of itself does not establish that they are of no value. It is possible that a downward ejection will occur in which the droplets are still moving with their initial acceleration. Recognizing this situation, as compared to drops falling out of their parabola, demands consideration of the radiating effect in the pattern.

Summary

Having defined the point at which a group of stains converge or originate from, how does the analyst use this information? Lacking specific knowledge of the crime, this process helps identify the general location at which some type of impact occurred. This is particularly useful in placing the subject or victim at specific locations within the scene at specific moments of bloodshed. By differentiating two distinct points of origin from what appears to be a single spatter pattern, it also identifies an additional impact that was not otherwise evident.

This information is used to refute or corroborate claims by those involved in the incident. Imagine a situation in which a subject claims self defense and that all blows delivered to the victim were struck as the victim made a standing attack against the subject. In evaluating the Area of Origin, let us assume we find evidence of blows delivered from a point low on the ground. Finding this, we have physical proof with which to dismiss, at least in part, the subject's statement.

In the past, analysts generally used the stringing or graphing methods to identify this area. Automation advances now allow one to complete the analysis in a more refined and efficient fashion. Even with such advancements, the analyst still recognizes that any Area of Origin estimation is exactly that *an estimation*. It establishes what is impossible and what is possible, as it relates to the origin of a given impact pattern.

Notes

1. Carter, Alfred, BackTrack/Strings User Manual, Forensic Computing of Ottawa, 1992, p. 17.
2. Wolson, Toby, Documentation of Bloodstain Patterns, presentation to the Int. Assoc. Bloodstain Pattern Analysts (IABPA), Colorado Springs, CO, 25 Sept 1992.
3. MacDonell, Herbert L., *Bloodstain Pattern Interpretation,* Laboratory of Forensic Science, Corning, NY, 1982, pp. 44–53.
4. Balthazard, Victor, Piedelievre, R., Desoille, Henri, and Derobert, L., *Etude Des Gouttes De Sang Projecte,* XXII Congres De Medicine Legale, Paris, France, June 1939, p. 25.
5. MacDonell, Herbert L., Balthazard Was Great But He Didn't String Us Along, *International Association of Bloodstain Pattern Analysts,* Volume 11, Number 1, March 1995, p. 10.
6. Registered Trademark of Forensic Computing of Ottawa, Ottawa, Canada.
7. Registered Trademark of Forensic Computing of Ottawa, Ottawa, Canada.
8. Registered Trademark of Forensic Computing of Ottawa, Ottawa, Canada.
9. Registered Trademark of Forensic Computing of Ottawa, Ottawa, Canada.

Considerations When Evaluating Impact Patterns

7

Impact patterns offer significant information in many bloodstained scenes. Area of Origin as discussed in the previous chapter allows the analyst to isolate, under good conditions, where the source of these patterns was in three dimensions. Even when this is not possible, an impact pattern often allows the analyst to generally set parameters of what is possible and what is not possible given the stains present. Impact patterns often help us understand the orientation of the event to the scene.

Methods of Description

As described in the Pattern Diversity chapter, the first aspect of any effort is in recognizing the pattern and assigning it to some particular classification. Impact patterns consist of only circles and ellipses that radiate outward. On any given surface the further out in the pattern, the more elliptical the stains will be. As far as the authors are concerned, impact spatter is impact spatter. Mist is a specific form of impact that when present may allow a more refined understanding of the potential source event, yet it is still just an impact pattern. But there are still mindsets employed when labeling stains that one may encounter. Thus, it is important to explain these ideas. For a number of years spatter were categorized based on the alleged velocities associated with their creation. Bloodstains routinely became categorized into three velocity groupings. These categories refer to the nature of the impact causing the stain, not the speed of the escaping spatter. The lawyers and scientists, who authored the NAS Strengthening Forensic Science report in 2009, seemed oblivious to this basic mindset claiming in that report: "One can tell if blood is spattered quickly or slowly."[1,2] This simply is not the case, velocity associations were associated with the force directed at the blood mass, not the speed of the droplets. These velocity spatter classifications were known as low-, medium-, and high-velocity impact spatter, often abbreviated LVS, MVIS, and HVIS, respectively.[3]

Low-velocity spatter were those patterns we now associate with drip, drip trail and drip patterns. The only force applied to the blood to produce the spatter stain is the force of gravity, which produces the larger drops. The resulting stains were generally considered to be 4 mm or larger in diameter.

Medium-velocity impact spatter (MVIS) were patterns in which the preponderant stain size was generally 1–4 mm in diameter, created as a result of some application of force.[4] This type of spatter would now simply be referred to as Impact Spatter. Historically, they were reported as resulting from a force of up to 25 ft/s and were often associated with blunt force situations. The droplets produced are smaller than that found in low-velocity events, thus the resulting stains are smaller as well. This increased breakup (the smaller drops) is due to the additional force applied in the event.

High-velocity impact spatter (HVIS) were patterns in which the preponderant stain size was reported as 1 mm or less in diameter. These stains were reported as being

 DOI: 10.4324/9781003408734-7

associated with a force in excess of 100 ft/s.[5] High-velocity patterns often include smaller submillimeter-sized stains as well; descriptively, such stains were referred to as mist-like (see Figure 9.3). The mist stain has effectively replaced this term, but with clear and important modifications.

There were several problems associated with the "velocity" method of description, leading to it being discarded in mainstream BPA. The first relates to the reported "velocity" figures. The figures of 5′ per second, 25′ per second, and 100′ per second and their association with a specific size of spatter have never been fully explained, nor has the foundation of such association been defined. These associations were effectively anecdotal. The reported velocities also left obvious gaps (e.g., what happens between 25′ per second and 100′ per second). A second more problematic issue becomes evident when discussing stains among analysts and lawyers. Quite often, when using a velocity label as a description, others assume it to mean a definition of the event. That is, by classifying something as a high-velocity incident, they believe the analyst means the underlying event must be a gunshot. This is not a valid association. A third issue related to velocity labels is that many analysts who use these terms tend to set absolute ranges of spatter size for the so-called low-, medium-, and high-velocity spatter patterns. Thus, a medium-velocity pattern will consist of 1–4-mm-sized stains, high velocity consists of stains 1 mm and smaller. Analysts generally quote the work of MacDonell and Bialousz as a basis for such ranges. In fact, there are no absolute size ranges evident in the writings of either MacDonell or Bialousz. Nor is it clear if they intended that absolute parameters were in force based on their experiments.

It is also evident that MacDonell and Bialousz never intended to assign event status to "medium velocity" or "high velocity" labels. MacDonell and Bialousz simply provided a basic method of definition or description for spatter patterns, and generally associated certain events with the patterns. Of some note, when discussing the differences associated with HVIS and the current Mist term, the range of spatter size initially defined by MacDonell (0.1 mm) for HVIS, was discarded at some point and replaced with the 1 mm. But 1-mm-sized stains occur across the spectrum of what would be considered MVIS and HVIS. To add more ambiguity, if one were to evaluate and sample the typical MVIS pattern, the analyst would find there are literally hundreds of very small stain (0.15–1 mm) present and in far greater numbers than the 1–4-mm sized stains. Unfortunately, the larger stains tend to capture the eye and the smaller stains were ignored. Thus, the defined ranges are not particularly valid in and of themselves. What remains evident in MacDonell and Bialousz's work was the correlation that as we increase the amount of force applied to the blood source, we see a corresponding decrease in spatter size.

An alternative to the velocity label approach was the descriptive label. As a result of their observations, Parker et al. chose to refer to spatter patterns based on a descriptive label. Their intent was to describe the stain based purely on the physical characteristics the analyst observed. Parker et al. classified these groupings as either "fine" or "coarse."[6] Their fine category included spatter with diameters of 3 mm or less, while coarse spatter were those stains with diameters of over 3 mm.[7] The approach was not radically different from MacDonell's; it simply didn't interject any association to the underlying event.

Another advocate of a descriptive approach was Judy Bunker. Ms. Bunker did not adopt Parker's terminology, but she taught students to identify all patterns present, documenting each with a detailed physical description based on observable characteristics. The purely descriptive approach has merit. Bunker's intent was for the analyst to first observe the stain and make conclusions about it later.[8]

There are some general statements regarding source events and the associated spatter. Larger spatter are typical of situations in which blunt trauma is the cause of the bleeding injuries, but not singularly. Very small spatter are more typical of gunshot injuries, explosive force or high-speed machinery but once again not singularly.

The problem with these generalizations lie in the variables that define spatter size. There are two primary variables that define the size of the resulting spatter. The nature of the force of disruption is but one. The second is the volume present in the blood mass being broken up. In the first, as we increase force, we see a decrease in spatter size. In the second, as we decrease the volume, we see a decrease in spatter size. Both variables are unknowns at the crime scene, and they tend to create contradictions associated with the generalizations. There can be a low force situation that produces very fine spatter based on a small volume at the point of disruption; just as there may be high force circumstance involving a large volume that will produce much larger spatter. Thus, a blunt force blow directed to a victim in the beginning of an attack (when less blood is present to be disrupted) may produce spatter that are radically different from spatter created when subsequent blows land and more blood is exposed.

It should be evident, the MVIS/HVIS mindset was discarded for all these reasons. Although it and other descriptive methods played a role in understanding and describing spatter patterns, in the end these methods presented too many contradictions and ambiguities.

Preponderant Stain Size versus Range of Stain Size

Unfortunately, within any given pattern, there may be a wide range of individual spatter sizes and literally hundreds, if not thousands, of individual spatter present. It is physically impossible to accurately sample the pattern in these instances and provide a quantitative result.

In the past, particularly when using the MVIS/HVIS mindset, one would simply try and define the preponderant size of the stains present. Imagine a pattern of 15 individual spatter stains with the following sizes:

Individual Stain Size	No. of Stains
0.25–1 mm	3
1.1–4 mm	10
4.1–6 mm	2

If the analyst chose to use the smallest stain (e.g., 0.25 mm), the pattern might be categorized as fine, mist, or high velocity. On the other hand, if the analyst chose to use the largest stain present (e.g., 6 mm), the stain might be categorized as coarse or low velocity.

To properly understand a pattern, the analyst looked at the preponderant individual stain size (using a visual approach) within the pattern. In the example, that happens to be of 1–4 mm size (i.e., 66% of the pattern fits this range). Using a velocity label, that would identify the pattern as "medium velocity."

The problem associated with defining preponderant stain size is that it tends to be subjective. When deciding the preponderant stain size, analysts use an eyeball method. They simply evaluate the stain and pick the size that appears to be the most prevalent. As indicated larger stains tend to jump out at the analyst and smaller stains tend to be overlooked.

The immediate solution to this issue is simply to define the range of spatter present in the pattern. The analyst locates those stains that appear the smallest and measures them, then locates the larger stains and provides this range in the description.

Impact Droplet Size

As previously discussed, the size of a spatter stain is a basic consideration, which may at least point us in the general direction of defining the nature of the force involved in creating the stain. For drops projected into the scene, the volume of blood present in a droplet is the primary determining factor for the subsequent stain size. What forces are required to cause subdivision of a blood source to create droplets? As MacDonell, Parker, and most other researchers discovered, blood and stable blood drops do not spontaneously degenerate from one volume to another.

It is the energy transferred by an impact to the mass of blood that ultimately causes the blood to break up at the source. The blood mass is ejected along any avenue of escape, as it cannot be compressed. As it does this, it follows a general sequence of becoming sheet-like structures, that devolve into ligaments, which then break up into smaller droplets. These smaller masses of blood stabilize through the effect of surface tension creating individual stable drops. Once these masses have undergone this ejection and become stable droplets, no further force beyond gravity and air resistance are present to affect them. To further divide and become smaller droplets, some additional force must act on them. Air resistance and gravity certainly affect the flight path of these droplets, but will not break stable drops up into smaller droplets.

Once set in motion by an impact, no spontaneous subdivision will occur as it is infrequent droplets will collide in flight.[9] The force of impact and the volume present define the drop size which establishes the stain size we ultimately see.

As one might expect, based on the size of impact spatter stains, impacts result in significantly smaller droplets than those resulting from drips and similar actions. We have found impact droplet diameters of 2 mm and smaller.[10] Remember, this is the actual droplet in flight and not the resulting stains. Droplets of this size generally have volumes of 0.005 ml or less.

This issue of the droplet's size is significant in our consideration of impact spatter, for one reason. The smaller the droplet, the quicker it attains and holds a true spherical shape. As a result of the very small droplet diameters found in impact spatter events, considerations in determining impact angles and defining the area where the impact occurred are generally reliable. There is a confidence that the collapsing droplet was generally spherical in shape when it contacts a surface. At the source of the impact, we may see major droplet oscillations or unstable spine and sheet-like structures. These effects dissipate in less than 0.01 second. Away from the source of impact, the shape of droplets in flight are not distinctly affected by other extraneous forces (e.g., wind or air currents) and they travel as near-perfect spheres.

Pattern Configuration and Dispersion in Impacts

Patterns associated with impact spatter have a radiating effect (see Figure 7.1). The convergence of this radiating pattern, as one would imagine, is the point where the impact occurred to the blood source. This provides significant information to the analyst when

Figure 7.1 An example of the radiating effect observed in spatter patterns. The origin of the spatter is approximately 40 cm from the floor; note how the pattern expands outward the farther from this point it is deposited. This particular pattern was the result of a stomping death. (Photograph courtesy of Lt. William Gifford, Anchorage Police Department, Anchorage, AK.)

trying to isolate where the spatter event occurred. The analyst can use various aspects of this pattern to understand where the source of the pattern was. If a complete Area of Origin is possible, this is the best situation. But a complete AO analysis is not possible in every situation. Nor is it always necessary. The analyst can evaluate the pattern in several other ways, which include visual orientation, using the area of convergence alone, and dispersion characteristics of the pattern. Each provides at the very least some limited information indicating the location of the source event.

The nature of how the pattern and individual stains within the pattern manifest themselves is dependent upon the relationship of the target with regard to the impact site. For targets that lie parallel to the primary direction of force in the impact, we are likely to see elliptical stains within the pattern. On targets that lie perpendicular to the force, we will see the more spherical stains. Figure 7.2 provides a basic understanding of this relationship. But impact events tend to disperse droplets onto multiple surfaces that have varying orientations to one another. Impact patterns should be considered as a whole.

There will be situations where the analyst seeks only to understand a basic aspect of where the source event occurred. AO may not be significantly probative, but a general corroboration of some basic belief is helpful to the analyst. In such situations, visual recognition and correlation of the impact and directional angles on the varying surfaces will provide a relatively clear understanding of the general area of the source event.

There may be situations where the primary pattern is deposited on a substrate that affects the impact angle. This is particularly true of fabric surfaces. In these circumstances, defining impact angle is inappropriate, as will be discussed in the Fabrics chapter. Under such circumstances, if one cannot define impact angles effectively, then a complete AO analysis is impossible. But in some circumstances, such as tightly woven cotton in bedding or sheets, the directional angle is not affected. Given symmetrical stains, the long axis of

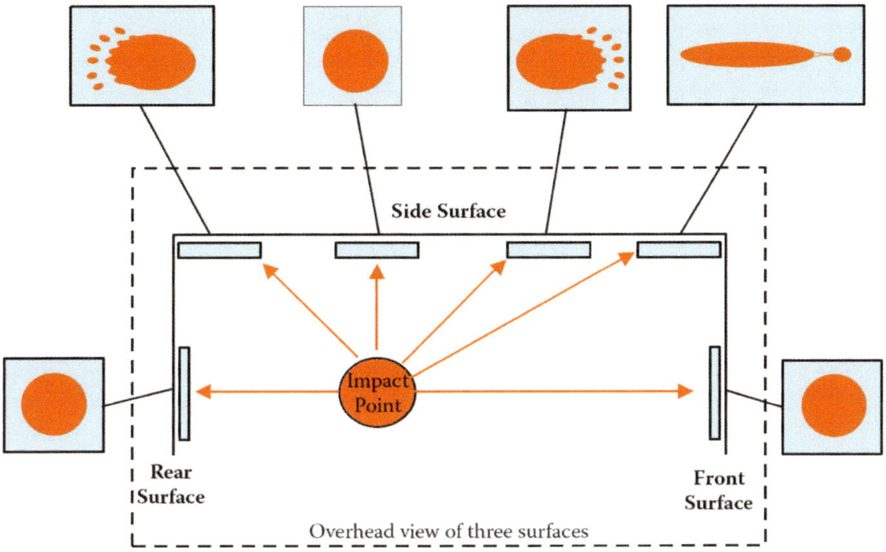

Figure 7.2 Stain shape and the relationship of the target to the impact. The stains we find in any given event may exist on any number of target surfaces. Thus, the position of the target in relation to the actual impact will produce any number of different impact angles. By considering the resulting stain shapes, the stain positions, and the overall pattern configuration, we can distinguish the probable impact area.

the stains can functionally be evaluated and a two-dimensional area of convergence isolated. This provides information relative to the pattern that may well be probative, even if the three-dimensional area of origin remains unknown.

A final consideration is dispersion characteristics. As impact patterns disperse and radiate outward, it should be evident that the closer a given target is to the source event, the more stains one will find in any given area (e.g., per square inch or per square foot). As the droplets travel outward from the impact site, the distance between the stains increase as they follow their individual radiating paths. Dispersion characteristics of the pattern can be helpful in deciding how close or distant it may have been to the stained surfaces. Once again, this information may be limited but in a specific context it may allow the analyst to refute or corroborate some belief about the event.

This is best understood if we limit our discussion to stains in a single dimension. Using Figure 7.3, imagine an impact creating droplets. These droplets travel along a single plane, as represented in the figure. Based on the radiating effect, the droplets passing the cross section labeled A are closer together than when passing section B.

If we replace the cross section with a vertical target and include a three-dimensional droplet pattern originating from the impact, the resulting stains will reflect this dispersion. Figures 7.4 and 7.5 both show examples of this effect. The two patterns, produced by similar gunshot events, were each at different distances from the event. As a result, the closer target (Figure 7.4) shows a level of dispersion lower than that evident in the distant target (Figure 7.5).

Always consider that this is not the only factor affecting the dispersion of droplets and their resulting stains. You will recall that air resistance has an inverse effect on droplets based on their size. Figure 7.6, a *Tracks* model, demonstrates how droplet size can distinctly affect the distance it travels. All the droplets in the model were projected at the same speed and along the same initial path. As a result of their size, they followed distinctly different

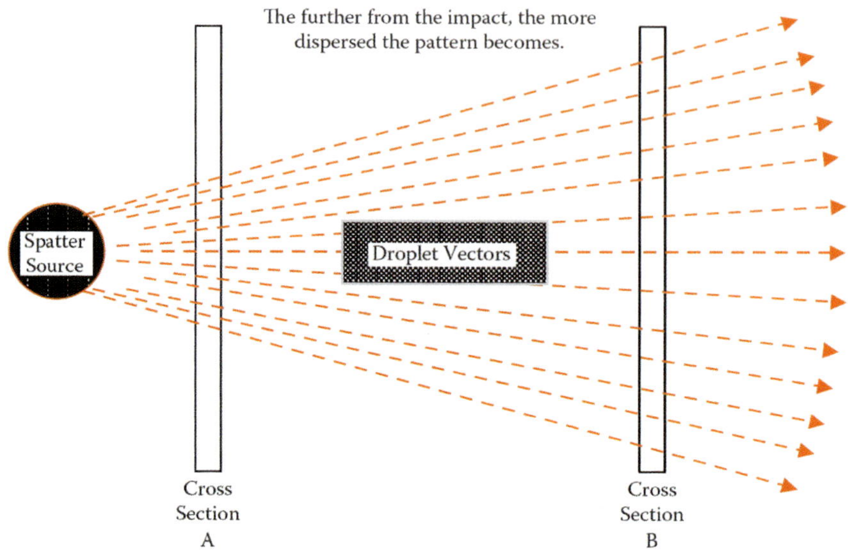

Figure 7.3 The radiating effect of a spatter pattern results in a higher level of dispersion in the stains farthest from the impact. The figure is idealized and does not consider the effects of air resistance or individual droplet parabola. These factors also play a role in the dispersion present in a pattern.

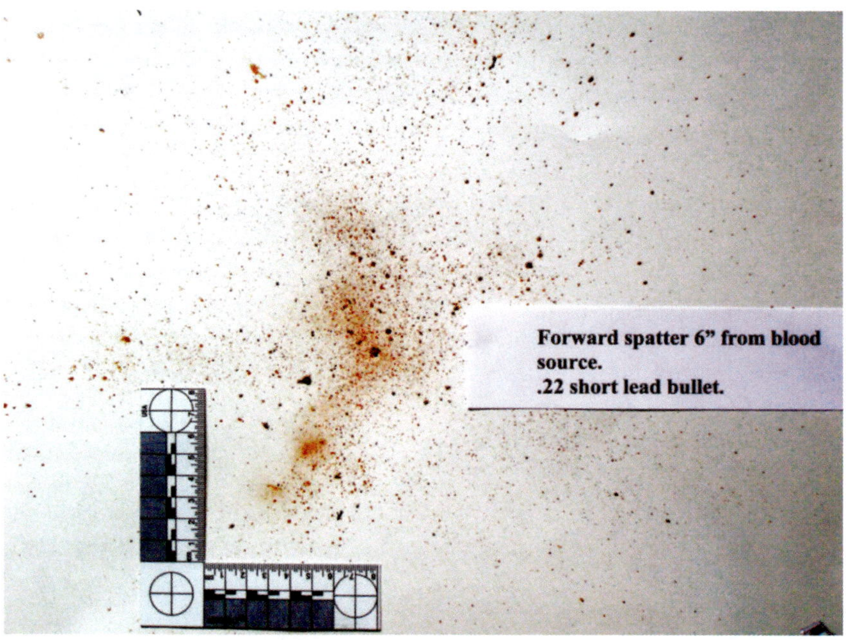

Figure 7.4 This pattern shows the level of dispersion in a gunshot event when the target was 6″ from the blood source. Note the misting effect and the high number of stains present in the central area of the pattern. Compare this to the dispersion in Figure 7.5.

Figure 7.5 This pattern shows the level of dispersion present in a gunshot event when the target was 18″ from the blood source. The pattern has a more dispersed nature with less of an evident misting effect.

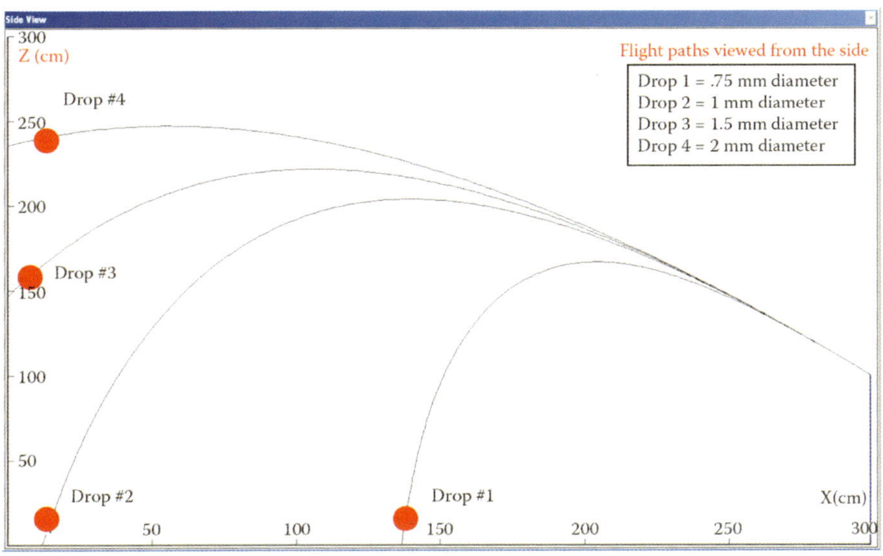

Figure 7.6 Tracks model showing four droplets projected along the same initial path at the same speed. Note that all four follow different trajectories, with the smallest droplet (Droplet #1) traveling the shortest distance. This is primarily the result of the effect of air resistance on the droplet, which is inversely proportional to the mass of the droplet.

flight paths. Smaller droplets simply do not carry as far as larger droplets. This issue is a consideration when viewing the dispersion of patterns, particularly those associated with gunshot events. Smaller stains will not travel any great distance.

In terms of the overall dispersion of the pattern many factors affect the overall pattern shape. The way the weapon contacts the impact site and the shape of the weapon are factors as well. These factors may limit how the blood ejects from the impact site. The angle at which the weapon strikes a surface, or the nature of the weapon's shape or configuration may preclude blood ejecting in a symmetrical manner. Impact spatter may eject in a very tight dispersion cone, in other situations the spatter may eject in wide dispersion cones.

The analyst should never accept, as it is not evident based on empirical data, that a given weapon will produce a singularly distinctive pattern each time it is used. Minor variations in the weapon's contact or application may result in significantly different effects when recreating the stain.

In some instances, this difference may provide the analyst with a better understanding of "how" the weapon was used. If the weapon creates a distinctive pattern when used one way and another pattern when used some other way, this may become evident during recreation attempts. If present, such evidence certainly points to the specific application of the weapon.

Dispersion characteristics also relate to our consideration of the presence of spatter on the subject. It is not an absolute fact that a subject involved in a spatter-producing event will have spatter on his person.[11] This is a common concern in court: How could a suspect have been in a scene where there was a significant spatter event and not be spattered themselves? Lack of spatter may result from being behind the radiating pattern, where the level of dispersion is often quite small. Spatter ejects along any avenue of escape, and often this is away from the attacker based on the application of the weapon from that direction. Additionally, the way the event takes place may limit the spatter dispersion in some manner, such as in the case of intervening clothing or bedding. Lack of spatter alone does not prove lack of presence or participation in the event. This idea is consistent with Sagan's well-known adage, "Absence of evidence is not evidence of absence."[12] In terms of this consideration, the analyst first considers where the spatter was distributed within the scene and associated surfaces. This generally will assist in understanding whether the dispersion cone was tight or broad. Then given consideration of "where" an attacker could have been, this may allow some decision on whether the attacker was or was not exposed to the spatter dispersion cone.

Dispersion characteristics of a pattern often provide the analyst with information regarding the general distance the spattered target was from the impact. They also verify that a given item was in proximity to the event in the first place. The pattern's configuration helps establish the relationship a given target had with regard to the impact. All of this, as Eckert and James explained, "[helps] determine the relative position of the assailant and victim at the time the blows were delivered."[13]

Figure 7.7 depicts the T-shirt taken off a murder victim. There are several significant spatter patterns present on the back of this T-shirt, two of the primary patterns are labeled B and C. Figure 7.8 shows the spatter present on the right back shoulder (labeled C). The pattern clearly shows a radiating effect to the right and the presence of a significant number of individual spatter stains. The further away from the origin, the more dispersed the pattern becomes. The pattern resulted from a blunt trauma beating with a bat like object. Figure 7.9 shows a pattern labeled B. Evident in and around the larger stains in this pattern are numerous very small stains. The larger stains were first thought to be spatter from a second victim, but DNA analysis verified these stains to be consistent with the victim wearing the T-shirt. When submitted to the lab, the lab chose to check the smaller stain

Figure 7.7 The back of a T-shirt from a murder victim. Stain Pattern C is a spatter pattern radiating out from the shoulder. Stain Pattern B is spatter and possible cast-off stains on the mid-back. Both stain patterns furnish information based on the dispersion characteristics.

overlaying the larger. These stains were determined to be consistent with the second victim's blood and establish that at the time of her beating, the T-shirt was in close proximity and oriented with the wearer's back to the event.

Spatter produced in stomping situations is rather complicated. Although boot stomps and similar events have not been studied in great depth, they generally produce our concept of the "complex" pattern. Such patterns have characteristics above and beyond the typical impact event, but significant spatter is often produced.

Boot stomps and similar impacts to a mass of blood already exposed in the scene will present a similar appearance as a typical impact. However, the total ejection from this event is quite different. Aspects of the blood distributed are "spatter" like; some of the blood is broken up into small droplets and widely dispersed in a radiating pattern. But due to the volume involved there will be streaming ejections, producing spine like structures. Such situations produce what is clearly a complex pattern, with characteristics consistent with both impact and gush/splash patterns. Blood volume, avenues of escape, the amount of pressure applied at a particular point, and the nature of the sole of a shoe all have some influence on the resulting pattern.

Note that in such instances spatter is ejected out and away from the stomping foot, thus when spatter is produced, we should expect the spatter to be on the non-involved shoe as well as any clothing covering the lower non-involved leg. We should not expect directional spatter on the stomping foot, unless both feet were involved. Another obvious

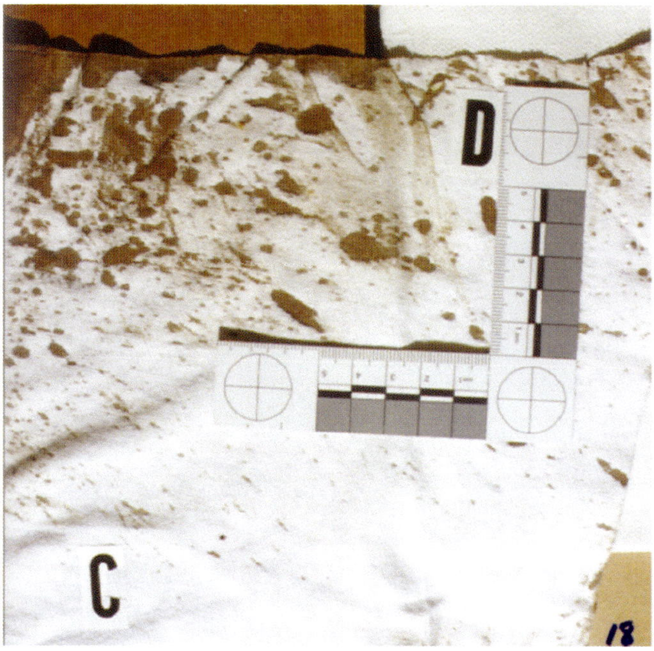

Figure 7.8 A closer view of the spatter associated with Pattern C in Figure 7.7. This pattern could be correlated to a scene pattern on surrounding objects based on the dispersion and the radiating effect, which effectively placed the victim in a general position when this event occurred.

Figure 7.9 A close-up photograph of Stain Pattern B from Figure 7.7. The small spatter intermixed in the larger stains was initially thought to be associated with a single event. DNA analysis established that the source of the smaller stains was the victim's wife, who was also beaten. Due to the small size of the stains and their orientation only on the mid-back of the T-shirt, this placed the T-shirt wearer in close proximity to his wife, with his back turned to her as the spatter-producing event occurred.

consideration is that if a shoe is involved in a stomping and producing spatter, that shoe will by necessity have been bloodied through contact.

Spatter Resulting from Gunshots

Gunshot Spatter: Forward Spatter and Back Spatter

Impact spatter often results from gunshot injuries. When a projectile strikes a target, sufficient force is transferred to the wound to create spatter, including mist, the fine aerosol spray of blood droplets. As MacDonell illustrated, the droplets resulting from such an impact radiate out in a three-dimensional pattern, as all impact patterns tend to do. This radiating effect is what we refer to as a dispersion cone.[14]

Spatter ejected from the exit wound follows a cone-shaped pattern in the general direction of the projectile, known as forward spatter. The entry wound also produces spatter, creating what is known as back spatter. See Figure 7.10. Obviously, forward spatter patterns only occur when there is an exit wound of some nature, while back spatter is always possible. Figures 7.11 and 7.12 graphically depict the development of the cone of forward and back spatter as a bullet passes through a bloody target. In addition to spatter, the force of the projectile may drive tissue and bone out of the wound as well.

Regarding the conical shape of both forward and back spatter patterns, consideration of dispersion holds true here also. The closer to the target, the more spatter we are likely to find. In spatter patterns found very close to the target, the lack of dispersion combined with the size of the droplets involved (e.g., the mist-like stains) may produce a fine spray-paint effect. See Figure 7.13. The area contained within the pattern will appear as if someone used a spray can to create it. Further away, the pattern becomes more dispersed with fewer mist-like stains, eliminating this effect. Refer back to Figure 7.4, which shows this mist-like pattern. In Figure 7.5, the stains are more dispersed as a result of being further from the blood source.

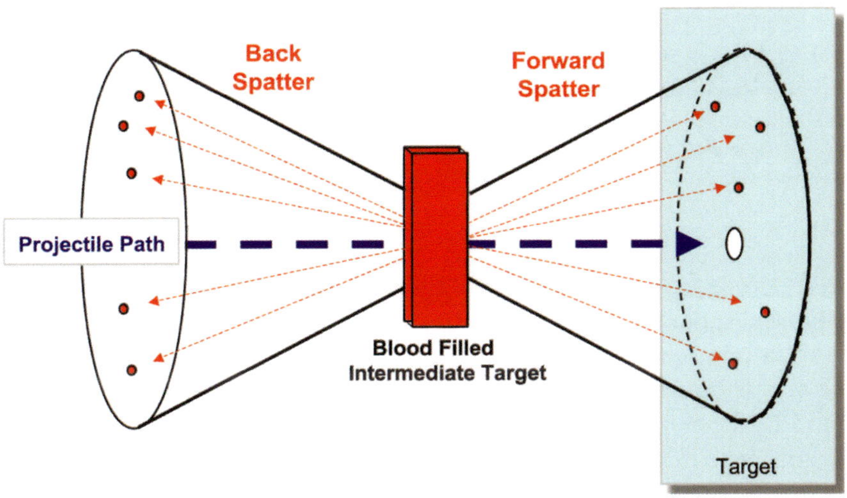

Figure 7.10 Spatter resulting from a perforating gunshot produces both forward and back spatter. The individual spatter project out from the source (e.g., the body) in a cone-shaped pattern, but generally do not travel very far because they are quite small.

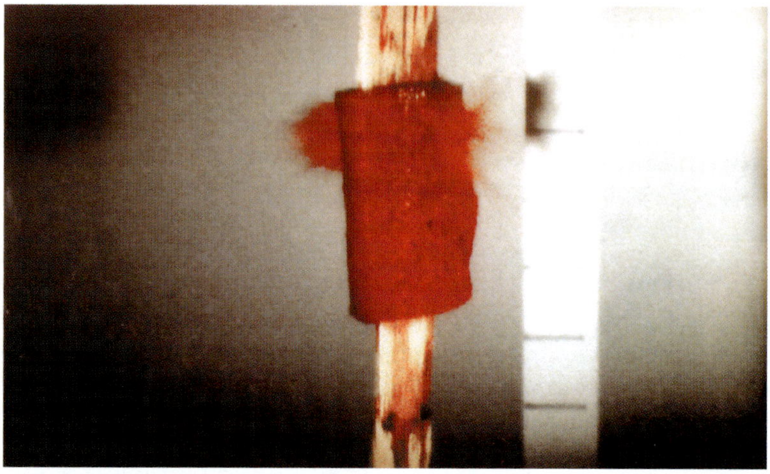

Figure 7.11 Forward and back spatter production. In this photograph, the development of the cone of spatter is evident. The bullet has passed through the sponge, moving right to left. Clouds of small droplets (both back and forward spatter) are ejected from the blood source.

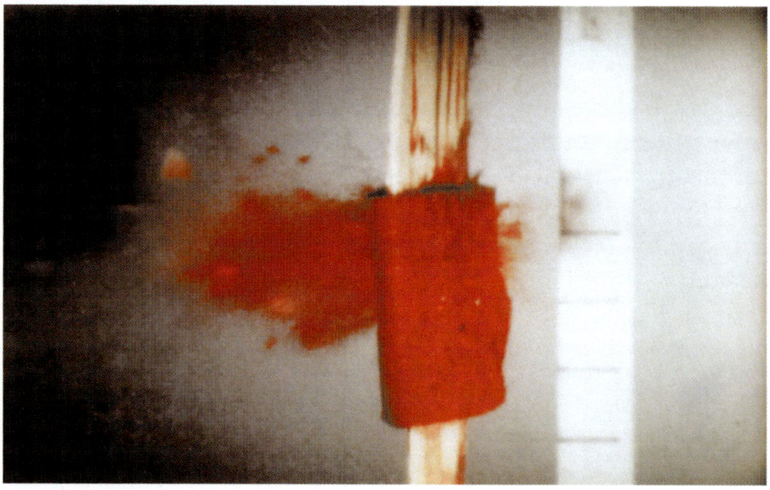

Figure 7.12 In this photo taken milliseconds later, the cone of forward spatter is well developed and extends out from the target. The cone of back spatter has peaked and is dispersing.

Figure 7.14 shows four frames from a high-speed video of a bullet striking a blood-filled sponge. In Frame A, the bullet has struck the sponge and a red-colored cloud is beginning to develop. In Frame B, the cloud is much more evident, yet in both Frames A and B the true nature of this cloud (e.g., the size of the involved droplets) is not really evident. In Frames C and D, the cloud has expanded and become more dispersed and the tiny nature of the blood droplets making up the cloud is more evident. We can literally see the mist. The stains created from this mist are referred to as atomized or misting stains, submillimeter-sized stains. Such stains are created from such small volumes, that no matter what their angle of impact, they will not demonstrate any ellipse and will appear generally circular. This is a result of their small Weber number, the surface tension of these small volumes

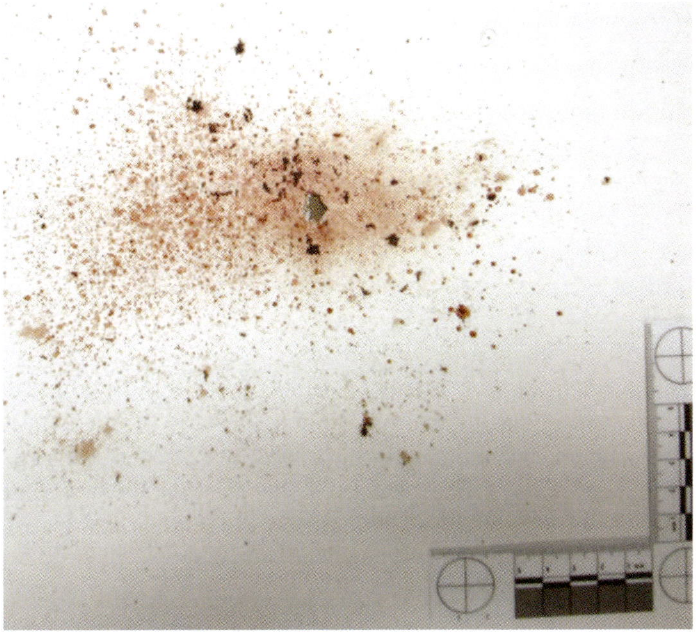

Figure 7.13 An example of a misting stain. Along with larger spatter, gunshot events often create very small stains (<0.1 mm). When the stains are close together, the result is a pattern that looks "spray painted" on. Based on the small mass of these submillimeter-sized stains, they cannot travel far and are usually found on objects close to the event.

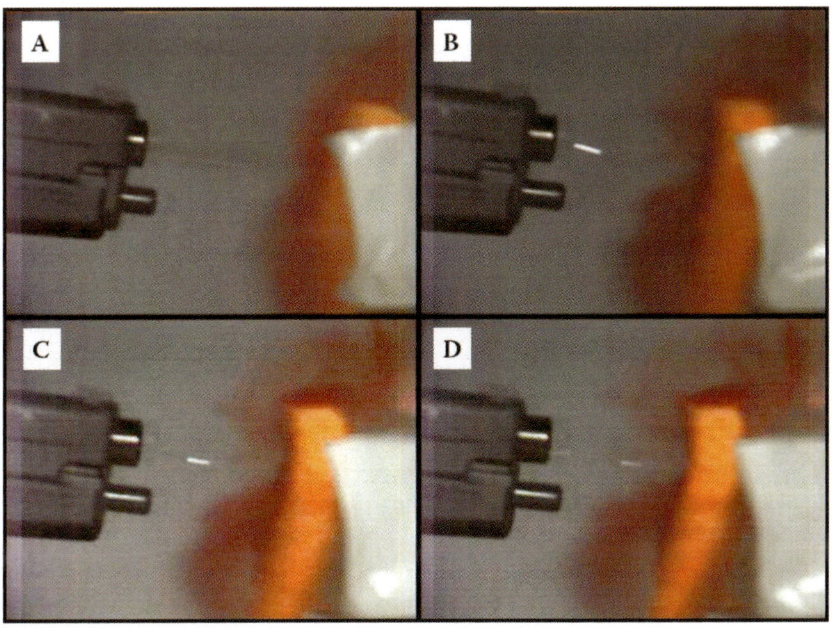

Figure 7.14 These four images were taken from a high-speed video of a bullet being fired into a bloody sponge. In Frame A, the bullet has already struck the sponge and a cloud of small droplets is developing off the surface. Through the four-frame sequence, the cloud ejects backward toward the weapon, becoming more dispersed with time. (Photographs produced in cooperation with Matt Noedel and the Midwest Forensic Resource Center, Ames, IA.)

prevents the normal collapse of the droplet we observe in large drops. Like a droplet of mercury, they simply land retaining their spherical shape and then dry.

Size Ranges of Gunshot Spatter

MacDonell and Bialousz in their original work, clearly made the point that gunshot spatter included larger stains beyond the aerosol or atomized stains. They stated:

> In addition to the fine particles, larger droplets, but essentially all under 1/8 in. diameter, will also be produced.[15]

Laber and Epstein placed the size of the mist stains as 0.2 mm or less, adding, "Beyond this, the size of spatter produced by gunshot overlaps many other impact spatter patterns."[16]

Anyone conducting experiments will routinely find spatter from gunshot with 1–2 mm diameter stains intermixed among the mist-like stains. Once again, the presence of a wide range of spatter size in any pattern, including those created by gunshot, should not surprise the analyst.

Presented with mist-type stains, a pattern with a preponderance of 0.1 mm and smaller stains, this simply points the analyst toward an explosive force source, which in many cases associated with crime will be gunshot. It does not define gunshot to the exclusion of other explosive force, but in a given context (considering the full methodology of BPA) may allow the analyst to opine the pattern was produced by gunshot. One interesting observation regarding "mist" claims, it is common that the decision to categorize stains as mist is often based on a visual mindset. This is highly subjective. The stains must be examined and measured. Keep in mind that no one, no matter what their eyesight, can visualize the circles associated with 0.1 mm stains. If the analyst can see the circles and ellipses, the stains may well be small (0.2 mm), but they are not mist. Use a loupe or magnification technique and measure the stains before claiming mist is present.

Kinetic Energy, Wound Cavitation, and the Creation of Gunshot Spatter

In reading current references, there remains a lack of agreement on what forces cause spatter, particularly back spatter, during a gunshot injury. There seem to be many beliefs regarding the creation of gunshot spatter.[17,18]

Dr. Martin Fackler, through his efforts to understand missile-caused wounds, disputes portions of these beliefs, or at least some of the misconceptions that are often attributed to them. First, Dr. Fackler points to collapse of the temporary wound cavity as the probable source of most gunshot-oriented spatter, both forward or back.[19] Second, he established that kinetic energy potential does not define wound cavitation, nor is spatter produced by the sonic shock wave which in some instances precedes the projectile. In his experiments, Dr. Fackler discharged his weapons from 9 m to 25 m from their targets. This eliminated "hot gasses" as the source of such wound cavitation.[20] This is not to say that in instances of close or contact wounds, hot gasses do not play a role in adding to tissue disruption. Yet, as Dr. Fackler points out, it does not cause an increase in wound cavitation itself.[21]

Dr. Fackler defines temporary wound cavity disruption by saying, "After passage of the projectile, the walls of the permanent [wound] cavity are temporarily stretched radically outward."[22] He likens the disruption process to that of a splash being made in mud. The potential for tissue disruption depends upon both the projectile mass and projectile velocity, while the utilization of this potential is based on construction, shape, and its interaction with tissue.[23] Dr. Fackler's findings dispute the widely held belief that a higher-velocity projectile always produces a larger temporary wound cavity. Dr. Fackler also established that a bullet creates both the entrance and exit wounds milliseconds before the highest level of temporary cavity occurs.[24]

What we are left to consider, based on this information, is that the permanent wound cavity created by the physical crushing of tissue as the projectile passes through tissue also causes a large temporary wound cavity. This temporary stretching of the surrounding tissues then collapses back to the permanent wound cavity. This collapse of the temporary wound cavity often occurs after both the entrance and exit of the projectile, leaving two holes from which blood may be ejected by the force of the collapse.

Interestingly, the shock waves produced by a gun can affect spatter dispersion. Depending upon the speed of the bullet, the bullet can overtake the shock wave and even when this does not happen, there are residual shock waves produced, reflecting off associated surfaces as well as one that appears to be associated with the muzzle. In video efforts, these shock waves have been observed effectively reversing direction of the fine mist droplets. This may explain why back spatter events tend to be less symmetrical than forward spatter patterns as well as the reason for finding fewer mist-like stains.

Double-Shot Impact Events

An interesting effect related to gunshot spatter and the temporary wound cavitation effect occurs when two or more successive wounds are received with a short interval of time between the shots. During the intervening period, the first wound bleeds filling the wound track with blood and tissue. When the second shot impacts the body, the tissues surrounding the first wound compress as a function of the temporary wound cavity causing the blood-filled wound to "squirt" out its contents. The resulting spatter is much heavier than that normally associated with a single gunshot and tends to travel further. The closer the two wounds are to each other and the longer the interval between the shots, the more dramatic the effect.

Figure 7.15 is a photograph of a ballistic gel block used to videotape this effect. Two small cavities were produced and filled with blood. The ballistic gel was then shot with a .22 caliber weapon from left to right. The alignment of the weapon was such that the bullet path crossed the cavity on the left, but only came in proximity to the cavity on the right. Figure 7.16 is a digital capture from the high-speed video taken of this shooting event. In this frame, the bullet has struck the ballistic gel, passed through the blood-filled cavity on the left and passed to the right side of the second cavity. A very pronounced streaming ejection is evident emerging from the cavity on the left and a second ejection is forming from the cavity on the right. Figure 7.17 shows the same event milliseconds later. The mass of blood ejected from the left cavity is breaking up into spatter. The ejection from the right cavity is now as pronounced as the cavity that was struck by the bullet directly.

As one would expect, the blood-filled cavity that the bullet passed through ejects spatter. This ejection is in a volume greater than that associated with a single gunshot. It is a form of streaming ejection. The cavity where the bullet simply passed by and was compressed also ejects blood and spatter in a streaming fashion. What this simulation demonstrates is that compression of a wound track by a force such as a temporary wound cavity is sufficient to eject spatter in volume from a preexisting wound.

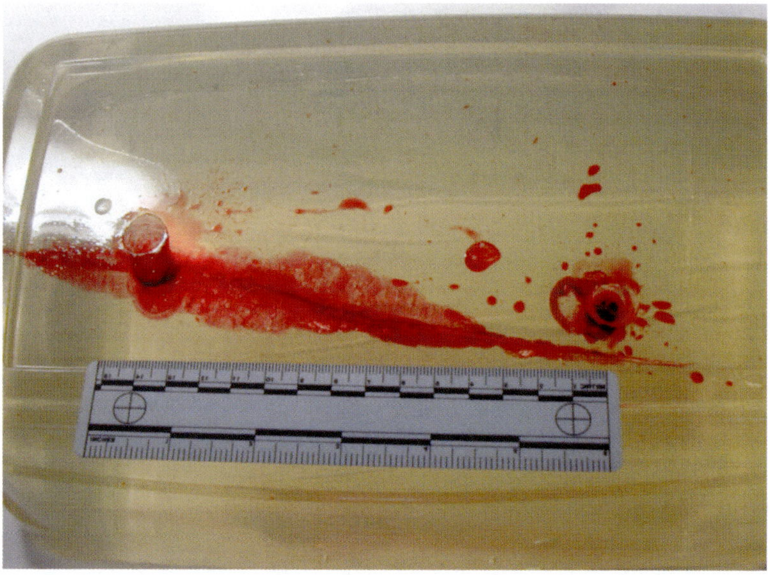

Figure 7.15 This photograph depicts ballistic gel used to demonstrate a double-shot impact. Two defects were produced and filled with blood. A bullet was shot through the defect on the left, while the path of the bullet simply passed close to the defect on the right. Figures 7.16 and 7.17 show the resulting ejection of blood. (Photographs produced in cooperation with Matt Noedel and the Midwest Forensic Resource Center, Ames, IA.)

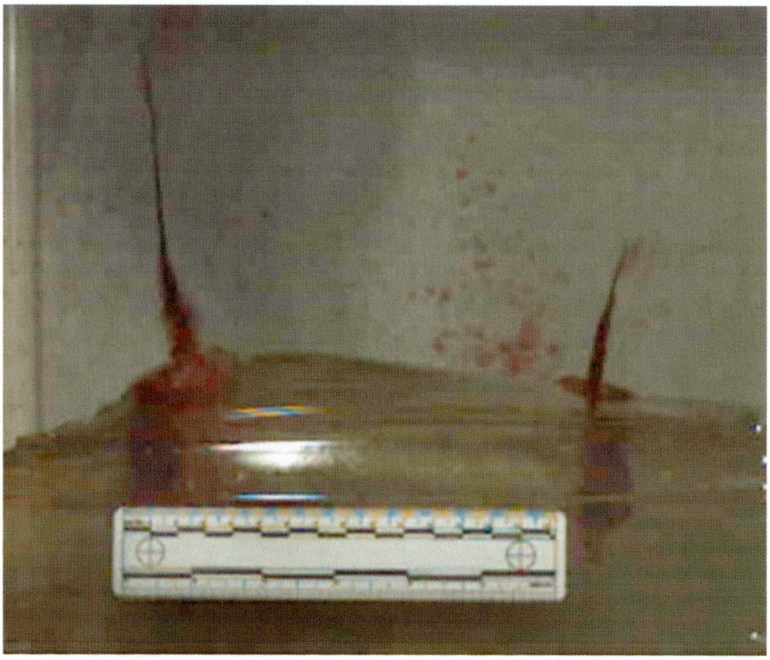

Figure 7.16 This photograph depicts the ejection from Figure 7.15. In this first image, a large streaming ejection is evident from the defect struck directly by the bullet. However, a streaming ejection is also developing from the defect on the right, where the bullet merely passed nearby. (Photographs produced in cooperation with Matt Noedel and the Midwest Forensic Resource Center, Ames, IA.)

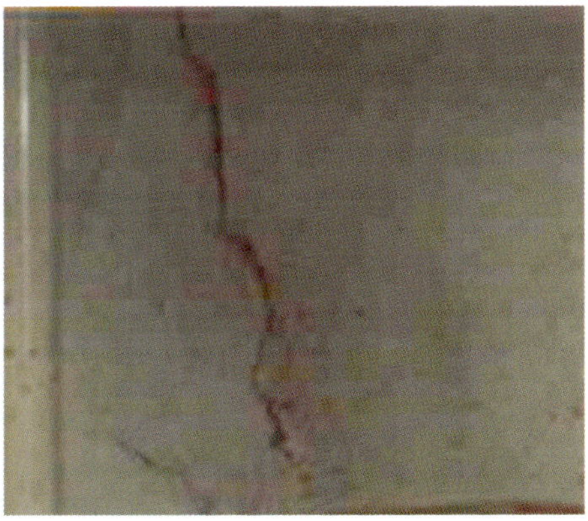

Figure 7.17 This second image shows the volume of blood ejected from the second defect. This sequence demonstrates that the temporary cavity produced by a bullet passing near a defect has the potential to eject significant volumes from a preexisting defect. (Photographs produced in cooperation with Matt Noedel and the Midwest Forensic Resource Center, Ames, IA.)

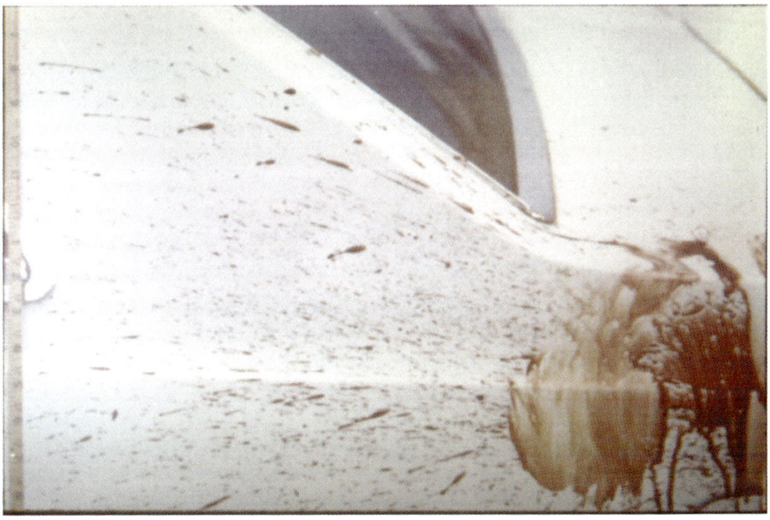

Figure 7.18 A case example of the volume of spatter ejected as a function of a double-shot impact. The victim in this instance, was shot once in the chest, remained upright leaning against the car, and was then shot a second time. Spatter was ejected in volume as far forward as the front door. Double-shot impact events produce spatter far in excess of a typical back-spatter event.

Figure 7.18 demonstrates a case example of how much back spatter can be produced in these situations. The victim was shot once in the chest with a 30–06 rifle. He collapsed to a leaning position against the side of the vehicle, creating the contact staining. While he stood in that position the wound track filled with blood, the victim was then shot a second time in the chest. The level of back spatter ejected from the two wounds in intense and atypical of the back spatter associated with a single shot (even that associated with a rifle). Spatter were found as far forward as the front door of the car.

Figure 7.19 depicts the wound associated with a second case example involving a double-shot impact event. The victim was shot twice in the chest with a 9mm handgun. The two

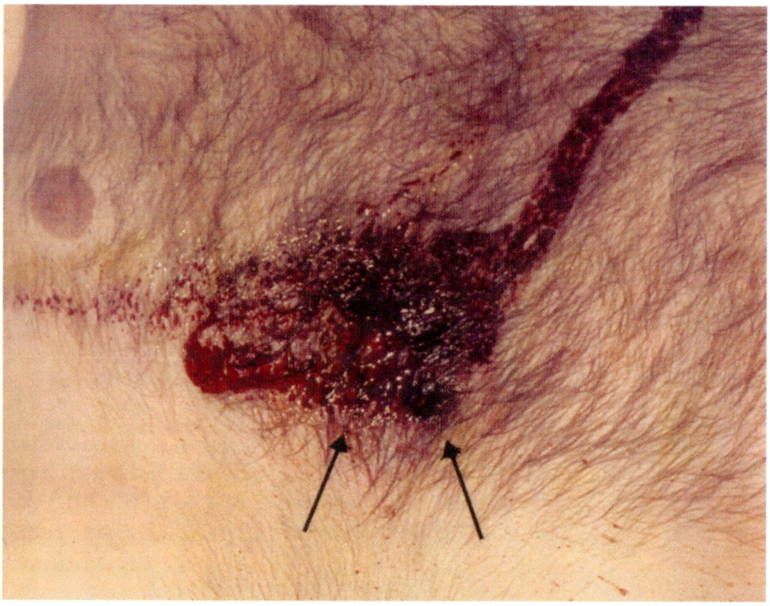

Figure 7.19 Another double-shot impact event. The victim in this instance was shot twice in the chest with a 9-mm semiautomatic pistol as he lay on his back in bed. Note that the two wounds to the chest are extremely close together (the arrows point to the wounds). The subject claimed a minute or less elapsed between the first and second shots.

wound tracks are within inches of each other. Based on the information provided by the shooter, at least one minute passed between shots allowing blood to collect in the first wound. Figure 7.20 shows the volume of back spatter ejected to the wall and ceiling by this event.

Gunshot Spatter Pattern Shapes and Dispersion

In considering typical gunshot spatter impact pattern shapes (unlike those associated with a double-shot impact event); forward spatter patterns when present tend to be more symmetrical than back spatter patterns. This is most likely due to the primary force of the impact being transmitted in the direction of the projectile. Back spatter patterns tend to be less defined.

In considering where or in what direction the spatter is ejected, one might imagine that the ejection of the cone of spatter from the gunshot event would align to the path of the bullet. What occurs is the cone of spatter will eject generally perpendicular to the surface from which it is ejected. For example, Figure 7.21 shows a bullet striking a simulated head from left to right. In this high-speed photograph, the path of the bullet can be seen (the gray streak on the right side of the simulated head). This simulation placed emphasis on the left side of the target (the entry side had a large blood reservoir) and less so on the right side (the exit side); as a result, the back spatter cone is far greater than the forward spatter. This is simply a function of the simulation setup and is not typical of a normal single gunshot event. What this and other similar efforts define is that the cone of spatter is ejected generally perpendicular to a surface and does not specifically align with the bullet path. The back spatter event (left side of the simulated head) best demonstrates this, but the forward spatter cone (right side) also demonstrates this effect as well.

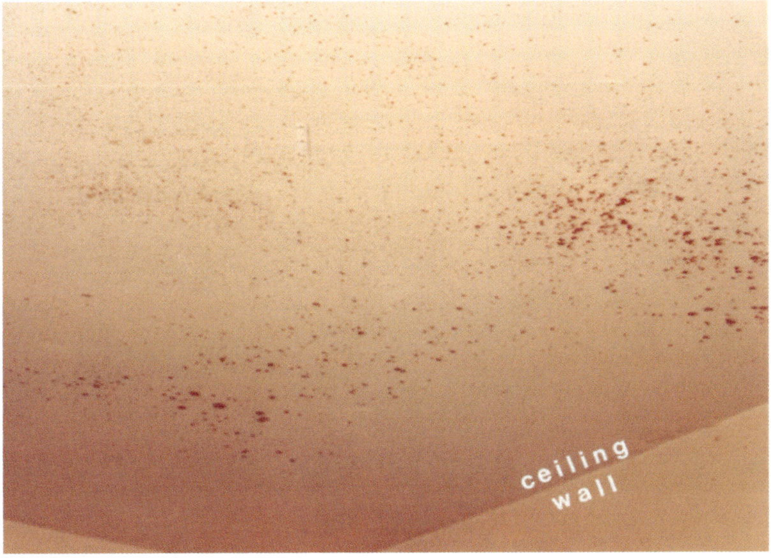

Figure 7.20 A view of the ceiling above the victim of the two chest shots shown in Figure 7.19. The ceiling was 6′ from the point where the victim lay facing it. The result of this double-shot impact event is a pattern well in excess of the typical volume, size, and distribution found in back spatter events.

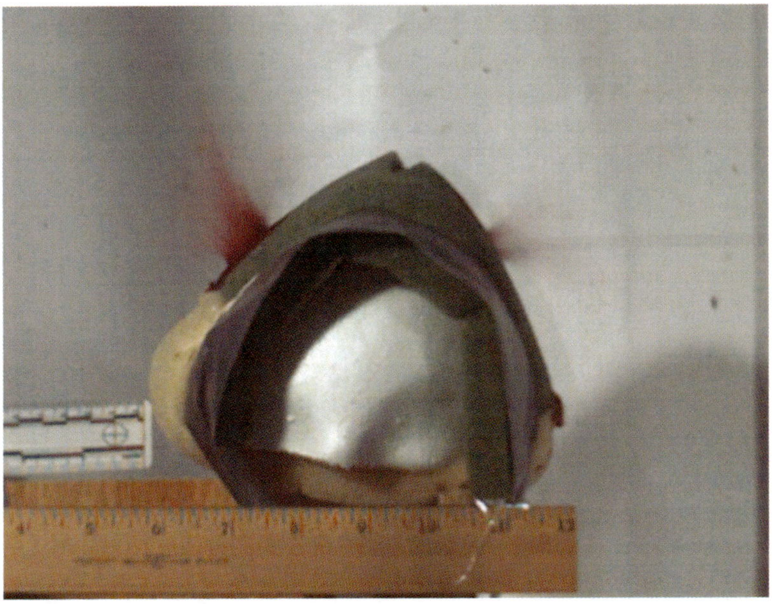

Figure 7.21 This image demonstrates that the cone of spatter produced by a gunshot event ejects generally perpendicular to the surface of the defect and not specifically in line with the bullet path. The bullet passed through the simulated head on a horizontal trajectory, left to right (a gray ghost image of the bullet is evident on the right side). The spatter, however, ejects perpendicular to the surface of the simulated head. (Photographs produced in cooperation with Matt Noedel and the Midwest Forensic Resource Center, Ames, IA.)

Figure 7.22 Heavy spatter resulting from several gunshot wounds to the head. The pattern in and around the door handle/armrest was associated with the final perforating shot.

As with other spatter patterns, the presence of spatter resulting from a gunshot cannot be absolutely predicted. In some instances, the pattern may be heavy and obvious. See Figure 7.22. In other instances, the spatter may be extremely limited or nonexistent. Due to the dynamic nature of the event and the variations possible, there is no way to predict based on the wound alone where or how much spatter will be present. Characteristics of the wound (e.g., skin, tissue, bone), positioning of clothing, hair or other items can all impact the presence of spatter or lack thereof. The small size of spatter in these instances may also prevent the analyst from finding them to begin with. Dark surfaces, heavy weave, carpet, coarse soil or vegetation can make it difficult, if not impossible, to locate these very small stains.

Of note to the analyst, the presence of back spatter on the subject or weapon is in part dependent upon the manner and distance at which the weapon is held. Due to their small mass, the small droplets associated with mist will carry only about 4–5 ft horizontally. Air resistance, if you will recall, is inversely proportional to droplet size. No matter how fast they are moving the droplets rapidly lose speed and fall. There are also indications that in some instances these aerosol drops of blood dry almost instantly while suspended in air. This too may be a factor in the lack of staining on surrounding objects. The small, dried specks simply fall to the ground and fail to adhere to any surfaces. Even if observed, movement by wind or airflow can alter their location.

It is possible that while suspended in the air, these droplets may move on air currents, which may result in deposits farther than 4 ft. This demands that an active air current of some kind is moving in the scene. In this instance, the droplets would likely "rain" down and be found only on the top of exposed horizontal surfaces.

Within the 4–5 ft range, a gunshot event may deposit small stains on the weapon, shooter's body, or clothing. Larger stains (0.5 mm and larger) may be deposited from greater distances. Figures 7.23 and 7.24 depict back spatter on the inside surface of a crossbow, while Figure 7.25 shows back spatter on the hand of a suicide victim. The size of such stains, combined with their highly dispersed nature, can make them easily overlooked both by the shooter when cleaning up and by the investigator during subsequent contact.

Back spatter is often deposited on the inside of the weapon's muzzle, external surfaces of the weapon, or on internal surfaces exposed during operation. See Figures 7.26, 7.27, and 7.28. During operation of a semiautomatic pistol, spatter can be deposited on aspects of the internal lower receiver, the recoil spring and recoil rod. These areas should always be examined, particularly if there is some belief a weapon was cleaned after discharge.

If the analyst wishes to establish that spatter was deposited into a barrel or to determine how far it was deposited, one of the more functional means of doing this is a technique developed by Martin Eversdijk. Using silicone products like *Forensic Sil*,* Eversdijk developed a functional method of documenting not only the position of the spatter in

Figure 7.23 An example of back spatter found on the inside facing of a crossbow. The owner was shot in the head while holding the bow in a firing position and pointing it at a police officer. The officer fired his weapon and struck the subject in the forehead. The presence of the back spatter supports the officer's contention that the victim was aiming the bow at the time he was shot. (Photograph courtesy of Lt. Johnny Kuhlman, Oklahoma City Police Department, Oklahoma City, OK.)

* *Forensics Sil* is a Registered Trademark of Armor Forensics.

Figure 7.24 A second photograph of the crossbow, with a scale of reference. Several other small spatter stains are evident to the right side of the photograph. (Photograph courtesy of Lt. Johnny Kuhlman, Oklahoma City Police Department, Oklahoma City, OK.)

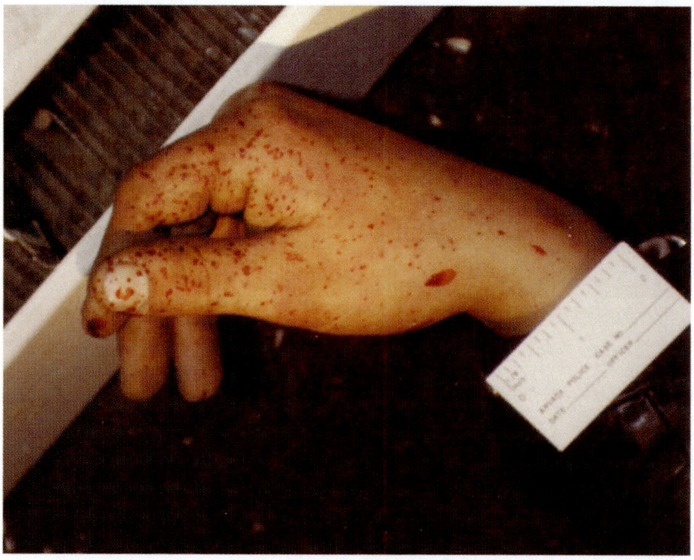

Figure 7.25 Back spatter present on the hand of a suicide victim. Also, note the large elliptical stain at the base of the wrist. This is a good example of why targets capable of motion during an event represent a poor medium in terms of directionality. The large stain cannot be reconciled as being from the initial deposit of back spatter as it shows opposite directionality. This deposit must have occurred after the wrist rotated away from the body. (Photograph courtesy of John Graham, Arvada Police Department, Arvada, CO.)

Figure 7.26 Blood deposited inside the barrel of the firearm. This photograph was taken using a fiber-optic light inserted in the breech. (Photograph courtesy of Don Blake and Bill May, Norman Police Department, Norman, OK.)

Figure 7.27 Spatter deposited on the trigger and frame of a semiautomatic. By contrasting the condition of the weapon, it is reasonable to predict that the hand that held the weapon was similarly stained.

the barrel, but a method that recovers the spatter stain as well. These products extrude the casting material using a trigger mechanism, which allows the material to be injected into any space. The product is automatically mixed in an extrusion tube, which eliminates concerns of improperly mixing the casting material. This allows the analyst to inject the casting material directly into the barrel of the weapon. To preclude any possibility of forcing blood further into a barrel, the material is injected from the breech end. Once the

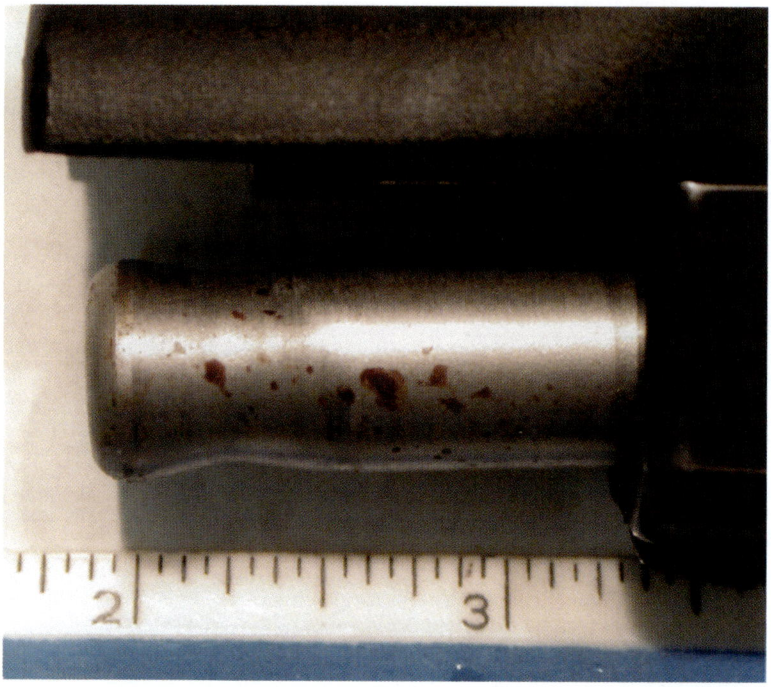

Figure 7.28 Spatter deposited on the internal surfaces of a semiautomatic. As the weapon is fired, the slide moves back exposing the recoil spring guide of the weapon and the outside of the barrel to the spatter event. If a suspect cleans the weapon after the event, these areas may be forgotten. (Photograph courtesy of Paulita McGuire, Indiana State Police, Indiana.)

barrel is filled, a small tab of material is left in place at the breech end to assist in removing the cast. The normal cure time for such products is typically 3–5 minutes, but it is better to allow a longer cure in this circumstance. Once the material sets, if needed, the muzzle end of the cast is clipped flat allowing the entire cast to be pulled out the opposite side. See Figure 7.29. When the cast is removed, the casting material captures the actual spatter in the barrel and these spatter become an integral part of the outer surface of the cast. The spatter can be examined, measured from the muzzle end to determine how far into the barrel they were deposited and as they are on the surface of the cast, they can also be tested (DNA/serology) like any other bloodstain. See Figure 7.30.

Evaluating Expectorate Blood

In instances in which a victim receives bleeding injuries to the mouth, nose, throat, or lungs, and continues to breathe, a form of radiating spatter may result. The spatter resulting from the expiration of blood can range from heavy large stains and on rare occasion to light mist-like stains comparable to those found in gunshot situations. See Figure 7.31. At times these stains can mislead the analyst and their similarities demand proper evaluation. For example, Figure 7.32 depicts expectorate blood on the sleeve of a husband who came to the aid of his wife after she was ambushed and shot multiple times by her ex-husband. A pattern of this nature could be mistaken for classic impact spatter.

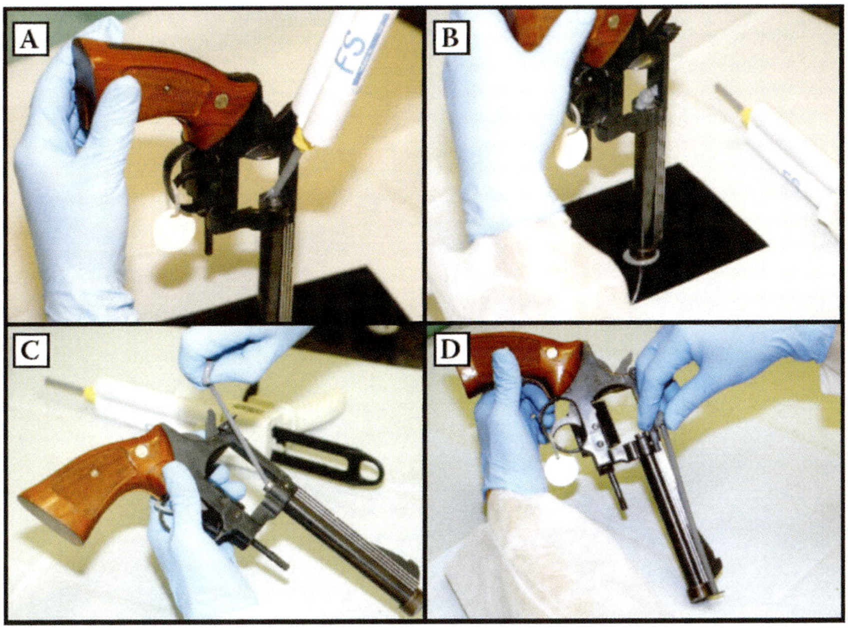

Figure 7.29 Using forensic casting material to document the depth of spatter in a barrel and to collect the spatter. In Frame A, the material is injected into the breech end. In Frame B, the material is allowed to cure and a small tab is left to assist in removing it. In Frame C, using the tab the cured material is recovered from the barrel. Frame D shows the complete cast. Figure 7.30 shows the result. (Photographs courtesy of Martin Eversdijk, Amsterdam Police Laboratory, Amsterdam, the Netherlands.)

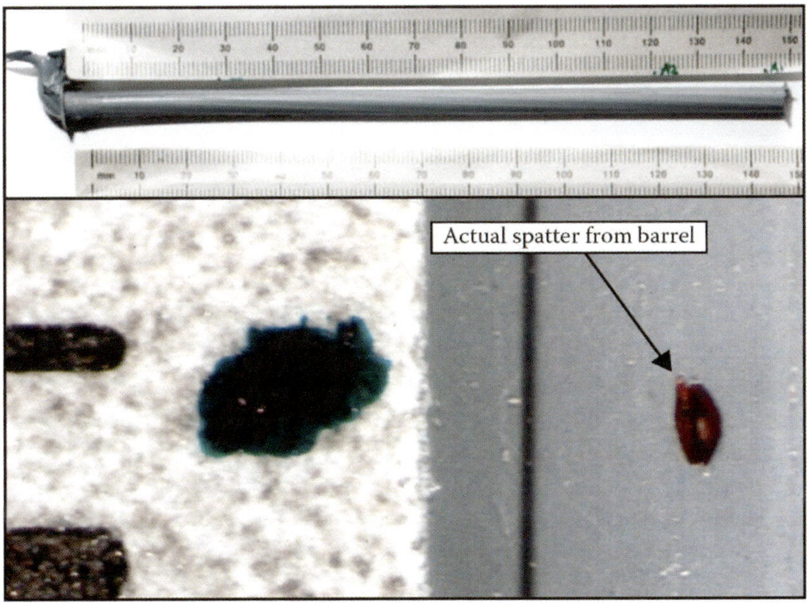

Figure 7.30 The top photograph shows the entire cast after recovery from the barrel. The lower photograph depicts a single spatter on the casting material. This is the actual stain caught in the cast. It can be evaluated like any other stain (e.g. DNA/serology). (Photographs courtesy of Martin Eversdijk, Amsterdam Police Laboratory, Amsterdam, the Netherlands.)

Figure 7.31 An example of expectorate blood. In this instance, the subject was wearing these cutoff jeans as he stood over a victim, whose throat was cut. Stains are evident on both the inside and outside of the pant leg. On fabric, the specific characteristics (e.g., vacuoles, mucous strands) may not be apparent and the pattern may appear as impact spatter. (Photograph courtesy of Donald R. Schuessler, Department of Public Safety, Eugene, OR.)

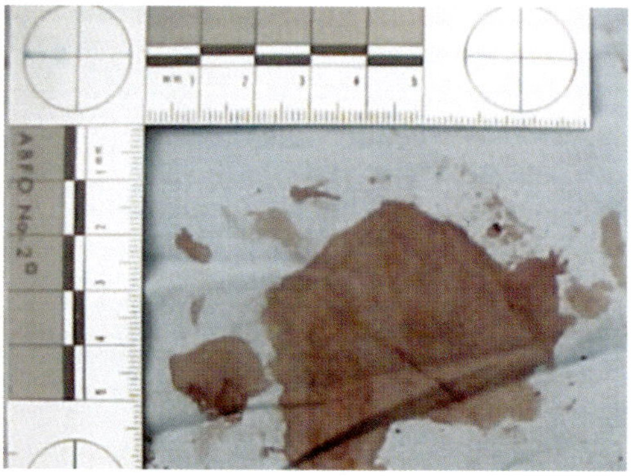

Figure 7.32 Expectorate blood on the sleeve of a man who came to the aid of a shooting victim. Initially of concern, the presence of similar spatter in the scene clearly from the expectorate event helped in understanding the true mechanism of creation of this pattern. Taken by itself without scene context, the pattern on the fabric is certainly consistent with spatter associated with a gunshot.

Oftentimes, expectorate blood that mimics typical impact spatter is less vivid in its coloration. It may give an indication of being watered down. There may be evidence of air bubbles in the pattern, or traces of popped air bubbles. See Figures 7.33 and 7.34. Additionally mucous strands may be present, spine like interconnections between individual spatter stains. See Figure 7.35. Keep in mind that the absence of these characteristics does not absolutely exclude expectorate as a mechanism. Lacking the distinct characteristics associated with expectorate, such a pattern will simply appear as a typical impact

Figure 7.33 After being stabbed multiple times in the back, head, neck, and chest, the victim began projecting spatter through his breathing (expectorate blood). The spatter landed on various articles to include this container. (Photograph courtesy of Sandra M. Roberts, Eugene Police Department, Eugene, OR.)

Figure 7.34 A close-up of the spatter from Figure 7.33. Note the presence of the bubble rings in the pattern. (Photograph courtesy of Sandra M. Roberts, Eugene Police Department, Eugene, OR.)

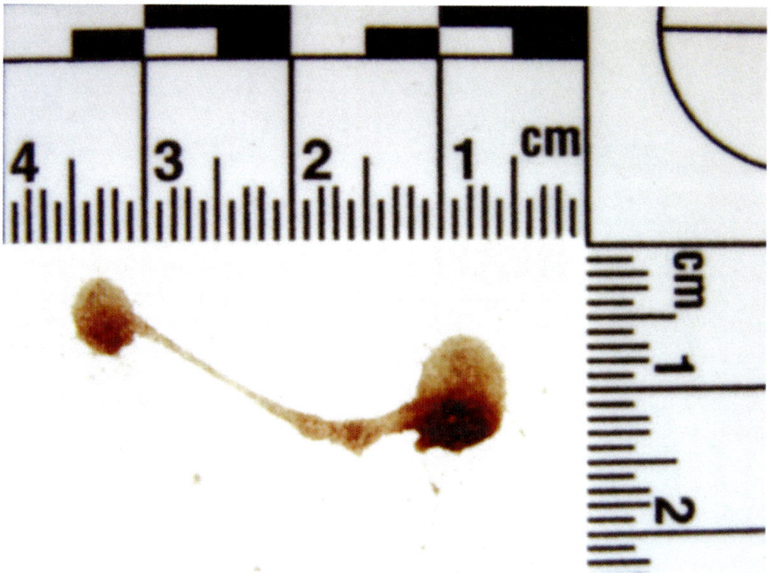

Figure 7.35 Mucous strands intermixed in spatter. This is another characteristic of expectorate stains.

pattern. Expectorate blood is often recognized by correlating the position of the spatter with blood observed in the nose, mouth, or airway. If no evidence of blood is found in the airway, either at the scene or during the autopsy, expectorated blood is generally excluded as a possible source for the spatter.

Dispersion of expectorate tends to be a less symmetrical than that found in typical impact spatter events. Often, multiple events from different breaths may be deposited onto the same surface.

Fly Spots

Although not true spatter, another pattern that can be confused with impact spatter is the "fly speck" or "fly spot" pattern. Flies present within the scene will feed on blood found there. This blood is both tracked about, regurgitated and excreted by the flies. In the instance of the tracking pattern, the marks are extremely small, but small characteristics caused by the insect's legs may be evident on close examination.

In the case of regurgitation, the specks are remarkably symmetrical. See Figure 7.36. If the pattern is the result of excreted blood by the fly, subsequent movement can produce a stain with an elliptical component that may mimic directional spatter. Often, the analyst finds these patterns in warm areas where the flies rest, such as high in window corners or along walls where the sun strikes. Such stains will usually test positive for blood with a presumptive test. Obviously, care should be exercised in evaluating any abnormal patterns that meet these criteria. The recognition of the pattern as a fly artifact will usually be evident by considering the pattern in concert with all the other evidence.

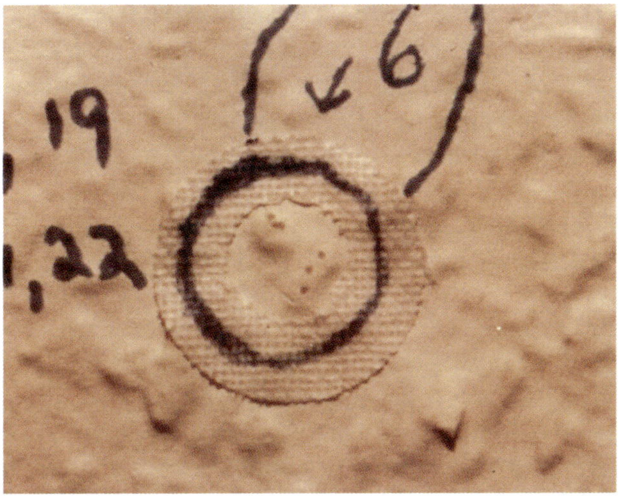

Figure 7.36 An example of fly spots. Fly spots in some situations can easily mimic impact spatter. Oftentimes, it is only after considering the entire scene context that the source of fly spots will become evident.

Summary

We encounter impact patterns frequently at crime scenes. The analyst's first issue is to properly describe and categorize them. Whatever method one chooses to use, its focus should be in describing those characteristics observed by the analyst.

In considering patterns, the analyst must recognize that spatter size ranges widely from event to event and may range widely in any given event. Overlap in stain size is often encountered. This demands a cautionary approach by the analyst in evaluating any spatter pattern.

A major point of concern when considering a spatter pattern, as MacDonell put it, "a few bloodstains do not a pattern make."[25] Whichever stains are chosen for evaluation, the analyst should be confident that they have a relationship to one another.

The radiating effect of the spatter pattern certainly helps the analyst recognize this relationship. Considering the dispersion found on an item can also help establish relationships with the scene. Spatter dispersion increases the farther away from the target the item is. Even lacking an ability to perform a formal AO analysis, aspects of dispersion and convergence associated with directional angles may offer some insight into where the source event was in relationship to the pattern.

Unfortunately, the spatter pattern is impossible to predict in all instances. The presence of hair, clothing, bedding, and other items, or the characteristics of the weapon and how it was used, may affect the deposit of the spatter in the overall pattern.

In situations of gunshot, spatter may form both as forward spatter, that which is projected outward in the direction of the bullet from the exit wound, and back spatter that is projected back toward the weapon from the entry wound. This latter situation often results in deposits of spatter on the subject or other parties present or in proximity to the wounding. When attempting to distinguish forward from back spatter, the analyst may look for a less symmetrical pattern and fewer stains in the latter. Beyond that, little distinction is evident based on the stains themselves.

Back spatter patterns in gunshot situations may effectively place the individuals at or near the victim. The tiny mist stains associated with gunshot events rarely travel horizontally more than 4–5 feet from their source. When found on someone's clothing, it is difficult to refute that person's proximity to the event.

Both expiratory blood and flies produce patterns that an analyst can misidentify as impact spatter. The analyst, if knowledgeable of such stains when considering the full context of the scene, is less likely to make such a mistake.

Keep in mind that the analyst does not "predict" the presence of impact spatter on clothing, the suspect or even in specific areas in the scene based solely on the type of event. What they do is contrast the condition of the actual scene in terms of where and what level of spatter is present and then using that condition make valid predictions about what level of spatter may be present on anything in similar proximity. In doing so, they must still consider issues such as the extent that the pattern radiates outward, the size of spatter present, and the dispersion level observed. Rarely, if ever, can absolute statements be made in the way of predicting the presence of back spatter. Statements such as "all gunshots to the head produce back spatter" or "a .22 head shot will not produce spatter" are not empirically valid statements.

Notes

1. Committee on Identifying the Needs of the Forensic Sciences Community, National Research Council, *Strengthening Forensic Science in the United States: A Path Forward*, electronic edition; National Academie Press, Washington, DC, 2009, p. 178.
2. MacDonell, H. and Bialousz, Lorraine F., Flight Characteristics and Stain Patterns of Human Blood, National Institute of Law Enforcement and Criminal Justice, LEAA Report PR-71, U.S. Government Printing Office, Washington, D.C., Nov. 1971, pp. 20–21.
3. MacDonell, H. and Bialousz, Lorraine F., Flight Characteristics and Stain Patterns of Human Blood, National Institute of Law Enforcement and Criminal Justice, LEAA Report PR -71, U.S. Government Printing Office, Washington, D.C., Nov. 1971, p. 21.
4. Bevel, Tom and Gardner, Ross M., *Bloodstain Pattern Analysis Theory and Practice, A Laboratory Manual,* TBI Inc., Oklahoma City, OK, 1990, p. 18.
5. Parker, N.L. et al., Summary report of the Bloodstain Analysis Research Group, Florida Department of Law Enforcement, Tallahassee, p. 65.
6. Parker, N.L. et al., Summary report of the Bloodstain Analysis Research Group, Florida Department of Law Enforcement, Tallahassee, p. 66.
7. Parker, N.L. et al., Summary report of the Bloodstain Analysis Research Group, Florida Department of Law Enforcement, Tallahassee, p. 57 and 66.
8. Bunker, Judy, Presentation to the Rocky Mountain Association of Bloodstain Pattern Analysts, 22 Apr 1991.
9. MacDonell, H., *Bloodstain Pattern Interpretation,* Laboratory of Forensic Science, Corning, NY, 1982, p. 3.
10. Ryan, Robert T., The Behaviour of Large, Low-Surface-Tension Water Drops Falling at Terminal Velocity in Air, *Journal of Applied Meteorology,* Volume 15, Number 2, Feb. 1976.
11. Pex, James O. and Vaughan, Charles H., Observations of High Velocity Bloodspatter on Adjacent Objects, *Journal of Forensic Sciences,* Volume 32, Number 6, Nov. 1987, pp. 1587–1594.
12. Beard, Kenneth V., Personal communication, Sept. 1989.
13. Gardner, Ross M., Deformation Levels of Blood Droplets Created by Impact, presented to the International Association of Bloodstain Pattern Analysts, Nov. 9, 1996, Albuquerque, NM.
14. MacDonell, Herbert L., *Bloodstain Patterns,* Golas Printing, Elmira Heights, NY, 1993, p. 61.

15. Eckert, William G. and James, Stuart H., *Interpretation of Bloodstain Evidence at Crime Scenes*, Elsevier, New York, 1989, p. 55.
16. Sagan, K., *Shadows of Forgotten Ancestors*, Ballantine Books, 1992, p. 387.
17. MacDonell, H., *Bloodstain Pattern Interpretation,* Laboratory of Forensic Science, Corning, NY, 1982, p. 70.
18. MacDonell, H. and Bialousz, Lorraine F., Flight Characteristics and Stain Patterns of Human Blood, National Institute of Law Enforcement and Criminal Justice, LEAA Report PR-71, U.S. Government Printing Office, Washington, D.C., Nov. 1971, p. 21.
19. Laber, Terry L. and Epstein, Barton P., Experiments and Practical Exercises in Bloodstain Pattern Analysis, Minnesota B.C.A. Forensic Science Laboratory, St. Paul, MN, Nov. 1983, p. 38.
20. Laber, Terry L. and Epstein, Barton P., Experiments and Practical Exercises in Bloodstain Pattern Analysis, Minnesota B.C.A. Forensic Science Laboratory, St. Paul, MN, Nov. 1983, p. 38.
21. Pex, James O. and Vaughn, Charles H., Observations of High Velocity Bloodspatter on Adjacent Object, *Journal of Forensic Sciences*, Volume 32, Number 6, 1987, p. 1588.
22. Fackler, Martin L., Col (Ret) M.D., Presentation to the Int. Assoc. Bloodstain Pattern Analysts, Colorado Springs, CO, Sept. 24, 1992.
23. Fackler, Martin L. Col(Ret) MD, personal communications, Aug 1995.
24. Fackler, Martin L., Col (Ret) M.D. et al., Emergency War Surgery, U.S. Government Printing Office, Washington, D.C., 1988, pp. 13–32.
25. MacDonell, Herbert L., *Bloodstain Patterns*, Golas Printing, Elmira Heights, NY, 1993, p. 64.

The Physically Altered Bloodstain Principle

8

The final aspect of predictability in BPA is the principle of the Physically Altered Bloodstain (PAB). The PAB principle recognizes that once exposed, blood is affected by environmental conditions. It can be articulated as follows:

> *Exposed blood will react to environmental conditions (e.g., airflow, temperature, humidity, variations of surface) in a relatively predictable manner.*[1]

Once exposed, blood reacts in a relatively consistent fashion to environmental conditions producing physical manifestations in the resulting stains. The specific aspects of these changes that provide some level of predictability include the minimum length of time required to achieve some level of coagulation or drying, and estimates of the volume present in an accumulation. In terms of drying and coagulation, this may allow the analyst an ability to define the minimum period since the blood was first exposed. In terms of accumulations, this may allow an estimate of the total volume involved, even when the accumulation occurs on a permeable surface. The PAB principle has also been referred to as physiologically altered bloodstains (PABS).[2] The nature of the PAB principle demands that generally to achieve a conclusion, some level of case specific experimentation is required, as in almost all instances it is dependent on environmental conditions.

PAB is also an area where additional validation research is ongoing. If there is an Achilles' Heel in the discipline of BPA, it is the PAB principle. For several years, various aspects, primarily coagulation, were based on anecdotal observations with limited consideration of other associated variables. PAB is not, however, without directed effort. This effort includes considerations of extreme temperature effects, issues relating to the development of the perimeter stain, and considerations of substrate effects on coagulation.

Coagulation Issues

Coagulation is a highly complex chemical process involving 13 different clotting factors present in the blood. This process is known as the Fibrin or Coagulation Cascade, the second aspect of the process known as hemostasis, the body's response to injury.[3] This cascade consists of a series of reactions within the 13 factors. The first aspect of hemostasis is the creation of the platelet plug, which attempts to clog the injury site to stem the blood loss. This plug may be sufficient in and of itself to plug the injury. In instances of a larger injury, coagulation supports the stabilization of the platelet plug.

Coagulation changes soluble fibrinogen, a glycoprotein molecule produced in the liver, into insoluble fibrin to produce a stable matrix that captures cellular material. This matrix, which we visualize as a gel-like substance (the clot), works alongside the platelet plug to further stabilize the platelet plug injury and limit blood loss. Coagulation occurs along two

 DOI: 10.4324/9781003408734-8

primary pathways. The first pathway is known as the tissue factor (TF). This TF pathway was previously referred to as the extrinsic pathway or tissue plasmogen activator (TPA). TF is produced and present in sub-endothelial cells. When injury occurs, this TF is exposed to liquid blood containing the various clotting factors, which initiates the Fibrin Cascade. This chemical process is very complex, the ultimate effect is a chain of reactions: prothrombin > thrombin > fibrinogen > fibrin.

The specific pathway is as follows:

1. Damage results in TF interacting with the liquid plasma.
2. The TF reacts with Factor 2, which then acts with Factor 7, then together they act on Factor 10 through an enzyme, protease. This then reacts with Factor 5.
3. Factor 10 binds with Factor 5, resulting in the protein prothrombinase.
4. The prothrombinase and Factor 2 react and create prothrombin. Calcium ions in the blood and the prothrombin lead to the creation of thrombin.
5. The thrombin binds with fibrinogen (a soluble protein) making fibrin, which is insoluble.

The contact pathway, previously called the intrinsic pathway, is more complex and involves both collagen and TF. The cells that lyse (the cell membranes break) expose the various proteins and compounds inside. Both TF and collagen are exposed as a result and the blood in contact with the injured skin which initiates the various factors. That both TF and collagen are involved may explain the faster pace in which blood clots in contact with skin. Once again a series of factors interact, but the clotting chain ultimately produces that found in TF process (prothrombin > thrombin > fibrinogen > fibrin).

What the analyst is presented with through these various interactions is clotted blood. The condition of the clot may provide the analyst with an indication of the minimum time that has passed between bloodshed and the observed condition of clotting.

In defining these observed conditions, Anita Wonder described three stages for clotting. She referred to these as: clot initiation, clot firmation, and clot retraction. According to Wonder, initiation starts between 10 seconds and 1.5 minutes after bloodshed. Wonder described clot firmation as the point where, if disturbed, no flow-back will occur in the volume. Wonder reported times of 5–20 minutes after the bloodshed. Clot retraction, which Wonder defined as serum separation from the fibrin mass, begins anywhere from 30 minutes to 1.5 hours after bloodshed.[4,5] Serum separation is the condition observed when the plasma flows out from the gelled material. This fluid is often straw colored or light pink. Figure 8.1 shows a pool with evident serum separation and clotting.

Laber and Epstein looked at coagulation, specifically considering substrate effects. Some basic conclusions derived from that effort included recognition of the first indications of clotting occurring between 15 minutes and 19 minutes, although they were not specific in what constituted this first indication. Subsequently, both small and large clots were observed in their effort at 22–25 minutes. Heavily congealed blood appeared at approximately 43 minutes. They noted that substrate is a factor in coagulation, specifically when clotting occurs on skin. In the instance of skin, the first indication of clotting occurred as early as 6–8 minutes, with clotted material appearing between 6 minutes and 12 minutes. The presence of heavily congealed blood occurred in these situations at 27–35 minutes.[6]

Figure 8.1 A pool of blood showing both clotting (the congealed dark mass) and serum separation (the pinkish colored flow that moves out of the pool to the left and down the wood floor).

In 2005, James, Kish and Sutton discussed clotting in their book, *Principles of Bloodstain Pattern Analysis, Theory and Practice*, not to be confused with *Bloodstain Pattern Analysis Theory and Practice: A Laboratory Notebook* (Bevel and Gardner published in 1990). They did not provide specific times for clotting conditions, but offered the following considerations:

> [B]lood clots and serum separation stains surrounding them, as well as the degree of observed blood drying should be recognized as important information at crime scenes.[7]

They cautioned that "Estimates of this interval should be made with extreme caution." They continued, stating, "estimation of degree of clotting and drying time of blood should be reproduced experimentally using freshly drawn human blood of similar volume placed on an *identical* surface with *similar* environmental conditions during the experiment as were observed at the scene."[8] It is important to note their specific use of the two words similar and identical. Using the same substrate is critical in any coagulation study, and the analyst must attempt to match the environmental conditions as closely as possible. Yet, the analyst must recognize that no one can match the exact environmental conditions.

Rossi and James looked at clotting with regard to the presence of saliva. Using an observation like Wonder's Clot Initiation, they saw times associated with initial clotting of 9–25 minutes depending upon the various subjects. Of particular interest to their effort was the effect of saliva mixed with the blood. Under this condition, initial clotting times were significantly reduced. They reported observed times of 2–9 minutes. They commented that not only should accelerated times be expected in saliva/blood mixtures, but also mixtures involving cerebrospinal fluid.[9] The latter appears to be a consideration, but this relationship has yet been defined through research.

Rossi et al. also commented on the many medical factors that may affect clotting times. These include: anemia, hemophilia, low platelet counts, liver disease, vitamin K deficiencies, leukemia, and chemotherapy.[10] All of which can or could effectively slow or retard the coagulation cascade. It is important to recognize in any coagulation study what the analyst is seeking is not to define "the time," but rather to define the *minimum* time required to achieve a specific condition of coagulation. By retarding the cascade these concerns simply increase the time required for the coagulation process to occur. It is not that such issues should not be pursued, but any condition that slows coagulation is effectively irrelevant to the purpose of a coagulation study. If the minimum time required for coagulation to some particular level is 40 minutes, anything beyond 40 minutes doesn't alter the minimum time conclusion.

The true probative value of any coagulation experiment is a matter of investigative context. Based on experimentation, if the observed condition of the scene clot would require a period significantly less than that believed or claimed, such information may refute that belief. Once again, any conclusion demands experimentation and a cautious and conservative approach. Effectively, these times are never measured or compared as a matter of mere minutes. So even the presence of cerebrospinal fluid and saliva, which clearly reduce the time for coagulation, may not impact the ability to deal with a specific probative issue in a given investigation.

Environmental issues of temperature, volume, humidity, the presence of saliva or cerebrospinal fluid, and the surface on which the blood was shed must be considered when designing any experiment. What is evident based on all these authors' efforts is that anecdotal claims as to typical coagulation times previously relied upon, may not be valid. Additional studies on coagulation continue and analysts should offer conclusions relating to coagulation times cautiously.

Case Example – Clotting

A young mother and her child were found shot. The child was shot twice, once in CV-2 and once in the head. In the hallway where the child was found, were two pools of blood. One pool of blood soaked through the carpet, carpet pad and into the sub-floor. This established a passage of time between rolling of the victim from this position and the second shot. In the second pool was a congealed mass showing the handprint of the child. See Figure 8.2. This second pool indicated the child was moved after being shot, bled in this second location while her hand was in this position long enough for coagulation to occur.

Drying Time of Blood

Drying time and the presence of a perimeter stain (previously referred to as skeletonization) are both important alterations of blood that may aid in the analysis. Investigative concerns may exist for drying times of both spatter stains and larger volumes of blood. In terms of the perimeter stain, Pex and Hurley established that within 50 seconds after blood is deposited, a ring always appears on the edge of a disturbed bloodstain. Of particular interest was the fact that if disturbed immediately after deposition, no ring was present.[11] Empirical data from repeated experimentation suggest that in some instances this ring may appear earlier or may take as long as 75 seconds to appear. This condition may occur even if the disturbance involves a vigorous wipe of the perimeter stain.

Figure 8.2 An area of congealed blood, which holds a child's handprint. As the accumulation occurred, the blood congealed while the hand was present in it. The hand was then removed. This type of pattern defines evidence that points both to the nature of the object that caused the pattern and to issues regarding timing and sequencing aspects. (Photograph courtesy of Edmond Police Department, Edmond, OK.)

The condition of this perimeter stain can be quite probative in instances where the disruption producing it occurs from a shoe or hand/finger. In these circumstances it may be possible to isolate what caused the disruption (e.g., a shoe mark or fingerprint). Claims that stains were disrupted at the time of discovery of the victim(s) by individuals who were moving about the scene are common and may be valid. But the presence of no perimeter stain or a minimal perimeter ring will indicate the disruption occurred very close to the time of deposition. Given a situation where the time of discovery is clearly distant from the assault, such a condition may be probative.

Case Example – Perimeter Stain Condition

An individual suspected in a murder of a rival, explained the presence of his shoes' marks in the scene by claiming he walked in as an unknown individual was standing over the victim in his final position at the far side of the room. He claimed he was present for a mere moment at the door and then fled.

Multiple shoe marks in blood were present that were associated to his shoes. The marks occurred deep into the scene, beyond his claimed positioning. Additionally, the shoe marks interrupted spatter and other stains on the floor, with no perimeter stains. Just as damning, was the presence on spatter on top of the shoe marks. See Figure 8.3

Complete or partial drying of blood is dependent upon environmental factors and the surface on which we find the blood. A single drop may take anywhere from 10 minutes to 90 minutes to dry completely, depending upon those factors. The perimeter effect consistently occurs not later than 75 seconds after initial bloodshed. See Figure 8.4.

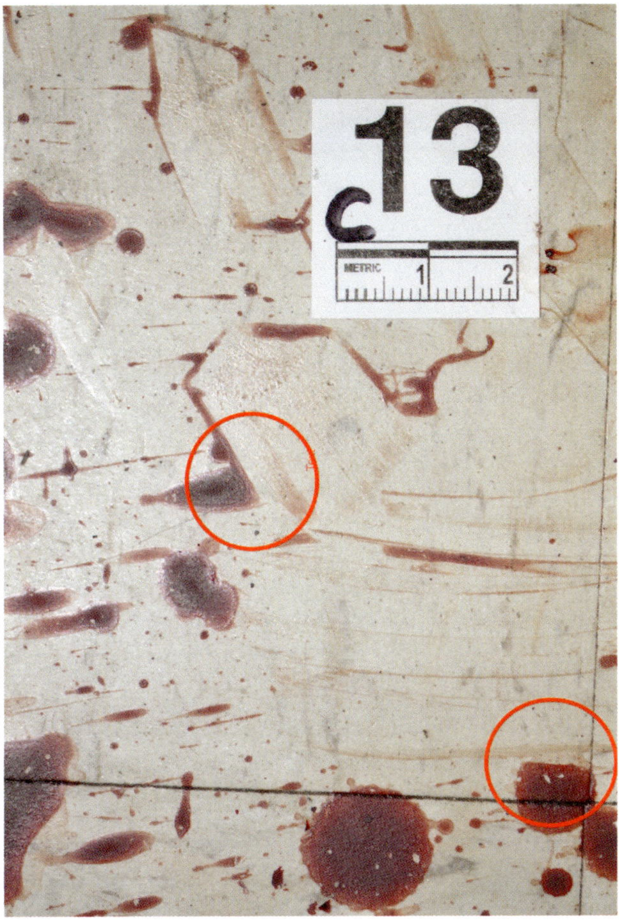

Figure 8.3 A suspects shoe mark disrupts several spatter associated with the assault with no evidence of any perimeter stain in the disturbed spatter. In other areas, spatter were deposited on top of the shoe mark, effectively refuting his claim of having simply walked into the scene as the assault was ending.

In terms of individual bloodstain drying times, environmental factors are always important. Drying is a physical process, different from coagulation. Minor variations of air temperature alone have a limited effect on drying time of individual drops; although extremes in temperature should be considered. Significantly colder temperatures will increase drying time, while significantly hot temperatures will reduce drying time. Surface temperature of the substrate can certainly affect drying time. In case of a hot surface temperature drying time is reduced. Airflow across the stain or drop has a dramatic effect. The Florida Department of Law Enforcement's (FDLE) early experiments indicate that increasing air flow across the surface where a single blood drop was deposited can change the drying time from as much as 90 minutes to as little as 20 minutes. Another consideration in the FDLE effort was the volume present and how it is distributed on the surface. Single drops of blood produced from a consistent volume were dropped from increasing heights, which of course resulted in an increasing stain size. The larger stains dried faster than the smaller stains. The obvious correlation is the more surface area exposed for a given volume, the quicker the drying time.[12]

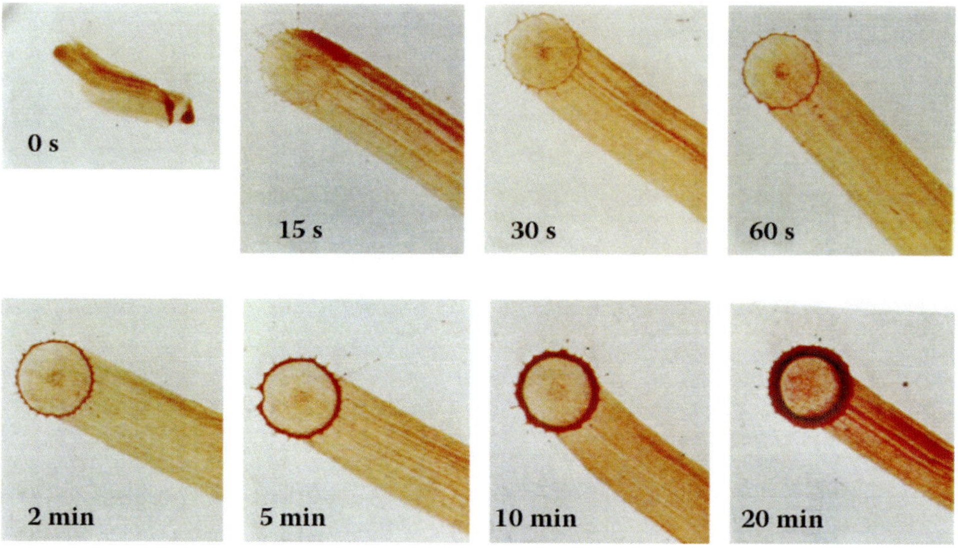

Figure 8.4 The perimeter stain effect in a blood droplet. Each droplet was deposited and then wiped through at the time indicated. Note that 30 seconds after deposit a ring is present on the outside boundary of the stain. At 60 seconds, the ring is readily apparent. This ring effect is known as a perimeter stain or skeletonization and allows us to distinguish between stains that were disturbed immediately after deposit and those that were disturbed later on. The time necessary for this ring to appear is independent of environmental factors. The time it takes the droplet to dry completely, however, is dependent upon the environmental factors.

To establish actual drying times of droplets or other volumes, the analyst must know the scene's environmental conditions. By matching critical conditions during experimentation, the analyst may be able to define with some level of confidence the drying times of stains of different sizes. Like coagulation, this merely offers a range of times required. Without the appropriate investigative context (e.g., a claim that is significantly greater than that observed experimentally), the experimental range may have no probative value. Even lacking this situation (a distinct time difference), the perimeter stain effect does help define whether a specific pattern or stain was or was not disturbed immediately after its creation.

Drying time of volume accumulations (e.g., large pools) also requires experimental effort. In many instances, the pool simply isn't dry at the time of police discovery. So, not being completely dry, and having to evaluate how dry it was, represents a subjective call up front. Caution is demanded in such circumstances.

Volume Estimations

Drying and clotting issues revolve around the passage of time since the bloodshed, which may be probative. Another common issue raised about blood pools is the total volume represented by the pool or saturation. This is particularly true in instances where no body is present. At issue is whether the wounded individual could survive based on the volume of blood lost in the scene. There are instances in which cases were adjudicated without a body, in part, because the blood volume made it clear without medical intervention the individual would die.

Such estimations are exactly that estimations. Any attempt to define this volume must consider the surface characteristics, including its absorbency, the depth of the clot or congealed blood on the surface, and the leveled nature of the surface. Such estimations require corroboration using experimental methods, which consider all the factors listed above.

Lee and coworkers developed four methods of considering this estimation. All the methods are based on the dry weight of bloodstains. They established a dry weight constant (0.4167 ml/mg) which they simply multiplied by the weight of the dry blood crust present in the stain. The most difficult part of their process is eliminating the weight of any substrate the dried crust may be present on. Two of their four methods included:

- Removing and weighing the actual blood crust present on nonabsorbent surfaces and multiplying that number by the constant.
- Weighing the substrate and blood crust, determining a weight of the substrate alone, subtracting this weight, and then multiplying the remainder by the constant.[13]

A second method developed for estimating blood volume is a wet volume estimate in which the analyst attempts to create a similar-sized pool. The first step is to measure the overall dimensions of the stain in question, defining the total area within the stain. It is important to use the same substrate in an unstained area. The analyst then pours whole blood onto the surface in a slow and methodical fashion; all the while monitoring the overall stain size as it develops. See Figure 8.5. The analyst then measures the test stains (several tests

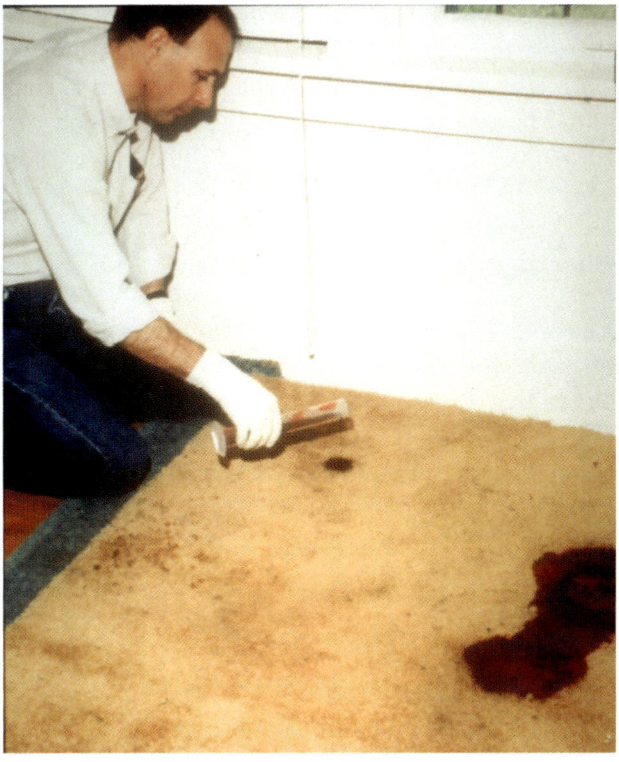

Figure 8.5 To use a wet volume estimation, the analyst creates a similar sized stain on a similar surface. Blood is added slowly and the analyst monitors the size of the stain. This is done in order to match the general dimensions of the questioned stain as closely as possible.

Figure 8.6 The analyst then measures the total area evident in the standards. This figure is compared to the area in the questioned stain. This comparison in correlation to the known volume of blood used to create the standard, allows the analyst to roughly estimate the volume of blood in the questioned stain.

may be necessary) defining their total area. See Figure 8.6. The comparison of the area encompassing the standards will provide the analyst with a rough estimate of the volume of blood present in the questioned stain. The weight of the bleeding victim on the surface, as well as anyone who might have stepped into the pool can further disperse and enlarge the pool and this should be considered in the analysis. Although certainly not absolute, these estimations of blood volume, wet or dry, may at times serve the investigative process.

Other Alterations Encountered

Altered stains occur across every type of pattern and with the single exception of the Wipe, they are not a "classification" for any stain. They are merely adjectives associated with some pattern. The Wipe holds this "classification" distinction due primarily to the probative value it offers, however, a wipe in a taxonomy mindset can be considered a complex pattern. There must be some preexisting stain that has been wiped.

Alterations of a stain or volume will occur as a result of both clotting and drying. Additional alterations such as dilution, disruption, color change, and even some form of interruption of the pattern (a void within the pattern) may occur. Such alterations may be helpful or significantly probative.

Voids

Voids are typical in many situations; however, they are most often encountered in spatter situations. A void is an area that lacks bloodstains within an otherwise continuous bloodstain pattern. Between the source (e.g., an impact, spurt or cast-off event) and the location where we find the pattern, it is always possible that some secondary target was present or

Figure 8.7 After receiving one gunshot wound, the victim fell to a position on the floor and the hat displaced from his head. A second shot was fired, producing a distinct ejection of spatter (a double-shot impact event). The voided area behind the cap and the spatter on the inside surfaces helps us understand the location of the item when the event occurred.

intervened. Spatter directed at the primary target is deposited on these secondary intervening targets. This will result in a lack of spatter in a specific area of the pattern on the primary target.

This effect, historically referred to as ghosting, may indicate the shape of this intervening object. If this object is recognized, this may help identify specific movements, or lack thereof, of these objects in the scene. As an example, consider Figure 8.7. A double-shot impact spatter event occurred to the right of the baseball cap when the victim was shot with a .45 caliber pistol. Based on the voided area on the wall behind the cap and the spatter on the cap itself, it is apparent the cap acted as a secondary target, creating the lack of staining in the otherwise continuous pattern on the wall. Figure 8.8 depicts another void in spatter.

In correlating the secondary object that resulted in the void, there is an expectation that it will be similarly stained (both in type and quantity) as was the target area around the void. The only exception would be if this object were cleaned or significantly altered in some other fashion.

It is important to caveat the consideration of voids by saying there is a big difference between a spatter pattern that is interrupted and then continues, as compared to a pattern that simply stops. See Figures 8.9 and 8.10. Voids may occur in either instance, but in the latter the analyst may be less confident that they are viewing an actual void. If the pattern ends abruptly with a distinct linear demarcation, it is a true void. See Figure 8.7. If no distinct demarcation is present, it may simply be the natural end of the spatter pattern.

Voids are often functional for establishing the position of the victim, assailant or other objects at the time of the event. For example, voids on the clothing may establish that a hand or arm was in some specific position at the moment of staining. In such instances, arms and legs may shield the clothing and skin behind them from the spatter event. The resulting voided areas may provide the analyst with a clearer picture of the individual's body position. Voids can and do appear in other forms of stains, particularly pools and contact stains.

Figure 8.8 A void in a spatter pattern. This photograph shows two important considerations with regard to voids. First, the sharper the demarcation, the more confident the analyst will be that they are viewing a void. The correlation of the void to the magnifying glass is quite apparent. Second, the item that created the void should have similar staining unless it has been cleaned or significantly altered. If the magnifying glass was the object creating the void, it will have the same level of spatter as the pattern itself. (Photograph courtesy of Sandra M. Roberts, Eugene Police Department, Eugene, OR.)

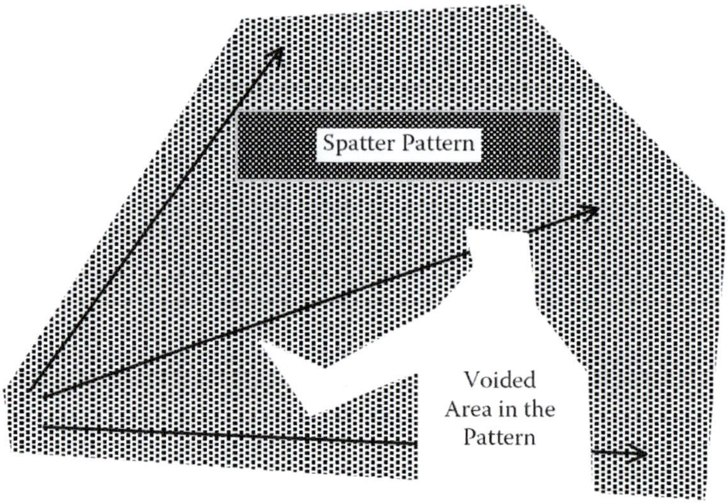

Figure 8.9 Interrupted pattern with an obvious void. Voids of this nature are easy to identify as the pattern clearly continues beyond the limits of the voided area. Often, the shape of the void will assist the analyst in identifying the object that intercepted the spatter or blood.

Case Example – Void

Figures 8.11 and 8.12 show a case involving a woman killed and left on a floor. The resulting voids in the blood show two distinct things: (1) The position of the victim's arms and torso when she was left on the floor. (2) The victim remained static in this position for an extended period after blood loss, producing the demarcated outlines. The husband claimed to have arrived as the suspect departed, suggesting the assault had just been completed; but also claimed to have moved her in resuscitation attempts.

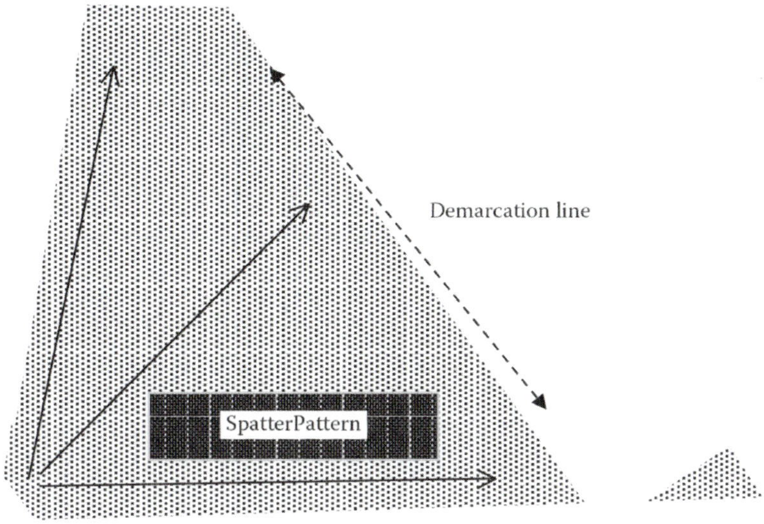

Figure 8.10 A less evident void situation. In this instance, we may still be viewing a void (e.g., the area to the right of the demarcation line) but it is also possible we are simply viewing the natural termination of the spatter pattern. In a pattern of this nature, decisions relating to whether this is a void require a level of caution.

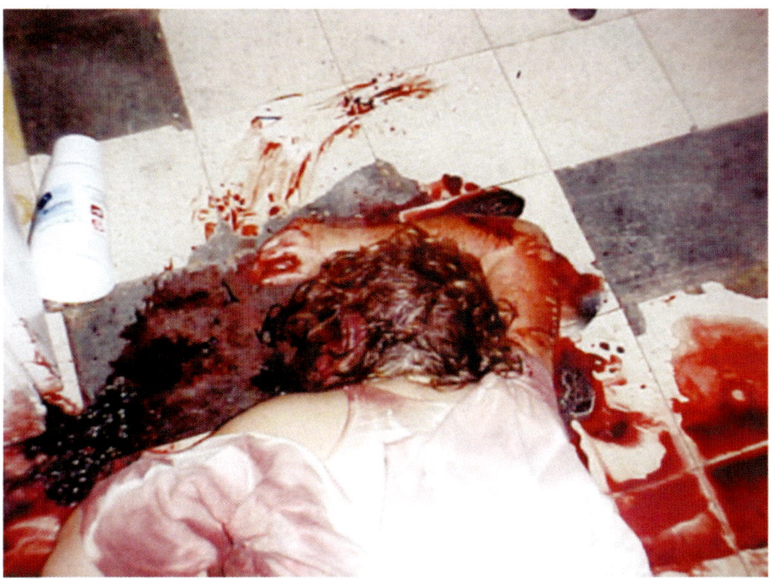

Figure 8.11 A victim present on the floor surrounded by a pool of blood.

Dilution

Another form of alteration occurs when whole blood becomes diluted. The most often encountered occurrence of dilution is with water, after blood is washed up. For example, Figure 8.13 depicts several diluted blood drops and a diluted flow in a sink. Their presence indicates that after wounding of the victim(s), someone utilized the sink to wash something (e.g., themselves or some object).

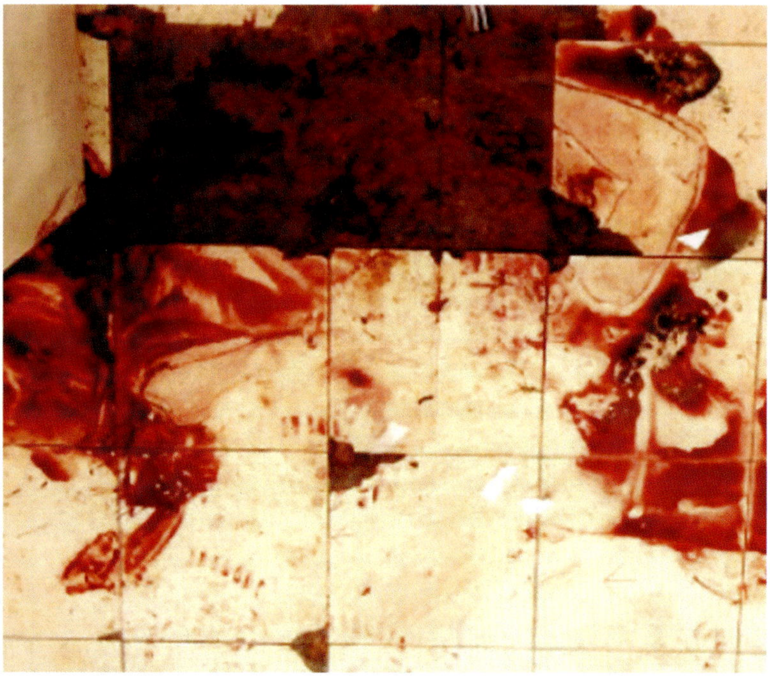

Figure 8.12 When the victim in Figure 8.11 is removed, there is a clear void present, particularly the right arm. The pool developed subsequent to her positioning.

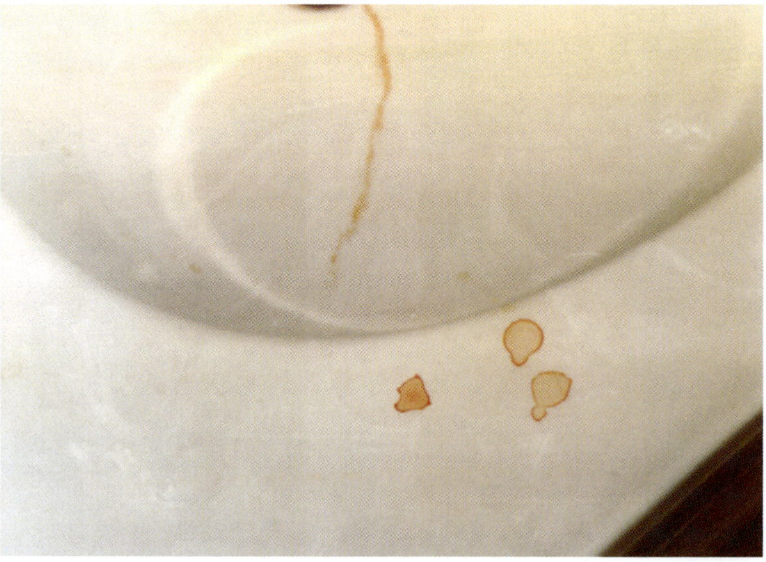

Figure 8.13 Diluted bloodstains. The stains in and around this sink were created by blood mixed with water. Alteration by dilution can aid in understanding that some other action (the diluting event) occurred subsequent to bloodshed.

In instances of spatter events associated with abdominal or head wounds, the individual spatter produced and directed to the varying surfaces may not be whole blood. The spatter, upon inspection, may be a mixture of blood and abdominal fluid or cerebrospinal fluid.

Color Change

Freshly exposed blood can range significantly in color. Arterial blood tends to be a bright red, whereas venous blood may appear a duller red due to less oxygenation. Once exposed to the environment in the scene, bloodstains tend to darken further due to oxidation. This color may range from the typical reds to a chocolate brown. The longer the stain is exposed to the air, the more likely that darkening will increase. Although this color alteration, when present, will suggest this elapsing of time, it is simply an indicator of time passage to the analyst. There are no known protocols that allow this factor to be evaluated further. There are simply too many variables that may lead to this darkening. Situations may be encountered where a given stain discolors significantly greater than other scene stains, even though the context suggests all depositions occurred essentially at the same time. These situations appear to be related to contaminants on the substrate, but this condition is not clearly understood. Color change, as shown in digital images or photographic prints, should be used cautiously to assess any bloodstain situation.

Summary

The Physically Altered Bloodstain principle is an important aspect of bloodstain pattern analysis. It is not called upon in many cases, but it should be understood. Remember that PAB conclusions are always limited. These alterations are a function of scene environmental conditions that are not always fully understood or known. Caution is always in order. Coagulation and drying time conclusions will almost certainly require experimental effort. What such experimental efforts may provide the analyst is always a minimum range of time required to produce the observed scene condition. The presence or absence of a perimeter stain is the obvious exception. Perimeter stain issues are encountered often and the condition of the perimeter stain or lack there of can be quite probative even without experimental effort.

Volume estimations require caution and most decidedly will require some level of experimentation. Even under the best of circumstances, such efforts will only provide a broad estimate of the volumes involved. But in certain investigative instances, such estimations may be quite probative.

Other alterations, such a dilution or the creation of a void, may prove probative and can be illuminating in a given context. In other situations these conditions may have no probative value whatsoever.

Notes

1. Gardner, R. and Griffin, T., Foundations of the Discipline of Bloodstain Pattern Analysis, A Response to the Report of the National Academy of Sciences, *Journal of Forensic Identification*, Volume 60, Number 4, 2010, p. 480.
2. Wonder, Anita K., *Blood Dynamics,* Academic Press, San Diego, CA, 2001, p. 100.
3. https://my.clevelandclinic.org/health/symptoms/21999-hemostasis, 18 Sept 2022.
4. Wonder, Anita K., *Blood Dynamics,* Academic Press, San Diego, CA, 2001, pp. 104–109.
5. Wonder, Anita K., What is Blood, *International Association of Bloodstain Pattern Analysts News,* Volume 2, Number 2, May 28, 1985.

6. Laber, T.L. and Epstein, B.P., Substrate Effects on Clotting, *The Canadian Society of Forensic Science Journal*, Volume 4, Number 34, 2001, pp. 209–214.
7. James, S.H., Kish, P., and Sutton, T.P., *Principles of Bloodstain Analysis Theory and Practice*, Taylor and Francis, Boca Raton, FL, 2005, p. 189.
8. James, S.H., Kish, P., and Sutton, T.P., *Principles of Bloodstain Analysis Theory and Practice*, Taylor and Francis, Boca Raton, FL, 2005, p. 192.
9. Rossi, C., Holbrook, M., James, and Mabel, D., Medical and Forensic Aspects of Blood Clot Formation in the Presence of Saliva – A Preliminary Study, *JBPA*, Volume 28, Number 3, Sept 2012.
10. Rossi, C., Holbrook, M., James, and Mabel, D., Medical and Forensic Aspects of Blood Clot Formation in the Presence of Saliva – A Preliminary Study, *JBPA*, Volume 28, Number 3, Sept 2012.
11. Pex, O.J. and Hurley, N.M., Sequencing of Bloody Shoe Impressions by Blood Spatter and droplet Drying Times, Presentation to the Int. Assoc. Bloodstain Pattern Analysts, 1990 Training Conference, Reno, NV, Nov. 1990.
12. Parker, N.L. et al., Florida Department of Law Enforcement Summary Report of the Bloodstain Pattern Analysis Research Group, Mar 1982, pp. 68–81.
13. Lee, Henry C., Gaensslen, R.E., and Pagliaro, Elaine M., Bloodstain Volume Estimation, *International Association of Bloodstain Pattern Analysts News*, Volume 3, Number 2, October 1986, pp. 47–55.

Bloodstained Fabric Issues
Fabrics as a Substrate

9

Bloodstain Pattern Analysis on fabrics has evolved over the past several years. The fabric aspect was seldom mentioned, if at all, in early "basic" bloodstain classes. Any mention generally considered the location of bloodstains on clothing with little or no consideration about the nature of the fabric itself. Even today fabric considerations are mentioned in a limited fashion in many Bloodstain Pattern Analysis texts. For example, MacDonell mentioned the effect of a fabric's surface and the consideration of differentiating spatter from blood transferred through contact.[1] *Principles of Bloodstain Pattern Analysis: Theory and Practice* devoted several pages to bloodstain patterns on clothing, primarily from the standpoint of location and documentation.[2] Wonder discussed collection of clothing from people (victims, witnesses, and subjects alike), garments recovered from the scene, the examination of clothing including the use of magnification, and case examples involving such efforts.[3] Previous editions of this book identified and discussed issues associated in evaluating stains on fabric, as well as its problematic substrate characteristics. With the development of numerous Bloodstain Fabric courses of instruction, more attention is now directed specifically at this substrate.

Within this history and these developments there appears to be a continuum of sorts with respect to Bloodstain Pattern Analysis and fabrics. At one end of that focus is consideration of the location of bloodstains on fabric, particularly clothing. The other end of this continuum currently delves deeply into the construction of fabrics. We will attempt to answer both of these considerations without losing sight of the primary issue, recognition of what type of bloodstains may be present on fabrics encountered and what, if anything, they may tell us. It is imperative the analyst is aware of the potential effect of fabric construction in general, as fabric substrates tend to mute class characteristics of basic patterns and in some instances may prevent effective analysis. Current "BPA Fabric" courses effectively concentrate on the limitations imposed by a fabric substrate; the focus is not on what one can do in such instances, but rather what one can't do.

The issue of whether a detailed understanding of the construction of fabrics is necessary raises an obvious contradiction to the science. When considering typical bloodstained scene substrates (walls with varying types of paints, floors of different construction) it is understood that some substrates encountered may mute, alter or completely prevent standard BPA evaluation. For example, soil surfaces, rough or irregular surfaces, and hydrophobic surfaces can eliminate the ability of the analyst to pursue aspects of BPA. One does not "study" the various substrates of walls, floors or other objects in detail. One simply recognizes whether a pattern is present on that substrate and whether that pattern has appropriate characteristics that can be evaluated. This is an extension of Woodward's concept of Invariance Under Intervention. Under some circumstances, the basic predictability of BPA can be affected by *any* substrate, to include fabric substrates.

If one does not pursue detailed study of all the various paints encountered, or the varying materials of construction encountered in the typical scene, the authors then pose

DOI: 10.4324/9781003408734-9

a simple question: What purpose does understanding the detailed construction of a given fabric offer the analysis? Either a "pattern" is present on the substrate or it is not. That pattern either demonstrates the characteristics considered in Pattern Diversity, Stain Shape and Vector Correlation, or PAB or it does not. So, in the estimation of the authors, any approach for fabrics is no different than any other substrate, but it is important to recognize that fabric construction has a tendency to interfere or limit the analysis. Granted, analysts have always recognized problematic scene surfaces (including fabrics) and it is certainly important to recognize particularly problematic fabric surfaces (e.g., waterproofed fabric, silk). But claiming that every time a fabric is involved, the analyst must recognize its specific characteristics is a statement that flies in the face of typical BPA analysis. Fabric substrates are no different than any other substrate; the question is whether a pattern is present and if present to what level can it be evaluated.

Fabrics are a common substrate. It is rare indeed for the analyst not to be confronted with various items of bloodstained clothing and other fabric surfaces during an analysis. These are likely to include either the victim's or suspect's clothing, as well as any bedding. The analyst is just as apt to encounter bloodstained draperies, and furniture surfaces. Fabrics do not significantly alter the analyst's approach to Bloodstain Pattern Analysis, but they do create specific concerns:

- Understanding that a fabric may mute class characteristics, preventing a complete understanding of the pattern type
- Differentiating contact from spatter stains
- Understanding the limitations of directionality and impact angle imposed by fabrics
- Recognizing issues associated with pattern transfers
- Evaluation approaches
- Documentation procedures
- Overcoming poor collection/documentation procedures

Overview of Textiles and Fabrics

Whatever else one might say, fabrics are diverse and complex substrates. The behavior of fluids on a fabric is a function of several factors including composition, construction, texture, condition, and absorptivity. The two types of fiber encountered are natural and synthetic. See Figure 9.1. Individual fibers (two or more stands) make a thread or yarn, which is then used in different fabric construction arrangements.

Cotton is perhaps the most common natural fiber in terms of everyday clothing construction. Wool likely comes in second after cotton. The short lengths of these natural fibers are spun together to create extended complex threads. As a general consideration, such natural fibers tend to retard capillary action as they are more porous, which typically will increase absorption of fluids.

Synthetic fibers are constructed chemically or manmade. As such, the length of these fibers may be indefinite. Again, as a general consideration, synthetic fibers that are not designed specifically as liquid repellant enhance capillary action due to unbroken, contiguous surfaces. Being non-cellular they tend to be less porous retarding absorption.

The construction of a fabric will be one of two basic types, a knit or a weave. See Figure 9.2. A knit is made up of a series of what is referred to as interlocking loops.

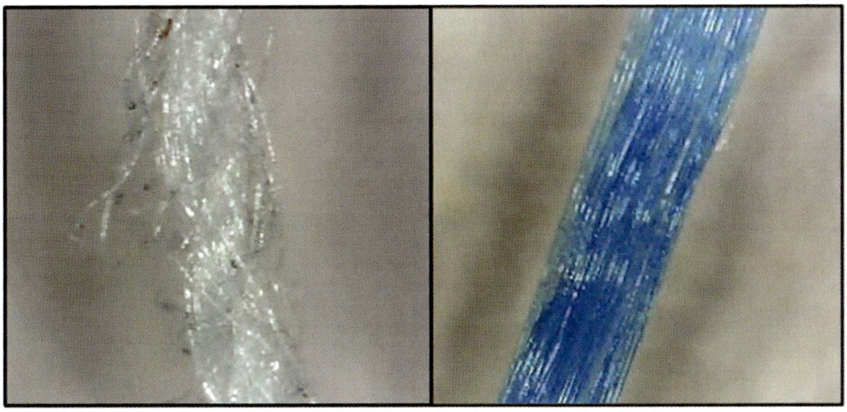

Figure 9.1 Fiber composition. The fiber on the left is a natural fiber, cotton. The fiber on the right is a synthetic fiber, nylon.

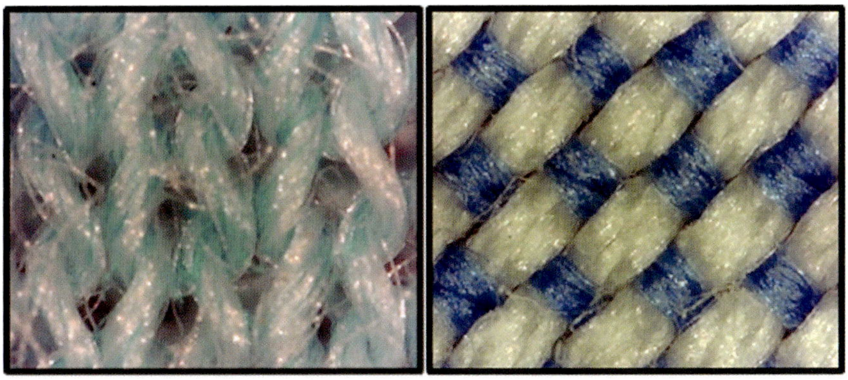

Figure 9.2 Fabric construction. The image on the left shows an example of knit construction. The image on the right shows an example of weave construction.

A weave is made up of a series of alternating interlocked threads. There are many and varied forms for the construction of both knits and weaves.

For knits, these include the basic warp and weft knits as well as a multitude of subtypes, including Jersey, Rib, and Interlock. For weaves, this includes plain weave, satin weave, twill, open and closed weaves.

A fabric's feel, texture, or finish can be manipulated during manufacture. Examples of such manufactured processing include physical actions (e.g., crushing, rolling), chemical treatment, or coatings such as applique. The results of these processes can create variations in the same basic fabric's finish, producing a dull or lustrous finish or perhaps a fine or coarse finish.

During or subsequent to manufacture, fabrics can be treated in any number of ways for various purposes. For example, a fabric can be treated to make it flame-retardant, stain-resistant, or water-repellent. Even laundering techniques can alter the nature of fabric (e.g., starches and fabric softeners). There are more than 70 different "treatments" for fabrics.[4]

If this multitude of variations in fabrics were not enough, the condition of the fabric at the time of deposition can create significant variation. Examples may include whether the clothing is new, worn, or laundered. Even whether a fabric is wet or dry at the time of deposition can alter its behavior. Either of these two conditions may affect wicking or diffusion characteristics in addition to other actions.

Let's consider this multitude of variations with a simple hypothesis associated with evaluating bloodstain patterns on fabrics:

> *Hypothesis: Fluid (specifically blood) deposition is affected item to item with respect to its composition, construction, texture, treatments, and condition. Thus, to analyze a bloodstain on a fabric, one must understand all of these variables.*

Were this hypothesis correct, to accomplish Bloodstain Pattern Analysis on any given fabric requires the analyst to consider *specific details in depth* for each of the listed aspects, effectively demanding that one become a textile expert. Even then, the unknown variables associated with the condition aspects of the specific fabric involved would effectively eliminate even the "fabric expert's" ability to understand the fabric. Who is prepared to say they know the full history of a fabric recovered at a scene? This entire proposition seems a bit harsh to be a valid claim and would effectively demand that Bloodstain Pattern Analysis on fabrics should not be accomplished by anyone.

As noted earlier in this chapter, this proposition flies in the face of everyday efforts associated with Bloodstain Pattern Analysis. Fabrics are certainly not the only substrates on which bloodstains occur. Nearly every other scene substrate involves significant variability with respective to their characteristics. Factors such as the effect of surface texture, treatments (e.g., paint, finishes), its ability to absorb, and general composition (linoleum versus concrete), as well as the condition of the typical substrates routinely encountered are dealt with on a case-by-case approach. Ultimately, the bloodstain pattern analyst is considering a single issue: Does the stain observed present sufficient characteristics associated with the predictable aspects of Bloodstain Pattern Analysis to allow evaluation to some degree?

The authors emphasize that the same identical approach is used for fabric substrates and is appropriate. Evaluate the stain, know the general limitations imposed by fabrics, be aware of and know critical substrates and the issues associated with them (e.g., silk and water-resistant fabrics). Then, within those acknowledged limitations, determine the characteristics of the stain under consideration and evaluate those characteristics in a methodical manner.

Significant Considerations When Dealing with Fabric Substrates

Taxon Concerns – Are Physical Characteristics Muted?

As described, fabrics are complex substrates. They may absorb or not, and there may be varying degrees of surface irregularities present. Any of these issues may affect the deposition behavior of blood. See Figure 9.3. One of the most significant issues encountered is a muting of the physical characteristics of the resulting stain. A specific taxon may be absent,

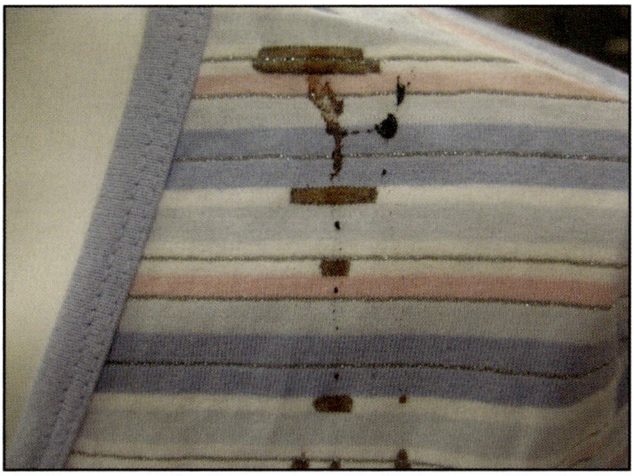

Figure 9.3 Fabric characteristics can significantly affect the behavior of the blood as it makes contact. In this instance, a flow interacts quite differently with the varying stripes in the knit. It fails to absorb on most and then easily absorbs with evident capillary action on the white dyed area. (Photo courtesy of Gillian Leak, Principle Forensic Services, UK).

making specific classification difficult. The analyst may be left classifying the pattern in a very general fashion, or if all taxa are absent, preventing any analysis.

It is common on fabric that classifications are highly ambiguous (e.g., spatter or linear spatter) and or even functionally impossible (e.g., this is a bloodstain). That said, if the taxa are fully present or present in some form then classification may be possible. For example, Figure 9.4 shows a stain that is some form of linear spatter.

The bottom line for the analyst is simple. If a "pattern" is present with obvious characteristics, evaluate it as you would any other pattern; but expect ambiguous patterns. With ambiguous or limited taxa, the results are ambiguous classifications. Do not force a classification in these situations. If there is any reason to believe that the fabric itself is creating the evident characteristics, conduct experimentation as required using the same fabric if allowed. In this instance, unstained areas of the item that are not probative are cut out for experimentation (obviously away from the item of evidence).

Concerns in Differentiating Contact from Spatter Stains

Blood has a medically recognized adhesive quality; because of this property it should not surprise us to find bloodstains on those who discover victims of crime. Such witnesses may touch, cradle, or move the victim in some fashion, which in turn will result in the transfer of blood onto clothing and skin. The likelihood of such transfers are not lost on the perpetrators; they too may claim incidental contact as the source of stains found on their clothing. The reader may recall Dr. Ziemke's reminder about the necessity of recovering the clothing of those who discover the victims of beatings and shootings.[5] That concern is still true today, since the analyst can often corroborate or disprove statements of such individuals based on the bloodstains present on their clothing.

One of the more critical considerations made using clothing is distinguishing contact from spatter. In certain situations, it is possible for very light contact of a bloody object to mimic or appear to be a spatter stain. When presented with such a stain, it is necessary for

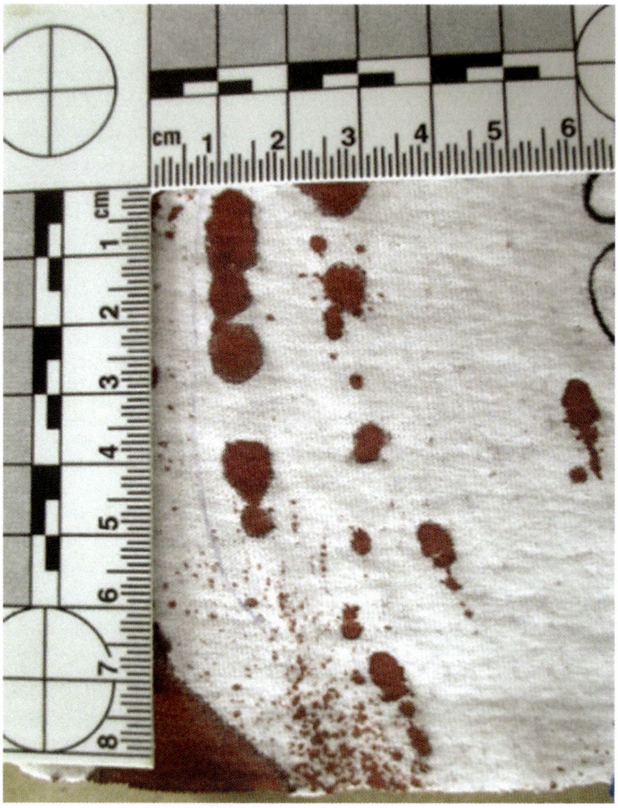

Figure 9.4 If the taxa are fully present or present in some form, then classification may be possible. It is a general expectation that the taxon may be muted on fabrics. This stain on white cotton is some form of linear spatter, but lacking further characteristics, no further classification is possible.

the analyst to distinguish between spatter (created by dynamic events) and stains created by mere incidental contact. When contact stains are found as a large pattern or a large saturation, the differences between contact and spatter are easily distinguished; however, in the case of very limited contact, the resulting stains are often quite small, and their true source may not be immediately evident.

Even in the case of these very small contact stains, the analyst can often make this distinction. Remember, the two stains result from distinctly different actions. One is projected onto the target, while the second merely contacts the target. When viewing the contact stain under low magnification, the analyst will observe blood traces on the upper weave of the fabric. As these light stains are generally the result of a limited smearing action with lateral motion, the deeper areas of the weave are often left untouched by the blood source. Even in situations in which there is compression contact with saturation, areas within the weave may still be free of blood. Figure 9.5 demonstrates this appearance under 60× magnification on 50/50 cotton/polyester material.

Spatter, however, strikes the clothing in a projected fashion. The small droplets impact into the weave, rather than brushing against it. When viewing a spatter stain, the lower levels of the weave are likely to exhibit staining. Figure 9.6 shows two projected spatter stains on 50/50 cotton/polyester material. Note the distinct differences between the stains

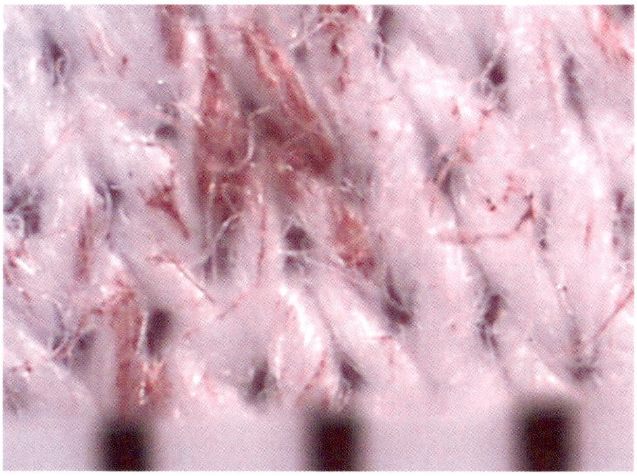

Figure 9.5 Contact staining of white cotton under 60× magnification. Direct compression contact by a bloody item generally creates a saturation stain that is easily recognizable as contact, but light contact may mimic spatter. Under magnification, however, the contact stains appear primarily on the top surface of the weave because the nature of contact is very often some form of lateral motion (e.g., a swipe).

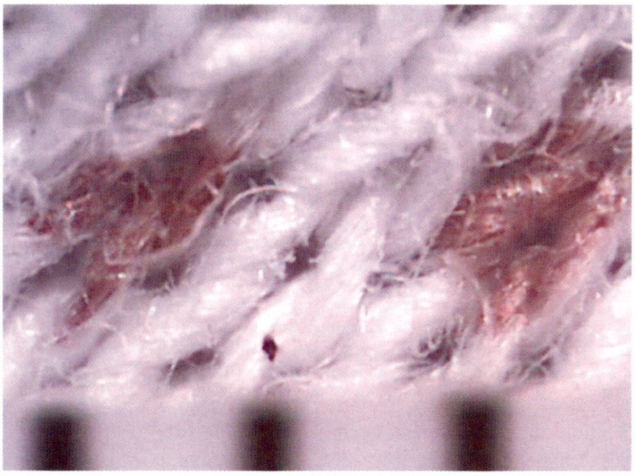

Figure 9.6 Spatter stains on white cotton under 60× magnification. Spatter is projected into the cloth, so not all of the stains will be on the top surfaces. Note the two stains have blood deep into the depths of the weave. Not every spatter stain on cloth will demonstrate this appearance, but the presence of several will validate that the pattern is spatter and not contact.

in Figures 9.5 and 9.6. Pex and Vaughan found this type of examination particularly useful in situations involving gunshots.[6] In their efforts to distinguish between contact staining and back spatter situations, they found that small back spatter droplets were often evident deep in the weave of a fabric. Keep in mind that with regard to small mist-type stains, the droplets can dry before impact and will not adhere as a liquid droplet would.

The problem with this examination approach is that some contacts may be sufficient to produce larger stains on top of the weave (unlike those observed in Figure 9.5) and not every spatter drop impacts deep into the junction of a weave. Spatter drops are so small that they can and do impact on upper surfaces of the threads. The stains resulting from such impacts will appear on upper surfaces as well, so there can be distinct similarities between staining seen in contact and that seen in a spatter event. To make the call with some level of confidence, the analyst should seek several clear examples of spatter that have been deposited into a weave juncture. In the weave, differences between the two stain types will become clearer. The presence of such stains deep in the weave allows the analyst to distinguish the stains as projected stains rather than some form of contact. In a contact stain, lateral movement is very often seen in any transfer. It is very difficult to make contact with a straight in and out movement with no lateral motion, thus the staining occurs on the surfaces of the weave. In an actual scene environment, evidence of this lateral movement is almost always present.

As always, it is important to contrast the condition of the scene with the clothing items. Is there evidence of an event capable of creating spatter or light contact in the scene, one consistent with that found on the clothing? Ultimately, if clear examples of contact or spatter are not present, the analyst should not be afraid to say, "I can't resolve the issue."

Contact and typical spatter stains are not the only form of stains encountered in such an analysis. Figure 9.7 is an interesting example of a stain that appeared on cursory view to be related to a spatter event. The stain certainly gives the appearance of meeting the spatter definition. The stain was deposited and recovered from the right chest area of the T-shirt, taken from a suspect in a shooting. See Figure 9.8. The victim was shot once in the head. During an examination using proper lighting and magnification, it became apparent the stain consisted of coagulated material. A subsequent DNA analysis was also interesting. The physical stain (not the substrate) when tested consisted of two DNA sources, the primary contributor was the suspect and the secondary contributor was the victim. Although no specific mechanism of creation could be identified for the stain, what was evident from the examination was that the stain did not result from spatter created as a result of the gunshot.

Figure 9.7 A stain on a piece of cotton T-shirt that upon initial evaluation was considered to be spatter. The T-shirt was recovered from a man who was suspected of killing another person with a shotgun.

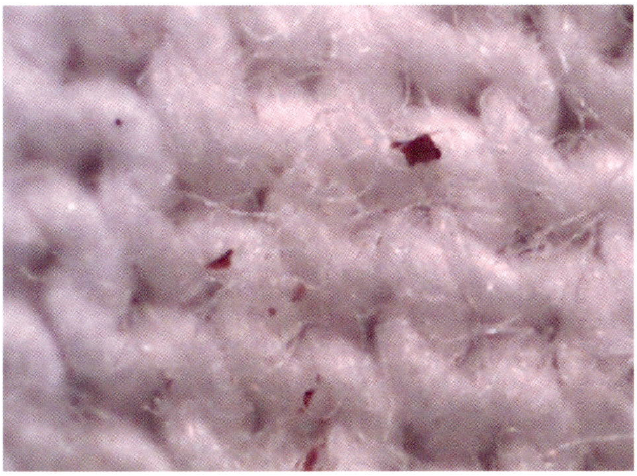

Figure 9.8 The "spatter" from Figure 9.7 viewed under 60× magnification. The stains were clearly the result of some mechanism that occurred after the blood coagulated. DNA analysis added another wrinkle to the initial evaluation. The physical stains were tested (not the substrate) and found to be a product of two DNA sources. The primary source was the owner of the T-shirt, and the secondary source was the victim. Whatever the actual method of creation, these stains were not the result of a spatter event involving fresh blood.

Recent claims suggest that the analyst must use a magnification of at least 100× to be confident when trying to distinguish contact from spatter. Empirical experience shows any magnification 20× and up is usually more than adequate to see the described differences.

In 2005, an interesting issue regarding contact versus spatter was raised in a highly contentious murder trial. A group of eight stains was found on the T-shirt of a murder suspect, which he claimed were the result of contact when he discovered the victims. At trial, a concept known as "painted fibers" was raised by the defense experts and presented to the court as proof that the stains were in fact, the result of contact and could not have originated from a spatter event. The defense relied so heavily on the idea that the word "painted fiber" was used in excess of 100 times in front of the jury by the defense experts.[7] This was all well and good with one exception, before that moment, no one in Bloodstain Pattern Analysis had discussed, written anything on, or even heard the term "painted fiber." Thus, this evidence had never been exposed to any peer-review or additional research.

The first concern is: What is a painted fiber? There have been discussions with respect to two different appearing stains and whether each can be called a "painted fiber." A painted fiber can appear either coated with blood or with a small "drop" that appears to encapsulate the fiber. That is distinct from simple blood deposition. See Figures 9.9 and 9.10. The best physical description of the painted fiber is that a small, often spherical droplet is attached to or wrapped around a single fiber. This fiber is usually disassociated and separate from other staining. Empirically, the presence of "painted fibers" is a given, but what probative value does such a stain provide?

In the court case, the primary issue revolved around a series of small stains that appeared to be spatter. As typical of these types of situations, some of the stains appeared spatter-like, others less so. Given the presence of a painted fiber, the underlying premise was *if* a painted fiber is only produced through contact, and there are painted fiber(s) present; *then* the stain's source must be contact. The claim, if true, had significant value to the jury.

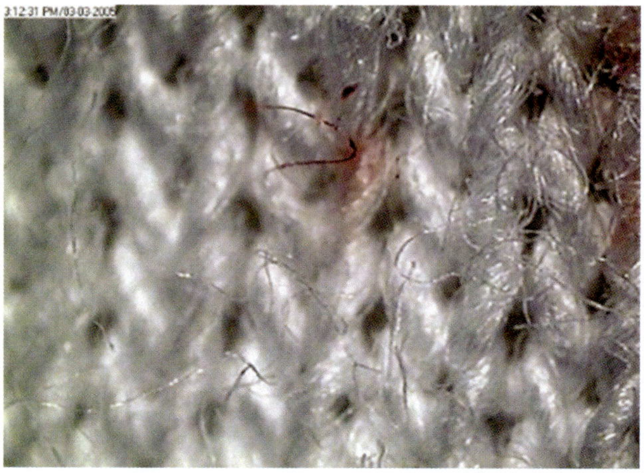

Figure 9.9 Painted fibers occur in a variety of situations including simple contact and spatter deposited on fabric, particularly when the fabric is wet or damp. As in this instance on white cotton, the fiber can appear coated with blood.

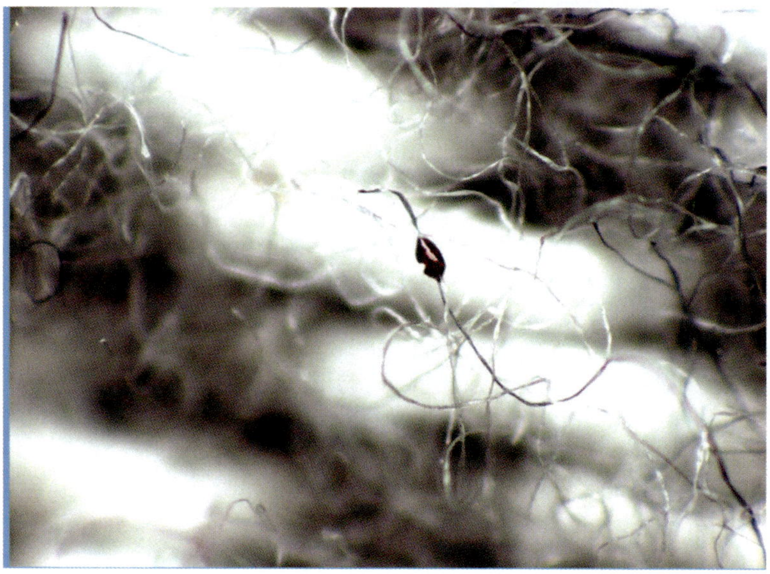

Figure 9.10 Another form of painted fiber on cotton, where the fiber is effectively encapsulated by a small volume of blood.

Since the introduction of this claim, the authors have routinely experimented, attempting to create painted fibers through various mechanisms. This includes introducing the idea into the course curriculum in an advanced BPA class. The painted fiber premise is presented as a background for students learning to develop experimentation. Students are presented with the hypothesis (Painted Fibers are only produced by contact) and then left as a group to devise experiments to prove or refute this hypothesis. The results produced by the student groups over nearly a ten-year period are straightforward. Painted fibers are produced through multiple mechanisms, including contact and spatter-type events. Contact circumstances produce more painted fibers, but painted fibers are repetitively produced in spatter events. Regarding the spatter events, the condition of the cloth is a factor. Slightly moist or wet cloth routinely produces more "painted fiber" stains. On dry cloth, they are less likely to be produced. Thus, we are left to consider that in terms of "is it contact or not," the painted fiber has no probative value. Perhaps some additional probative aspect may be determined in time. But for now, the painted fiber issue appears moot.

Another basic consideration regarding spatter on fabric is how difficult it is to see. On white or light-colored fabrics, the contrast of these small stains usually makes them evident on any initial visual inspection. But if the fabric's color is dark, visual examination of the clothing without proper lighting and magnification may make the stains all but impossible to see. The larger the fabric's weave, the more difficult it will be to see. See Figure 9.11. Keep in mind that blood does not fluoresce, and oblique white lighting will likely provide the best visual examination environment. IR photography of dark clothing may provide a functional means of bringing out patterns on such fabrics. See Figure 9.12

In one study conducted by Holbrook regarding spatter and contact issues, Holbrook concluded: "[T]he overall appearance of individual stains including size, shape, and penetration of the weave/knit, should be considered together with the characteristics of the fabric on which they are deposited."[8]

Concerns Regarding Directionality and Impact Angle

Surface characteristics are always a concern for the bloodstain pattern analyst, since target surface affects the collapse of the droplet and thus the symmetry of the resulting stain. The presence of the typical spatter stain offers additional information relating to the direction and angle of the spatter's deposition onto the substrate. Fabric substrates can easily interrupt such considerations. As was previously discussed, a symmetrical outflow of fluid

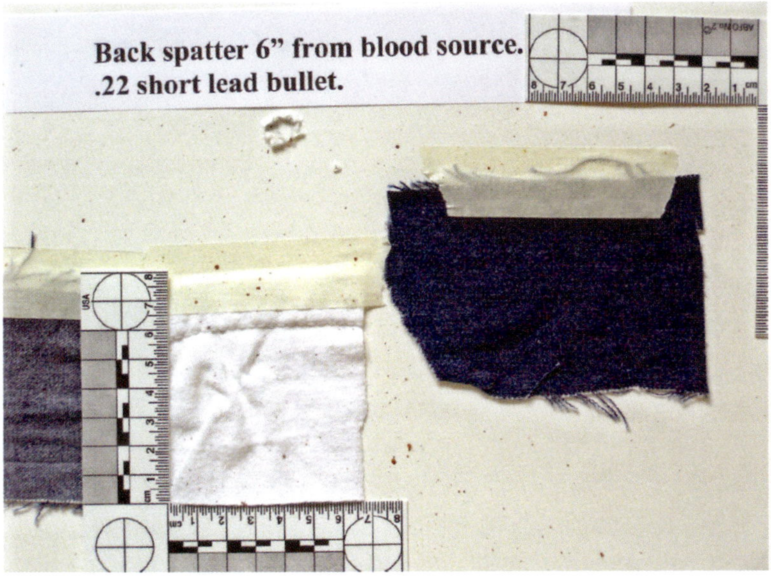

Figure 9.11 The very nature of a fabric can mask the presence of spatter. In this instance, white cotton, a cotton blend, and jean material were exposed to a back spatter event from a weapon. The staining on the dark jean material is difficult, if not impossible, to see without the assistance of proper lighting and magnification.

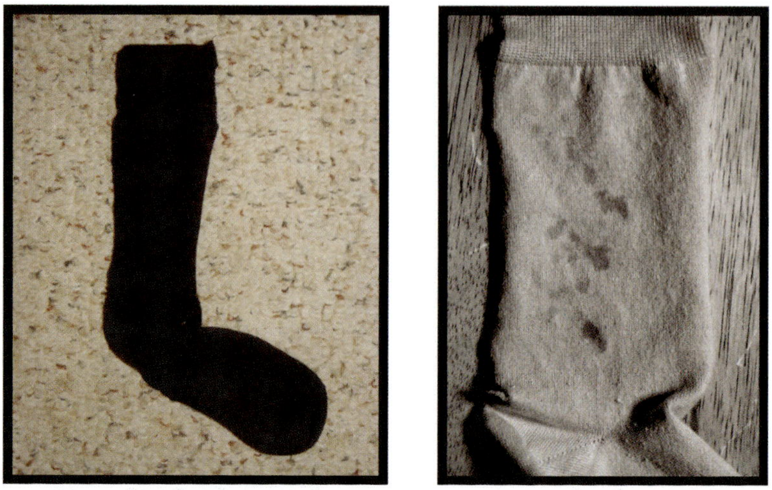

Natural Light Photograph **IR Light Photograph**

Figure 9.12 IR photograph produced with a Fuji UVIR camera body with 60 mm Nikon lens and a Peca 89A filter. Camera was set to manual mode and focus. Shutter speed set at 1/15th s and *f*/4.8. ISO 400 using ambient soft light and no flash. (Photograph courtesy of Jason Guffey, Central Piedmont Community College, Charlotte, NC.)

Figure 9.13 When considering bloodstained cloth, caution is always required. These stains on the jean material might suggest that the drops struck at a slight angle, but in fact, they are a product of a 90° impact. When angles of impact are reported on clothing or fabric, they are described in a general fashion (e.g., struck at an acute angle, struck generally perpendicular).

during the droplet's collapse leads to regular or symmetrical stain shapes. Asymmetrical outflow leads to irregular stain shapes. Various fabrics can present some interesting effects on stain shape and certainly limit the analyst's ability to understand impact angle or directionality. Distortion of the stain on a fabric's surface is a function of the texture of the fabric as well as its absorption characteristics.[9] That distortion makes impact angle determination suspect. For instance, in Figure 9.13, the stains on the jean material have what

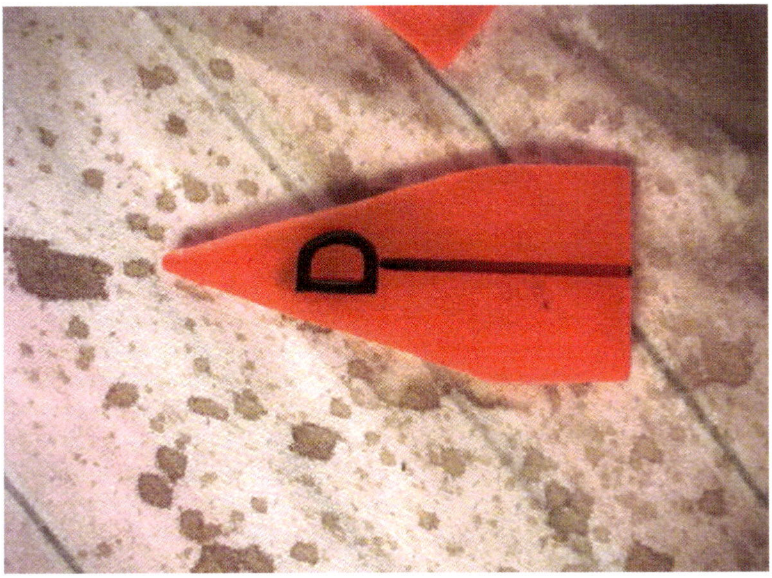

Figure 9.14 In considering directionality, these stains offer little assistance based on the shape of the stain alone. First, many are asymmetrical. Second, and more importantly, there are few if any tails or satellite spatter to define which side of the ellipse is the leading edge and which side is trailing edge.

appear to be elliptical shapes; however, the drops fell at 90 degrees to the fabric. Certain fabrics, particularly those made up of a high percentage of polyester or nylon and those treated with a water-resistant chemical and or a fabric designed as water-resistant, can produce spatter stains that appear rectangular. Particularly in the latter, the stains may not even remain on the clothing. Droplets can fall to the fabric's surface, dry and then fall off the surface without being absorbed into the material.

If the shape of the stain affects impact angle calculations, it can also affect directionality issues. Lacking in many spatter stains on fabric are satellites or scalloped edges. These characteristics normally assist the analyst in recognizing directionality. On fabric, the stains can appear as merely ellipses. See Figure 9.14. In these instances, a single stain may provide no indication as to directionality. Yet, analysis of directionality in these instances is not completely lost, as the radiating nature of the pattern may provide an indication for the pattern's directional deposition. See Figure 9.15.

First and foremost, it should be stated that the error of impact angle calculations on even the best fabric substrate is much higher than that encountered in other typical crime scene substrates, sufficient in some instances to completely disguise the true angle of impact. Error rates may be as high as +/– 15 degrees. This is not always the case. For example, Figures 9.16 and 9.17 demonstrate spatter deposited onto a shirt. Not only are typical spatter stains produced, but the basic characteristics of impact (stain shapes becoming more elliptical the farther out in the pattern) are also evident. Finding typical elliptical shaped stains *with tails and satellites* tells us the droplet *is* collapsing in a generally normal fashion. Thus, the general directional and impact angles are viable. On such surfaces, the typical variation of shapes (e.g., circular to elliptical shapes) found in impact events, all of which are a function of the angle of deposition will likely be present. But given the error associated, the best advice is to present any claims of impact angle on fabric as "the

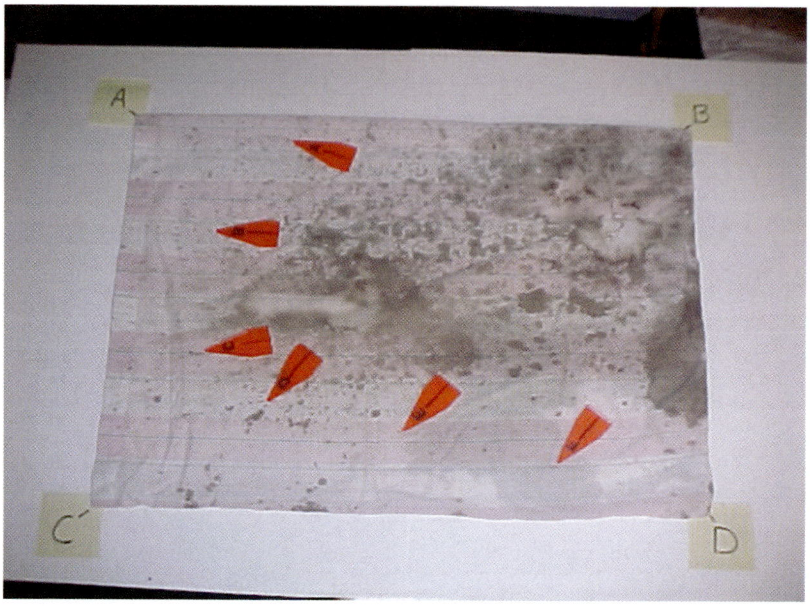

Figure 9.15 Directionality on fabric or cloth is not always a complete lost cause. If we consider the overall pattern in Figure 9.14, along with the major axis of the involved ellipses, we can still visualize the radiating nature of the spatter pattern, which effectively defines the directionality.

Figure 9.16 A spatter deposited on a cotton shirt. A classic elliptical shape formed along with a tail and satellite. This demands the droplet was collapsing in a typical fashion.

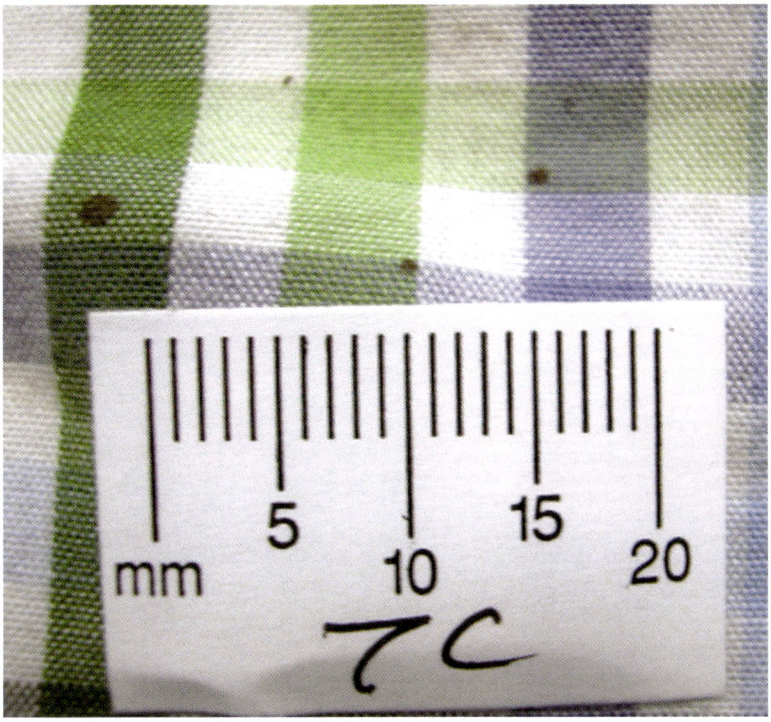

Figure 9.17 Several spatter from the same pattern and shirt as shown in Figure 9.16. There is a typical range of shapes evident in the pattern (the further out in the pattern, the more elliptical the stain), which lends confidence that the droplets are collapsing in a typical fashion.

drops struck acutely" or "the drops struck generally perpendicularly." Consider stain shape cautiously, as you would on any other surface. Claims of specific impact angles on fabric should be avoided.

If there is any question about why an "ellipse" is present, the analyst can evaluate the fabric's behavior. This is easily accomplished using a small non-probative area. Cut this section out and direct the spatter onto the surface, as well as dropping passive stains onto the surface. The bottom line is to be wary of fabrics/clothing and be very careful of conclusions regarding impact and directional angle issues.

Pattern Transfer Concerns

When the bloodstain pattern analyst evaluates items for pattern transfers, they generally look for curves and distinct demarcations (some line or lines that should not occur naturally), which may suggest an object of some nature. Bloodied clothing wadded up and in contact with itself can produce linear demarcations that may mimic pattern transfers. The patterns produced may include distinct curves that, on first blush, may suggest knives or other objects as their source. Therefore, when evaluating pattern transfers on clothing or fabric, caution is always necessary.

The best method of resolving this issue is to cautiously consider any single-dimension patterns (e.g. a single demarcation) and always look for multiple landmarks (e.g., multiple demarcated lines at angles to one another) in the pattern transfer. In the latter situation, the analyst can be more confident that it is a pattern and unrelated to clothing rubbing

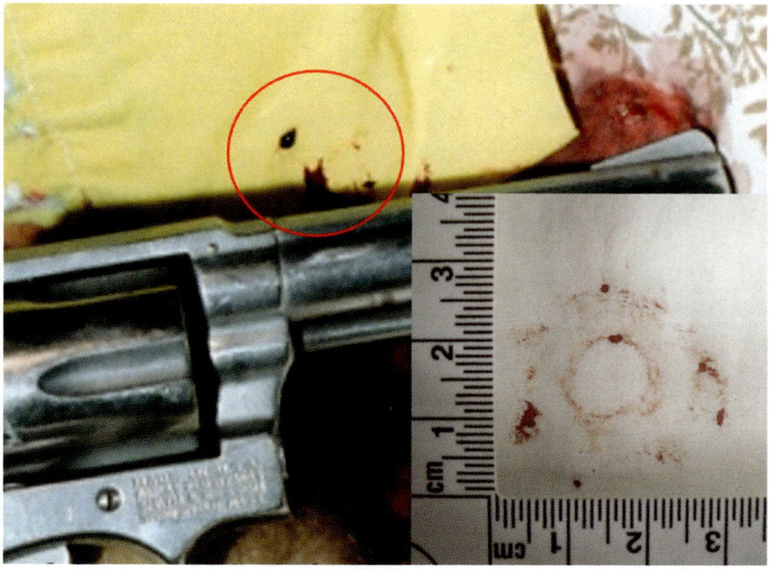

Figure 9.18 When evaluating a given pattern transfer, the analyst should be able to recreate it under similar conditions. In the scene of an alleged suicide, a small circular pattern was found on the associated pillow case. It consisted of an inner voided circular feature, surrounded by a bloodied circular feature. Additionally at the top of the pattern were two vertical protrusions. Using the muzzle of the weapon, a similar pattern was produced.

against itself. The analyst should then be able to reproduce the pattern through the alleged source mechanism.

Case Example: Pattern Transfer

In a reported suicide, a small pattern transfer was noted on the pillow case above the weapon, as it lay *in situ*. The pattern consisted of an inner circular feature void of blood, surrounded by a bloodied circular feature. There were also two small vertical protrusions at the top of the pattern. Using the weapon (after all additional forensic work was done) the barrel was bloodied and produced the insert pattern by bringing the muzzle portion of the barrel into contact with the fabric in a vertical orientation. See Figure 9.18. The probative issue was simple. If the weapon was merely dropped to its position on the pillow by the victim, how did the end of the barrel make contact with the pillow case in the required vertical orientation?

Additionally, the analyst should consider enhancing the pattern with a chemical treatment. Such enhancements, as discussed in Chapter 12, may bring out additional details in the pattern transfer, making its source more evident. Latent (invisible or only slightly visible) pattern transfers on any substrate often provide greater detail than a patent (visible) pattern transfer. The transfer of a pattern on a light-colored fabric may be obvious, yet enhancement may help visualize unseen portions of the pattern. When examining dark fabrics, begin with a strong white-light source, directing the light at different angles onto the fabric. Rotating the fabric during the examination may help as well. This can change the viewer's perspective, assisting in discerning the pattern.

Evaluation Approaches

The various colors associated with the fabrics encountered often create issues during any evaluation. Certain color combinations can make it difficult to visualize patterns or pattern transfers. Dark clothing can make it nearly impossible to see the stains. Ultimately, when stains are present, the analyst is responsible for capturing photographs of the stain or pattern, both for further analysis and for presentation to the jury. Chemical enhancement techniques, typical crime scene ALS techniques, the use of infrared cameras, and even the digital manipulation of colors in a digital photograph can assist in this effort.

The varying nature of the dyes utilized in the creation of fabrics makes it impossible to provide general approaches (e.g., what light source to apply). White light, oblique lighting, and as described, altering the angle of the light onto the surface is always the first step. After that, various light source techniques should be attempted. Remember that blood will not fluoresce. Blood, however, absorbs both UV and IR light, which causes the stain to appear darker. A particular fabric dye may behave differently, fluorescing or lightening under some particular light source.

This is true in IR wavelengths. The bloodstain (as well as gunpowder depositions) will darken dramatically under IR, whereas many dyes associated with fabrics will lighten. In some instances, even printed, dyed, or woven patterns on the fabric can be removed using IR. See Figure 9.19. Unfortunately, the dye's behavior to IR wavelengths is an absolute crapshoot. It is not singularly a function of the material of the fabric; it is the dye that defines this behavior. For example, some jean material will lighten significantly, whereas other jean material will not lighten at all. See Figure 9.20.

Applying Good Fabric Preservation and Documentation Procedures

BPA, as with most evidence, relies heavily on its recognition at the earliest possible moment of an investigation. In many instances, recognition of this evidence by the first responding officers and on-scene medical personnel – EMTs/Paramedics is critical. If bloodstain evidence, however seemingly minor, is overlooked or ignored by these individuals, the investigative information that a bloodstain pattern analyst may derive may be limited to nonexistent.

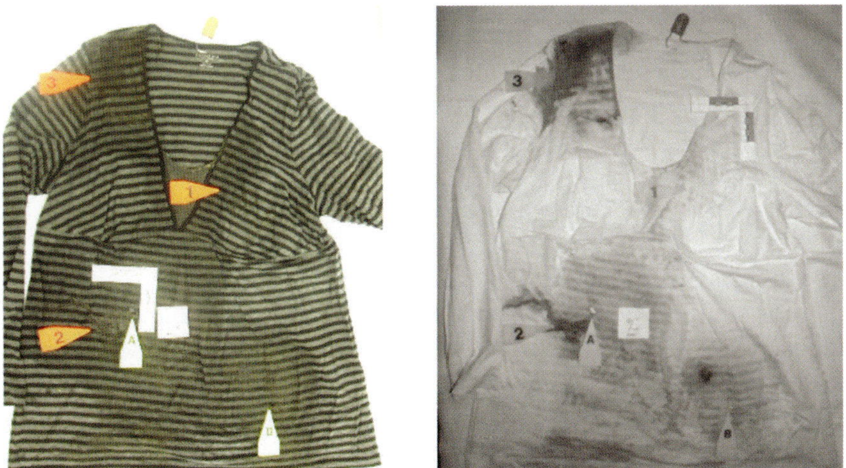

Figure 9.19 Given the variety of colors and patterns associated with clothing items, infrared photography can be of some help when trying to visualize bloodstain patterns. The photograph on the left is the item under natural light, the photograph on the right is under infrared light.

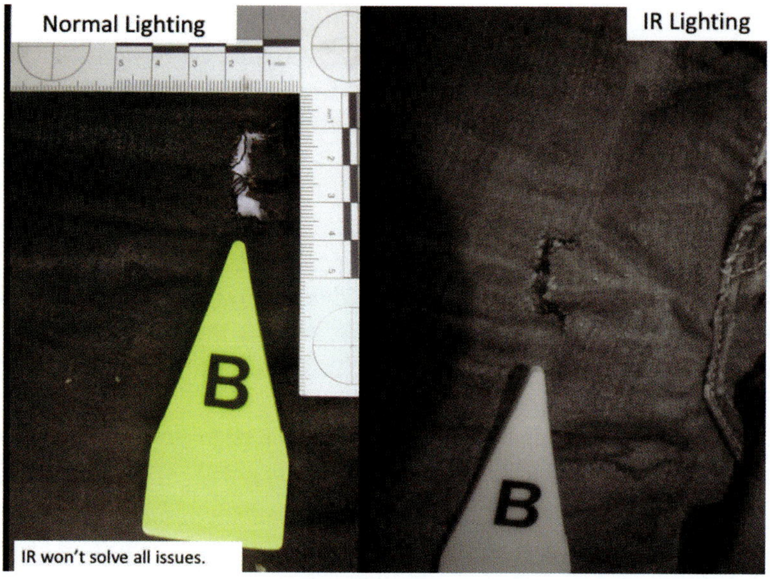

Figure 9.20 The behavior of a fabric under infrared light is very much an issue of the dyes utilized. Jean material is an excellent example. Some jean material dyes will lighten significantly, while others, as in this case do not lighten at all.

The first responding law enforcement officers have many immediate responsibilities upon arriving on the scene. During cross-examination by a defense attorney at a homicide trial, one of the authors was asked, "What is the most important duty of the first officer on the scene of a violent crime?" The question was unexpected, and this witness stopped to think for a moment. In fact, there are many duties, of nearly equal importance including not compromising possible evidence. This is not the time to examine the scene for minute bloodstains.

The importance of bloodstains and bloodstain pattern evidence is a topic that should be introduced in all basic law enforcement academies. This may not be possible or time-limited given the number of training hours required at an academy and the large number of required courses and skills to be instructed. At many basic police academies, bloodstain pattern evidence receives only 60–90 minutes of instructional time. Besides briefly introducing BPA, the instructor(s) do emphasize the importance of blood on a scene in general and on clothing worn by victims, witnesses, and suspects. This level of instruction helps, but it is not sufficient.

Bloodstain pattern evidence should be a topic of continuing training and instruction at some point after a law enforcement officer is working in the field. At a minimum, such instruction should address bloodstain evidence on clothing and the responsibilities of the officer(s). If, or once, there is sufficient law enforcement personnel on scene to ensure safety, scene security, and medical attention, at least one officer should preliminarily look for any possible evidence, protect it, and document it in some means such as a note and/or photographs. Unless it is contrary to an agency policy, this officer should photograph the scene around any injured or deceased with bloodletting wounds before any alteration occurs. Such photographs can capture clothing conditions as well. Even minimal medical aid may produce post-incident artifacts in the scene or on the wounded person's clothing. These can include simple drips when the body is rolled or manipulated or even dynamic

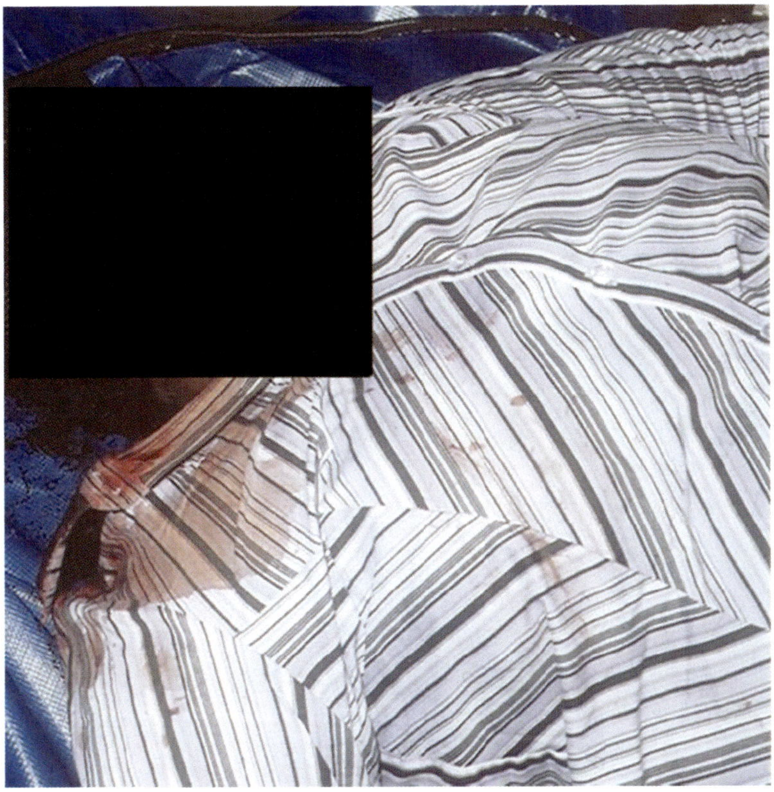

Figure 9.21 The condition of the clothing at the scene is important to document. This condition will change with time and manipulation of the victim. Compare the condition of the blouse in this photograph with its subsequent condition shown in Figure 9.22.

stains such as cast-off. On more than one occasion, such photographs by on-scene officers have provided vital information to a bloodstain pattern analyst. Another valuable resource for such images would be those obtained by Body Worn Cameras (BWC). BWC may capture aspects of stained clothing or even stains on the body that are ultimately lost during manipulation and transport.

Prior coordination between the various agencies with death scene responsibilities is also helpful. Among the issues to discuss would be the importance of in situ photographs, for comparison to the bloodstains observed on the deceased's clothing after collection, the importance of photographs before a deceased is moved in any way that can produce bloodstain artifacts on the clothing and the deceased. See Figures 9.21 and 9.22. This includes photographs of the deceased's hands and lower arms to document the presence or absence of bloodstains, the feet, bare or clothed, and footwear, especially on the plantar surfaces of bare or clothed feet and the bottom of the outsoles. Recognizing that movement and transport by the ME's office can destroy significant patterns or produce additional patterns is key, and any actions taken to resolve these concerns is important.

In situ, the deceased's clothing and body routinely hold bloodstain pattern evidence. If these stains are not documented before movement, they are often lost as a result of substantial saturation that occurs as the body is bagged for transport. The best advice for resolving these concerns is to document the deceased entirely on the scene before movement. This includes those aspects of the deceased that are not oriented up.

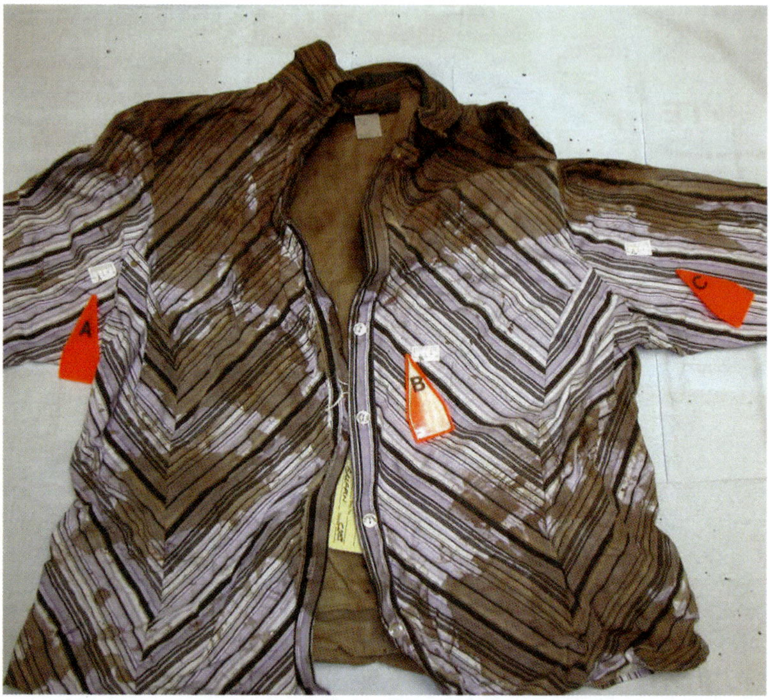

Figure 9.22 The condition of the blouse in Figure 9.21 after transport of the victim and examination at the morgue, is anything but pristine. New patterns (primarily saturation stains) are produced and mar other patterns of interest. It is imperative that the clothing be documented in situ at the scene.

At the actual point of removal of the body, care should be taken to capture the underside of the body and any staining that may be present. Although not a direct function of the clothing examination, this "after-removal" photograph is often forgotten. Basic crime scene procedures will generally ensure photographs of the body in place at the scene, but when the body is released, most of this effort is already accomplished. Thus, when the body is removed, no one considers photographing the area where the body was removed from. The presence of stains, patterns, or the lack thereof can significantly influence the Bloodstain Pattern Analysis.

Collection of bloodied apparel before damage is obviously the most effective approach in resolving these concerns. Other than initial assessments by first responding officers and medical intervention by paramedics, it is generally established that manipulation of the deceased is restricted to medical examiners and their associated investigators. Thus, the idea of prior discussions and coordination between law enforcement and medical examiners may provide a means of preserving and documenting bloodstain evidence on and around the body, as well as ensuring the ME's needs are met. If not previously done in a prior formal or informal setting, the scene investigator realizing there is critical evidence present, can certainly seek permission from the appropriate medical authority to remove such items on scene. Ultimately, all of those on a death investigation scene have the same goals – solving the crime and working to bring closure to the family of a deceased. Thus, coordination that meets everyone's needs is essential.

Clothing removed from the victim by EMS on a scene or by doctors at the hospital is also important. Obviously, such clothing has been manipulated and damaged, but

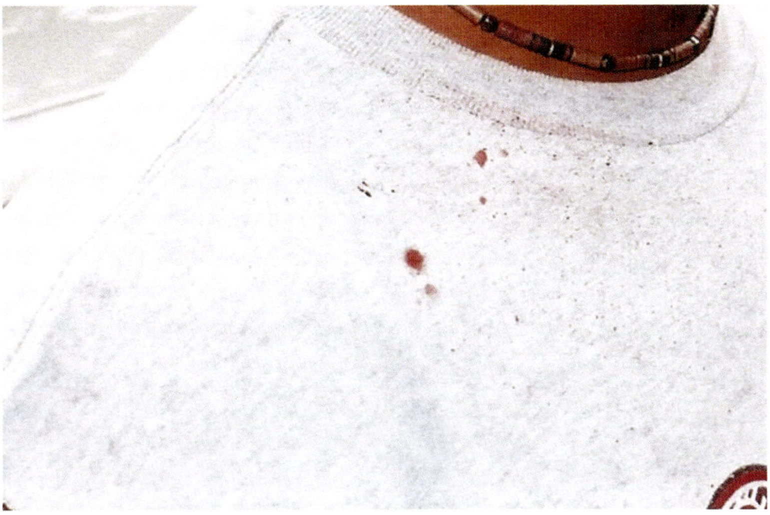

Figure 9.23 This individual was involved in a shooting incident. The pattern on his T-shirt suggests it is spatter; however, despite the obvious bloodstains, the initial responders did not take this clothing and it literally walked away from the scene. Without the clothing to examine and validate the nature of the stains, no conclusions can be drawn.

the bloodstain patterns themselves may still be evident and of value. There is no reason, other than laziness, for these clothes not to be recovered. Unless they are stolen, crime scene investigators must pursue and collect the clothing. Excuses that the clothes were mixed with the trash or "thrown away" should never be accepted. The victim's clothing can provide critical evidence, and every reasonable effort to recover the clothing should be pursued.

Stains on the clothing of the deceased is not the analyst's only concern. The clothing of individuals suspected of involvement in a bloodied scene and those who discover victims should be pursued as well. See Figure 9.23. Obviously, bloodstain on suspects can be highly probative, corroborating or refuting specific claims. It is not uncommon that the individuals discovering a victim come under some initial suspicion from the investigative team. Once again, the presence or absence of bloodstains on their clothing may support or refute claims made by them. When involved parties are sent on their way without consideration of staining on the apparel and without detailed documentation, significant issues often arise at later points in the investigation. Involved party clothing should be examined carefully and when necessary, seized for subsequent examination under more controlled conditions.

It should be unnecessary to say, but appropriate crime scene photography procedures are employed both on scene and during subsequent evaluation of fabric and clothing items. The technique known as Roadmapping (discussed in Chapter 13) is the most effective way to demonstrate the condition and position of the various patterns present on the clothing during off-scene evaluation. In roadmapping, some form of large readable label for each pattern is introduced and conspicuous in each photograph. Overall photographs are taken with all these labels to show the overall orientation of the clothing or item. See Figure 9.24. Each close-up photograph also has a label, thus preventing any confusion about what area or pattern is depicted in the photograph. If macro-photographs are necessary, the large label may need to be replaced with smaller millimeter labels. See Figures 9.25 and 9.26.

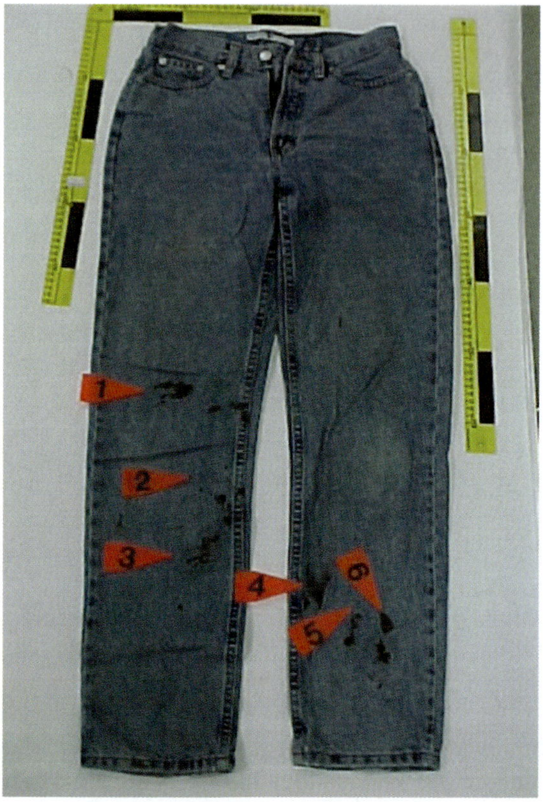

Figure 9.24 Using the "Roadmapping" documentation method, overall photographs are taken of an item of clothing, which include appropriate scales and labels indicating where specific stains are on the item.

In addition to appropriate photography of any bloodstained item, sketching of clothing can be important as well. The sketching aspect of off-scene documentation may follow one of two procedures. The analyst can capture digital images of the item (all sides or surfaces) and then print the digital image. This print is then annotated. Such annotated prints are effective additions to the analyst's notes. This is a technique is common in crime laboratories.

Hand-drawn sketches are still a functional method of documentation as well. Concerning such sketches, some analysts have suggested that they should be "to scale," but this claim is based on a lack of understanding of what "to scale" means in crime scene documentation and what it imposes on the analyst. In crime scene work, a "to scale" drawing means exactly that; any measurement taken from the sketch will match a measurement in the actual scene based on some indicated scale. A level of error is always assumed, but the nature of that error is usually quite low. If the "to scale" sketch does not match or a significant error is evident, opposing counsel can attack the sketch. For this reason, most hand-drawn crime scene sketches are prepared in a "not to scale" format showing landmarks, evidence, and general measurements, but without the claim that each aspect of the sketch will match exactly. In this latter approach, counsel is without challenge.

Consider this in the situation of a bloodied piece of clothing. Bloodying and subsequent drying often leave such fabric items misshapen. The stretched and disfigured fabric

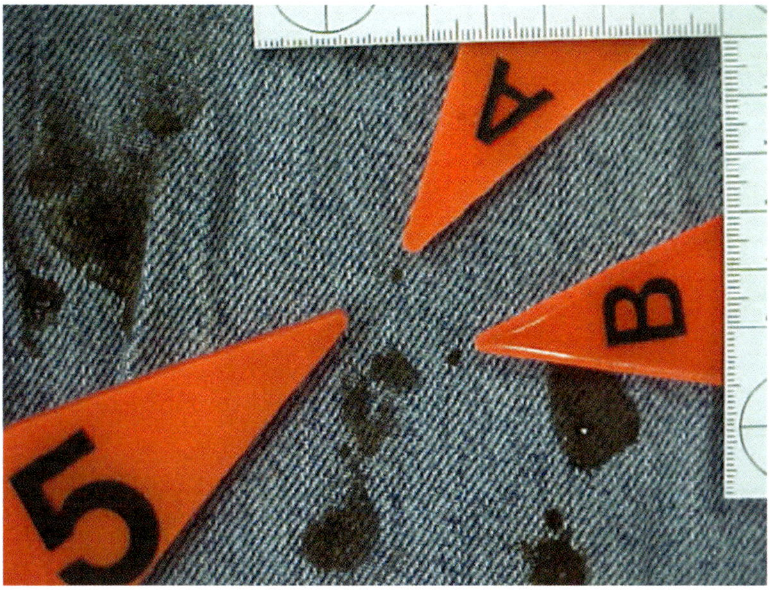

Figure 9.25 A photograph of Stain #5 from the item depicted in Figure 9.24. As this pattern is made up of several very small stains, additional close-up photographs are necessary.

Figure 9.26 A close-up photograph of Stain #5A from Pattern 5 depicted in Figure 9.25. The larger orange label is too big for the close-up photograph, so an additional small millimeter scale is introduced alongside the stain and annotated so the viewer knows which stain they are looking at.

can also be torn or cut, with ragged edges. The ability of anyone to functionally draw "to scale" each surface of such an item is asking a bit much. The task is impossible, given the expectation of what a "to scale" drawing is. Thus, clothing sketches should not be represented as "to scale"; they should be demonstrative. Just as with annotated photographs,

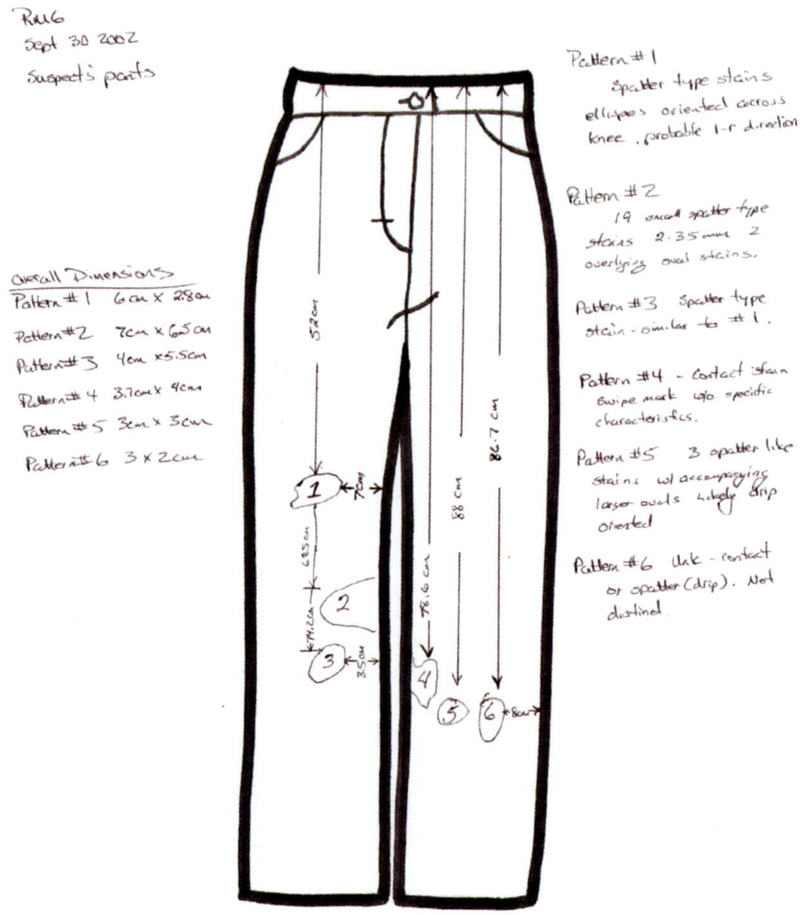

Figure 9.27 Clothing sketches should be detailed and accurate, but it is nearly impossible to create a "to scale" clothing sketch of a bloodied item. Clothing sketches should not be represented as such.

the representation of any annotation should be as accurate as possible. But recognize any measurement on a misshapen bloodied item is unlikely to be "exact" when reproduced.

Whether using a hand-drawn sketch or an annotated photograph, patterns of interest should be identified, and their overall dimensions or generalized position demonstrated in some fashion. See Figure 9.27. Measurements from associated landmarks (e.g., "extending six inches to the left of the inseam") should be annotated to assist others in understanding where the pattern was observed. All sketches should be supported with overall and close-up photographs showing the specific details.

Overcoming Poor Collection/Documentation Procedures

If appropriate on-scene documentation efforts are not accomplished and a request is neither made nor granted to recover the clothing, how can the analyst overcome marring or recognize what patterns are post-incident? Depending upon the level of marring, it may be possible to pull information from such clothing. If the saturation staining is not excessive, underlying patterns may still be visible. As a rule however, marring by saturation cannot be overcome. Once destroyed, there is no functional means to remove or look through the saturation and recover the pattern information. This is an area ripe for research, and perhaps in the future, some combination of chemical(s) and Alternate Light Source wavelength will make it possible to see through the saturation staining and evaluate any patterns beneath. For the moment, prevention of post-incident saturation is the best course of action. If this is not possible, post-incident patterns can often be recognized by the analyst. This is done by using any in situ photographs that are available and comparing them to the current condition of the clothing.

To make this distinction, the analyst should seek and find any photographs, videos (e.g., BWC footage) of the area on the clothing in question. Remember that these areas are often captured in other crime scene photographs, such as close-ups of items of evidence around the body. The intent of the photographer may be to capture some other aspect of the scene, but in the process, patterns on the clothing are captured as well. The photographs used by the analyst should include only those taken on the scene before significant manipulation of the remains. It is understood that these photographs will not allow detailed views, but analysis of the stain from the photograph is not the issue; recognition of its existence is. The photographs are examined from their varying perspectives to locate stains or patterns that were present on the scene and associate them with landmarks (e.g., seams, defects, and buttons). The stains are then located on the actual object, using these landmarks. If the pattern is present on the clothing in the crime scene photographs, then the pattern can be examined on the item with a level of confidence that the stain being examined was, in fact, present before the removal and manipulation of the victim.

Recognizing that the bloodstain pattern analyst is not always the party responsible for the crime scene processing, it should be evident their analysis can fall victim to the efforts of the crime scene processors. For this reason, there is an inherent responsibility to properly train the crime scene processors on documentation of bloodstain patterns and particularly bloodstained clothing. The analyst shouldn't be a passive player. They should do everything in their power to see that proper training is given and the crime scene investigators employ appropriate techniques. Failing to take such steps may result in receiving evidence in an inappropriate form that will ultimately handcuff the analysis. See Figure 9.28

Understanding Context – Two-dimensional and Three-dimensional Issues

There is a somewhat unique consideration involving clothing items, as analysts such as Taupin effectively stated, "Clothing (and textiles) need to be placed in context."[10] This statement appears simple and direct enough but consider the way clothing items (as well as other fabric items) are collected and subsequently examined. Clothing items are removed on-scene or at the morgue, containerized, and then at some subsequent point, laid out for more effective documentation. If accomplished properly, each item is laid out flat on a large surface, and both inside and outside aspects are photographed. This is the only effective approach to capture the entire object and any associated staining. In effect, the analyst is capturing a 2D view of any stains present.

In a scene, surfaces are generally static and mostly unchanging, so understanding the context in that instance is relatively straightforward. However, the human body can move

Figure 9.28 This is the actual collection procedure used by an investigative organization looking at a homicide. All of the clothing of the victim was removed at autopsy and stuffed into this single paper bag. Not only are cross-contamination issues possible, but also post-incident artifacts in the form of additional contact staining and/or saturation are likely to result from this abysmal effort.

in many orientations as well as garments worn, and those orientations can change significantly, both with regard to the surrounding scene and to any bloodstain events.

Worn clothing always represents a complex three-dimensional surface. Viewed solely in a two-dimensional aspect, it is more difficult to understand the orientation of varying surfaces to the bloodstaining event. The location of bloodstains, especially spatter events (e.g., impact or cast-off) on clothing, can be highly probative. Understanding the associated orientation of some aspect of the clothing with regard to a bloodstain event may force positional issues, demanding a specific position of the wearer at the moment of deposition. This is more difficult to understand when viewed solely in a 2D position. This is also limiting when trying to document some evident orientation for the jury. There are two techniques used to resolve this problem. These include using a mannequin when viewing the clothing or having the clothing worn during subsequent analysis. Both techniques are functional. The mannequin technique has certain limits in terms of understanding and documenting the dynamic range of movement possible. The second technique of wearing the clothing is far more insightful, but creates issues the analyst must deal with.

Mannequins (partial or full) allow the analyst to better understand and document general orientations of the pattern discovered. See Figure 9.29. The varying orientations

Figure 9.29 A three-dimensional examination of clothing using a partial mannequin (torso).

Figure 9.30 A surrogate in full PPE wearing bloodstained clothing for a three-dimensional examination of the bloodstains on the garment (shirt).

possible of how the item lay while worn on the body become more evident. The limitations of the mannequin are the inability to bend or twist a torso, neck, or head, or explore potential orientations of limbs (e.g., hands, arms, or legs). Nonetheless, this technique is functional and helpful.

The second technique involves a person wearing the clothing. A surrogate of similar height, weight, and build dons the clothing. See Figure 9.30. This allows nearly any possible orientation of the body, clothing, and stains to be explored and recognize any interrelationships that may be present. This technique does create certain issues. First and foremost, it can only be pursued after all DNA collection efforts have been performed.

Second, it requires appropriate personal protection equipment (PPE) and procedures for the surrogate. The surrogate must use full PPE to ensure protection, particularly the face, when placing pull-over shirts or hoodie-type clothing items on. In terms of documentation for the court, this approach has no equal. Any potential position can be explored and documented, allowing refutation or corroboration of certain claims.

General Considerations Involving Fabric Collection and Documentation

There are a variety of basic considerations to keep in mind when evaluating clothing and fabric items:

- Clothing and other fabric items containing or believed to contain bloodstain evidence should be collected and packaged per agency policy, *assuming* the policy meets essential criteria. Each item should be packaged separately in a "breathable" container such as paper or cardboard on the scene when possible. If there is any concern that the item is wet, it should be returned to an appropriate drying chamber and then repackaged.
- If the item is obviously saturated, containerization may require something other than a breathable package to prevent blood from soaking through the packaging and contaminating surrounding areas during transport. Take whatever steps are necessary on scene to prevent cross-contamination of the item itself (e.g., one area of saturation affecting unstained areas) during transport. If the item is highly saturated, recognize the realities of limiting such cross-contamination. Back at the lab, carefully remove the item, place it in a drying chamber, and once dry, repackage.
- As with all crime scene documentation, the best image capture approach is a "raw" format or uncompressed images. JPEG images will generally suffice for sharing with counsel or display to a jury, but NEF or TIFF images allow more effective image processing during analysis if required.
- A tripod and macro lens are generally necessary when capturing stains. A wide-angle lens and hand-held camera will likely be necessary when taking photographs of the complete item.
- Ring lights are highly effective in close-up work, but don't forget that adjusting the orientation of the light to the fabric may help with the contrast aspect. Thus, a handheld strobe or light may be effective as well. This is particularly true when trying various ALS wavelengths during photography.
- Bedding context may be highly probative. Such items may be oriented wrong sides up, wadded, or overlapping. Ensure that how the bedding is found in situ is effectively documented before collection. Blankets, sheets, or pillows may be oriented oddly (wrong side facing upward), overlapping, or wadded in some fashion. As discussed, understanding the context and orientation of an item is important at later stages when laying out the bedding for evaluation. This may demand marking an item on scene to help with such orientations (e.g., up, down, SE corner).
- Visible labels of fabrics or clothing items should be documented. Such tag information regarding the composition of the fabric and/or if the fabric has been subjected to any manufacturing treatments may be helpful at later stages of the analysis.

- Be particularly careful when employing enhancement reagents. First, remember that enhancement is accomplished after all other means of documentation and testing have taken place. Be aware that such reagents can produce specks that could be misinterpreted as "spatter." Be sure that one is documenting true circles and ellipses. Particularly, when using luminol, it is certainly not recommended to measure a luminol reaction and use such measurements to generate some claim of a particular impact angle. Consider the fact that you are attempting to measure a light source, something that is luminescing as a chemical reaction occurs. Due to the very nature of spatter on fabric, remember the best one can be confident of given a "well-formed" stain on fabric, if the spatter fell acutely or perpendicularly.

Summary

Recognize that fabric presents limitations. Taxa are often muted or may be completely absent. This will undoubtedly limit the bloodstain pattern analyst in certain areas. Still, such targets do not eliminate the ability to evaluate the basic patterns on a case-by-case basis. A major concern of the analyst is properly documenting clothing stains at the scene, so their value is not lost. Additionally, the analyst must try to educate those encountering suspects and subjects, teaching initial responders to be cognizant of bloodstains and to seize bloodstained clothing and footwear whenever practical.

Quite often, a simple question is posed to the Bloodstain Pattern Analyst. Are the observed bloodstains the result of contact/transfer that can result when a bleeding victim is cradled or medical aid rendered? Or are the bloodstains the result of a dynamic mechanism associated with the creation of spatter? This question can be resolved in many instances by examining the stains under magnification. As with any conclusion, there must be clear examples that support the conclusion. Lacking clear examples, it is only appropriate to report that the question cannot be resolved.

Basic considerations such as directionality and impact angle must be considered cautiously when dealing with cloth and fabric based on the nature of the target surface. Various fabrics will all but eliminate the ability to distinguish this information. The analyst should always understand the basic nature of the cloth or fabric being evaluated before defining any specific conclusions.

When conducting Bloodstain Pattern Analysis on fabric, especially clothing, we suggest using a multistep methodology. First, use a macro-examination approach. Where on the clothing are the bloodstains and bloodstain patterns? This can be challenging when dealing with dark fabrics. Examine the item using a strong, white light source, directing the light onto the fabric from different angles. Supplement this examination using ALS and/or IR photography, progressing to some type of magnification technique, depending on what is available to the analyst. Use of enhancement chemical, if necessary, is always a last step.

Second, compare the bloodstains on the fabric to the presence or absence of bloodstains in photographs of the object or subject at the scene, especially the earliest photographs available. Be confident the stains being considered are not post-incident artifacts.

Third, consider the context of the stains relative to the clothing's orientation when worn. Are the bloodstains located in a limited portion of the item, such as on the cuffs of sleeves or the cuffs and lower legs of pants? Is there some orientation demanded between the clothing and the stain's source event? Does the location and subsequent stain or pattern classification support or refute specific theories or statements of associated parties?

Fourth, are there any specific fabric aspects that must be considered? Is there any indication that the fabric's construction or condition somehow alters the characteristics observed?

Fifth, based upon the physical characteristics present and the preceding considerations, to what degree can one offer conclusions relating to the stains and patterns present?

Remember that fabrics as a substrate are more likely to impede the analyst rather than help them. This is the nature of the beast.

Notes

1. MacDonell, H.L., *Bloodstain Patterns*, Revised Edition, Golos Printing, Inc., Elmira, NY, 1997, pp. 83–88.
2. James, S.H., Kish, P.E., and Sutton, T.P., *Principles of Bloodstain Pattern Analysis: Theory and Practice*, CRC Press, Boca Raton, FL, 2005, pp. 106, 107, 283–293.
3. Wonder, A., *Bloodstain Pattern Evidence: Objective Approaches and Case Applications*, Academic Press, 2007, pp. 240–249.
4. https://sewguide.com, 7 Nov 2023.
5. Lochte, Theodore, *Gerichtsarztlichte und polizeiartliche Technik*, 1914, Wiesbaden, Germany, English Translation, p. 19.
6. Pex, James O. and Vaughan, Charles H., Observations of High Velocity Bloodspatter on Adjacent Objects, *Journal of Forensic Sciences*, Volume 32, Number 6, Nov. 1987, pp. 1587–1594.
7. State v. Camm.
8. Holbrook, M., Evaluation of Blood Deposition on Fabric: Distinguishing Spatter and Transfer Stains, *International Association of Bloodstain Pattern Analysts News,* Volume 26, Number 1, March 2010, pp. 3–12.
9. Slemko, Joeseph, Bloodstains on Fabrics, International Association of Bloodstain Pattern Analysts News, pp. 3–11, Dec 2003.
10. Taupin, J.M. IABPA workshop "BPA on Clothing", International Association of Bloodstain Pattern Analysts, 6 Oct 2011.

A Methodology for Bloodstain Pattern Analysis

10

As previously discussed, meeting the rigors of science is a critical aspect for any analyst. One of the most significant aspects for achieving such rigor is employing a standard protocol or methodology. In an environment where lawyers routinely attack the tenets of every modern forensic discipline, it is imperative that practitioners of any forensic discipline know and apply a functional methodology. Across any discipline, there may be no single "absolute" method; in fact, different functional and valid methods may exist. Although these methodologies may vary in minor ways, they will all follow a predictable path, that of the scientific method. Their existence and application enhance the probability that the analyst, no matter what level of expertise will arrive at a defensible conclusion.

In the past, the discipline of Bloodstain Pattern Analysis has suffered needlessly in terms of reputation within the legal system. In many ways, bloodstain pattern analysts have been their own worst enemy. Examples include the failure to articulate a defined taxonomy for bloodstain classification, and a failure to understand fully the application of these classifications. Worse yet, individuals claiming themselves experts in the area of Bloodstain Pattern Analysis have literally stated the discipline is an art and subjective.[1]

How is it possible that a discipline with a 170-year consistent history can be labeled an "art" and suffer from such misconceptions? The answer in large part is attributed to not fully understanding or applying an accepted methodology. Analysts are taught all of the skills applied in Bloodstain Pattern Analysis – recognition of patterns, impact angle determination, point of convergence, area of origin determinations — – but not everyone leaves a technical course understanding how and in what order to apply these skills. Thus, the inability of analysts to explain how the "analysis" was performed leaves the perception that the discipline is without foundation. If the rank and file analyst can't explain in simple terms how they go about the analysis, then how can the discipline hope to escape a tag of "unscientific"? The problem, however, has never been the lack of appropriate procedures, but rather a failure to articulate these procedures.

As a discipline, Bloodstain Pattern Analysis defines conclusions specific to the events that occurred during the bloodshed incident by using information derived from size, dispersion, shape, volume, pattern characteristics, the number of bloodstains, and their relationship to each other and the surrounding scene. This is done by asking and answering questions using the scientific method.

The Scientific Method

Often in attacking an analysis, opposing experts and lawyers imply that only the "scientist" can apply the scientific method; mere mortals such as analysts, investigators, and police are incapable of understanding the mysterious tenets of this thing called "the scientific method." In such arguments, one can be sure that the counselor will use great emphasis

DOI: 10.4324/9781003408734-10

and intonation each time he uses the word "science." More often than not, these same individuals will have no clue of what the scientific method is or exactly what it entails. Nevertheless, by furthering the myth that the scientific method is something mysterious and beyond the average human's understanding, they hope to prevent the presentation of critical information. So what is "the scientific method" and how does it relate to the process of crime scene reconstruction and Bloodstain Pattern Analysis?

First and foremost, the scientific method is not some mysterious, complex, or magical thing. It is, in its simplicity, a defined process used to resolve complex problems. The scientific method might best be described as a circular path that begins with a question, which leads to an answer, which begets another question. In effect, any research effort using the scientific method creates an ever-expanding and self-correcting body of knowledge related to some specific issue.

The scientific method is not, however, a mere collection or regrouping of information. Having the information in front of you and not recognizing what it means is a classic failure of many in the criminal justice system. As we will discover, it is the analysis of the data collected through the scientific method that leads to knowledge. Mere ownership of information means nothing.

The scientific method is certainly not some distant concept for the elite, unattainable by the non-scientist. It is imperative to dispel this particular idea, as it paints any individual employed as a "non-scientist" as inferior and incapable of achieving a valid conclusion. As Leedy commented in his book *Practical Research, Planning and Design:*

> Everywhere our knowledge is incomplete and problems are awaiting to be solved. We address the void in our knowledge, and those unresolved problems, by asking relevant questions and seeking answers to them. The role of research [also known as the scientific method] is to provide a method for obtaining those answers. By inquiringly studying the facts, within the parameters of the scientific method. ... [But] the word "research" has a certain mystique about it. It suggests to many people an activity that is exclusive and remote from everyday life .... Although this concept of research may seem somewhat remote or academic, many of us rely on a truncated form of it each day to dispose of less formal matters than those solved by the more elaborate methodology of pure research.[2]

Thomas Huxley, a nineteenth-century student of science, wrote in this regard:

> The method of scientific investigation is nothing but the expression of the necessary mode of working the human mind. It is simply the mode by which all phenomena are reasoned about, rendered precise and exact. There is no more difference, but just the same kind of difference, between the mental operations of a man of science and those of an ordinary person. As there is between a baker or butcher weighing out his goods in common scales and the operations of a chemist in performing a difficult and complex analysis.[3]

Henry Rhodes, who clearly believed in the scientific method for criminal investigation, stated that it is a fallacy to suggest that the academically trained are more likely from start to finish to be able to handle a criminal investigation than are the police. Trained observation, discrimination, and a sense of the value of evidence are necessary skills that police experience provides.[4]

Robert Pirsig described an insightful correlation of the day-to-day use of the scientific method in his book, *Zen and the Art of Motorcycle Maintenance.* This odd tale of self-discovery devotes an entire chapter to explaining the scientific method. Pirsig, in describing a mechanic's dilemma of identifying the source of a mechanical problem, wrote:

> [The] solution of problems too complicated for common sense to solve is achieved by long strings of mixed inductive and deductive inferences, that weave back and forth between the observed machine and the mental hierarchy of the machine found in the manuals. The correct program for this interweaving is formalized the scientific method.[5]

Pirsig's interesting association is quite accurate. Humans tend to resolve problems using an odd mixture of empirical observation combined with rationalism in order to make sense of complex or abstract ideas. This mixture is itself the very nature of the scientific method, which seeks to blend empirical knowledge (that gained through observation and deduction) with rationalism (that gained through application of reasoning and inductive thought). However, the scientific method provides structure to the process. Different authors describe the scientific method differently, but this structure generally includes six basic steps. See Figure 10.1.

The first step in the scientific method is to identify the problem. This ensures that a focus is established and all efforts are directed toward the intended goal. The most significant part of defining a problem is in breaking a larger problem down into smaller sub-problems. As we introduce the crime scene analysis concept of "Event Analysis" later in this book, the importance of dealing with discrete, small problems will become quite evident. However, it is just as important in Bloodstain Pattern Analysis.

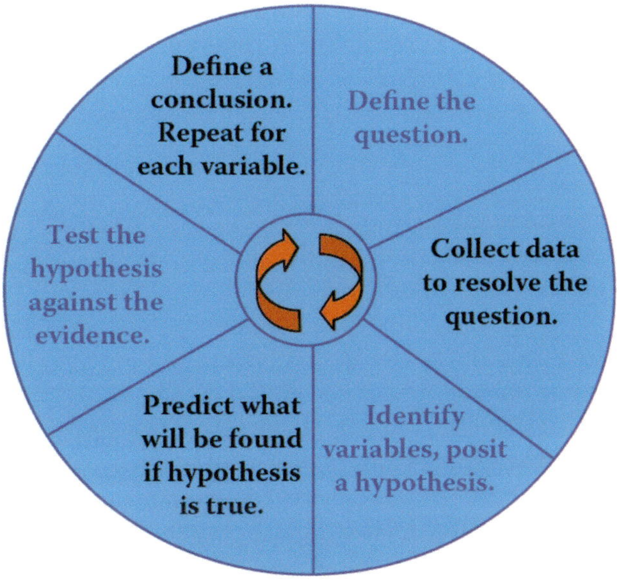

Figure 10.1 The scientific method is a structured process of asking and resolving distinct questions using several steps. Various authors describe these steps in slightly different fashions, but all follow a similar path. The answer to one question often begs another question, thus the process is circular. The result of the scientific method is an ever-expanding, self-correcting body of knowledge.

Consider this step in relation to a simple bloodstain pattern. If your stated problem is "What does this stain establish?," the problem is far too broad to begin with. Smaller, more discrete, issues must be answered before the larger problem is resolved. These include:

- Is the stain actually blood?
- If it is a bloodstain, where in the established classification of Bloodstain Pattern Analysis does it fit?
- What do the directional aspects of the stain indicate?
- What interrelationships exist between the pattern and other patterns or evidence?
- Given the classification, what mechanisms were present in the scene to create the stain?
- Is one mechanism more consistent than another as an explanation, and if so why?

In order to resolve the larger question of "What does this bloodstain establish?," one must first deal with and answer the smaller questions.

The second step of the scientific method is to collect and gather data that might help establish an answer to the question. In pure research and certainly in physical science, this step entails the laymen's icon of the scientific method; that is, the scientist performing an experiment. However, data is collected through means other than experimentation alone. In the instance of crime scene or Bloodstain Pattern Analysis, empirical evidence is present in the scene photos, scientific reports, and other information collected during the investigation. Occasionally, it may be necessary for the crime scene or bloodstain pattern analyst to design and conduct an "experiment" to help aid in the discovery process. Unfortunately, creating objective experiments is an area where great care should be taken to design the experiment properly. In Chapter 16, we discuss issues involved with designing and conducting such experimentation. As previously mentioned, having the data and understanding the data are two entirely different things. In this step, the analyst must also look for and recognize interrelationships of the data.

The third step of the scientific method is to posit some hypothesis regarding the problem. What might the solution be? A hypothesis is an educated guess and vastly different from a mere guess.

The fourth step is to make predictions relative to the hypothesis. If the hypothesis were true, what would one expect to be present in the scene or the data available? If the hypothesis were false, what predictions could be made? Predictions are nothing more than simple counterfactual statements; "if this, then that" statements. Such predictions will also point the analyst to data they might otherwise have missed.

The fifth step is to test the hypothesis by comparing our expectations (predictions) for a given hypothesis against the observed data. If correct, a hypothesis will demand that certain things occurred, and therefore evidence of these occurrences will exist. If we put forth a hypothesis that a given bloodstain was created by a specific mechanism, then somewhere in the data we should find evidence supporting the availability and presence of this mechanism in the scene. Furthermore, we may be able to create a similar mechanism and, with its employment, create a similar type stain. It is important to recognize that in terms of finding answers to problems, the scientific method is far more effective for disproving a given hypothesis than it is for proving a single hypothesis. Often referred to as premise negation, the scientific method will allow us to state with some level of authority that certain things could not occur, but it will not always allow us to identify exactly what did occur.

After testing a given hypothesis, the analyst is left with one of three possibilities. The hypothesis is supported by the data indicating it is possible, the hypothesis is rejected and excluded as a possibility, or the data is inconclusive. If rejected, the hypothesis must be altered or an alternative hypothesis considered. Remember that in the scientific method, the data defines the conclusion; the conclusion does not define the data. This important nuance is lost on any number of professionals, scientific background or not, and manifests itself in behavior where individuals begin to rationalize why certain items of evidence or data are unimportant or "don't matter."

The final step in the scientific method is to conclude something from the information examined. What one concludes is dependent upon the data and interrelationships. After exhaustive study, we may conclude that we cannot answer the question. On the opposite end of the spectrum, we may conclude that we have a clear answer to a given problem. One never really knows until after the analysis. This is perhaps one of the more important aspects of applying the scientific method – recognizing what we do not know. As Aristotle once commented while comparing himself to an antagonist, "I am better off than he is, for he knows nothing, but thinks he knows, while I neither know, nor think I know." In this latter particular I seem to have slightly the advantage of him."[6] The analyst must always remain cognizant of what they do not know.

The greatest misunderstanding of science and the scientific method lies in the conclusion. One of the greatest myths of science is that "scientific certainty" defines something absolutely, that science provides unwavering answers. This simply is not the case. To understand this issue better, we must elaborate on science and its role. Science seeks to explain our world; it does so by seeking theories that explain observed phenomena. As Nordby stated, "The goal of the scientific explanation since the time of Galileo has been to expose nature's general covering laws."[7] Granted in BPA, we understand that these scientific generalizations (e.g., covering "laws") may have ranges in which they are valid following Woodard's mindset, but we are still seeking to expose what the rules of BPA define. Stephen Hawking also commented on this aspect in relation to the science of cosmology, stating, "The eventual goal of science is to provide a single theory that describes the whole universe."[8] But Hawking went on to explain that the basic theories by which science understands our universe are in fact inconsistent with each other. And if inconsistent, then one must be wrong in some aspect. Science knows and understands this, but accepts the theories as the best explanation possible at the moment, while seeking a combined theory that will ultimately incorporate them all, but eliminate the inconsistencies.

Science gives definitive answers but not ultimate truths. Paul Leedy described this idea in depth, defining what he called regions of data. The first region was primary data, which lies closest to truth. Leedy felt primary data is "the most illuminating, the most truth manifesting," while secondary data, although helpful, lies further from the realm of truth. In cautioning the research student to be wise enough to know that what one makes from the data is truly only a glimpse of ultimate truth, Leedy stated:

> The English word for fact comes from the Latin. The Latin origin is in the word facere, meaning "to make" – what the situation makes or manifests to the observer. The etomology provides the first clue as to the nature of data, they are manifestations of the truth rather than the truth itself. No one has ever looked upon the truth itself – pure, undisguised, naked truth. ...The mind yearns to understand the Ultimate. As a means to that goal, we have chosen the pathway of research [the scientific method]. But it always ends at the farthest reaches of data,

> which are at the brink of a canyon in whose depths lies the inaccessible Ultimate Truth. … Truth is forever just beyond what is represented by the data and hence, just beyond human grasp.[9]

The scientific method provides us with an understanding, a recognition, of what the data represents. At times, the data may be very precise, providing clear and specific information in some area of an analysis, but more often, it will allow us only a glimpse of what happened. Just as physicists accept that the theories of relativity and quantum mechanics are the best understanding of our world given our current knowledge and technology, we must accept that a conclusion derived from the scientific method brings us only to the best explanation given the data.

But does applying the scientific method to an investigation elevate the investigative process to the status of "science"? Perhaps Nordby answers this best, referring again to science's goal of discovering covering laws. He commented that science, specifically in a deductive nomological approach, starts with a phenomena and works backward toward a covering law. Once established, such laws help predict and explain other phenomena. However, there are no covering laws that define acts such as murder. Crime lies at an intersection of many laws and causes. In such instances, explanations come first and they are derived by reasoning backward. Nordby then writes:

> Reasoning backwards analytically requires reading natural signs, formulating tentative explanations, and methodically testing them. … These explanations do not become "less scientific" because they fail to invoke established laws of nature, or "more scientific" because they appeal to covering laws. The "scientific" status of an explanation is independent of its connection with the laws of nature.[10]

We recognize that BPA operates within the idea of Invariance Under Intervention mindset, there may not be specific covering laws in BPA, but we understand where the predictability lies. The scientific method allows us to seek answers using a standard process given this predictability. At its core, the scientific method relies on an ability to define discrete objective questions, seek and find answers to those questions, and ultimately apply the answers in order to solve larger, more complex questions. If posing and answering questions is the heart of the scientific method, that begs the question, what does the bloodstain pattern analyst ask? The ultimate question in Bloodstain Pattern Analysis is: How was this stain or pattern created? To achieve an answer to that, the analyst first asks and answers a number of far more discrete questions.

A Practical Methodology for Applying the Scientific Method

If the bloodstain pattern analyst does not know what questions to ask and in what order to ask them, how then can they functionally apply the scientific method? A practical approach that solves this problem can be articulated in a nine-step methodology. The bloodstain pattern analyst should:

1. Become familiar with the entire scene.
2. Identify the discrete patterns among the many bloodstained surfaces.

3. Categorize these patterns based on an established taxonomy.
4. Evaluate aspects of directionality and motion for the pattern.
5. Evaluate angles of impact, areas of convergence, and areas of origin.
6. Evaluate interrelationships among patterns and other evidence.
7. Evaluate viable source events to explain the pattern, based on all of the evidence.[11]
8. Define a best explanation of the events.
9. Peer review the effort.[12]

The scientific method is a broad-brush mindset, applied to solve any complex problem in any discipline or pursuit. It is the underlying foundation of any discipline's protocol, but there must be a discipline-specific protocol to set the scientific method in motion within that discipline. A methodology/protocol is required for every single discipline defining "how" and in what order the analysis is conducted. These nine steps effectively solve that "how."

Step 1: Become Familiar with the Entire Scene

No scene analysis is ever attempted without first assessing the limits of the scene and knowing generally what is contained within it. This step involves answering the most basic questions posed in crime scene processing: What is the true extent of this scene? What areas will I consider and what items are present? If the analyst inappropriately limits himself or herself, critical stains and/or evidence may be overlooked.

To preclude this, before any other action, the bloodstain pattern analyst seeks to familiarize themselves with the scene using whatever means and data are available. This includes seeking out information relevant to post-incident artifacts. It is a rare situation when EMS and police activity fail to disturb the scene in some fashion. Whenever possible, the analyst should seek out initial responding officers and any initial documentation (e.g., photos exposed by initial responders) to compare it to the condition of the scene as viewed by the analyst.

CASE EXAMPLE: BECOME FAMILIAR WITH THE SCENE

Figures 10.2, 10.3, and 10.4 depict one room in a major crime scene. The overall scene consists of a home in which three victims were beaten in three different rooms. To illustrate methodology, the example will consider a single room.

The room depicted is a sparsely furnished bedroom. The victim found in this room showed signs of life and was removed by EMS. She subsequently died as a result of multiple blunt force injuries. In surveying the room from the entryway at the northeast corner, we observe multiple bloodstained surfaces in the northeast quadrant, including the floor. The victim was removed from a position behind the doorway and luckily a photograph that shows her position was taken by an initial responding officer. See Figure 10.5. This photograph when compared to the crime scene photographs show that many of the stains we observe on the floor, as well as the bloodied clothing, are actually post-incident artifacts created as EMS worked with the victim. By comparing Figure 10.5 with Figure 10.2, it is evident that the white print comforter has been repositioned in the room, which may explain the lack of bloodstains farther to the west. This is critical information, which will aid us in the next step of the process.

Figure 10.2 An overall of one room in a homicide scene. In the initial step of assessing the scene, the analyst looks at what is present, considers issues of what is and is not evidence, and what may be post-incident artifacts. Note the clothing in the lower right corner; it appears cut off. On this initial look, it appears to be a post-incident artifact created by Emergency Medical Services (EMS).

Step 2: Identify Discrete Patterns

Bloodshed situations often involve multiple events that are ongoing in an area. Because of these events, stains and patterns can be deposited on top of or around other stains. These patterns tend to merge, making it difficult, if not impossible, to distinguish between them. In this step, the analyst answers these simple, but often elusive, questions: Where are the patterns I intend to evaluate? Which of these stains belong together?

How does the analyst make this assessment? A critical consideration is the stain's size, shape, position, and directionality. A group of similar sized elliptical stains that appear to radiate out from a central point, and a series of linearly positioned circular stains with consistent directionality help the analyst determine initially what stains he believes entail a particular "pattern." Note that it is difficult for some to separate aspects of Step 3 from this consideration. An experienced analyst upon seeing a pattern is likely to recognize what type of pattern it is immediately, but any initial classification is exactly that, an initial hypothesis. Over time, these beliefs may or may not hold up. Stains the analyst thought were related may prove to be two or more patterns overlying each other. Patterns that appeared on first blush to be one thing might actually turn out to be another. These initial

Figure 10.3 Another view of the room and artifacts present.

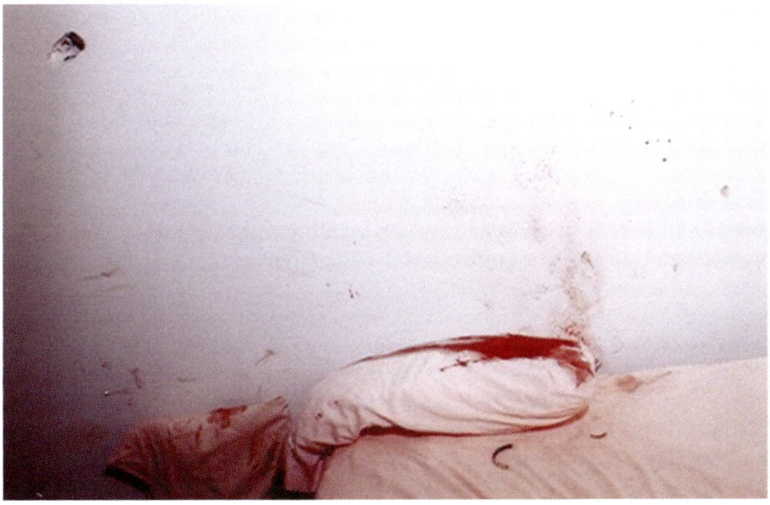

Figure 10.4 A better view of the bed. Bloodstains of various types, as well as the wall defects are evident.

beliefs are simply initial assessments that focus and guide the examiner in the subsequent steps of the analysis. Recognize that even for the experienced analyst, recognizing something as a "pattern" and deciding what kind of pattern it is are two distinct actions.

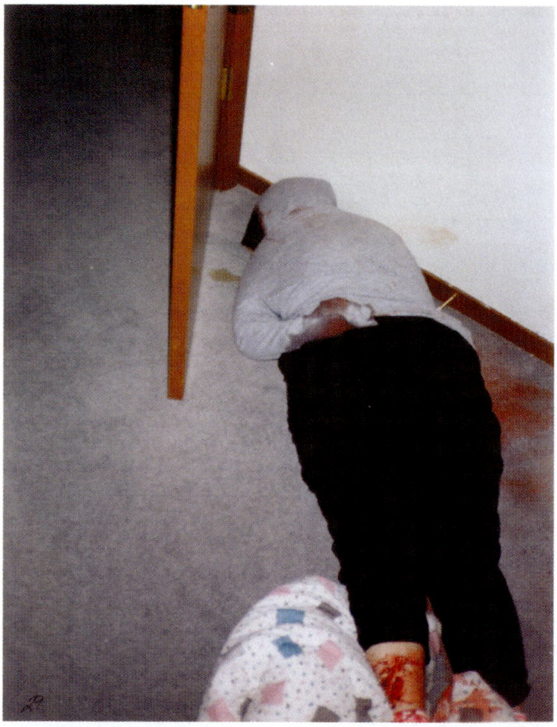

Figure 10.5 This photo taken by initial responding officers before the victim was removed is significant. In it, we can contrast the condition of the clothing, the floor, and the position of the white comforter in comparison to the way it was found and documented by the crime scene team (see Figures 10.2, 10.3, and 10.4).

CASE EXAMPLE: IDENTIFY DISCREET PATTERNS

For purposes of the example, the concentration will be on the primary stains, ignoring many of the smaller stain groups. Looking at Figure 10.2 through Figure 10.5 allows us to eliminate a number of the contact stains observed in the final scene. As these are post-incident, they serve no value to the analysis.

In Figure 10.2, there is a large stained area on both the vertical and horizontal aspects of the box spring. Beneath this area is a large blood accumulation just in front of the brown slipper. This is Group 1, an area for consideration.

In Figure 10.3, excluding the numerous post-incident contact stains between the door and the bed, there is a large stained area at a point consistent with the victim's final resting position. This is Group 2.

In Figure 10.4, there are a large number of very small circular and elliptical shaped stains evident on the wall above the mattress. This is Group 3 (see Figure 10.6).

To the right of these stains are additional stains surrounding what appear to be physical defects in the wall (see Figure 10.7). This is Group 4.

When the pillow from the floor that is leaning against the east wall is removed, we find an additional pattern present on the wall as well as similar staining on the back side of the pillow. See Figures 10.8 and 10.9. This is Group 5.

On the top surface of the pillow present on the bed is a large stained area (see Figure 10.10). This is Group 6.

Above Group 4 and to the right of Group 3 is another grouping of larger stains. This is Group 7. See Figure 10.11.

There are now seven different groups of stains that require consideration. The next step in the process is to classify these stains.

Figure 10.6 A close-up of Stain Group 3, which consists of a large number of small circular and elliptical shaped stains on the wall. Their size, shape, dispersion, and directionality suggest they are associated with one another, yet we readily recognize they may be from multiple similar events.

Figure 10.7 A close-up of Stain Group 4. All around this group are small circular and elliptical stains associated with Group 3, but the pattern of concern is the staining in and around the defect.

Figure 10.8 When the pillow leaning against the east wall is removed during scene processing, an additional pattern is located on the wall behind it. It will be considered as part of Group 5.

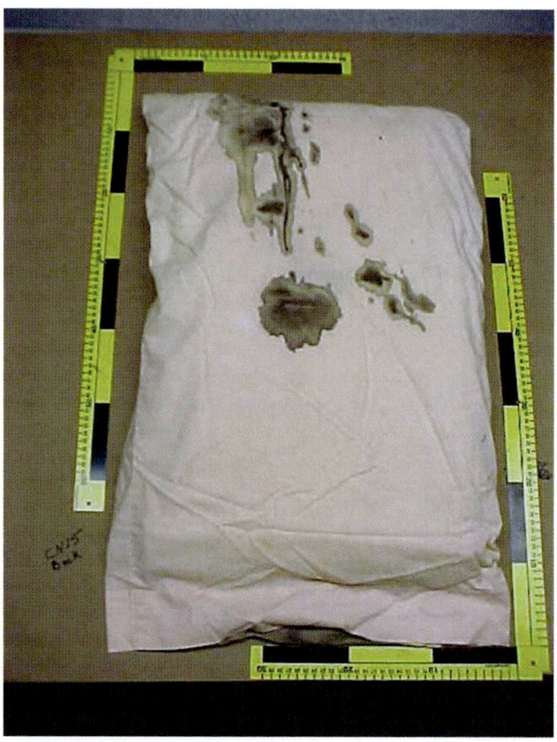

Figure 10.9 Not only was a pattern present on the east wall behind the pillow, a pattern was evident on the surface of the pillow that was facing the wall. It too is part of Group 5.

Figure 10.10 The top surface of the pillow found on the bed is stained as well. This is Group 6.

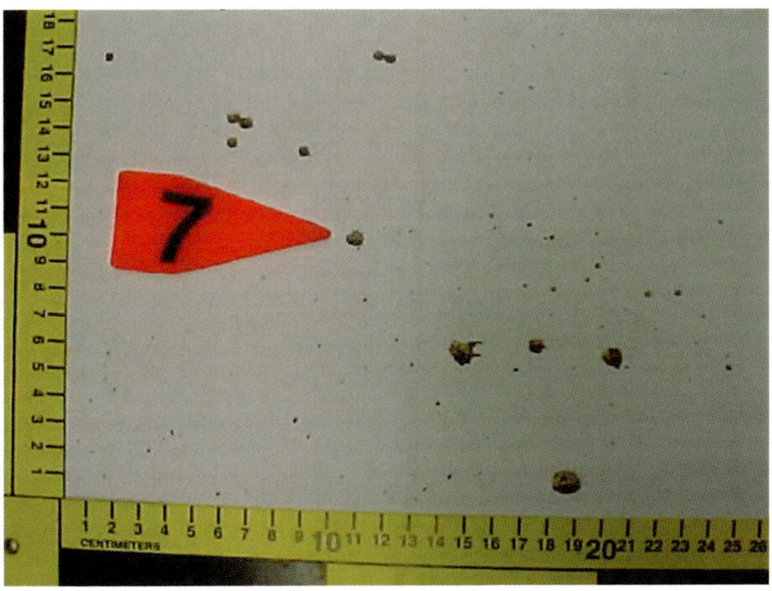

Figure 10.11 Above and to the right of Group 3 (the small circular- and elliptical-shaped stains) is another series of stains. Their size and directionality suggest they are a separate event, Group 7.

Step 3: Classify the Patterns

Classification is the core step in Bloodstain Pattern Analysis. Considering one pattern at a time, the questions are: What kind of stain is this? Is this contact or spatter? Is it impact spatter or spatter produced by a cast-off action? To find an answer to these questions, the analyst must ask questions that are more precise about the characteristics of the stain being considered. Classification requires some form of criteria against which the questioned stain will be evaluated. In other words, what are the physical characteristics of shape, size, dispersion, orientation, etc. that I am looking for to differentiate one pattern from another? As discussed in Chapter 4, criteria are not the same thing as definitions. For example, the typical definition of a cast-off stain is a stain created when blood is flung or projected from an object in motion or one that suddenly stops some motion. This definition does nothing to explain what characteristics the analyst used to conclude that the stain pattern was cast-off in the first place. There must be some set of articulated criteria against which each stain is examined.

Classification using established criteria is the key to a successful and objective analysis. Before the analyst can ever begin to consider a hypothesis in the context of the crime, they must know what kind of patterns are present. As Saviano pointed out, the focus is on the pattern itself with no attempt to infer the broader meaning of the pattern in the context of the scene.[13]

Using the decision map (see Figures 4.35 and 4.36 in Chapter 4) and taxonomy introduced in Chapter 4, we will apply this idea of classification to the stains and pattern in the following example.

CASE EXAMPLE: CLASSIFY THE DISCRETE PATTERNS

In Figure 10.2, Group 1 consists of the large stained area on both the vertical and horizontal aspects of the box spring. The stains are non-spatter, as they are not made up of small elliptical or circular shaped stains. The pattern is 15″–20″ in length on the horizontal surface, which connects and extends down the vertical surface. There is an accumulation of blood, without any specific shape, but conforming to the surface contour of the box spring as well as aspects that lead from one point (the horizontal area of the box spring) to another (down the side of the mattress). The fact that the two areas are connected supports an initial decision to discuss both aspects as one. An accumulation of blood is present on the floor beneath this pattern as well.

Applying the decision map to this group determines that it is a non-spatter stain. It has regular margins. There is no patterned deposit. There is both movement with surface contours, as well as aspects of accumulation and absorption. Based on these characteristics, the pattern is best classified as a complex pattern – a flow and saturation. The pattern beneath this grouping (just beyond the edge of the pillow on the floor) is distinct but follows a similar decision map path, which leads us to a saturation/pool.

In Figure 10.3, Group 2 consists of the large stained area to the far left of the bed, just inside the door but before the hallway. Once again, there are no spatter-type stains (e.g., circular or elliptical) present; there is an accumulation of blood conforming to the surface contour and absorbed into the carpet. The stain is not wholly blood, however, with evident particulate matter present to the south of it, which is visually consistent with vomitus.

Applying the decision map, the stain is a non-spatter stain. It has regular margins. There is no patterned deposit. It has aspects of absorption. Based on these characteristics, the pattern is best classified as saturation.

In Figure 10.4, Group 3 consists of hundreds of small stains evident on the wall above the mattress. A better view of these stains, as seen in Figure 10.6, shows that they are small elliptical and circular shaped stains, indicating they have been airborne. The stains are present across the east wall starting at the level of the mattress and centered on the mattress. The pattern extends significantly out to either side and up the wall. Individual stains in the pattern range from submillimeter to approximately 3 mm. There are literally too many stains to count. There is evidence of a radiating effect. The radiating effect, however, is not defined, so we may be looking at overlapping stains from different events. The stains are not oriented in any linear or curvilinear deposition. Also missing are any individual large volume stains.

Applying the decision map to Group 3 the pattern is made up of spatter. It is not linear or curvilinear. There is a radiating dispersion. There are no ringlets or vacuoles present. The majority of the stains are greater than 0.1 mm. Based on these characteristics, the pattern is best classified as an impact pattern.

In Figure 10.4, Group 4 consists of the area involving the defects. Figure 10.7 shows the nature of these defects. Although impact spatter (small elliptical stains radiating from a central area) overlay this pattern, the primary stains observed are without obvious elliptical or circular components. The staining outlines the defect as well as covers the depressed area, creating a pattern effect.

Applying the decision map to Group 4, the stain is not spatter. Although the pattern is more easily defined by the defect, the blood along the edges indicates a patterned appearance. Based on these characteristics, the pattern is best classified as a pattern transfer.

When the pillow on the floor is removed, it reveals Group 5 as seen in Figure 10.8. Starting approximately 20″ up the floor and extending to the floor is an accumulation of blood that has generally regular margins and leads from one point to another following the force of gravity. Additionally, when we remove the pillow, we find similar staining on the top portion on the back side (see Figure 10.9). The staining includes an absorbed accumulation of blood with generally regular margins that extends down the pillowcase following gravity.

Applying the decision map to Group 5, the stain is a non-spatter stain. It has generally regular margins and movement with surface contours without absorption (on the wall). There are aspects of movement and absorption on the pillow. Based on these characteristics, the pattern is best classified as a flow and the pattern on the pillowcase is classified as a flow and saturation.

In Figure 10.2, Group 6 consists of the stains on the pillow on the bed. Figure 10.10 gives a more complete overview of this group of stains. On the top surface of the pillow is a large accumulation of blood, which extends outward to the left. The accumulation has absorbed into the fabric of the pillowcase and underlying pillow.

Applying the decision map to Group 6 the stain is a non-spatter stain. It has regular margins, and both movement with surface contours as well as aspects of absorption. Based on these characteristics, the pattern is best classified as a complex pattern – a flow and saturation.

In Figure 10.4, Group 7 is the set of large stains to the right of Group 3. A closer view of these stains is seen in Figure 10.11. Once again, we see a series of elliptical and circular shaped stains. The pattern consists of at least 12 individual stains ranging in size from 3 mm to 10 mm. They are all generally circular and show evident directionality slightly down and left to right. The size of these spatter stains compared to the spatter associated with Group 3 is distinctly larger. They show no radiating effect. There is no evidence of individual large volume droplets (e.g., flows from individual stains). The stains do have directional angles that run generally parallel to the overall pattern (left to right) and there is a somewhat progressive change in the shape of the stains (the further right in the pattern they are found, the stains become more elliptical), although this is not distinct.

Applying the decision map to Group 7 the stains are spatter. There is a linear arrangement evident. There is no volume in the stains (no flows from individual stains). There is a progressive change in the shape of the stains, although it is not distinct. Based on these characteristics, the pattern is best classified as a cast-off pattern.

In these initial steps, the analyst first familiarizes themselves with the entire scene so they knows where the bloodstains are. Next, they attempt to distinguish discrete patterns among the various bloodstain surfaces. Once this is accomplished, they classify these discrete patterns using a viable (physical characteristic driven) classification system. These first three steps are always accomplished in sequential order.

Steps 4 and 5 are accomplished as needed and in varying orders, depending upon the situation. Certain aspects of these steps have already been considered when trying to define the discrete patterns in the scene, but are revisited to ensure validity of any initial assessment of the classification.

Step 4: Evaluate Aspects of Directionality and Motion for the Pattern

The questions posed in this step of the process are relatively simple. They include: From which direction did the stains being considered originate? Is there motion of deposition indicated by the stains? Information of this nature will be significant at later stages when considering the flow of events and actions in the crime scene. Evaluating directionality obviously aids in establishing area of origin for impact spatter patterns, so it is often evaluated simultaneously with Step 5. Nevertheless, direction and motion are not solely issues for spatter patterns; various drip, contact, and flow patterns require this consideration as well.

CASE EXAMPLE: EVALUATING DIRECTIONALITY AND MOTION

Group 1 was classified as a flow and saturation pattern. The accumulation of the pool/saturation led to the flow and the aspects of the flow follow gravity, moving down the side of the box spring.

Group 2 as a saturation pattern demonstrates no obvious direction or motion aspects.

Group 3 was classified as an impact spatter pattern. By looking at the individual directional angles of the stains around the periphery of the pattern, we can see they are elliptical and indicate motion upward and outward away from the central area of the pattern, which is situated nearly centered on the mattress. The shapes of the stains in the central portion are circular, indicating they struck at less acute angles (more closely to 90°). So, the pattern is radiating out and away from an area somewhere low on the wall, just above the top side of the mattress.

Group 4 was classified as a pattern transfer, which demands some form of contact. In this situation, the direction of contact is more a function of the depression associated with the pattern than of the contact pattern itself.

Group 5 was classified as a flow pattern and saturation. The flow follows gravity, but its immediate source is not evident.

Group 6 was classified as both a flow and saturation. As the pillow is positioned in the final scene, the flow extends from the primary saturation, left toward the doorway and wall.

Group 7 was classified as cast-off. The individual stains show evident directionality slightly down and left to right. This, of course, is based on the long axis of the stain and the evident tails on the right side.

Step 5: Evaluate Areas of Convergence and Origin

Using the mathematical techniques explained for Area of Origin, the analyst seeks to answer specific questions relating to impact spatter. These include: Can an impact angle be determined for this stain? If so, what is it? Is there an area of convergence for a group of related stains? If so, can the area of origin be defined? If so, where is it? The answers to these questions can weigh heavily on the ability of the analyst to evaluate subsequent hypotheses in Step 7.

Often, when evaluating spatter patterns, it will become evident that the spatter involved is not from a single source event. Area of convergence and origin evaluations may indicate that there was more than one source event involved. This will alter the analyst's initial thoughts of what is a "discrete pattern," often requiring the analyst to break out and discuss separately the different spatter patterns that overlay each other.

CASE EXAMPLE: EVALUATING AREA OF CONVERGENCE AND ORIGIN

Group 3 was the only impact spatter pattern located. In Step 4, it was apparent that the stains are radiating out from a position somewhere centered on the bed. But in evaluating the stains, no clear convergence point is evident. Taken in consideration with the sheer number of overlapping spatter stains present in the pattern, this indicates multiple events were involved. Individual stains can be examined for their impact angle to validate the radiating effect observed. For example, stains centered and just above the level of the mattress show 70°–80° impacts, while stains at the outer periphery of the pattern show impacts at acute angles down to 15°. Their directional angles all lead back toward the center of the pattern, but there is no evident single convergence for all of the stains.

Any attempt to identify an area of origin from the pattern would be limited in value because the pattern suggests multiple impact events overlying one another. Lacking clear convergence points, it would be difficult to associate specific spatter with a specific event. If desired it could be done, but it would be difficult and provide limited information in the context of this crime. However, the overall radiating effect does indicate that the forceful events were occurring relatively low and centered on the bed.

Using a tree/forest analogy, steps 1–5 have considered what the pattern tells us. We are looking at the individual trees, so to say. After completing these steps, the analyst must now begin to consider what the patterns define together, moving from a tree to the forest mindset. Only then can they seek to explain the patterns in the context of the unique incident. To accomplish this, Steps 6 and 7 are considered in order.

Step 6: Evaluate Interrelationships among Patterns and Other Evidence

In this step, the analyst begins to look at the patterns in relation to one another and in relation to other items of evidence. Whereas in Step 3 the focus was on the individual pattern, here the analyst looks at the scene with a broader perspective. The questions posed include: Are any of the patterns related? Whose blood is involved? Are they the result of a common bloodshed event? What do they define when considered together? For instance, one often finds a cast-off pattern that clearly emanates from a spatter pattern. There may be a number of spatter patterns present on different scene surfaces that considered in context actually define a single pattern. Consideration and refinement of these answers can also alter the initial conclusions made in Step 2, for what was thought initially to be several distinct patterns might in fact be a single pattern. Keep in mind, these relationships are many and varied (e.g., DNA, spatial position, bloodstained artifacts) and there is no way to concisely list them all.

CASE EXAMPLE: EVALUATE EVIDENCE/PATTERN INTERRELATIONSHIPS

Groups 1, 5, and 6 indicate that a free-flowing blood source was present in the room. Given their respective locations in relation to the victim's final position behind the door, they suggest a common source and sequential autonomous movement of that source from an area on the bed to the door. The first indication of the open blood source is present in contact with the pillow on the bed, with a flow pattern that extends in the direction of the wall and Group 5. Group 5 is very similar in its nature, but with no evident source above it or in contact with the wall. Group 1 shows the blood source was present low on the bed, but the source was not in contact with the bedding above Group 1.

Group 3 indicates multiple forceful impacts to a blood source while it was on the bed. The pattern transfer and defects in the wall support this. These defects show that blows were being directed at something. This occurred after the object being swung was already bloodied. The directional cast-off (Group 7) moves left to right and, taken in consideration with the orientation of the physical defects themselves, indicates the force was being applied from the north side of the bed (the left side).

The saturation/pool at the door (Group 2) is consistent with the final resting position of the victim's head and indicates she had a free-flowing blood source associated with her head. A check of autopsy information reveals that the victim from this room had numerous skull fractures, but no open lacerations or wounds. There are no connecting pools, associated drip patterns, or blood into blood patterns associated with the pool, indicating the source was on the ground and remained stationary after arriving here.

Step 7: Evaluate Viable Source Events in an Effort to Explain the Pattern

Using all of the information derived from the analysis in Steps 1 through 6, the stains and patterns must be considered in the context of the unique scene examined in an effort to answer the question: What caused these stains? This includes considering any unusual variables suggested by the scene (e.g., extreme temperatures, wet versus dry substrates). This step effectively leads us back to the direct application of the scientific method, but at broader investigative issues. As Saviano suggested, the analyst must develop and test various hypotheses and theories related to the patterns in an attempt to explain their creation and any specifics they define about these events.[14] Additional information that must now be considered includes autopsy results (e.g., what types of injuries were received and where).

In this step, when confronted with instances of unique issues or very specific crime scene conditions never before encountered, if the analyst is not comfortable with these considerations based on prior experience and training, this may require conducting specific research or experimentation in order to resolve a hypothesis.

CASE EXAMPLE: EVALUATE VIABLE SOURCE EVENTS

Throughout the entire scene (the additional rooms in the house), repetitive pattern transfers are present, all consistent with the size and shape of the physical defects seen in this room (see Figure 10.12). These patterns as well as the defects suggest a bat-like object. The pathologist tests this hypothesis against the wound analysis. The pathologist supports this conclusion. No other wounding or patterns are present to suggest any other weapon.

In considering Group 4 (the defects with edges bloodied by contact), it shows that a bloodied object was directed in a North-South orientation against the wall. The defects and resulting

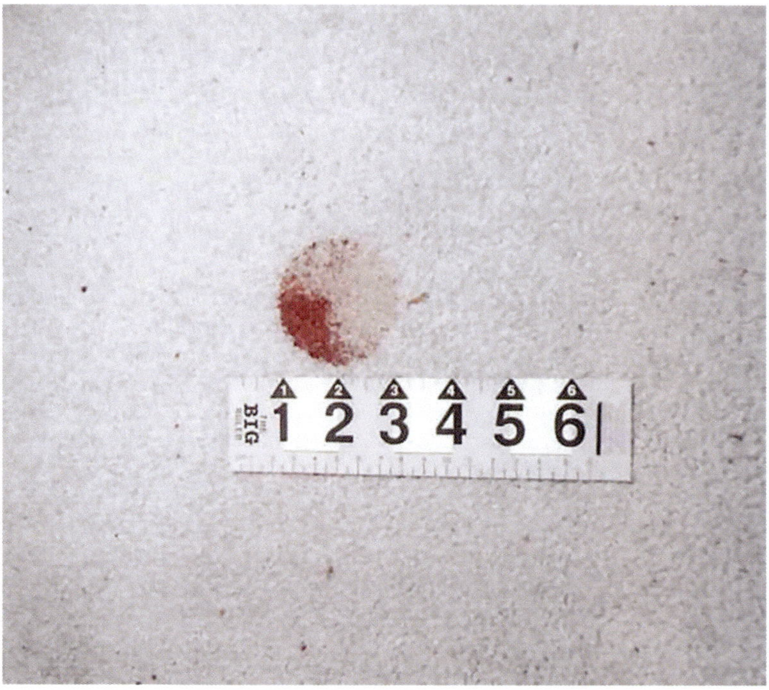

Figure 10.12 Throughout the scene, a number of repetitive pattern transfers are found similar to this pattern. In context, the wall defects and similar pattern transfers on the wall and bed, suggest a bat-like object was used as a weapon.

contact pattern themselves do not specifically define a "bat," but they are consistent with such a weapon.

Group 7 indicates that a bloodied object was moved left to right (N-S) and slightly downward at a location above the mattress. The lack of heavy staining on the victim's hands eliminates any defensive action by the victim as a source of this pattern. The cast-off pattern is certainly consistent with that produced by a bat or bat-like object and the presence of the impact spatter in the area, as well as the contact patterns, supports that the weapon would have been sufficiently bloodied to produce cast-off.

The blood source for Patterns 1, 5, and 6 are all consistent with a free-flowing blood source moving under its own power. This hypothesis is tested by verifying that the patterns are the victim's blood. The victim is verified as the source. Based on the sheer volume, it is apparent the source is flowing quite heavy, but there are no connecting transfer patterns or drag marks between them. This supports that the victim retained some mobility between these positions. At one point, the source was low to the ground in contact with and bleeding against the box spring. Lacking any other wounds, this places the victim's head at this location and would have exposed her back to any attack. Such a hypothesis, combined with our knowledge of the condition of the weapon (distinctly bloodied), allows us to immediately test it by considering her clothing. In examining the victim's clothing, a pattern transfer consistent with the class characteristics of the defects is located on her right rear shoulder (see Figures 10.13 and 10.14).

Pattern 3 indicates a number of forceful impacts to a blood source that was centered on the bed. The only wounds that had the ability to expose a blood source are to the victim's head (flows from the nose or mouth based on fractures to the skull). There are seven physical defects on the wall behind the bed. Given the deposition of blood flows and pools that begin on the bed, this evidence supports a belief that the initial bloodletting blows were delivered on the bed. Although no area of origin was calculated, the general convergence of the spatter indicates the victim's head was in a position centered and relatively low on the mattress. The first contact point of the blood source is the saturation and pool on the pillow, which demands the victim's head was either in contact with the pillow or close to it after bleeding began. There are no drips or similar flow patterns on the sheets beneath the pillow, which effectively eliminates the pillow being raised off the bed.

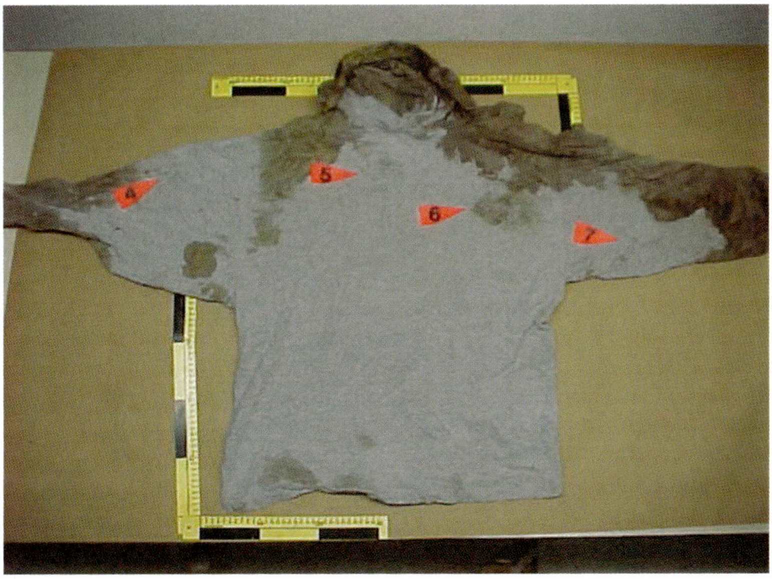

Figure 10.13 Using the scientific method, the scene context suggests a position in which the victim's back was exposed to her attacker at one point in the attack. This begs the question: Is there evidence that supports the attack to her back? The back of the victim's sweatshirt provides the answer, with numerous pattern transfers and directional spatter.

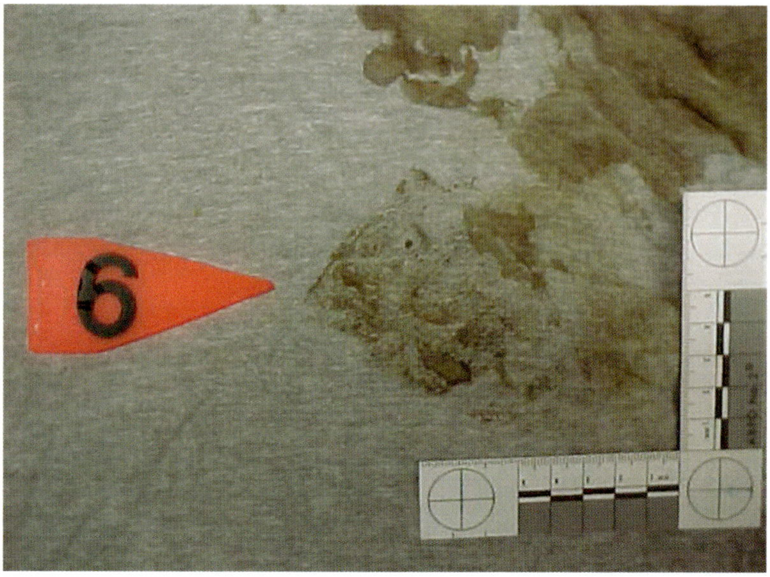

Figure 10.14 A close-up of a large pattern transfer on the back of the victim's sweatshirt. It has characteristics (size and shape) similar to those found in the defects in the east wall. A similar pattern was reproducible using a bloodied bat.

Step 8: Define a Best Explanation Given the Data

In Step 8, all of the data presented are used in a broader sense to reconstruct the order and nature of the events associated with the bloodletting. In effect, we answer the question: What happened and in what order did it happen? Saviano eloquently described how the analyst moves from individual hypotheses to larger more encompassing theories, testing each objectively along the way.[15] Thus, the data from the bloodstains define conclusions about specific actions, and through an understanding of individual actions, theories of what happened can be formulated and tested in this final step.

A critical consideration in this final step is order of actions, if possible. Sequencing and timing information from individual patterns are applied in an effort to define sequential relationships between stains. This effectively produces some level of order for the entire incident.

An area of concern in Step 8 is the injuries sustained by the various parties. The analyst should examine EMS, ER, and autopsy reports and photographs. The questions of concern are the following. What bleeding injuries were present? Were arterial sources exposed? What level of mobility did the victim have subsequent to certain injuries?

CASE EXAMPLE: DEFINE A BEST EXPLANATION GIVEN THE DATA

Based on all of the relationships discussed previously, the following is the best explanation of events given the data:

- The victim's initial bloodletting wounds were directed at her as she was on the bed.
- At one point, her attacker was to the north side of her bed, directing the blows N-S.
- The victim was positioned on the bed long enough to allow multiple blows to be directed at her.

- At some point after an initial blow or blows were struck, the victim began bleeding with her head in contact with the pillow. This occurred while the pillow was generally positioned as it is in the final scene.
- The victim subsequently moved toward the north creating the flow onto the wall. This movement may have been under her own power, or while being pulled. Her head was not drug across these surfaces.
- Subsequent to dismounting the bed, she was low (e.g., perhaps kneeling) with her back exposed to the attacker and her head in contact with the box spring.
- She subsequently moved, was moved by another, or fell to her final position and was not moved again until the arrival of EMS.

Step 9: Peer Review

Peer review entails presenting the analyst's conclusions to another trained analyst. This ensures that the analyst laid a proper foundation for the conclusions offered. A peer review may also lead to correlations that the analyst missed, or recognize subjective inferences or bias introduced inadvertently by the analyst.

The application of this methodology allows for objective analysis of the scene and bloodstains. Aspects of the analysis are deductive (e.g., classification and directionality). Later steps combine both deductive and inductive logic (e.g., associating specific actions to the stains). This blending of logic, using the scientific method, allows us to ultimately arrive at defensible conclusions. Are there questions we cannot answer regarding this scene? Absolutely. Nevertheless, those actions that were identified have a foundation. So long as that foundation is accurate and we have considered all viable possibilities, we can remain confident of the conclusions drawn.

Applying the Methodology in Different Environments

There are three basic environments in which the bloodstain pattern analyst will apply the foregoing methodology:

1. *Active Scenes:* Crime scenes where law enforcement is currently processing or has retained control of the scene subsequent to processing.
2. *Released Scenes:* Scenes involving ongoing investigations where law enforcement has documented the scene using standard processing protocols, but has released the scene back to the owner.
3. *Cold Case Scenes:* These scenes are similar to the released scenes; the only record of their condition is in the law enforcement documentation. Due to the time lapse since the event, the scene itself may have been significantly altered or destroyed.

Active Scenes

Active scenes offer the greatest opportunity for data collection to the bloodstain pattern analyst. If appropriate crime scene processing protocols are in use, then the scene context has not been destroyed by processing efforts. The bloodstain pattern analyst is able to enter and examine the scene in situ. The analyst follows the nine steps of the methodology as discussed, but their presence at the scene allows for more effective documentation beyond the basic crime scene documentation. It also ensures that stains and areas are not overlooked during Step 2 (Identify Discrete Patterns among the Bloodstained Surfaces). Just as important, the bloodstain pattern analyst can recognize and better understand spatial relationships of the scene, which can be important when considering Step 6 (Evaluate Evidence/Pattern Interrelationships).

In the active scene, it is imperative that the analysts understand their role in crime scene processing. Although this understanding is sometimes lost on the laboratory scientist, crime scene processing is not just showing up at the scene and bagging everything in sight. It has underlying direction and purpose. Crime scene processing protocols require that six activities occur. These are (1) assessing, (2) observing, (3) documenting, (4) searching, (5) collecting, and (6) analyzing. Since every action taken by the crime scene processing team ultimately alters the context of the scene, generally these activities are accomplished in sequence. The express purpose of this sequence is to engage in the least intrusive actions first, followed by the more intrusive (e.g., basic scene documentation is always accomplished prior to any physical search of the scene) and ensures that scene context is not lost as a function of scene processing.

If the bloodstain pattern analyst is present at an active crime scene, they should be trained in basic crime scene procedures and must recognize how their actions fit into this overall sequence. Thus, it is inappropriate for the bloodstain pattern analyst to show up at the scene and begin a procedure such as Roadmapping prior to basic scene documentation efforts. The analyst must understand crime scene concepts such as "going back," the process of stopping a procedure when new evidence is discovered and returning to earlier steps in order to document the new evidence. Most of the procedures the bloodstain analyst will attempt (e.g., Roadmapping, area of origin determinations, bloodstain enhancements) are undertaken only after primary scene processing is accomplished.

If properly trained and on-scene, the analyst can directly assist in creating better crime scene and bloodstain pattern documentation. Areas for the analyst's consideration include the following:

- Pointing out small but important stains and patterns that might otherwise be left out of the basic documentation.
- Reminding individuals to photograph the victim's clothing as completely as possible prior to placing the victim in a body bag.
- Reminding individuals to photograph the areas that are beneath the body or beneath objects that were displaced in the room.
- Evaluating any suspects that are on scene for bloodstains on their bodies or clothing and then documenting their condition.
- Recognizing the need for enhancement of crime scene surfaces for latent blood.

Without a doubt, the active crime scene is the best scenario for evaluating the bloodstain scene. Released scenes and cold case situations present issues that ultimately limit the bloodstain pattern analyst to varying degrees.

Released Scenes

It is not uncommon for the bloodstain analyst to be called in to consult after an active scene has been processed and released. Although the basic methodology does not change, the analyst must recognize certain limitations inherent in this situation.

First, the analyst is no longer master of their own fate with regard to Step 2. The analyst can only analyze those patterns and stains that were documented, whether intentionally or not. The analyst should request and examine all photographic documentation by all parties. This includes photographs taken by first responders and ME investigators. It also includes any video footage taken of the scene. Keep in mind that oftentimes the crime scene processors will have a number of photographs that were not submitted to the DA or included in the final report. These photographs may have been considered of poor quality, redundant, or of no value in understanding the scene. Yet in these photographs, stains and patterns may unintentionally be documented that were not immediately documented by the crime scene processor for whatever reason.

The analyst should request all of the scene diagrams created by the original investigators. The crime scene sketches provide context to the photographs, eliminating some ambiguity. They often provide direct information regarding locations and distances associated with the evidence and particularly the bloodstain patterns.

The analyst should also request a complete list of recovered evidence. This list will identify what objects were taken from the scene and what is available for direct examination. Physical items taken out of the scene should be examined whenever possible, and the analyst should not merely rely on photographs of these items.

The scene itself remains a functional source of information. Granted, the context of the scene has been significantly altered during processing and certainly after release by authorities. Bloodstains, however, are a resilient type of evidence. Even after washing of surfaces, spatter stains may still be evident. Flows, pools, and saturation stains may be evident for example by examining beneath carpeting, additionally enhancement techniques for latent bloodstains is possible. The analyst must consider any post-processing activity and how it may have affected or produced bloodstains in these aged scenes. However, a scene examination may still offer valid and helpful data.

Of interest to the analyst are spatial relationships in the scene. Crime scene analysts have long understood that spatial relationships are not immediately evident from examining two-dimensional photographs. A simple and purely innocent choice of lens or the angle in which a lens is oriented can make a very small space look quite large or hide the evident relationship of one surface to another. Thus, a scene visit can be helpful if for no other reason than to understand the spatial relationships present in the crime scene documentation.

Another source of data for the bloodstain pattern analyst are any notes prepared by the crime scene investigators. These notes are prepared during the observation, documentation, and collection phase of processing. They should include detailed descriptions of the scene as observed by the crime scene investigator. These notes are ultimately synopsized into a report, but in any ethical organization, they are retained as a part of the case file because they are an original and contemporaneous record of the investigator's observations and actions. Details regarding stains and patterns the investigator did not consider pertinent for inclusion in the final report may be present in the notes. These additional details may functionally aid the analyst in understanding the nature and extent of various patterns.

After all is said and done, when dealing with a released scene the bloodstain pattern analyst may be left considering testimonial evidence as to the nature of stains and patterns. The analyst will often be presented with areas that were poorly documented with minimal photographic coverage. Obvious staining may be present, but the extent and overall nature of that staining will not be evident based on the photography. Detailed notes about these areas may not have been produced and the only functional source for information is the crime scene technician's memory. If the crime scene technician is considered an objective and knowledgeable source, the analyst may have to rely on their memory. This reliance should always be considered cautiously.

Cold Case Scenes

There is generally little difference between released scene circumstances and the cold case situation. The additional limitations in the cold case situation may consist of the following:

- No physical scene to visit. If enough time has passed, the scene itself may no longer exist or may have been significantly altered in some fashion.
- Investigative personnel who had direct knowledge of the condition and circumstances of the scene investigation may not be available for interview or too much time has passed and their memories are limited and of no value.
- There may be no physical evidence still in the possession of authorities associated with the case.

Beyond these limitations, the cold case is approached and resolved based on the crime scene documentation available.

Summary

Not long ago a Texas appeals court judge wrote:

> We are dubious of the claim in this record that blood spatter evidence can determine the aftermath of a violent incident of bloodshed and try to determine the locations of individuals before, during, and after bloodshed and to try and determine perhaps the sequence of events that occurred.[16]

The sheer ignorance of this statement still amazes anyone who has ever read it, yet it is a persistent mindset held by many in the legal profession. The proponents of this misinformation are furthered in their cause by a failure of this discipline to adequately prepare rank and file analysts to explain how they conduct their analysis using a structured approach.

A practical methodology is necessary for Bloodstain Pattern Analysis and the analyst must be able to articulate how that methodology is applied. The purpose of such a methodology is to ensure that the analyst follows an accepted procedure based on the scientific method. The failure to utilize a practical methodology will invariably affect the validity of the conclusions drawn.

One practical method of applying the scientific method in Bloodstain Pattern Analysis involves using the following steps:

1. Become familiar with the entire scene.
2. Identify the discrete patterns among the many bloodstained surfaces.
3. Categorize these patterns based on an established taxonomy.
4. Evaluate aspects of directionality and motion for the pattern or stain.

5. Evaluate angles of impact, points of convergence, and areas of origin.
6. Evaluate interrelationships among patterns and other evidence.
7. Conduct supporting experiments when necessary and evaluate viable source events to explain the pattern, based on all of the above.
8. Define a best explanation given the data.
9. Peer review.

These steps effectively lead the analyst through the crime scene data, following the scientific method, to supportable conclusions regarding what the bloodstain data establishes. The basic methodology used by the analyst is no different when applied across the three different situations the analyst might encounter: active, released, and cold case scenes. The only difference to the analyst is in the limitations that the latter two situations may present to the analysis.

Notes

1. Akin, L., Interpretation of Blood Spatter for Defense Attorneys and Investigators: Part I, unpublished.
2. Leedy, P.D., *Practical Research, Design and Planning*, Macmillan Publishing Co., NY, 1988, p. 3.
3. Osterburg, J.W., *Criminal Investigation: Reconstructing the Past*, Anderson Publishing, Cincinnati, OH, 1992, p. 12.
4. Rhodes, H.T.F., *Some Persons Unknown*, John Murray Publishing, London, 1933, p. 5.
5. Pirsig, R.M., *Zen and the Art of Motorcycle Maintenance*, Bantam Books, New York, NY, 1988, pp. 92–97.
6. Drake, H.L., *Plato's Complete Works*, Littlefield Adams and Co., Patterson, NJ, 1959, p. 43.
7. Nordby, J., *Dead Reckoning, The Art of Forensic Detection*, CRC Press, Boca Raton, FL, 2000, p. 122.
8. Hawking, S., *A Brief History of Time, from the Big Bang to Black Holes*, Bantam Books, NY, 1988, pp. 10–11.
9. Leedy, P.D., *Practical Research, Design and Planning*, Macmillan Publishing Co., New York, NY, 1988, p. 3.
10. Nordby, J., *Dead Reckoning, The Art of Forensic Detection*, CRC Press, Boca Raton, FL, 2000, pp. 123–124.
11. Gardner, Ross M., Defining a Methodology for Bloodstain Pattern Analysis, Journal of Forensic Identification, Volume 56, Number 4, July/Aug 2006.
12. Gardner, R.M., *Practical Crime Scene Processing and Investigation*, CRC Press, Boca Raton, FL, 2004, pp. 267–268.
13. Saviano, J., Articulating a Concise Scientific Methodology for Bloodstain Pattern Analysis, *Journal of Forensic Identification*, Volume 55, Number 4, 2005, p. 466.
14. Saviano, J., Articulating a Concise Scientific Methodology for Bloodstain Pattern Analysis, *Journal of Forensic Identification*, Volume 55, Number 4, 2005, pp. 467–469.
15. Saviano, J., Articulating a Concise Scientific Methodology for Bloodstain Pattern Analysis, *Journal of Forensic Identification*, Volume 55, Number 4, 2005, pp. 467–469.
16. 2000 West Law 190193 (Tex. App. El Paso), West Law 2000.

Anatomical Considerations in Bloodstain Pattern Analysis

11

MICHAEL FERENC

Introduction

To interpret a bloodstain pattern, the analyst should understand a modicum of biology. The details of blood, blood clotting, and associated bleeding disorders form complete medical subspecialties and are far beyond the scope of this chapter or the reader's patience. However, a brief overview of the subject may still be beneficial.

Blood Cells and Plasma

Human blood is a complex fluid with both liquid and suspended solid components. The solid components account for approximately 45% of the blood volume and consist of the following cellular components: red blood cells, white blood cells, and platelets. These cells are derived from precursor stem cells in the red bone marrow. The liquid component, plasma, accounts for approximately 55% of the volume.[1,2,15]

The red blood cell, or erythrocyte, measures approximately 7–8 μm (micrometer or one-millionth of a meter) in diameter and is approximately 2 μm in thickness. The red blood cell contains no nucleus but does contain hemoglobin. Red blood cells circulate for approximately 90–120 days until they wear out and are removed by scavenger cells in tissues like the spleen and liver.

Hemoglobin, the oxygen carrying protein in red blood cells, contains iron, which is responsible for blood's red color. The hemoglobin also is the basis for many of the presumptive blood tests used at crime scenes. The simplest, oldest, and least specific methods are based on the medical guaiac test. When hydrogen peroxide is dropped on a blood-stained swab or piece of paper impregnated with guaiac (gum guaiac is a green-brown tree resin), the iron in hemoglobin, acting as a reducing agent, promotes a blue color change to the guaiac (more modern blood testing techniques instead use an antibody-based color change reaction specific for human hemoglobin).[1]

Erythrocytes are normally biconcave disks. However, as a bloodstain dries on a target surface, the red blood cells first crenate (become shrunken and rounded with scalloped or spiky surfaces) and then hemolysis occurs (the cells rupture and break apart).

In medicine, an estimate of the number of red blood cells can be done in several ways. One way is to calculate the amount of hemoglobin per unit volume of blood. Normal values are approximately 12.5–15.0 g/dl (grams per deciliter or one-tenth of a liter) for women and approximately 14.0–17.5 g/dl for men. This is equivalent to approximately $4.5–5.1 \times 10^6$ red blood cells per microliter (or one millionth of a liter) for women and $4.5–5.9 \times 10^6$/μl for men. Another method is to centrifuge for several minutes a small glass capillary tube

DOI: 10.4324/9781003408734-11

filled with blood. The heavier solid components (the cells) are pulled to the bottom and the liquid part is pulled to the top. As a percentage of the height of the volume of the blood in the tube, the red blood cells normally are 38–45%. When expressed as a percentage of the total blood volume, this number is called the hematocrit. The hematocrit is approximately three times the hemoglobin value in normal circumstances. A very thin pale band forms between the plasma and red blood cells accounting for usually $<1\%$ of the total volume. This is called the buffy coat and contains the white blood cells and platelets.

Platelets, or thrombocytes, are another cellular component that lacks a nucleus. Platelets are 2–4 μm fragments that are released from a special bone marrow precursor cell (megakaryocyte). They survive only about 9–12 days in the circulation. Platelets are involved in the initial stage of hemostasis (the stopping of blood loss). On the inside of a blood vessel at an area of injury, platelets clump to form a small plug and release chemicals that trigger other parts of the clotting process. The average person has 150,000–450,000 platelets per microliter of blood. Spontaneous bleeding usually does not occur unless the platelet count falls below 20,000 per microliter.

White blood cells come in several subtypes such as lymphocytes, monocytes, and granulocytes. Each subtype serves a different role in fighting infections and regulating the immune system. White blood cells are nucleated and, thus, are the source of DNA for forensic analysis of bloodstains. The average person has approximately $4.4–11 \times 10^3$ white blood cells per microliter of blood.

Plasma is a straw-colored, viscous liquid and accounts for approximately 55% of the blood volume. It is composed of over 90% water with the remainder being salts, carbohydrates, proteins, lipids, and dissolved gases. A blood clot consists of red blood cells, white blood cells, and platelets enmeshed in the fibrin formed when coagulating plasma proteins interact. The remaining clear pink-yellow liquid is referred to as serum. Serum separation is often observed in a crime scene and appears as this pinkish-yellow fluid surrounding a large volume of coagulated blood.

Viscosity is a measurement of a fluid's resistance to flow. The viscosity of blood is due to both the resistance of the plasma (plasma alone is 1.5–1.8 the viscosity of water) and, primarily, the percentage of red blood cells. Viscosity of plasma and red blood cells is approximately three to four times that of water. A decrease in red blood cells below the normal values (anemia) reduces viscosity. For example, an otherwise healthy person can survive with a hematocrit as low as 21%. The opposite extreme of too many red blood cells (polycythemia) results in increased viscosity (a hematocrit of 60% is associated with a viscosity about eight times greater than water).[8]

Coagulation and Hemostasis

Hemostasis is the process by which the body prevents or limits blood loss after injury. The three initial components of hemostasis are vascular smooth muscle spasm, formation of a platelet plug over the damaged endothelium (the inner surface lining of a vessel), and formation of a fibrin plug from coagulation factors in the plasma.[2,3]

The plasma contains a series of dissolved proteins (the classic major factors are numbered I through XIII), most of them made by the liver. When triggered, a complex coagulation cascade occurs, ending in the formation of a fibrin clot. The pale pink-yellow fibrin clot is a meshwork of interlinked proteins that blocks off or narrows the damaged vessel. If a particular factor is absent or abnormal, the formation of fibrin is prevented or significantly reduced. For example, deficiency of factor VIII is a common cause of hereditary hemophilia. The coagulation cascade is sometimes described in terms of an intrinsic pathway or an extrinsic pathway. Some of the factors are unique to one or the other pathway. Substances released by the platelet plug and by the damaged blood endothelium activate the intrinsic pathway, while substances released from the disrupted tissues at the injury site activate the extrinsic pathway. The separation of these pathways is somewhat artificial, as there is quite a bit of interaction between the two (a factor VIII deficiency hemophiliac, whose intrinsic pathway is defective, can still form fibrin clots through the extrinsic pathway). The result by either route is a fibrin clot. The clotting process at an injury site begins almost immediately, and a small superficial cut can form a clot in approximately 1–3 minutes.

Competing with clot formation is the process of blood lysis (dissolving). Injured tissue also releases a substance (tissue plasminogen activator, TPA) that changes another plasma substance, plasminogen, into plasmin. Plasmin dissolves fibrin, allowing the body to break down portions of the blood clots and reestablish blood flow to the tissues. Examples of this effect can sometimes be seen at a crime scene. Very rapid blood loss where the escaping blood has had little or no contact with TPA will often result in typical clot formation. A slower rate of blood loss, where the blood has time to interact with TPA, may form pools of nonclotted liquid blood.

Individuals treated with anticoagulants for medical reasons are often referred to as being on "blood thinners." The blood is not actually thinned. Rather, the patient is given a drug that blocks or retards some part of clot formation. Examples include aspirin, which interferes with the release of a thromboxane, a substance needed to cause platelets to clump together. Coumadin blocks vitamin K. Vitamin K is needed by several of the coagulation factors to function. Heparin binds to a substance, antithrombin III, which blocks the conversion of several coagulation factors to their active forms. This in turn reduces fibrin clot formation. However, anticoagulants ("blood thinners") do not alter blood viscosity, the number of red blood cells, or the behavior of a bloodstain on the final target surface.

The Circulatory System and Shock

The average person has approximately 70 ml of blood for each kilogram of total body weight.[16] The variation of size and weight among different people means that the typical range of total blood volume is approximately 4–8 liters (8–16 pints). Editor use citation 1[1]

Circulation is normally a closed-loop system. Blood is propelled from the left ventricle of the heart (with 90–140 mmHg systolic [contracting] pressure, 0–8 mmHg diastolic [resting], at a rate of approximately 4–8 l/min) into the aorta (approximately 120/80 mmHg).[4] The aorta is one of several elastic arteries, so called because of the increased elastic fibers in its walls that expand and recoil helping to propel the blood along. The arterial tree branches out into smaller and smaller muscular arteries. The contraction or relaxation of the muscular wall of these arteries, and particularly arterioles (the final <0.5 mm diameter artery branches just before capillaries) can regulate the amount of blood flow to a particular region and the overall blood pressure.[3, 10]

The capillaries are the smallest (approximately 7–9 μm wide and approximately 1 mm long) but most plentiful vessel. In the capillaries, the now slowed blood is able to diffuse oxygen and nutrients to the tissues and take away carbon dioxide and waste materials. This is also the leaky part of the circulatory system where liquids can move in or out of the circulation.[15] Any excess fluid that leaks out is collected by a parallel one-way system of vessels called the lymphatics. The contents of the lymphatics eventually empty back into the circulation near the heart.

The capillaries join to form first venules and then larger and larger veins. Inflammatory processes also can make the venules leaky, allowing liquids and white blood cells to move out into the tissues. The veins have thin walls with a few muscle cells that allow for some contraction or dilatation. Veins are referred to as capacitance vessels and, under normal conditions at rest, can contain up to two-thirds of all the blood in the circulatory system. By merely standing up, each leg has the capacity to store an additional 350 ml of blood.[11] Many of the larger veins in the limbs have valves so that flow proceeds only one way, back to the heart. The superior vena cava and the inferior vena cava are the final large veins that empty into the heart's right atrium (pressure range approximately 0–8 mmHg). From the right atrium, the blood is pumped to the right ventricle, which pumps the blood through the capillary bed of the lungs. There it is returned to the heart into the left atrium and then into the left ventricle to start the circuit anew.

The passage of blood through vessels can be described as turbulent or nonturbulent flow.[3, 9,,12] Turbulent flow is movement of blood in a non-parallel direction in the vessel, producing irregular whirlpools or eddies that often can be heard with a stethoscope. This type of flow occurs at branching points in vessels, as well as in diseased vessels. Nonturbulent flow is roughly parallel movement of blood in a vessel and is the most efficient means for perfusing tissues. Sometimes this efficient nonturbulent flow through larger blood vessels is described in terms of Poiseulle's law. Poiseulle's law concerns the laminar flow of an incompressible, uniform fluid (that is, a Newtonian fluid such as water) through a cylindrical tube with a constant circular cross section. Laminar flow is where concentric fluid layers move in one direction with the innermost layer traveling the fastest (and the most efficiently). The mathematical details of this law are beyond the scope of both this chapter and this author. The relevant part of this law to bleeding issues concerns the resistance to flow through a tube. The key point is that if blood in a vessel is exhibiting laminar flow, then resistance to flow is inversely proportional to the fourth power of the radius of the lumen. If one assumes that at a constant vascular pressure, a cut vessel of 1 mm diameter loses 1 ml of blood per minute, then a cut vessel of 2 mm diameter would lose 16 ml of blood per minute, and a vessel of 4 mm diameter would lose 256 ml of blood per minute. Poiseulle's law and laminar flow appear to be useful in approximating the behavior of blood inside healthy larger arteries. However, blood vessels are not straight, are not uniform, and do not exhibit steady flow. Furthermore, blood is a complex fluid with liquid and solid components, not a Newtonian fluid. Any limited predictive value laminar or turbulent fluid dynamics has regarding blood loss ends at the tip of the cut vessel.

The condition known as shock occurs when there is insufficient blood perfusion to the tissues and organs to meet their physiological needs. Hypovolemic shock results from loss of circulating blood volume such as results from rapid bleeding. Other forms of shock include cardiogenic shock (e.g., a heart attack resulting in inadequate cardiac pumping) or septic shock (e.g., infectious process resulting in massive dilatation of most of the blood vessels all at the same time).

Sometimes hypovolemic or hemorrhagic shock is clinically separated into four levels.[16] The first is rapid blood loss (over a matter of minutes to a few hours) of less than 15% of the total estimated blood volume (EBV). This degree of blood loss is usually accommodated by an increase in cardiac output and constriction of the blood vessels without significant changes in heart rate or blood pressure. The second level occurs when blood loss is between 15% and 30% of the total EBV. This condition is associated with elevation of the heart rate to approximately 100 beats per minute (tachycardia), narrowing of the pulse pressure (the medical records will often say the pulse is "thready"), increased respirations (tachypnea), a slight decrease in blood pressure (hypotension), and anxiety. Any blood loss over 30–40% of the total EBV is associated with increasing tachycardia (over 120 beats per minute), greater tachypnea, worsening hypotension, decreased urine production, and mental confusion. Finally, when blood loss exceeds 40% total EBV, worsening of the vital signs, minimal urine production, sweating, constriction of the pupils, and lethargy will result.

The rapid loss of 30–40% of the total EBV in an otherwise healthy person is survivable, but requires the maximal use of all compensatory mechanisms such as increased heart rate, shifting of fluids from outside the circulation, and vasoconstriction. However, rapid loss of over 40% without immediate therapy is usually fatal. With this degree of blood loss, the circulation system cannot meet the needs of the body, and irreversible tissue death with multi-system organ failure ensues. Fluid resuscitation and aggressive medical therapy at this point may delay the inevitable but cannot halt the process.

When rapid blood loss is not replaced, initially the hemoglobin and hematocrit concentrations remain the same. Only after the circulatory volume is reexpanded with crystalloid or colloid fluids is the degree of bleeding reflected in the decreased hemoglobin and hematocrit (each 3% drop in the hematocrit is roughly equivalent to the number of red blood cells in a pint of blood). With blood loss over a prolonged period, such as days, a person can compensate for a loss of 50% of the total EBV through the shifting of extracellular fluids and increased fluid retention.[7]

Nontraumatic Causes of Bleeding

One of the issues an investigator must consider at a scene involving bloodshed is the following: Is this a crime scene? A victim may be rendered susceptible to rapid, oft-fatal blood loss from natural processes or incidental minor trauma. Here is a small sample of conditions that might be commonly encountered and mistaken as a crime scene:

From the Head: Epistaxis, or nosebleeds, can sometimes result in severe and even fatal bleeding with the blood loss amounting to several pints of blood expelled in minutes. It can be a spontaneous process, due to hypertension, or due to drug ingestion such as cocaine. Simple lacerations in the very vascular tissues of the scalp can also lead to rapid blood loss with ensuing shock and death, especially if the victim is impaired such as from alcohol intoxication.

From the Lungs: Blood expelled from the lungs is called hemoptysis. It can result from medical disorders including infectious diseases or inflammatory disorders such as pneumonia, tuberculosis, or autoimmune diseases. Cancers can erode into blood vessels causing sudden and massive hemoptysis. In asphyxia or drug intoxications, pulmonary edema and bleeding within the lung tissues can result in copious amounts of bloody-looking fluid in the mouth and nose that may even cover the face.

Blood can be expelled from the airways during a cough at very high velocities and become atomized in the process. In one study[5] looking at the efficacy of paper masks, simulated coughing resulted in velocities ranging from approximately 5 m/s to slightly over 13 m/s (roughly 16–42 ft/s). Another study[13] looking at cough-generated infectious aerosols found cough speeds in the range of 1.5–28.8 m/s (roughly 4.9–94 ft/s). Taking either set of values, a cough can propel blood at enormous speeds and should be considered when interpreting an impact pattern. Depending upon the rate of bleeding, the blood may or may not be visibly diluted by mixing with lung secretions or saliva.

From the Gastrointestinal Tract or Due to Liver Disease: The liver is key to normal blood clotting and anything that diminishes its function makes bleeding more likely. Alcohol abuse is one of the most common causes for cirrhosis of the liver in the United States. Not only does cirrhosis impair liver function and the coagulation process, the scarring that accompanies cirrhosis elevates the pressure in certain blood vessels increasing the risk of rupture. Most of these vessels are along the gastrointestinal tract. Rupture of such dilated vessels in the esophagus (esophageal varices) can result in vomiting of blood that collects in the stomach (hematemesis). This projected blood may appear fresh or may be altered by contact with stomach acid. When altered, it is referred to as "coffee-ground" hematemesis. See Figure 11.1. If blood instead passes through the bowel to the rectum, it may appear as bright red blood (hematochezia), or more likely as a dark brown to black, tarry, foul-smelling material (called melena and caused by the intestinal liquids and contents altering the blood).

Peptic ulcers (ulcers of the stomach or first segment of the small bowel) can also result in rapid gastrointestinal bleeding. Inflammatory processes, cancers, and malformed blood vessels are other causes.

From a Gynecological Source: Menstrual blood flow can vary widely from woman to woman, but will generally range from less than 30 cc to up to 80 cc of dark red, nonclotting blood over a three- to five-day period. The "blood" is mixed microscopically with sloughed endometrium (the lining of the uterus) and mucus. A normal childbirth can be associated with an estimated blood loss of 500–1,500 cc. If the placenta is positioned too low in the womb or partially on the cervix (placenta previa) in the third trimester, the woman can suffer sudden profuse vaginal bleeding that can be fatal to both mother and child.[6,14]

From Venous Insufficiency: Varicose veins that become disrupted from minimal trauma or from chronic medical ulcers can be a source of extensive, even potentially fatal, bleeding. In chronic venous insufficiency, the one-way valves in the leg veins fail. The superficial venous drainage in the legs proceeds not only upward but also inward through perforating veins to the deep venous trunks (the pressure in one draining deep calf vein, the posterior tibial vein, while standing at rest can be approximately 80 mmHg). Normally during exercise, the pressures in the deep venous trunks is elevated (in the calf this is as great as 250 mmHg, and in the thigh, 115 mmHg), but at the same time the pressure in the more superficial veins drops markedly nearing zero (this effect is referred to as

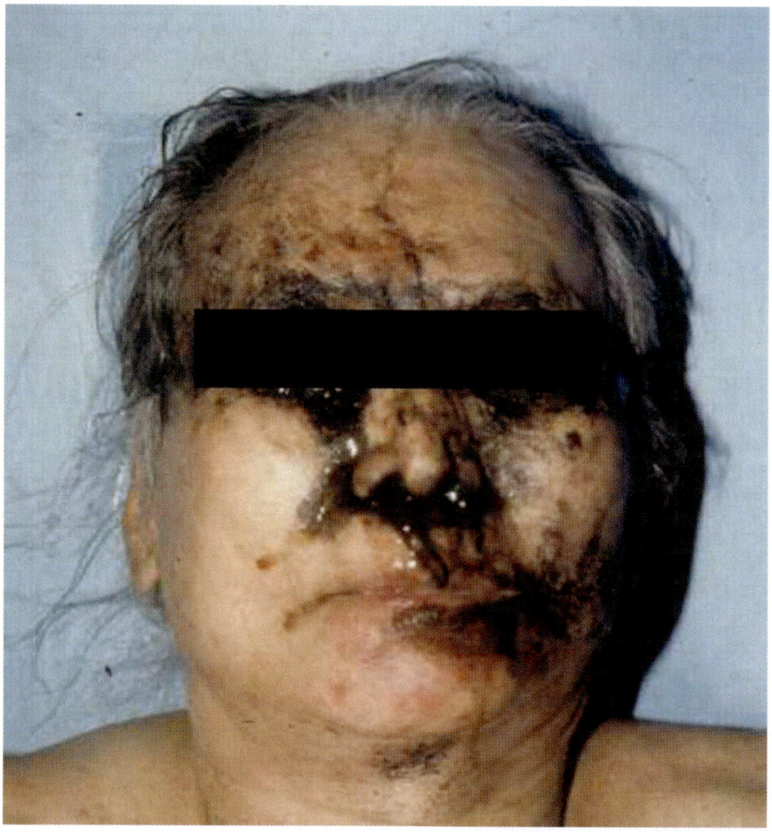

Figure 11.1 Coffee ground hematemesis. Blood that collected in the stomach and was subsequently vomited will have a distinctly brown appearance, caused by the stomach acid. (Photograph courtesy of Dr. David Winston, Pima Forensic Science Center, Tucson, AZ.)

the musculo-venous pump).[11] With chronic venous damage, the elevated pressures in the superficial veins can potentially lead to enormous rapid blood loss. Because of the increased pressure, blood loss of this nature can mimic some of the features of a streaming ejection, such as those expelled from an artery. See Figures 11.2 and 11.3.

From decomposition: Decomposition, in addition to distorting and discoloring the body in ways that mimic antemortem injuries, can result in postmortem liquids that may be confused with antemortem hemorrhage. Decomposition is the combination of autolysis (the normal breakdown of dead cells) and putrefaction. The putrefactive processes will break down tissues and produce gases that tend to expel liquids from facial mucus membranes. The autolytic processes also break down tissues and blood, making the decomposition fluids dark red to brown. The superficial skin frequently detaches and is filled with this liquid. The appearance can be a rather dramatic process that suggests a markedly bloody face with "blood" flow on the sides of the face. Furthermore, if a limb or the head is in a dependent position, decomposition of that lower congested region can lead to pools of suspicious-looking dark red liquid looking very much like blood dripped into blood or, in concert with putrefactive gases, possibly like projected blood.

Figure 11.2 The condition known as venous insufficiency syndrome has the potential to deposit a significant volume of blood into the scene. The patterns may appear "spurt"-like even though they are ejected from veins. (Photograph courtesy of ID Officer Rex Sparks, Des Moines Police Department, Des Moines, IA.)

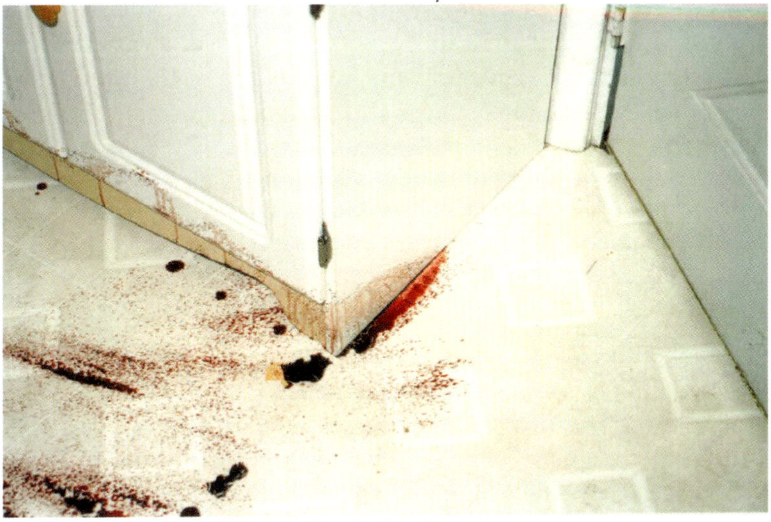

Figure 11.3 Another example of the patterns produced by venous insufficiency syndrome. Note the serpentine pattern on the vertical surface of the cabinet. These are quite similar to those produced by streaming ejections such as arterial spurts. (Photograph courtesy of ID Officer Rex Sparks, Des Moines Police Department, Des Moines, IA.)

Traumatic Pathology

The analyst should be familiar with some aspects of the pathology of injuries in order to understand the forensic pathologist's report. First, it is important to understand that a forensic pathologist describes the injuries assuming the victim is in what is called normal anatomic position. That position is person standing erect, legs together, with his arms by his side, palms facing forward. The injuries are described from the viewpoint of the victim right and left, front and back, etc.

The categories and examples of injuries[17] briefly described in the following, with a few exceptions, are limited to those physical injuries most likely to be of concern to a blood pattern analyst.

Firearm Injuries

A firearm projects more than a bullet from the muzzle. Following the bullet are the rapidly expanding gases created by the burned gunpowder, the residue of the burned gunpowder itself, and any unburned or partially burned gunpowder granules. The bullet and these gunpowder-related products will affect the appearance of the entrance wound.

If the muzzle of the weapon is fully pressed against the skin (a contact wound), then the bullet and all of the gunpowder-related products, including the large volume of gases, are forced into the wound. The gases expand the tissues, typically disrupting the skin and forming an irregular-shaped to stellate (star-shaped), large, gaping-open wound. Reapproximating the wound edges may reveal the original, roughly round, bullet hole as well as a circumferential abrasion pattern (an abrasion is a superficial scrape of the skin). There may also be an abrasion pattern in the shape of the muzzle (a muzzle imprint). Darkening inside the wound edges or on adjacent bone represents gunpowder residues. If the contact wound is over an area where the gases can easily expand inward and sideways, the entrance wound might consist of a simple round hole with a surrounding round abrasion rim, possibly with a muzzle imprint. The inner edges of the wound will show darkening from gunpowder soot and may show a localized pink discoloration of the skeletal muscle caused by carbon monoxide in the gases.

If the muzzle of the weapon is held away from the skin at a distance up to 18–24″, then, in addition to the bullet hole, there is usually a surrounding pattern of gunpowder residue on the skin. This pattern consists of irregular-shaped punctate abrasions around the wound (gunpowder stippling). The closer the muzzle is to the skin, the smaller and more intense this pattern is. These abrasions are caused by unburned or partially burned gunpowder, as well as any other debris propelled from the barrel, that impact the skin. A careful examination can often reveal occasional embedded particles of the gunpowder-related material in a few of these abrasions. If the muzzle is less than 4″–6″ from the skin, there is usually a diffuse, shiny, gray-black discoloration around the wound representing the addition of gunpowder soot being propelled into the skin. The closer the muzzle, the more compact and intense the soot pattern and the less likely one can see the pattern caused by the concomitant gunpowder stippling.

If the muzzle is still further away – 18″–24″ – then only a round to oval entrance wound surrounded by a rim of abrasion, created as the bullet enters, is seen.

An exiting bullet tents and tears the skin outward, typically causing a slit-shaped or irregular to stellate-shaped wound. Usually no abrasion rim is seen; although if the exit wound is shored up against a firm surface or very tight clothing, some abrasion may be seen. If the bullet in the body tumbles or breaks up, or if the weapon is a high-powered (high velocity) rifle, the exit wound can be a large gaping defect larger than its corresponding entrance wound. In general, no absolute statements can be made about any correlation between bullet size or type and the relative sizes of entrance and exit wounds.

It is essential to remember that the gunshot wound residue patterns just described are deposited on the first surface encountered. Therefore, a forensic pathologist considers it vital that he or she examine the clothing for soot, stippling, and even searing from the heat of the muzzle blast before offering an opinion about the range of fire. In addition, if a bullet tumbles in flight, breaks up, or ricochets, the characteristics of the entrance wound are often dramatically altered.

Shotguns have some additional wounding characteristics. Because of the large amount of gases produced by a shotgun (seen as well with high-powered rifles or large-caliber handguns), contact shotgun wounds usually show massive disruption of the underlying tissues. A contact shotgun wound to the head is likely to reduce the head to an unrecognizable pulp-like mass.

If the shotgun cartridge load consists of pellets, then at a range of less than 4–5 ft from the muzzle, the pellets enter as a single mass creating a round to oval entrance wound. The closer to the upper limit of this distance, the more likely one is to notice scalloping or irregularity along the margins of the wound representing the pellet mass starting to separate. Beyond 4–5 ft, the pellets continue to disperse from the central mass creating individual wounds. Eventually, the pellet mass spreads out such that only multiple, roughly round pellet wounds, each with minimal abrasion rims, are produced. Shotgun loads consisting of a single relatively large projectile (rifled slug) often produce large gaping entrance and exit wounds.

A smooth bore shotgun discharge also differs from most other firearms in that the shell contains a component (usually some form of wadding or plastic shot shell cup) between the projectile and the gunpowder charge. With contact wounds and range of fire up to 10–12 ft, the wadding can be projected inside the entrance wound. Still further away, the wadding may still impact the surrounding skin, leaving its own abrasion imprint.

Shotgun exit wounds can consist of small separate pellet defects or gaping irregular-shaped defects in the case of masses or clumps of pellet.

Firearm injuries can result in injuries and findings that are not directly along the wound track. For example, a gunshot wound may not involve the orbital region. However, there may be diffuse periorbital hemorrhage (appearing as black eyes or "raccoon eyes") resulting from the transmitted force of the blast. It does not represent a separate blunt injury. A neck, nasal, or oral injury may result in aspirated blood in the lungs that should not be interpreted as a separate injury.

Because of the great variability in the appearance of entry and exit wounds for different firearms, a competent forensic pathologist will often wait until the completion of his or her internal examination (after the pattern of soft tissue and bone injuries are determined) before designating each wound as an entry or exit wound. It behooves the analyst to read the wound descriptions and to review the photographic documentation before accepting the pathologist's opinion.

The variability in gunshot wounds also prevents anyone from making broad predictions as to the form or nature of the blood loss. Considerations must be given to the regions and organs injured by the bullet, as well as the degree of internal hemorrhage.

Sharp Force Injuries

A sharp-edged object that cleaves rather than tears the skin and underlying tissues causes sharp force injuries. Knives are the most commonly encountered sharp force instrument, but many other items such as sharp-edged kitchen utensils, scissors, slotted screwdrivers, and broken glass can be sharp force weapons. Implicit in this observation about other weapons is that there may be an intermediate type injury between sharp force and blunt force (described later) caused by a "dull" sharp instrument or a "sharp" dull instrument (as examples, a "dull" knife or a "sharp" axe).

Skin is a relatively elastic surface that is stretched along specific lines of tension. Therefore, sharp force wound edges have to be reapproximated to provide an estimate of the true wound size and shape (and through this, possible dimensions of the weapon). Typically, there is no abrasion rim surrounding the wound unless there is a defect on the blade, a knife is thrust to the hilt, or a weapon of intermediate sharpness is used.

If the depth of a sharp force wound is greater than its largest surface dimension, it is referred to as a stab wound. If the skin wound is longer than it is deep, then it is referred to as an incised wound (that is, a slashing or cutting wound). Some sharp force wounds may have both a stab (thrusting) component as well as an incised component as the thrusted weapon is also pulled parallel to the skin surface. The depth of the wound track may be reported as a range. Descriptions of injuries in the anatomic position cannot reveal if the victim was inhaling or exhaling, twisted, or bent over. Additionally, the compressibility of tissues should be considered. These are all factors that may influence the pathologist's ability to accurately estimate the true wound depth or even the true direction of the wound path.

The wound edges may give additional information about the weapon, such as squaring of one wound angle consistent with a single-edged knife, or waviness of the wound edge or surrounding skin suggesting a serrated blade. Defects or irregularities in the cutting surface may be reproduced in the wound edges. The adjacent skin can show evidence of multiple parallel superficial wounds (sawing) or linear abrasions (in suicides interpreted as hesitation marks).

Wounds on the hands or arms may be positioned such that they suggest that the victim was defending himself or trying to ward off the attack. The forensic pathologist may categorize these injuries as defensive wounds.

Once again, the variability of internal trauma and internal bleeding caused by sharp force injuries should make one very hesitant about making broad predictions concerning the rapidity, volume, or character of external blood loss based solely on the type of wound.

Blunt Injuries

A relatively broad-based force that tears and crushes tissues produces blunt injuries. These are the type of injuries one would see in a victim attacked with a bat, beaten with hands or feet, or resulting from a fall or motor vehicle collision. Externally, these injuries consist of lacerations, contusions, and abrasions.

Abrasions are scrapes or scratches of the most superficial layers of the skin. An abrasion can vary from a barely perceptible pink-tan mark to a dark red-brown discoloration. Postmortem drying and the draining of blood away from the skin cause initially pale abrasions to appear much darker and more prominent. In a victim who survives, the leakage of a small amount of blood and cellular fluids from the deeper skin can form a dried crust within a day (a scab).

Contusions are bruises. Bruising occurs when an applied force ruptures blood vessels, and the blood, with the impetus of circulatory pressure, extends into the surrounding tissues. An acute bruise can appear as a wide color spectrum besides the typical "black and blue" depending upon the depth, size, and density of the hemorrhage. Over several weeks as a bruise is repaired and reabsorbed, it undergoes color changes to green-yellow and eventually brown. If the age of a bruise is both relevant and in doubt, great care should be exercised in color-based examination alone. Benefit can be gained by microscopic examination of the bruised tissues by a forensic pathologist. It is very easy to be wrong based upon gross appearances.

Lacerations are tears. Lacerations occur when a blunt force shears or tears the skin and underlying tissues. A laceration is more likely to have irregular or uneven edges with abrasion along the margins. The subcutaneous tissues beneath the skin at a laceration often are undermined (producing a cul-de-sac-like region of separation in the soft tissue). Within the laceration's edges or in the undermined regions, bridges of the soft tissues are found. These tissue bridges represent intact stronger bands adjacent to weaker torn tissue. When the wound is gently pulled apart or lifted, string-like bands of tissue form a bridge from the skin edge to the deeper tissues.

Multiple factors can influence the extent of bleeding from a blunt injury. A laceration to a region with a large vascular bed (the scalp and face) will cause more rapid and voluminous blood loss than a less vascular region (the sole of the foot). Limb injuries expected to result in copious bleeding that are distal to a disrupted central vessel may produce minimal bleeding. Hypotension (low blood pressure) from internal injuries may limit bleeding from even severe external injuries. Finally, the forensic pathologist may determine that some of the injuries are postmortem and resulted in minimal or no bleeding at all. One should be hesitant to make overly specific predictions about bleeding and would be wise never to make any predictions without the full medical or autopsy reports.

The shape and patterns of injuries can provide information about their source. A crescent-shaped laceration may suggest a pipe or a hammer. For instance, several evenly spaced pairs of uniform rectangular-shaped lacerations may indicate the prong end of a hammer. Repetitive abrasion and contusion patterns in the skin can be suggestive of footwear injuries from stomping. As with pattern transfers in blood, there is no limit to the nature of pattern injuries. The forensic pathologist will be very interested in reviewing any pattern transfers at the scene. This may aid the pathologist in understanding the patterned injuries and recognizing potential sources of the pattern. As with blood transfer patterns, the detail from these pattern injuries usually provides only class characteristics. However, occasionally individual characteristics allow for a match or near-match of the patterned injury to a specific weapon or object (provided one-to-one photographic documentation could be provided to the criminalist).

Under the correct circumstances, a broad-based blunt force may leave little or no external injuries while producing fatal internal results in the head or torso. Even a fracture to the femur (thigh bone) can be associated with over 1,500 ml of local internal bleeding that can result in shock and death.

The head can be lethally injured with minimal external findings. A skull fracture can be associated with bleeding and contusions within internal membranes and in the brain. Contusions to the brain can occur without skull fractures or even discernible external blunt injuries to the scalp. When a moving head strikes a stationary firm object and the force from the impact is transmitted through the head, contusions to the brain are often seen on the opposite side (contrecoup contusions). Moreover, impacts to a head at rest without a skull fracture can occasionally produce contusions on the same side (coup contusions).

These latter contusions occur with an intact skull. They should be distinguished from contusions found underneath skull fractures (fracture contusions).

Of the several protective membranes surrounding the brain, the thickest and outermost layer is the dura mater. There are thin veins in the upper midline region of the brain that bridge across from the dura to the brain which can be sheared from a blunt impact. The bleeding from these veins (subdural hematoma) can collect slowly over hours and even days before causing symptoms. Bleeding from arteries intimately associated with the skull, typically resulting from a skull fracture, can produce a much more rapid collection of blood between the dura and the skull (epidural hematoma).

The Forensic Pathologist as a Resource

A forensic pathologist is a medical subspecialist who has training and experience in examining and documenting unusual or unnatural deaths, particularly deaths due to trauma. Basic issues or questions posed to the forensic pathologist might include the following:

1. Time of death? The observations at the scene of rigor mortis and livor mortis can form a rough basis for time of death. Rigor mortis is the stiffness (not contraction) that forms in all of the skeletal muscles after death. It is related to acidosis, which causes muscle cell contractile proteins to bind. Rigor mortis starts at about 1/2 to 1 hour after death and reaches a maximum by or before 12 hours. After about 24 hours, rigor mortis begins to dissipate as decomposition changes emerge. Livor mortis is the postmortem red to purple-blue settling of blood to the dependent portions of the body. Livor mortis is usually seen within a few hours after death and increases in intensity over the next few hours. Pressing the lividity during this time will reveal blanching. Turning the body over during the first few hours is like turning the sands in an hourglass. The lividity will redistribute to the new gravity-driven dependent portions of the body. However, at about 8–12 hours, the lividity becomes fixed. The drop in body temperature congeals the blood in the smallest vessels. This and the extravasation of blood into the tissues prevent any further blanching or shifting of the blood.

   There can be a wide range of variation to the onset and development of rigor mortis and lividity. The general "rules of thumb" noted previously should remain just that. Many factors, including obesity, muscle mass wasting, antemortem anemia, and strenuous exertion, can affect the onset and extent of both.

   Other postmortem changes such as declining core body temperature or the electrolyte content of vitreous humor can add information but are not generally accepted individually as reliable indices. With these tools, the forensic pathologist can provide a several-hour time range that may be useful to the investigator.
2. Is there an injury at all? The forensic pathologist can determine if antemortem injuries are present, or if postmortem changes and/or medical conditions are responsible for the apparent "bloodshed." Especially in decomposed remains, distinguishing antemortem from postmortem changes can be challenging. In some cases, bruises and lividity can be surprisingly hard to distinguish. The elderly are more prone to bleeding after incidental minor trauma that can nevertheless appear dramatic. The forensic pathologist can determine if these are trivial or significant injuries.

3. What is the source and extent of the bloodshed? The forensic pathologist can document the nature and extent of injuries that could account for the observed blood patterns. He may be able to distinguish features or injuries that limit the actions of a victim (for example, a gunshot wound to the thoracic spine rendering the victim unable to walk), or suggest the sequence of events at the blood pattern scene. Documentation of massive internal bleeding may explain the paucity of blood at a scene (instead of suggesting that there is another, yet undiscovered, crime scene). Documentation of blood in the airways may explain an otherwise confusing spatter pattern. The forensic pathologist can confirm the possible mixing of cerebrospinal fluid and blood in a head wound that would explain diluted stains at a scene. The forensic pathologist will certainly document damage to the larger vessels and may offer functional information on injuries to smaller vessels (taking into account that small, relatively minor, arteries can be responsible for significant bleeding), all of which helps explain patterns at the crime scene. An incompletely incised artery may explain an arterial spurt, while a fully transected artery that retracted back into the deep muscle may explain a more non-specific projected (gush) pattern.
4. Are all the injuries antemortem? The forensic pathologist should be able to determine if all of the wounds are associated with hemorrhage in the tissues. A lack of tissue hemorrhage could represent further injury after death.
5. Are the injuries hospital-acquired? The forensic pathologist can bridge the gap between the investigator and the health care providers. A review of the medical records can provide vital information to distinguish inflicted injury from that associated with therapy. As an example, bleeding that mimics contusions can surround venipunctures. Another common example is the injury produced by the rib spreaders used to open the chest for cardiac massage in emergency departments. This injury has been known to be erroneously attributed to various mechanisms of injury including gunshot.

The forensic pathologist should be a resource that every blood pattern analyst uses routinely. Moreover, as much as anything else, the forensic pathologist's job is to be able to explain his or her findings in basic understandable terms. If he or she cannot, get a second opinion.

The Author

The author, Michael Ferenc, M.D., J.D., is a forensic pathologist with over 36 years of experience who has worked in New York City, San Francisco, Maine, and Arizona.

Notes

1. Henry, J.B., Ed., *Clinical Diagnosis and Management by Laboratory Methods*, 12th ed., W.B. Saunders Company, Philadelphia, PA, 2001.
2. Kumar, V., Abbas, A.K., and Fausto, N., *Robbins and Cotran Pathological Basis of Disease*, 7th ed., Elsevier Saunders, Philadelphia, PA, 2004.
3. McCance, K.J. and Huether, S.E., Eds., *Pathophysiology: The Biological Basis for Diseases in Adults and Children*, 3rd ed., Mosby, Maryland Heights, MO, 1998.

4. Greenfield, L.J., Ed., *Complications in Surgery and Trauma*, 2nd ed., Lippincott Williams & Wilkins, Philadelphia, PA, 1990.
5. Inouye, S., Matsudaira, Y., and Sugihara, Y., Masks for Influenza Patients: Measurement of Airflow from the Mouth, *Japanese Journal of Infectious Disease*, Volume 59, 2006, pp. 179–181, Tokyo, Japan.
6. Goodnight, Jr., S.H. and Hathaway, W.E., *Disorders of Hemostasis & Thrombosis: A Clinical Guide*, 2nd ed., McGraw-Hill, New York, NY, 2001.
7. Glovicski, P. and Yao, J.S.T., Eds., *The Handbook of Venous Disorders: Guidelines of the American Venous Forum*, 2nd ed., Edward Arnold, London, UK, 2001.
8. Guyton, A.C. and Hall, J.E., *Textbook of Medical Physiology*, 9th ed., W.B. Saunders Company, Philadelphia, PA, 1996.
9. Mulholland, M.W. and Doherty, G.M., Eds., *Complications in Surgery*, Lippincott Williams & Wilkins, Philadelphia PA, 2006.
10. Dorhout Mees, E.J., *Cardiovascular Aspects of Dialysis Treatment – The Importance of Volume Control*, Kluwer Academic Publications, Hingham, MA, 2000.
11. Dodd, H. and Cockett, F.B., *The Pathology and Surgery of the Veins of the Lower Limbs*, 2nd ed., Churchill Livingstone, London, UK, 1976.
12. Berne, R.M. and Levy, M.N., Eds., *Principles of Physiology*, 2nd ed., Mosby, Maryland Heights, MO, 1996.
13. Hertzberg, J., Vansciver, J.M., and Miller, S., Flow Field of a Human Cough, Abstract to the Division of Fluid Dynamics Meeting of The American Physical Society, Chicago, IL, 2005.
14. Beckmann, C.R.B., Ling, F.W., Smith, R.P., Barzansky, B.M., Herbert N.P., and Laube, D.W., *Obstetrics and Gynecology*, 5th ed., Lippincott Williams & Wilkins, Riverwoods, IL, 2006.
15. Dailey, J.F., Ed., *Dailey's Notes on Blood*, 2nd ed., Medical Consulting Group, Springfield, MO, 1993.
16. Gutierrez, G., Reines, H.D., and Wulf-Gutierrez, M.E., Clinical Review: Hemorrhagic Shock, *Critical Care*, Volume 8, London, UK, 2004, pp. 373–281.
17. Spitz, W., Ed., *Spitz and Fisher's Medicolegal Investigation of Death*, 4th ed., Thomas, Springfield, MO, 2004.

Presumptive Testing and Enhancement of Blood

12

CHARLENE MARIE AND TOM J. GRIFFIN

Is this stain *blood*? Identifying the red, slightly viscous material oozing from a wound or pooling around a body is not difficult. But what about the brownish-green stain on the axe handle? Or the dark brown stains on the fence post? What about that sticky-looking, reddish-brown material on the interior driver's door? The true nature of blood and other stains found in the crime scene will be anything but evident to the analyst.

A significant part of identifying a stain as blood is based on context and observable physical properties. When it is obvious the stain results from an injury, there is little issue with a call of "blood." The presence of pinkish-red cerebrospinal fluid or a dark red pool showing serum separation at the edges of a volume indicates blood. Many times, however, the appearance of the blood has changed, and the association of the stain with a victim or suspect is not necessarily evident. The first step in deciding is to examine the stain visually. A good description and documentation of the target stain will often allow confident conclusions regarding the presence of blood on a target.

When the context and physical properties do not clearly indicate blood, presumptive tests are a viable and vital part of an analyst's toolkit. Even if the stains appear as blood, they may still need to be tested.

The next step is selective presumptive testing of the stain. Presumptive blood and bloodstaining testing are integral to evidence collection and scene analysis. What may be apparent at the scene may not be evident in the courtroom, so good documentation and appropriate testing are essential tasks in the analyst's scene work.

A presumptive reagent is a screening tool. A positive presumptive reaction indicates that the material tested is most likely blood. It implies that the sample should be collected and that further testing should be done. A responsible field report would incorporate all the information available, including the presumptive result. The report might read, "Presumptive testing indicates (or corroborates) the presence of blood." Similarly, if a presumptive test does not indicate the presence of blood, it is proper to record that "blood was not detected." That does not mean blood is not present; it means that using the available methods and reagents, we did not detect blood.

Presumptive Tests

There are several useful presumptive reagents. They each have their advantages and shortcomings. Some of the most sensitive and selective reagents have been abandoned because of their toxicity. The ideal reagent will offer good sensitivity and specificity, but not be toxic to the user. It will be easily prepared and easily used. The results will be unambiguous. Additionally, the reagent should have a good shelf life. There is no ideal reagent, but several good reagents are available that meet these standards.[1,2]

DOI: 10.4324/9781003408734-12

Most presumptive reagents are colorless dyes that become colored in the presence of blood and an oxidizer. All the presumptive tests discussed here detect hemoglobin (Hb). It is this iron-containing heme portion of the hemoglobin molecule that is detected, regardless of the dye that is used. The heme catalyzes the reaction with a presumptive dye and peroxide.

Most mammalian forms of the hemoglobin molecule have four heme groups per hemoglobin molecule. Figure 12.1 shows the heme structure. The iron ion (Fe^{+2}) in the middle of a porphyrin structure in the heme is responsible for oxygen binding, transport, and release. This is the most essential function of the heme group.

Catalytic presumptive tests using chemical reagents are based on the principle of the presence of peroxidase-like activity of hemoglobin in blood. This peroxidase-like activity catalyzes the oxidation by peroxide of the various dyes resulting in a rapid color change of the test reagent. Phenolphthalein, *o*-tolidine, leucocrystal violet, 3,3′,5,5′-tetramethylbenzidine (used in Hemastix™), leucomalachite green, fluorescein, and luminol are commonly used presumptive reagents. These reagents react with some plant materials, cleaning agents, metals, metal salts, and other sources of iron. For this reason, these tests are *presumptive* for blood, not confirmatory.

While any of these reagents can be used as a presumptive test for blood, it is important to consider the pros and cons of the various reagents before choosing a presumptive testing reagent.

Sensitivity, specificity, safety, and simplicity are the basic reagent criteria. Once the presumptive tests are done, the target stain must still be viable for genetic testing.

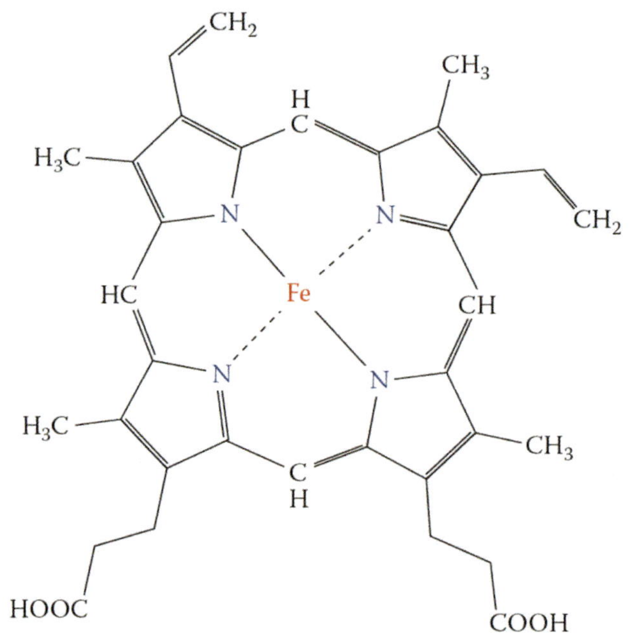

Figure 12.1 The heme structure. Phenolphthalein, *o*-tolidine, leucocrystal violet, 3,3′,5,5′-tetramethylbenzidine (Hemastix™), leucomalachite green, fluorescein, and luminol are commonly used as presumptive reagents. They react with the iron (Fe) ion present in the hemoglobin molecule.

Additionally, substrate and environmental conditions may be important considerations. What works in the laboratory may be cumbersome in the field. What shows up clearly on a light surface, may not show up on a dark one. Fortunately, there are several good choices and it is prudent to be familiar with more than one.[3]

Whenever organic dyes are used, an important consideration is the possibility of carcinogenic/mutagenic consequences. While known carcinogens are avoided, gray areas exist for some presumptive tests regarding their inclusion as suspected carcinogens. One of the ways to evaluate a particular dye for mutagenic behavior is the Ames test.[4]

The Ames test is an assay that places the material to be tested onto a specifically designed growth medium (usually doped with genetically modified salmonella). If the suspected material can grow colonies in the medium, it is considered a mutagen. The results are sometimes listed as "mutagenicity in salmonella." Since most carcinogens are also mutagens, if a material tests positive as a mutagen, the material is also considered a suspected carcinogen.

If a material is cleared via the Ames test, it is not mutagenic and is, therefore, *most likely* not carcinogenic. The Ames test is valuable because many materials that might not otherwise be tested can be screened with this method. A shortcoming of this screening test is that the material is tested against bacteria and then correlations are made regarding its effects on living beings.

Benzidines

Because of their sensitivity, Benzidine and its derivatives (see Figure 12.2) were used extensively as blood detection reagents. However, benzidine has been identified as a human carcinogen and is no longer recommended for use. A related material that is very commonly used is Hemastix™. The active dye in Hemastix™ is 3,3′,5,5′-tetramethylbenzidine (TMB).[5] This material, while structurally related to benzidine, has been tested via the Ames test and found not to be a mutagen. Additionally, the construction of Hemastix™ has the active materials embedded on a pad at the end of a plastic strip. The construction of this product contributes to the safety and the ease of its use for testing.

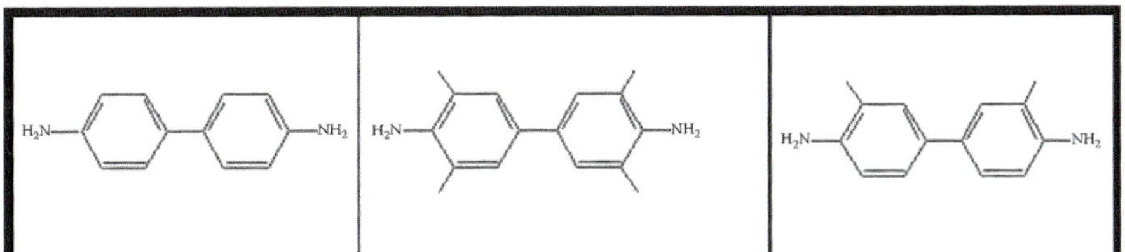

Benzidine	3,3′,5,5′-Tetramethylbenzidine (used in Hemastix™)	o-Tolidine
Carcinogen	Not a carcinogen	Suspected carcinogen
Benzidine-Based Dyes		

Figure 12.2 Benzidine and its derivatives. The derivative, *o*-tolidine (3,5-dimethylbenzidine), is a suspected carcinogen. It is still in use by some practitioners because of its excellent sensitivity.

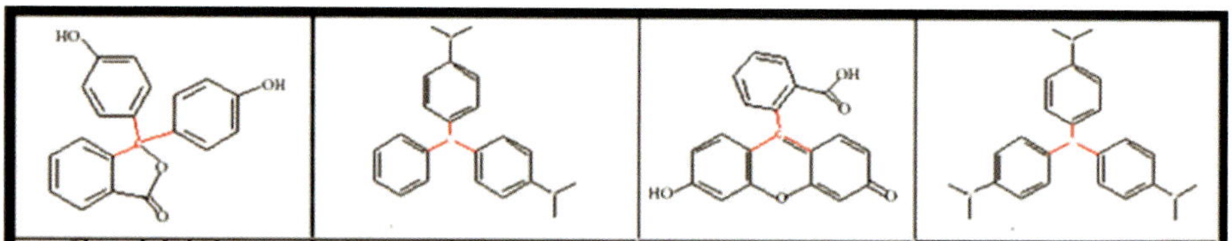

Phenolphthalein	Leucomalachite green	Fluorescein	Leucocrystal violet
Suspected carcinogen	Suspected carcinogen	Nontoxic	Nontoxic
Triarylmethane-Based Dyes			

Figure 12.3 Triarylmethane is the "backbone" or common structure of phenolphthalein, leucomalachite green, fluorescein, and leucocrystal violet.

Triarylmethanes

Phenolphthalein (Kastle–Meyer reagent), leucomalachite green, leucocrystal violet, and fluorescein are all in the triarylmethane group (see Figure 12.3). Like the benzidine group, they share a common basic structure. Structural variations give rise to their different color reactions and, in the case of fluorescein, cause the reaction to fluoresce.

Phenolphthalein was used for almost a century in commercial laxative formulations.[6] In 1997, the FDA concluded that phenolphthalein potentially *could* cause cancer in humans. The Ames test for phenolphthalein did not detect it as a mutagen, but mice studies showed evidence of carcinogenicity.

Fluorescein and leucocrystal violet are referred to in the literature as nontoxic.[7] Small changes in structure may yield surprising changes in toxicity. It is a good rule to be suspicious and, therefore, careful of *any* chemical testing reagent.

Luminol

Luminol is a chemiluminescent molecule. Chemiluminescence is the emission of light during a chemical reaction. Luminol (5-amino-2,3-dihydro-1,4-phthalazinedione, and also seen as 3-aminophthalhydrazide) produces light when it reacts in the presence of hemoglobin (see Figure 12.4). The molecule in the excited state (in Figure 12.4) emits light and then returns to a ground state. This is a fast reaction and because nitrogen (N_2) is expelled, it is not reversible. More luminol must be added for further chemiluminescence; however, this presumes sufficient hemoglobin is present for the additional reaction to occur. The structure of luminol is based on aniline and the phthalates (see Figure 12.5). There was some evidence that exposure to aniline by dye workers resulted in an increase in liver and bladder tumors. This gave obvious rise to concerns that luminol might be a suspected carcinogen. Subsequent research has shown this is not the case.[8]

Choosing a Reagent

Dye choices for presumptive blood testing include benzidines, triarylmethanes, and the chemiluminescent molecule luminol. It is common practice to choose phenolphthalein, leucomalachite green, or 3,3′,5,5′-tetramethylbenzidine (Hemastix™) when the stains are patent, or visible. Luminol, leucocrystal violet, and fluorescein are generally used to test for the presence of latent blood or enhance latent patterns. Note that nontoxic reagents are

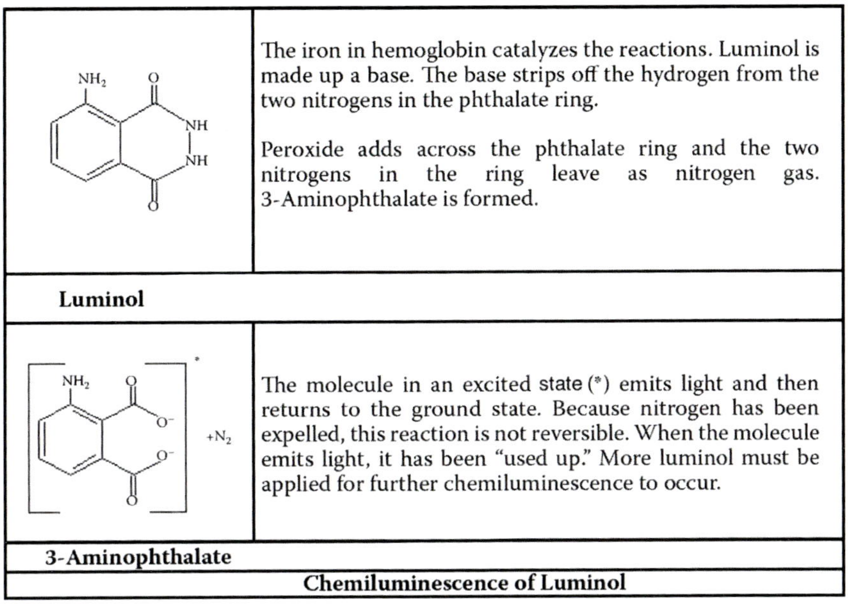

Figure 12.4 Luminol (5-amino-2,3-dihydro-1,4-phthalazinedione) produces light (chemiluminescence) when it reacts with the Fe ion in hemoglobin.

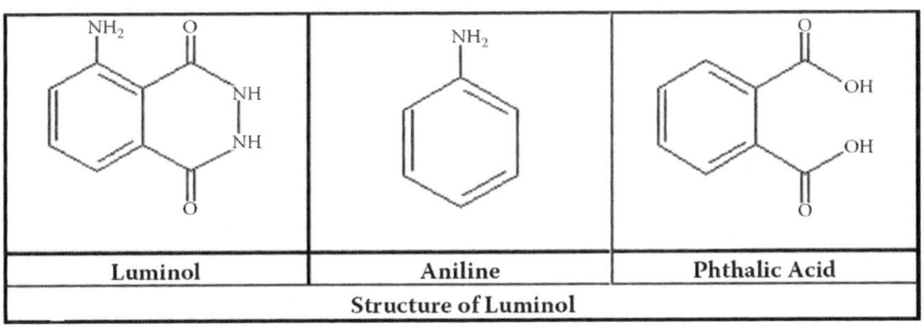

Figure 12.5 The structure of luminol is similar to that of aniline and the phthalates.

generally used to search for latent blood. Part of the reason is their excellent sensitivity, but it is also important to note that searching for latent blood often involves spraying a chemical reagent. It is much safer to spray a known nontoxic material.

Whichever testing reagents are used, it is important to validate the sensitivity and specificity of the reagents and to verify that the reagents are reacting properly on the day of use by conducting tests with positive (known blood sample) and negative (reagent blank) controls before the reagents are used in evidential testing. Lot numbers and expiration dates for the reagents should be recorded. Several commercially available products may be used. In forensic laboratories, such commercial products may be used or the reagents may be prepared in-house. Regardless of the source, most forensic laboratories will conduct various validation tests before adopting such reagents for standard use. Testing these materials yourself before committing them to evidential testing is still important.

A good understanding of the sensitivity, specificity, repeatability, and range of responses of the reagents on a variety of target substrates is important. The analyst should recognize the various sources of interference referred to as false positives (see Figure 12.6). Keep in

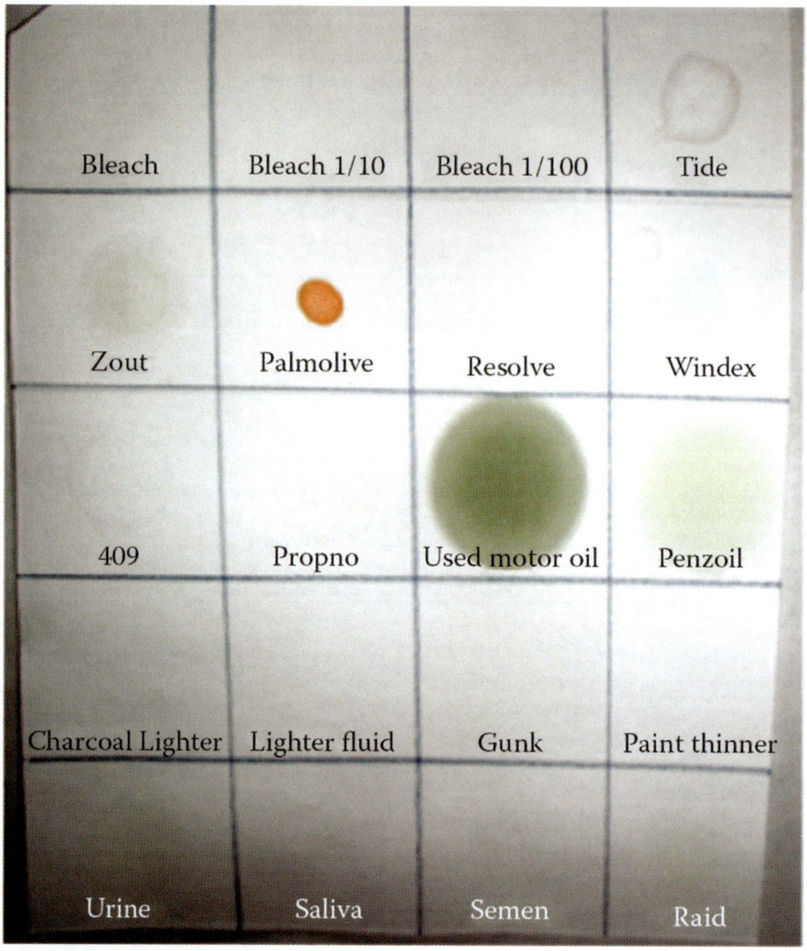

Figure 12.6 A stock card used to test for interference. The card includes samples of common cleaners, soaps, lighter fluid, motor oil, and other materials. Each substance is placed in a separate test grid to check for reactivity with the presumptive test. The cards can be prepared in advance and stored for future use.

mind that not all surfaces are created equal. An excellent example is black denim, which is a far more difficult surface to test than, say, white cotton. It is also essential to develop familiarity with the range of responses that might be seen. The analyst must understand how the reagent of choice will behave on a variety of substrates and under varying environmental conditions.

The analyst may also want to check the reagent on bloodstains of varying strength, especially if they are considering testing materials in the field and are unsure how old they are or how concentrated the stains are. One method for setting up a dilution series is to serially dilute whole blood with distilled water and then apply samples of the dilutions to cardstock (see Figure 12.7). The cardstock is divided into grids and each square labeled with the dilution to be applied. Several of these may be prepared, air-dried, and frozen until needed. This approach works well for reagents that will be sprayed onto a target. This is a fast way to perform a quality control (QC) check *and* test the reagent's sensitivity in a field setting if such testing is standard procedure.

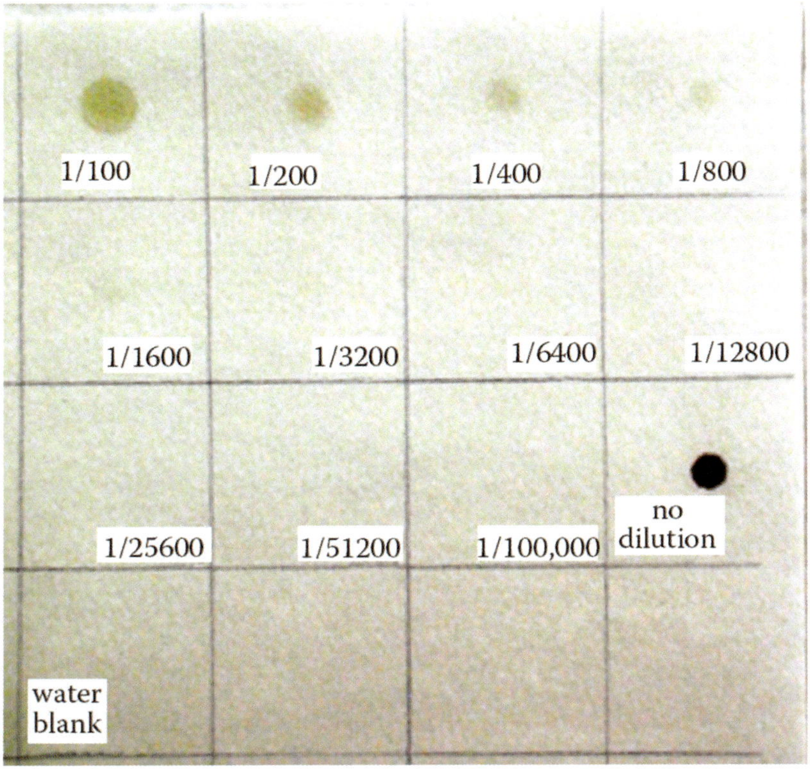

Figure 12.7 A bloodstain dilution series test card. The cards are used to test the chemical's sensitivity to diluted bloodstains. They are generally prepared in advance and stored in a freezer. They can be used when necessary to test the chemical's response.

When you conduct dilution or interference tests, it is a good idea to maintain a record of the results for both. These results become a guide of how your reagent behaves in your hands. Having a written track record of the reagent's performance is also very helpful. It is easy to photograph the test before and after treating the stains to document the effort.

As noted above, the reagent should be checked each day it is used. The reagent should be tested using a known blood source and a blank, which should include any solvent used to prepare the reagent. The QC check can be included in laboratory or field notes. A sensitivity check may be incorporated into the QC check by using the bloodstain dilutions card, as shown in Figure 12.8.

Just because a bloodstain dilution series reacts down to 1 in 100,000 on a piece of white cardstock or other test medium does not mean those results will always occur in a field application. Each of the presumptive searching reagents reacts differently on different substrates. As mentioned, black denim is a difficult substrate, and you will have a harder time finding very tiny and/or dilute stains on such a surface. You must use your eyes first. Before applying any chemical reagent, examine each item using different lighting arrangements and magnification. If a crime scene investigator has doubts or questions regarding evidential testing, they should contact the biological science section of the forensic laboratory that will ultimately be examining and analyzing the evidence.

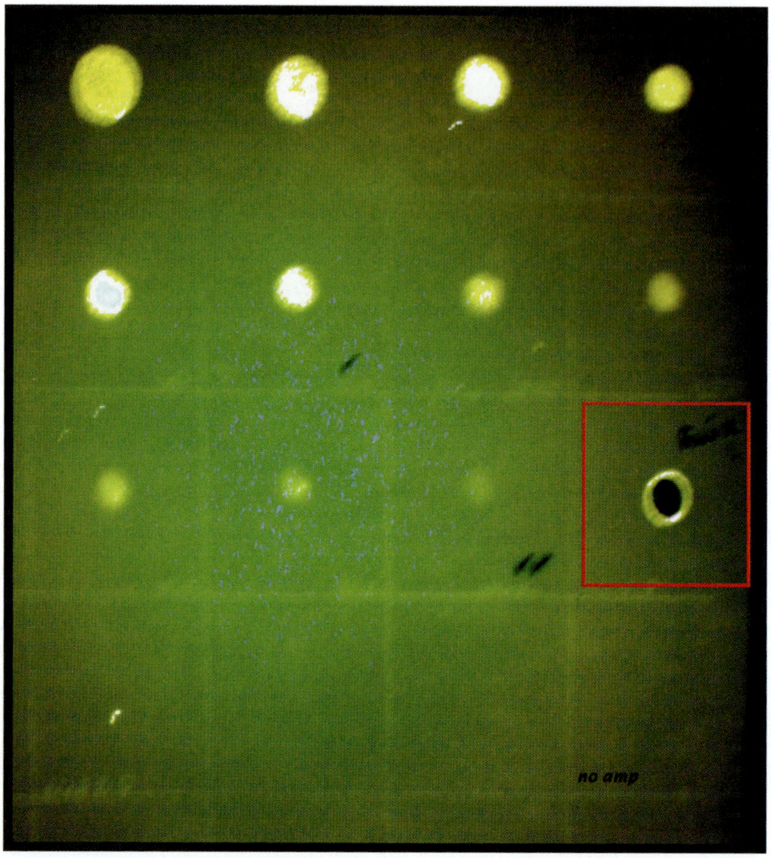

Figure 12.8 An example of a fluorescin reagent response using a bloodstain dilution card. 20 μl aliquots of blood dilutions were applied to the white cardstock grid. The box highlights an undiluted stain (whole blood). Note that the stain reacts only around the boundary, which is a typical response of fluorescin when sprayed on whole blood.

Genetic Testing Considerations

As with any presumptive testing, by its very nature such tests will use or reduce at least some of the blood present. In turn, this has the potential to limit the amount of genetic material ultimately available for DNA testing. While there is still some discussion about the effect of presumptive testing on DNA sampling, the dyes used and the formulas given in this chapter are all safe to some degree to use separately or in tandem and still allow genetic material to be recovered for DNA testing using Polymerase Chain Reaction (PCR).[9,10] Whenever possible, the best practice is to use one set of swabs to collect a sample from a bloodstain or pattern for DNA profiling. Then use another swab or set of swabs to collect a sample from a bloodstain or pattern for presumptive testing. The real challenge for presumptive testing and DNA sampling arises for samples that are already compromised. Specks of apparent blood or faint smears that might be blood are samples that must be handled carefully. These samples are the most susceptible to both contamination and destruction. As it turns out, they are also often the most probative samples.

Handling alone can cause the genetic material to diminish simply due to the mechanics of manipulating the material.[11] It is possible to deluge a tiny stain with a chemical reagent, making it impossible to find the sample later for further testing. Dilution, mechanical damage, and contamination are always possibilities and a significant consideration when dealing with already compromised samples.

Contamination of the sample may or may not be known or recognized. Contamination that is known can always be accounted for in the context of the scene (e.g., a suspect bleeding on a victim, an examiner sneezing on a sample, or the wind knocking a sample into another sample). Care must be taken to prevent or limit sources of contamination as much as possible, be observant of potential sources of contamination, and note any sources of contamination that are discovered.[12]

Most of the evidence we encounter is mixed with something from the environment. Often the evidence has dirt, fibers, foodstuff, or some other environmental association. Most of these contaminants will not further complicate genetic testing, but they can make the DNA analyst's sample cleanup more difficult. It is always important to describe the evidence context so that if information about environmental context will contribute to the genetic analysis, the DNA analyst can make sense of it. For example, bloodstains from the victim that are on the suspect's clothes will most likely yield a major donor profile from the victim and a minor donor profile from the suspect. The location of the bloodstain on the suspect's clothes is always a compelling association. These considerations make it even more critical to secure reference samples from all victims and suspects whenever possible, to assist in untangling any DNA profile mixtures.

As much as is possible, the analyst collecting the sample must not add anything further to the sample. Proper evidence handling, such as wearing protective gloves and clothing, using disposable or cleanable tools, and securing the samples properly will help ensure they arrive at the laboratory for testing without altering their original condition.

Formulations

Hemastix™, o-tolidine, phenolphthalein, leucocrystal violet, and leucomalachite green are the reagents used to detect patent blood. Formulas for their preparation, testing procedures, and interpretation of the results follow. Some forensic laboratories and forensic product providers may use alternate formulations to those presented in this chapter.

Hemastix™

Hemastix™ are plastic strips to which a reagent pad is affixed for testing stains or liquids for the presence of blood.[13] The dye and peroxide are both on the test pad. This makes use of Hemastix™ a one-step test. The oxidizer is diisopropylbenzene dihydroperoxide; the dye is 3,3'5,5'-tetramethylbenzidine (TMB). The reagent pad is moistened, and the embedded dye and peroxide react to heme to produce a dark green/blue-green color (see Figure 12.9). This single-use reagent strip is ready to use upon removal from the bottle and is disposable.

Hemastix™ Procedure

When the reagent pad is moistened with a small amount of distilled water, it contains all the necessary chemicals to perform the test. To demonstrate that a test strip is working appropriately, a known blood source (positive control) and reagent blank (negative control)

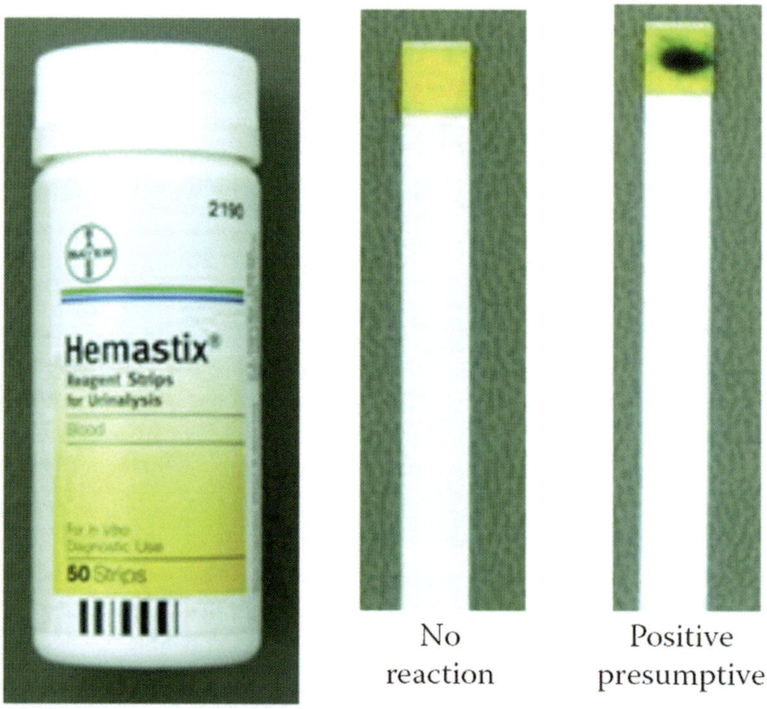

Figure 12.9 Hemastix™ are a commercially prepared presumptive test based on tetramethylbenzidine (TMB). A positive presumptive result is a blue-green discoloration on the test pad. The analyst should always ensure the package has not exceeded the expiration date. (Photograph courtesy of Sara Larsen, California Department of Justice, Redding Laboratory, Redding, CA.)

must be tested each day the reagent strips are used. These results should be included in notes, and it is good practice to record the lot number and expiration date listed on the bottle.

To conduct an indirect test, rub a moistened swab or a piece of small, folded filter paper on a suspected stain and then touch the swab or filter paper to the moistened reagent pad. To conduct a direct test, cut a very small portion of the suspected stain and touch it to the moistened reagent pad. A variation of the direct testing procedure is to bring the moistened reagent pad into contact with the stain as found. Alternatively in the laboratory, apply a small aliquot of stain extract directly to the pad. If available, test a small, unstained portion of the substrate material. A positive presumptive result for blood is a rapid color change to dark green/blue-green. No immediate color change is a negative result.

As with other reagents, exposure of the used test strip over time will result in a slow color change. This is not a positive presumptive result.

Preparing Phenolphthalein, Leucomalachite Green, and o-Tolidine

The dyes phenolphthalein, leucomalachite green, and o-tolidine are often used in a two-step procedure. First, the reduced dye solution is applied, followed by the oxidant. A 3% hydrogen peroxide (H_2O_2) solution is the most commonly used oxidant. This can be purchased or prepared from a stronger solution by diluting it in distilled water.

To prepare the dyes in the laboratory, phenolphthalein is heated in a basic (sodium hydroxide) water solution, and leucomalachite green is heated in an acidic (acetic acid) water solution. Zinc is added to both to catalyze the reduction. Be careful when disposing of used zinc because as zinc dries, it generates heat and can start a fire. Additionally, the used zinc may be considered a hazardous waste in some states. The other caution when heating these solutions is that hydrogen gas is formed. Take care not to seal the solutions too tightly. The hydrogen gas needs to vent. Preparation should be conducted in a fume hood.

Phenolphthalin Solution

While this presumptive test is commonly referred to as the "Phenolphthalein test," from a chemistry standpoint this is incorrect. As is detailed below, the actual compound or reagent reacting with the iron in the hemoglobin molecule is phenolphthalin. The testing reagent is prepared as in the preceding paragraph.

2 g phenolphthalein
20 g sodium hydroxide pellets
100 ml distilled water
20 g zinc

Boil gently (or reflux) on a heat block or in a heating mantle under a hood until the solution is colorless. This colorless alkaline (or basic) solution contains the reagent phenolphthalin, a reduced form of phenolphthalein. Store the stock solution refrigerated in a dark bottle with an additional amount of zinc to keep the solution in its reduced form.

Once the stock solution is prepared, a working phenolphthalin solution is created from it. Prepare a 1:5 dilution of the stock solution with ethanol (e.g., 2 ml stock + 8 ml of reagent ethanol). Add zinc to keep the working solution colorless (reduced).

As previously described, phenolphthalin and peroxide react with the hemoglobin in blood. The hemoglobin catalyzes the reaction and phenolphthalin undergoes oxidation back to phenolphthalein to produce the change from colorless to pink (see Figure 12.10).

Leucomalachite Green Solution

To produce the solution, leucomalachite green is heated in an acidic solution until it is colorless. Malachite green is the oxidized form. Leucomalachite green is the reduced form and the name by which this solution and test are commonly known.

0.5 g leucomalachite green
50 ml glacial acetic acid
75 ml distilled water
2.5 g zinc

Boil gently (or reflux) on a heat block or in a heating mantle in a hood until the solution is colorless. Store the solution refrigerated in a dark bottle with an additional amount of zinc to keep the solution in reduced form.

One study reported that minute bloodstains that had been previously subjected to presumptive testing with leucomalachite green gave poor partial short tandem repeat (STR) profiles due to degradation, especially when tested after long intervals of time had passed after the presumptive testing.[14]

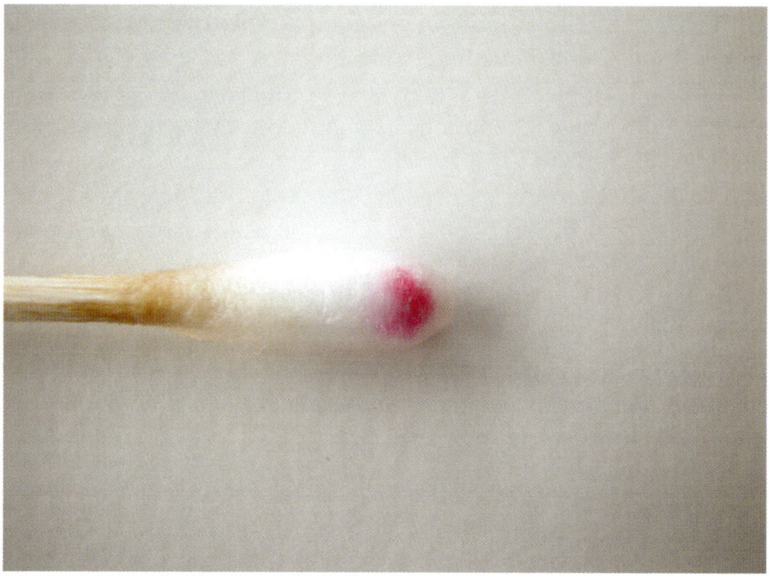

Figure 12.10 A presumptive positive phenolphthalein reaction for blood. The swab is moistened with alcohol and then rubbed against the suspected stain. A drop of phenolphthalin solution is placed on the swab, followed by a drop of peroxide. The pink color should develop immediately.

o-Tolidine Solution

To produce the solution in the laboratory, o-tolidine needs to be mixed in acidic ethanol; it requires no heating or zinc. Handle this reagent with care and use appropriate precautions. Do not spray this reagent. Always wear nitrile gloves to handle and mix the *o*-tolidine solution in a hood.

0.5 g *o*-tolidine
50 ml ethanol
50 ml glacial acetic acid

Mix the above materials in a flask and transfer to a dark reagent bottle. Store refrigerated. *o*-Tolidine is not widely used at this point, as it was found to be carcinogenic. It is however, very sensitive. It should only be used under laboratory conditions with appropriate safety procedures.

Testing Procedure Using Phenolphthalin, Leucomalachite Green, and the o-Tolidine Solutions

These are usually two-step chemical tests. A known blood source (positive control) and reagent blank (negative control) are tested each day of use with the presumptive test reagents to demonstrate that they are working appropriately. Some commercially available products, and forensic laboratories, may involve a three-step chemical test.

For an indirect test, rub a moistened swab or a piece of small, folded filter paper on a suspected stain. Cut a very small portion of the suspected stain for a direct test. If testing this way, you will need a spot plate, usually conducted in a laboratory. If available, test a small, unstained portion of the substrate material. Add a drop of the working solution to the swab, filter paper, or cutting. Add a drop of 3% hydrogen peroxide (H_2O_2).

The appearance of a rapidly developing pink (phenolphthalein) or blue-green (leucomalachite green and o-tolidine) color at this stage is a presumptive positive result for the presence of blood. A color change before the addition of the peroxide may indicate the presence of a chemical oxidant. No immediate color change is a negative result.

Interpretation

To obtain a reliable interpretation it is important to consider the following information:

Catalytic tests are very sensitive, but not specific. The appearance of a rapidly developing pink (phenolphthalein) or blue-green (leucomalachite green and *o*-tolidine) color is a positive presumptive test for the presence of blood. A color change before the addition of the peroxide may indicate the presence of a chemical oxidant. For Hemastix™, the appearance of a rapidly developing dark green/blue-green color on the reagent pad is a presumptive positive result for the presence of blood.

A positive color reaction alone in a presumptive test should not be interpreted as proof of blood. An appropriate conclusion is: "Stain A was chemically tested for blood. The presumptive blood test was positive." If visual characteristics and presumptive chemical tests are positive, then the results may be reported as: "The sample is apparent blood." If the chemical results are negative, the results may be reported as: "No blood was detected on the sample."

The major sources of "false-positive" reactions come from chemical oxidants and vegetable peroxidases. Fresh vegetable peroxidases can give reactions like blood, but the reactions are usually slower and weaker. Vegetable peroxidase activity fades rapidly upon drying. In addition, vegetable stains can generally be distinguished visually from blood. Color development before the addition of H_2O_2 may be due to the presence of a chemical oxidant. Copper and nickel salts are common chemicals responsible for false-positive reactions, as are some molds and bacteria found in bathrooms.

Searching for and Enhancing Latent Blood

When latent blood is suspected, three blood presumptive search and enhancement reagents are available: luminol, fluorescin, and leucocrystal violet (LCV). All three reagents are very sensitive, and each has advantages and disadvantages. They can be used sequentially if needed.

Luminol is a chemiluminescent reagent that produces a light-blue glow when it encounters blood.[15,16] See Figure 12.11. Luminol is easy to mix and use. The luminol glow is very characteristic but does not last very long and typically requires a darkened environment to visualize properly and for documentation requires appropriate camera setup (tripod mounted, long exposures). Various commercial luminol products are available, including Hemaglow™ and Bluestar™.

Crystal violet is a dye that is used as a general protein stain. General protein stains are less specific than presumptive blood testing reagents. However, the reduced form LCV is a presumptive blood reagent.[17] The colorless LCV solution turns purple on contact with latent blood. It can be used in available light and is very effective on light-colored surfaces.

Figure 12.11 A footwear pattern transfer enhanced by luminol.

Similar to phenolphthalein and phenolphthalin, fluorescein must be reduced to colorless fluorescin before use.[18,19] The reduced form is sprayed on a substrate and is then oversprayed with dilute hydrogen peroxide. When in contact with latent blood, it produces a bright yellow-green color when visualized with an alternate light source (ALS). The fluorescence is very bright and has good duration. Luminol and LCV require no overspray or ALS to visualize the result.

Fluorescein and LCV are triarylmethanes. They are structurally close to phenolphthalein and leucomalachite green, and they are thought to react with blood in a similar fashion. These processes, as described, generally will not degrade DNA that may be present in a substrate. Any processing, however, may reduce the amount of DNA present due to manipulating the material including diluting the sample.

Just as with reagents used to test patent blood, these three latent blood-testing reagents are subject to the same false-positive results in the presence of metal salts. Additionally, reactions are seen on dilute hypochlorite (bleach) stains, although the color, intensity, and duration differ from those seen on latent blood.

Leucocrystal Violet (LCV) Preparation

As previously described, the basic structure of LCV (see Figure 12.12) is the triarylmethane skeleton. To prepare the LCV solution:

Dissolve 10 g of 5-sulphosalicylic acid in 500 ml of 3% H_2O_2.
Add 4.4 g of sodium acetate.
Add 1.1 g of LCV.

Alternate LCV Reagent Preparation Method

The reagent may also be prepared in the field using a purchased 473-ml bottle of 3% H_2O_2. In this method, the solution is produced as follows:

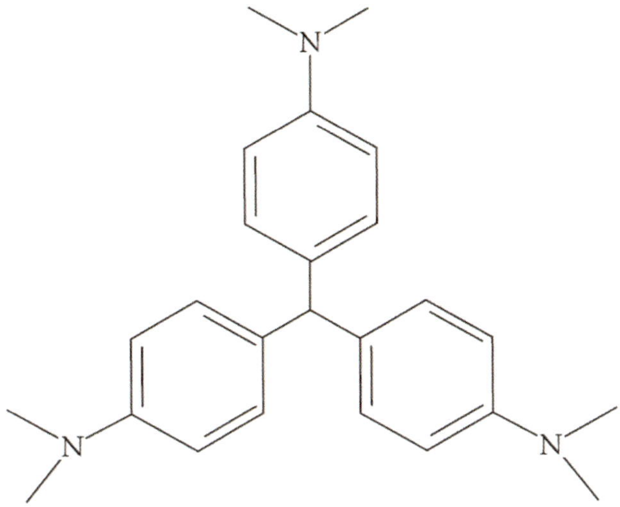

Figure 12.12 The structure of leucocrystal violet (LCV) is based on the triarylmethanes.

Pour a small volume of the H_2O_2 into a small cup and add 10 g of 5-sulphosalicyclic acid.
Dissolve 1.1 g of LCV into the above solution.
Pour this solution back into the bottle, replace the cap, and shake for 30 seconds.
Pour some of the solution from the bottle back into the small cup and add 4.4 g of sodium acetate. Mix until dissolved and pour back into the bottle. Replace the cap and mix thoroughly.

Note: 5-sulphosalicylic acid helps to fix blood to the substrate on which it is present. Discard LCV crystals if they become yellow.

Fluorescin Spray Solution Preparation

The materials and reagents used to prepare fluorescin include:

Fluorescein (e.g., Aldrich F245-6)
Zinc, granular
Ethanol
Glacial acetic acid
Hydrogen peroxide (H_2O_2)
Sodium hydroxide (NaOH)
Distilled water

Fluorescin can be prepared either by an acidic reduction in alcohol or a basic reduction in water. The alcohol preparation is useful for vertical and slick surfaces. Both preparations are oversprayed with H_2O_2 diluted to 3% in either ethanol or distilled water.

Fluorescin in Alcohol Preparation

To prepare the alcohol fluorescin solution:

Dissolve 0.1 g of fluorescein in 20 ml of ethanol in a small, lidded container.
Add 2 g of granular zinc.
Then add 1 ml of glacial acetic acid.
Warm for about 30 minutes. (You can use the warmth of your hand around the container.)
Decant or pipette 1 ml of the clear stock solution into 99 ml of ethanol to prepare the spraying solution.

Fluorescin in Water Preparation

To prepare the water-based fluorescin solution:

Add 0.1 g of fluorescein to 20 ml of distilled water in a small, lidded container.
Add 2 g of granular zinc.
Then add 1.0 g of sodium hydroxide (NaOH) pellets.
Warm for about 30 minutes.
Decant or pipette 1 ml of the clear stock solution into 99 ml of distilled water to prepare the spraying solution. The dilution of the stock solution can be varied to increase or decrease sensitivity in laboratory settings.

The alkaline, water-based fluorescin preparation just described results in a clear, but slightly lime-green colored solution (see Figure 12.13). This is diluted to produce a working solution and is then sprayed onto the target. In the presence of heme, this solution is oxidized and will fluoresce in a lime-green color. This fluorescence is visualized with an ALS. Generally, the technique works best when operating the ALS at the 445–455 nm range with a yellow barrier filter, but can also be visualized at 485 nm with an orange barrier filter (see Figure 12.14).

Before applying fluorescin, the substrate is always checked with the ALS to identify any natural fluorescence or other contaminants that may fluoresce. Any contaminated areas are marked to ensure they are not mistaken as part of the reaction for the latent blood.

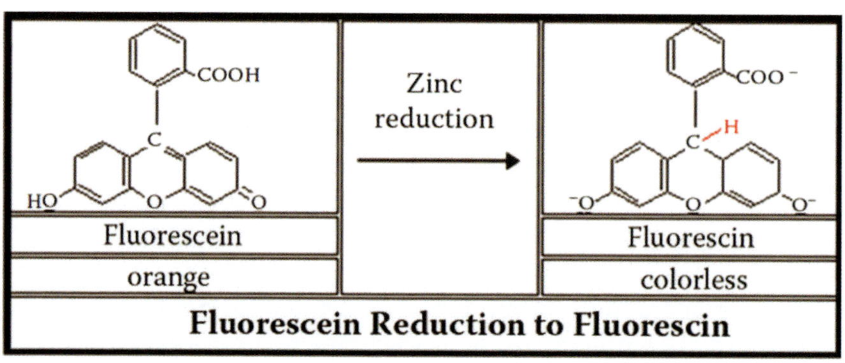

Figure 12.13 Fluorescin is prepared in either an acidic reduction in alcohol or by a basic reduction in water. The alcohol preparation is useful for vertical or slick surfaces.

Figure 12.14 A fluorescin reaction on carpet. To visualize the fluorescence, the analyst utilizes an alternate light source (between 445 nm and 485 nm) and either a yellow or orange barrier filter.

Luminol

The luminol reagents should only be mixed immediately before use. The materials and reagents used in luminol include the following.

Reagent Preparation

Sodium perborate is a solid material; thus, kits can be prepared and materials mixed in the field ahead of time.

0.5 g luminol
3.5 g sodium perborate, anhydrous
25 g sodium carbonate, anhydrous
500 ml distilled water
Dissolve luminol and sodium carbonate in 250 ml of distilled water.
Dissolve sodium perborate in 250 ml distilled water.
Combine and spray.

Alternative Reagent Preparation

In this preparation, the oxidizer hydrogen peroxide is used instead of sodium perborate.

Luminol (3-aminophthalhydrazide)
5% sodium hydroxide (NaOH)
3% hydrogen peroxide (H_2O_2)
Distilled water

To prepare the luminol solution:

Add 2 ml 3% H_2O_2 to approximately 50 ml distilled water.
Mix 0.05 g luminol into 10 ml of 5% NaOH.
Add this to the water/peroxide solution.
Bring the final volume up to 100 ml with distilled water.

Safety Considerations

Even when no safety data exists that incriminates any dye as a health hazard to humans, follow standard personal safety precautions. Spraying should be done, if possible, in a well-ventilated area and with suitable protection (laboratory coat or other protective clothing, gloves, mask, goggles, etc.). Respiratory protection is recommended when working in an enclosed space. Some agencies are addressing the issue of possible health concerns to humans who will be exposed to areas where chemical enhancement and searching with these reagents was conducted.

Hydrogen peroxide will irritate the eyes, skin, and respiratory system. Hydrogen peroxide is an oxidizer, which may make organic substances more flammable on contact. Alcohol also poses a possible flammable risk and appropriate ventilation is necessary. Before spraying, open flame sources, such as pilot lights, must be checked and eliminated. Keep in mind that any chemically treated evidence should be marked so that anyone encountering it later will be aware of the nature of the treatment.

Procedure for Using Luminol, LCV, and Fluorescin

Reagents should be checked each time before they are used. At a minimum, a known blood sample should be checked. Remember that undiluted blood may quench the reaction, although the borders of the stain will glow. (Refer to Figure 12.8.) These reagents are most useful when prepared shortly before spraying and should be disposed of when the application is complete.

Ensure that appropriate documentation and photographs are taken before any chemical enhancement procedure. If fluorescin is used, surfaces should be checked with an ALS before application of the reagent and any areas that fluoresced be marked. Spray the reagent lightly with as fine a mist as possible. A Chromist™ sprayer or similar device may be used on smaller areas. Plastic garden-type sprayers are helpful for larger areas. Very large areas should be sprayed in sections. Areas may be resprayed as needed; however, this presumes sufficient hemoglobin is present for the reaction to occur.

Photograph or Otherwise Document Areas Where Reactions Are Observed

As indicated, luminol reacts with many things, but through trial and use the analyst can often identify such false-positive reactions. The luminol reaction should be evaluated using three criteria: color, intensity, and duration. The color reaction should be a bluish-white to blue-green. White-blue and darker violets occur in any number of false positives. The intensity of the reaction should build over a very short period, and it should remain evident for some time, although it will fade. The length of this duration is very much a product of ambient lighting. At night outside, the reaction will remain for minutes. If indoors and the analyst turns the lights on or off, this will result in a shorter duration due to the eyes adjusting back and forth to the change in lighting. Immediate flashes of white or white-blue are indicative of bleach and cleaning products. Due to past abuses, some courts now prohibit testimony regarding luminol's use and any conclusions drawn from it. The best advice is to try to confirm the presence of blood, which, unfortunately, given the trace amounts involved, is difficult.

Protein Stains

Additional enhancement dyes that react with various proteins in body fluids and blood are available. One of the most common in use at crime scenes is amido black. The resulting enhancement is a black to black-blue color. Amido black can be applied both as a water-based or methanol-based application. Water-based procedures, which are most applicable for use at the crime scene, require some level of effort and should only be undertaken with prior experience (see Figure 12.15). Protein stains are not considered presumptive for blood, and any conclusions reached should be carefully considered in the context of the crime scene.

Photographic Documentation

Another concern for the analyst is documenting the luminol reaction. In the late 1980s, Fred Gimeno and Gary Rini perfected methods of photographing luminol reactions. Before these efforts, all luminol photographs were generally exposed in completely darkened

Figure 12.15 An amido black reaction on fabric.

rooms. The result was good luminescence but no frame of reference for the viewer. One could not tell where the reaction occurred on the evidence or in the scene. Gimeno and Rini produced images that included the luminescence and the evidence item using paint-with-flash (also referred to as paint-with-light) techniques similar to those used in spelunking and outdoor crime scene photography. Their efforts are documented in two articles published in the *Journal of Forensic Identification*.[20,21]

With film photography, exposure times of several minutes were often required to capture luminol reactions. With the advent of digital photography, photo documentation is much simpler. The sensitivity of the associated charged-couple device (CCD) that replaced film, allows for shorter exposure times. Using an ISO setting of 400, with an open lens iris (*f*/2.8 is good, but greater *f*/stops will work), exposure times required can range from as little as 15 seconds to perhaps a minute. If longer exposures are required, use of the bulb setting will accomplish the task. Based on the CCD light sensitivity, paint-with-light techniques are also simpler. Manual flash techniques, flashlight techniques, and even smartphone illumination for close-up work (e.g., footwear patterns) all accomplish functional photographs that capture the luminescent behavior and the associated scene context.

Fluorescein photography is much simpler, as an ALS is utilized. The ALS offers sufficient ambient light to demonstrate the background and the fluorescence. The only requirement is an appropriate barrier filter for the camera lens matching the goggles worn by the crime scene technician. As previously described, the filter and goggles may be yellow or orange, based on the wavelength of light the analyst utilizes.

Both luminol and fluorescein enhancements are best documented using a camera positioned on a tripod. It may be helpful to respray the target area during the documentation. The luminol chemiluminescence intensity may fade quickly relative to the time required to take a photograph, while the fluorescein reaction will remain much longer.

Digital images provide an alternative to using either the fill-flash or the paint-with-light procedures to document luminol reactions. One photograph is captured of the area of interest well-illuminated by ambient light and/or supplemental lighting. With the camera in the same position, a second photograph is captured of the same area of interest darkened and the resulting enhancement visible. Commercial software as well as certain Nikon cameras allow the photographer to overlay the two images to make both the enhancement and scene context visible.

With digital cameras, the analyst can test various exposures at the scene and immediately adjust the camera based on the result.

Interpretation

Luminol, LCV, and fluorescin are as described presumptive blood tests. These reagents are much more sensitive than patent blood presumptive reagents. They are useful in detecting patterns of older or indistinct bloodstains and detecting the remaining blood residue after a stain has been cleaned. These presumptive searching reagents are subject to false positives just like phenolphthalin and leucomalachite green.

A positive reaction is considered presumptive for blood. Because these reagents are so sensitive, a positive reaction may be reported as follows: "The target area glowed. Therefore, this may indicate the presence of blood. Further testing may be indicated. No inference as to species of origin may be made." If no apparent reaction is observed, the following may be reported: "No blood was detected." If the reaction differs from the control reaction (e.g., color, duration, intensity), then the result may be considered inconclusive and so reported. However, environmental conditions can cause ambiguous results. These processes are always presumptive.

A major consideration in any enhancement is validating the presence of blood. One of the best approaches when using enhancement chemicals is to spray lightly in areas of concern and *if* a response is indicated, stop and swab for DNA before attempting to fully enhance the area (Figures 12.16 and 12.17).

Confirmation of Blood

Running more than one presumptive test (e.g., serial test using Hemastix™, followed by leucomalachite green) does not establish a confirmation of blood. Presumptive tests are all *presumptive*. In response to the need to clarify the nature of the stain at the crime scene, various tests are now available that may provide confirmatory results for the presence of blood.

Immunoassay Confirmation of Blood

Immunoassay tests are designed to use the binding between an antigen and its homologous antibody. Commercially available tests include Hexagon OBTI™ and Hematrace™. Both tests were designed to screen for the presence of human hemoglobin utilizing an immunoassay technique. Because positive reactions have been observed with blood from humans,

Figure 12.16 A suspected crime scene in which there was evidence of bloodshed, but also indications that the subject cleaned up the area. This photograph shows the scene following the application of luminol. The blue-green luminescence shows various stains, including various contact patterns, a drag mark, blood trails, and spatter. (Photograph courtesy of Ray Clark, Oklahoma City Police Department, Oklahoma City, OK.)

Figure 12.17 The back stairs of the same crime scene from Figure 12.16. The drag mark was not evident in any fashion prior to the application of luminol. These photographs were taken in the dark, while a flashlight beam was bounced off the surrounding surfaces for several seconds. As a result, the photograph shows good luminescence and scene detail. (Photograph courtesy of Ray Clark, Oklahoma City Police Department, Oklahoma City, OK.)

higher primates, and ferrets, this test is used to confirm the presence of blood. It is not a valid confirmation of human blood. Additional testing by the crime laboratory (e.g., a species test) may be used to confirm that the blood is of human origin.

Immunoassay test cards should be checked before use. A small sample is extracted using the buffer solution provided. The amount of time for the extraction varies depending on how old the stains are estimated to be. Typical extraction times vary from 1 minute to 5 minutes. Older stains may require a longer extraction time.

A few drops of the extract are pipetted into a sample well-marked "S" on the immunoassay card (see Figure 12.18). A colored band will appear in the control (C) section to demonstrate the card is functioning properly, and a similar band will appear in the test section (T), if blood is present. The card may be read up to 10 minutes after testing. Bands appearing after 10 minutes should not be considered a positive result. If the control band fails to develop, the test should be conducted again using a new card.

Order of Testing Procedures

Presumptive testing, confirmatory testing, and enhancement of bloodstains present issues if they are not completed in an appropriate order. Presumptive testing (without direct application to the surface) is acceptable on the scene or at the laboratory. This allows the analyst some indication to move forward.

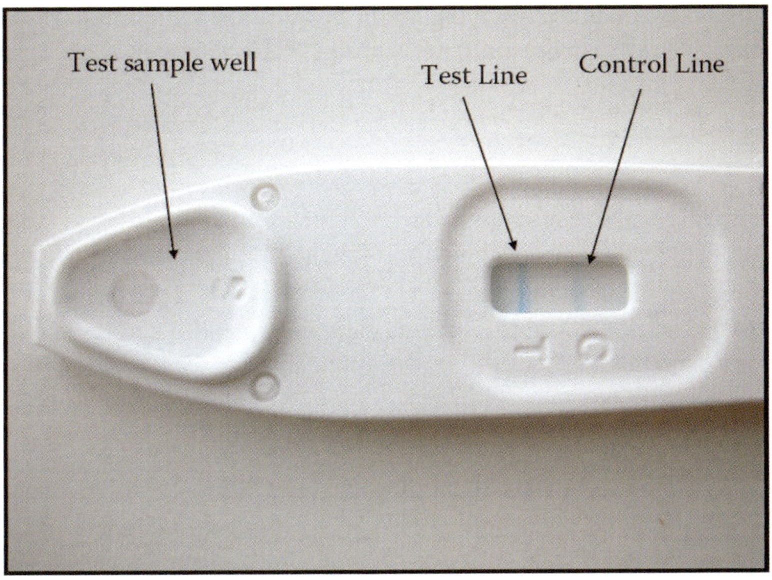

Figure 12.18 Immunoassay tests are also available to the analyst. Although not truly "human specific," these tests are more specific than other presumptive test techniques. After sampling the stain, a small amount of the test solution is placed in the sample well. A blue line will appear at the test (T) line if the test is positive. If a blue fails to appear at the control (C) line, the test should be repeated with another test kit.

Once a positive presumptive test is obtained, sampling the area in question for subsequent DNA testing should be accomplished whenever possible. If desired, confirmatory testing can then be completed on any remaining sample. Enhancement chemicals should always be the last procedure as they may interfere with DNA and other confirmatory tests. If there is any question about the order of procedures to implement, contact the supporting forensic laboratory for guidance.

In one case known to the authors, the questioned stains were submillimeter-sized and gave a positive presumptive result using phenolphthalin. The area was then enhanced using LCV. This effort was followed with a confirmatory test using HemaTrace®. The HemaTrace® yielded a negative result. Additionally, no samples of these stains were recovered for DNA testing.

This lack of a confirmation of human blood was significant. The order of procedures utilized in this instance was subsequently questioned. A literature search produced no papers on how LCV might affect HemaTrace® testing, post-LCV application. This led to experimentation where similar submillimeter-sized stains of known human blood were produced on a white T-shirt. The stains were allowed to dry and sprayed with LCV. After LCV drying, the stains were tested with HemaTrace®, and negative results were obtained. Thus, the order of procedures likely was a factor in the negative result.

Sample sizes utilized in testing are also important. The recommended sample size for HemaTrace® testing was reported as 3–5 mm^2. The bloodstains in the described case were less than the recommended sample size. In subsequent experiments conducted, HemaTrace® tests were negative after LCV application on smaller stains. Stains measuring 3.5 mm^2 and larger were also tested, all producing a positive result after LCV. These efforts indicate that submillimeter-sized human blood stains will test negative after direct LCV processing.

DNA testing is always preferred if the stains in question are limited. If sufficient stains are present, follow the order: presumptive testing, DNA sampling, confirmatory testing, and enhancement.

Summary

The various chemical reagents discussed offer the analyst multiple tools to check presumptively for the presence of blood and enhance latent bloodstain patterns. The critical concern for analysts is to understand the various capabilities and limitations of these tools, as well as to be comfortable and practiced in the application of whatever reagent they choose.

The Author

Charlene Marie retired as manager for the California Department of Justice, Santa Barbara Laboratory, with more than 30 years of experience as a chemist. She has held memberships in the California Association of Crime Laboratory Directors, the International Association of Bloodstain Pattern Analysts, the Association for Crime Scene Analysis, and the California Homicide Investigators Association.

Notes

1. Burdette, P.E., Presumptive Tests for Blood – A Comparative Study, CRE Report, No. 201, October 1976.
2. Cox, M., A Study of the Sensitivity and Specificity of Four Presumptive Tests for Blood, *Journal of Forensic Science*, Number 36, 1991, p. 1503.
3. Cox, M., Effect of Fabric Washings on the Presumptive Identification of Bloodstains, *Journal of Forensic Science*, Volume 35, Number 6, 1990, pp. 1335–1341.
4. Ames, B., Lee, F., and Durston, W., An Improved Bacterial Test System for the Detection and Classification of Mutagens and Carcinogens, *Proceedings of the National Academy of Sciences of the United States of America*, Volume 70, 1973, pp. 782–786.
5. Garner, D.D., Cano, K.M., Peimer, R.S., and Yeshion, T.E., An Evaluation of Tetramethylbenzidine as a Presumptive Test for Blood, *Journal of Forensic Science*, Volume 21, Number 4, 1976, p. 816.
6. Gaensslen, R.E., The Phenolphthalein Test, Section 6.3, *Sourcebook in Forensic Serology, Immunology, and Biochemistry*, U.S. Department of Justice, National Institute of Justice, August 1983, pp. 103–105.
7. Report on Carcinogens, Fifteenth Edition, National Toxicology Program, Department of Health and Human Services, 2021, http://ntp.niehs.nih.gov/go/roc.
8. Tice, R., Luminol – Review of Toxicological Literature, prepared for National Institute of Environmental Health Sciences, Contract No. N01-ES-65402, July 1997.
9. Tobe, S.S., Watson, N., and Deaid, N.N., Evaluation of Six Presumptive Tests for Blood, Their Specificity, Sensitivity, and Effect on High Molecular Weight DNA, *Journal of Forensic Science*, Issue 1, Number 52, 2007, p. 102.
10. Gross, A.M., Harris, K.A., and Kaldun, G.L., The Effect of Luminol on Presumptive Tests and DNA Analysis Using the Polymerase Chain Reaction, *Journal of Forensic Science*, Volume 44, Number 4, 1999, pp. 837–840.

11. Spear, T., Sample Handling Considerations for Biological Evidence and DNA Extracts, *TIELINE*, Number 25, 2004.
12. Shaler, R.C., Modern Forensic Biology, in *Forensic Science Handbook*, 2nd ed., Saferstein, Ed., pp. 525–535.
13. Technical Note for Hemastix™ Reagent Strips for Urinalysis, Diagnostic Division, Bayer Corporation, Rev. 9/95, 1992.
14. Alenazy, N.S., Refaat, A., and Babu, S., Comparison of the Effects of Two Presumptive Test Reagents on the Ability to Obtain STR Profiles from Minute Bloodstains, *Egyptian Journal of Forensic Sciences*, Volume 5, Issue 3, September 2015, pp. 103–108.
15. Quinones, I., Sheppard, D., Harbison, S., and Elliot, D., Comparative Analysis of Luminol Formulations, *The Canadian Society of Forensic Science Journal*, Volume 40, Number 2, 2006, pp. 53–63.
16. Perkins, J.D., Luminol – What's Glowing On?, *SWAFS Journal*, Volume 13, Number 2, October 1991, pp. 19–22.
17. Bodziak, W., Use of Leucocrystal Violet to Enhance Shoe Prints in Blood, *Forensic Science International*, Volume 82, 1996, pp. 45–52.
18. Maucieri, L. and Monk, J., Enhancement of Faint and Dilute Bloodstains with Fluorescence Reagents, *CAC News*, Volume 13, Spring 1992.
19. Cheeseman, R. and DiMeo, L.A., Fluorescein as a Field-Worthy Latent Bloodstain Detection System, *Journal of Forensic Identification*, Volume 45, 1995, p. 631.
20. Gimeno, Fred E. and Rini, Gary A., Fill Flash Photo Luminescence to Photograph Luminol Bloodstain Patterns, *Journal of Forensic Identification*, Volume 39, Number 3, 1989, pp. 149–156.
21. Gimeno, Fred E. and Rini, Gary A., Fill Flash Color Photography to Photograph Luminol Bloodstain Patterns, *Journal of Forensic Identification*, Volume 39, Number 5, 1989, pp. 305–306.

Documenting Bloodstains 13

There was a saying used in the Army CID regarding documentation: "If it's not written down, it didn't happen." Analysis and reconstruction demand this type of thinking whenever possible. No matter how good the evaluation and subsequent reconstruction, without documentation that allows us to share those efforts the evaluation may be for naught.

The Function of Documentation

Documentation serves several functions:

- Verification of scene integrity
- Providing a quality court presentation
- Allowing outside analysis
- Providing for independent defense evaluation

The most important function of documentation is simply maintaining a record of the scene integrity as the analyst finds it. We want to be able to state with some authority the condition in which we found the evidence and how that evidence relates to other items. The court expects and demands this of the crime scene technician.

Good scene documentation precludes investigators from including post-incident artifacts in the analysis. It ensures the inclusion of evidence that may not survive the processing attempts and provides sterling proof to the court of the scene condition. To this end, documentation cannot be haphazard.

Understanding the issues for maintaining scene and evidence integrity in Bloodstain Pattern Analysis demands an understanding of crime scene processing as a whole. This requires the analyst understand the six basic elements of crime scene processing: (1) assessing, (2) observing, (3) documenting, (4) searching, (5) collecting, and (6) analyzing. This book is not an appropriate vehicle for teaching all-encompassing crime scene processing techniques. For that reason, we offer this one comment: if the reader has not been exposed to training and experience in proper crime scene processing, that deficiency should be corrected before attempting to conduct a Bloodstain Pattern Analysis or crime scene reconstruction. A resource for this purpose is *Practical Crime Scene Processing and Investigation* by Ross Gardner and Donna Krouskup. Eventually, the court will need to know what the analyst knows. For court presentations, good documentation allows the analyst to clearly present often complicated information in a fashion that is understandable to the audience. This is particularly critical when considering presentations to the jury. In order to believe the analysis, the jury must understand the reasoning and evidence that support it.

Outside analysis is not uncommon either. If a given department has no specific training in Bloodstain Pattern Analysis or crime scene reconstruction, outside assistance may

 DOI: 10.4324/9781003408734-13

be brought in. The outside analyst is limited by the documentation efforts of those who process the scene. Good documentation makes the outside analyst's job easier.

The final function of documentation, that of supporting defense evaluations, is also important. No matter what others may say, police agencies serve but one master – the truth! Therefore, the analyst should have little concern over defense analysis of the analyst's work. If we expect others to discover the same truths we found, they must certainly have access to the same information we had. If the analysts applied themselves appropriately, then any analysis or reconstruction will stand the test of scrutiny.

Granted, with regard to opposing evaluations, it is easily argued that little in the legal profession deals with truth. Defense attorneys or prosecutors will raise odd and curious questions; certainly, they will not limit themselves to attacking only the evidence. Nevertheless, such attempts to muddy the water will be less successful if good documentation exists. Ours is an adversarial system in which defendants should and do have the right to their own impartial evaluation. Good documentation simply affirms the investigator's neutrality regardless of what side employs them.

Given these concerns for scene and evidence integrity, we will consider the following areas that define proper documentation:

- Collection and detection of bloodstains and traces
- Photography of bloodstains, including videotape recordings
- Sketching techniques
- Written reports for bloodstain documentation

Collection

It is not enough that the analyst recognizes a pattern as defining some specific action. Without knowing "who" was involved, this knowledge does little in defining a bloodstain pattern reconstruction. Bloodstain evidence does not stand alone; in particular, DNA evaluations are critical to understanding the relationship of bloodstains to the crime.

DNA analysis allows a particular pattern or stain to be correlated to a particular individual. Forensic laboratories primarily use short tandem repeat (STR) techniques. The sensitivity of STR technology today demands a level of care in collecting and storing any specimen in order for it to be of value.

Beyond documenting the stain's location, as a rule the analyst should be concerned with several issues in collecting stains for subsequent DNA analysis. These include:

1. Proper discrimination between stains
2. Sufficient identification of samples to patterns
3. Sufficient collection of samples prior to destructive testing
4. Precluding cross-contamination or additional post-incident contamination of samples

The first concern is really the investigative benchmark. Failing to discriminate if a given stain sampled belongs to a specific pattern will cause the DNA results to be useless to the analyst. In approaching each pattern for collection of samples, the analyst needs to be confident that the stains selected are representative of that pattern and only that pattern. If several spatter patterns intermix in a given area, one cannot arbitrarily select stains from the conglomerate. Always choose the stains carefully to preclude later confusion.

In addition to having confidence that the analyst is collecting stains from a single pattern, the recovered samples must be marked in a fashion that will guarantee relating them back to that pattern. Ensure that the evidence collection notes and report have a clear correlating label that identifies the stain group involved. The authors prefer establishing a number or letter for each stain group early in the scene processing. Any subsequent mention of the pattern includes a reference to this identifier. This identifying label is also listed on the evidence collection forms, along with other standard data. The following examples contrast this information with and without the use of the identifying pattern label.

Inappropriate Evidence Collection Example

Item #	Description
1	Stain, red in color, dried. Approximate dimensions 2 mm × 2 mm. Found on the west wall, about 2 ft from the floor. Stain swabbed and sealed. Package marked RMG 1200 2 OCT 23.

This description certainly tells where the item was found and the analyst can later identify the stain using the evidence mark (initials, time, and date), but it fails to associate the stain with a specific pattern or group in the scene.

Appropriate Evidence Collection Example

Item #	Description
1	Stain, red in color, dried. Approximate dimensions 2 mm × 2 mm. From Stain Group 2. Found on the west wall, about 2 ft from the floor. Stain swabbed and sealed. Package marked RMG 1200 2 OCT 23.

This description provides all of the previous information but adds a ready reference to the stain group the analyst identified in the crime scene documentation as Group 2. This simple addition to the documentation effort will eliminate subsequent controversy regarding whose blood is involved in which patterns.

Sufficient collection of blood samples is also an important issue. Remember that no single person's effort helps the bloodstain analyst more than those of the biological scientist. Don't hamstring that effort. Cutting corners while collecting samples at the crime scene may save a few moments, but it will prove costly in the end. Take the time to provide the laboratory personnel with sufficient samples to do their job.

The authors asked a group of practicing forensic scientists to comment on methods of collecting bloodstain samples. We felt we could obtain a consensus on the "best" method for collection. Unfortunately, there was no actual consensus. The methods of collection differed from organization to organization. Some experts believed liquid blood should never be collected from the scene, while others felt it an appropriate response. Some felt cutting of the actual stain from the object was a good method, while others felt swabbing was more effective. Since there is no consensus for the collection of samples, individual analysts must coordinate their specific collection techniques with their crime laboratory. This ensures that whatever method is used meets the needs of the supporting laboratory. Before scene processing, communication between scene investigators and the lab establishing the preferred collection method is essential.

No matter what method is used, collection techniques should eliminate cross-contamination as well as inadvertently adding DNA material. Depending upon the advice of the forensic laboratory this may require the collection of substrate and blank control samples. Substrate and blank controls may still be considered applicable to DNA analysis, as they allow the biological scientist to recognize when contamination is present that might interfere with the analysis. Another potential source of contamination comes from scene investigators or analysts themselves by breathing or releasing spittle onto the surface/sample to be collected. This is easily prevented by wearing a commercially available mask covering the mouth and nose when on scene.

Whatever method is used, it may be insufficient to collect a single sample from any given pattern. No matter the size, contamination issues or simply a degraded sample can result in an inconclusive biological report. The analyst should collect a fair sampling of stains (a minimum of two or three) from the pattern. When collecting more than one stain from a given pattern, the scene investigator or analyst should contact the forensic laboratory (ultimately receiving the samples for DNA analysis) as to whether multiple stains can be collected as one sample or if each individual stain should be collected separately.

To eliminate issues, the analyst should exercise specific care in the collection process, following the guidelines recommended by the forensic laboratory doing the DNA analysis. General guidelines include: First, use only new gloves and change gloves if the glove has any contact with the associated surfaces. Do not use standard "crime scene kit" tools for collection. Instead, use disposable scalpels or single-edge razor blades for cutting samples. Dampened sterile swabs for swabbing carefully, ensuring the sterile water container does not touch the swab. Use disposable pipettes and syringes if liquid blood must be collected. Scalpels can be thoroughly wiped with alcohol and allowed to dry between the collection of each sample.

Once recovered, the analyst must properly containerize the sample. Swabs should be allowed to air-dry before placing in any type of container. Commercially available containers include boxes in which the swab(s) are secured and swabs for which the sample and tip are enclosed in an attached plastic cover. If cuttings were collected, these should be allowed to air-dry before being packaged in a "breathable" container such as paper or cardboard. If the container has an adhesive seal requiring dampening, only use sterile water to coat the adhesive and close it. A more effective technique is to forego the adhesive and simply seal the container completely with tape. Alternatively, the portion of the packaging with the adhesive can be securely taped.

If liquid samples must be collected from the scene, the vials or tubes must be enclosed in a second container. Heat-sealed bags work well for this purpose. Should the tube be breached through spillage or accidental breakage, the heat-sealed bag may preclude the loss or degradation of the sample.

Items stained with blood, such as clothing, should be thoroughly air-dried. In the instance of heavily bloodied clothing, initial packaging must allow transport from the scene to an appropriate drying area/chamber, once again attempting to eliminate any cross-contamination. Once dried, individually wrap each in clean butcher paper. Use enough paper to encapsulate the entire item. If possible, do not fold the item directly over onto itself. It may be important to determine whose blood is on a particular item of clothing and where the blood is found on the item. Place an intervening piece of paper between the surfaces when the item is folded. This reduces the possibility of creating inappropriate pattern transfers and eliminates a source of cross-contamination if multiple DNA samples are present. Do not simply toss items of clothing into a bag or container, and certainly never toss multiple items of clothing into the same container. If the only containerization method available is to place the items into a clean paper bag on scene, consider how it is folded and placed within.

Just as the analyst prefers that the actions of those processing the scene will not disturb the physical stains, the biological scientist prefers that the analyst not intermingle or cross contaminate the samples. The destruction of biological specimens occurs due to many factors, several of which the analyst cannot control. It only makes sense that we eliminate sources of degraded samples whenever possible. The use of many chemical enhancement methods discussed in Chapter 12 represents a source for such degradation. Dr. Henry Lee demonstrated that many of these chemicals degrade standard ABO and enzyme serology tests for any forensic laboratory still using these procedures. Although his research was not all encompassing with regard to their effect on DNA, it certainly indicates a concern.[1] Current efforts suggest that some chemical enhancement methods do not have a negative effect on subsequent DNA analyses, except for the potential dilution of any sample that may be present on a surface. The best advice is to never arbitrarily spray enhancement chemical on any item of evidence until after collecting proper biological specimens or after coordinating with the biological science section of the forensic laboratory. If any questions exist on how to proceed, consult the supporting crime laboratory before acting.

Bloodstain Pattern Photography

Far too often, crime scene photographs of bloodstain evidence are meager, insufficient in detail, and generally lacking. Often photographers shoot from afar, never filling their viewfinder with the image of concern. The analysis process, and certainly opposing analysis, requires detailed photographs. The information presented in this section is important to the bloodstain pattern analyst, crime scene technicians, and investigators.

The challenge with bloodstain photography is recognizing the level of detail necessary for subsequent BPA. The technicians responsible for the scene photography may not understand the nuances of Bloodstain Pattern Analysis. What photographs are taken at the scene can easily determine if any BPA conclusion is possible. Sometimes they document too much of the wrong thing, or worse yet, far too little of the critical stains.

It is important to understand that not every drop of blood shed at the scene needs to be photographed. Common sense and good judgment are the best guides in making determinations of what must be documented. Prior training in the basic considerations of BPA and classification of stains will enhance the photographer's understanding of what must be captured. If a stain naturally causes questions in the technician's mind, it may be important in the ultimate analysis. If in doubt, however, take a photograph.

We offer the following guidelines as the minimum photographic documentation requirements:

- Document the entire scene in situ. Include overall photographs.
- Photograph evident pattern transfers, flows, and other fragile patterns as early as possible.
- Document all identified pattern groups or stains using both evidence-establishing shots and an overall photographs of the entire pattern with a scale.

- Take macro-photographs of all stains of interest. Include a reference scale in each.
- All macro-photographs should be taken at a 90° angle using a tripod.
- In instances of spatter patterns where area of origin determinations are necessary, document the individual spatter used in the determination.
- Use Toby Wolson's Roadmapping method or a similar procedure.[2]
- Use the camera effectively, orient the camera as necessary in either the landscape or portrait position.

The best photographic results in Roadmapping are produced by always using a tripod, even for pattern overall photographs. Granted current DSLR cameras allow capturing images handheld at lower shutter speeds, but the tripod ensures a quality focus without blurring. This may also be necessary if some panoramic photographs involving overlapping images is desired.

Documenting the scene is important no matter what the nature of the crime. The scene technician should photograph the scene immediately upon arrival, whether bloodstains are involved or not. As a rule, these photographs include overall or 360° photos of each room, including the ceiling and floor. Using a wider-angle lens (e.g., 28mm), the technician shoots across the room from each of the four corners at a minimum, or from any other configuration that captures the condition of the entire scene. All photographs of this nature should have some level of overlap with the next photograph.

Overall photographs serve little function in the courtroom. The lens used can cause perspective distortion. Nevertheless, these photos are immensely helpful to the analyst and investigator, particularly when verifying where some items were when the processing began. As every crime scene technician knows, a primary goal is to document the scene as found and provide documentation offering the ability to identify alterations, inadvertent destruction, addition, or movement of items within the scene during processing, collection, or other actions.

Questions regarding alteration of the scene will arise: Was the bloody fingerprint on the wall caused by EMS or the suspect? Overall photos help eliminate some of this confusion, which can plague the analysis attempt.

After taking the 360° or overall photos, the technician takes evidence-establishing and close-up photos of all obvious items of evidence. Establishing shots serve the function of identifying where in the scene a particular item is and any relationships it may have to other items of evidence. For example, Figure 13.1 through Figure 13.4 are Roadmapping photographs that take the viewer from an overall perspective of a grouping of major patterns (Patterns E, H, and I), to an overall of a specific pattern (Pattern I), and then to details within the individual stains that make up the pattern (Pattern I 1–4). Using these photos, the viewer understands where the pattern transfer in question is in relationship to the wall and floor. The photographer carefully moves through the scene, documenting the condition of all obvious items of evidence before anyone can disturb them.

As the analyst continues to process the scene using a standard processing model, photographs are made of all newly discovered evidence. The analyst can also take more controlled close-up photographs as the item is actually collected. This helps document the item's complete condition.

Having discussed the general requirements for photography, let's focus on the specifics of photographing bloodstains and bloodstain patterns. Pattern transfers in the scene and flow patterns are perhaps the most likely stains to fall into the "fragile" evidence category.

Figure 13.1 An overall photograph that establishes the presence of several patterns of interest on a kitchen floor (Patterns E, H, and I). The function of this photograph is to show an area and orient what patterns are present in this area.

Actions in the scene during processing can mar or obliterate them. Document these stains as soon as possible with close-up photography. Afterward, take all steps to protect the pattern from physical damage.

After all basic crime scene processing is complete, the bloodstain pattern analyst or scene technician creates a second series of photographs to support the bloodstain pattern issues using a technique such as Roadmapping (discussed in detail later in the chapter). These photographs capture all of the detail the analyst believes is necessary for the ultimate analysis and support of this analysis.

These photographs should also lead the viewer from the overall picture to the detail. Figures 13.5 through 13.9 depict a classroom exercise of what the basic objectives of this documentation should achieve.

Figure 13.5 is the overall photograph of an area. It defines five distinct patterns of interest. These patterns are labeled A through E. The labels introduced by the analyst will help clearly show the orientation of one pattern to the other and help orient this overall photograph with others from the scene.

Figure 13.6 is an overall pattern photograph of a specific spatter pattern. The purpose of the pattern overall is to show the entire pattern, even if all detail is not evident. The photograph is taken from whatever distance is necessary to show the complete pattern. In

Figure 13.2 An overall photograph of a specific pattern. The pattern overall shows both the pattern in its totality and acts as an evidence-establishing photograph for any subsequent close-ups.

smaller patterns, this photograph will act as a close-up as well. In larger patterns, close-ups may be necessary to see additional detail.

Figure 13.7 is an overall pattern photograph of Pattern D, a swipe. Once again, the purpose of this photograph is to document the entire pattern as it appears.

Figure 13.8 is an example of a close-up photograph intended to show the specific detail of Stain A1 from Pattern A. Note that it is annotated with a plumb line and a scale of reference. With the positional information obtained at the scene, this photograph could be evaluated using Area of Origin (AO) software.

Figure 13.9 is a close-up of the detail in Pattern D, specifically the presence of the bifurcations. This detail is important as it allows us to conclude whether the source of the swipe was hair or a hair-like object.

Each pattern of concern and the specific detail the analyst considers important must be captured in these photographs. Throughout the photography, each stain must retain a unique identity. A photograph with no landmark is of little value, as it makes it hard to be sure exactly what stain you are looking at or what the proper orientation of the stain is.

Overall pattern photographs (e.g., Figure 13.2) often suffice as functional evidence establishing photographs, so long as the sub-labels (e.g., A1) introduced into the picture are readable. However, in some cases, such as when dealing with small spatter stains and other small bloodstains, it may be necessary to use several establishing shots, each successively closer to the stain to show its orientation.

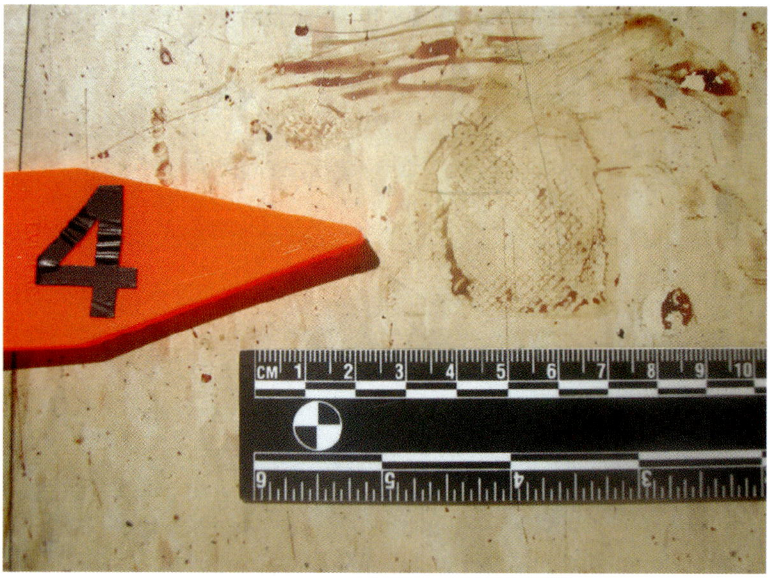

Figure 13.3 A close-up photograph of Pattern I-4, a pattern transfer, showing the detail evident in the pattern.

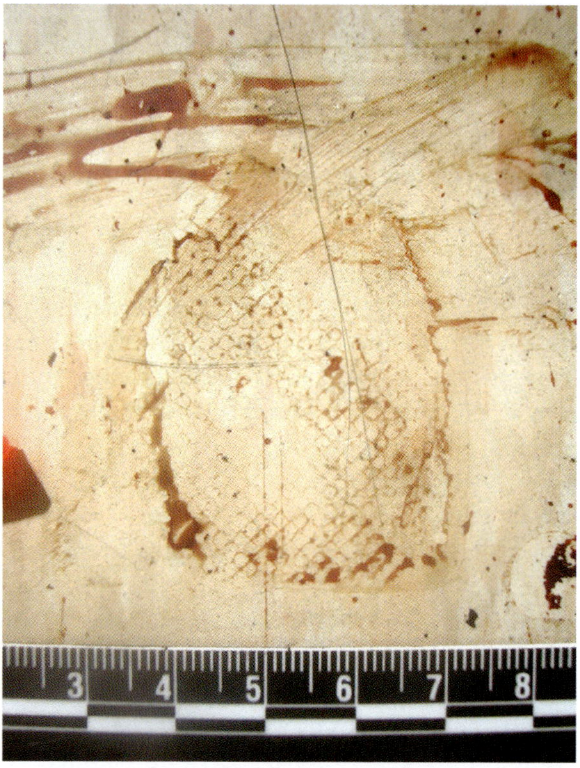

Figure 13.4 A second evidence close-up photograph filling the viewfinder's frame with the specific details of interest. This second photograph is often necessary when the pattern details are small. In this case, the pattern appears to be associated with a shoe. If that shoe were subsequently located, this photograph may offer the footwear examiner sufficient details to compare the two.

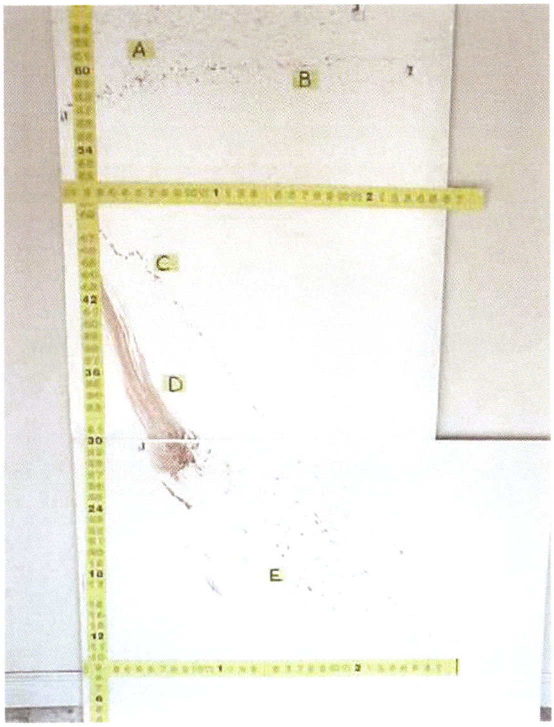

Figure 13.5 A training overall photograph that orients the position of the various stains to the scene. By themselves, such photographs rarely provide the detail necessary for proper Bloodstain Pattern Analysis, but they are still important as they allow us to orient each pattern of interest to the scene itself.

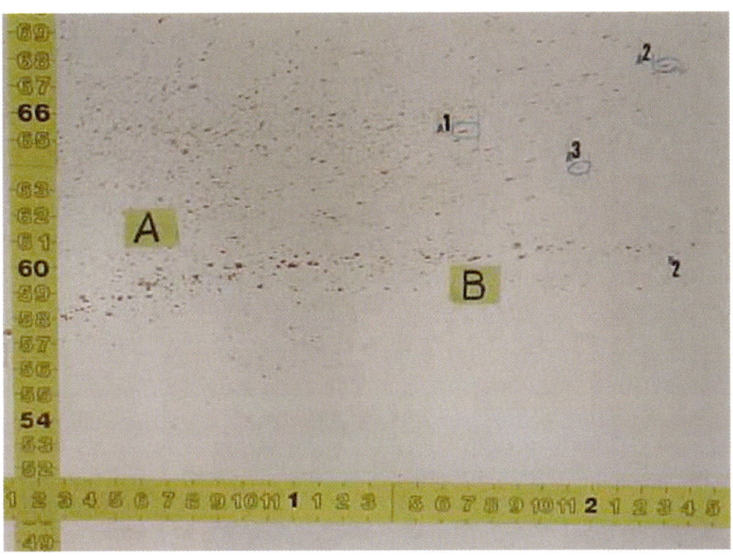

Figure 13.6 A training overall photograph of a specific pattern, Pattern A, an impact pattern. Similar to a scene overall, the pattern overall identifies the extent of a specific pattern. It may or may not show distinct detail, depending upon the size of the pattern of interest. Other patterns (in this case Pattern B) may also be depicted based on proximity, but the photograph's purpose is to highlight Pattern A.

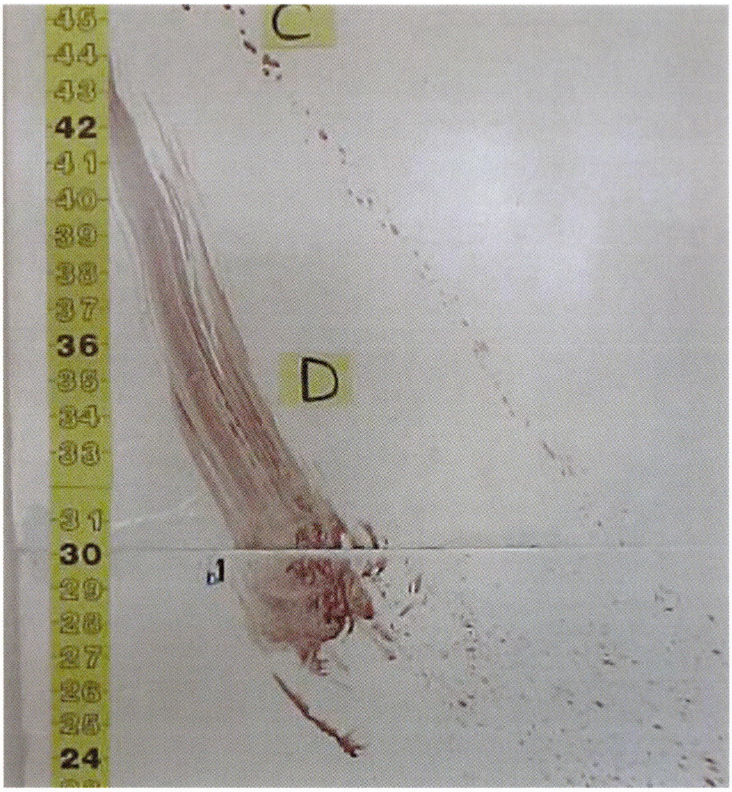

Figure 13.7 An overall pattern photograph of Pattern D, a swipe.

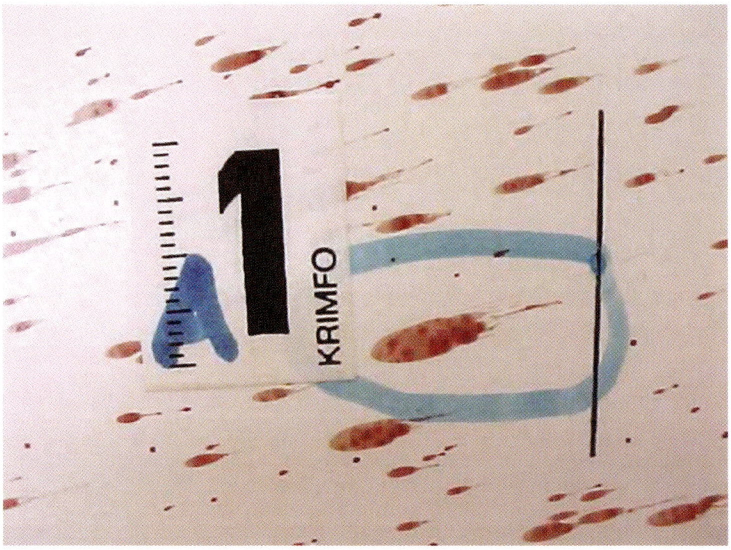

Figure 13.8 Close-up photographs of Stain A1, from Pattern A (Figure 13.6) the impact pattern. These types of photographs provide specific detail from the pattern or stain, allowing actual analysis. When conducting area of origin analysis, a close-up photograph of each individual spatter utilized is required.

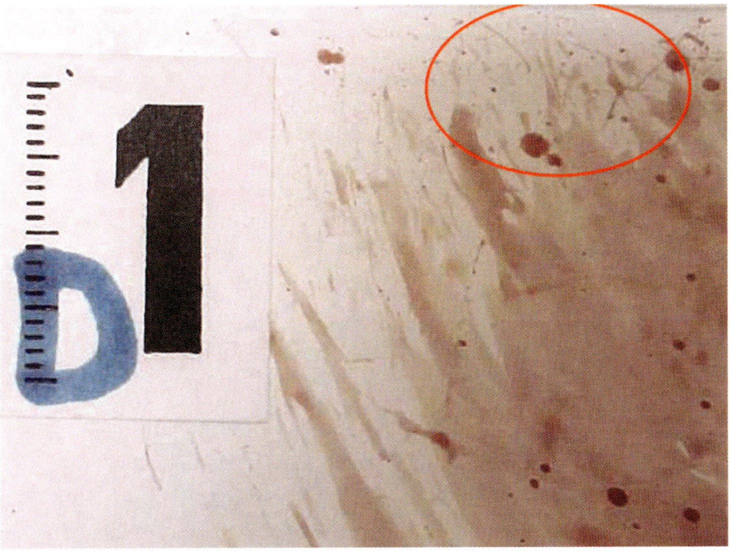

Figure 13.9 Close-up detail in Pattern D (Figure 13.7), the swipe. Important characteristics that allow the analyst to include or exclude some mechanism as the source of the stain must be documented. In this instance, the small bifurcations in the upper right corner of the swipe allow the analyst to identify the source of the pattern as hair or a hair-like object.

Without the evidence-establishing photograph, patterns and stains lose context, particularly when viewed by others who were never at the scene. Consider Figures 13.10 and 13.11. Without the establishing shot (Figure 13.10), the evidence close-up photograph (Figure 13.11) tells the viewer little about where the item is in the scene. The same is certainly true for close-up photographs of bloodstains found on evidence. The pattern evident in Figure 13.12 means little to the viewer without the accompanying photograph, Figure 13.13, which shows its location on the murder weapon.

The next step is to photograph the individual stains or details of interest. This means taking macro-photographs in which the viewfinder is filled with the detail present in the stain. Pattern transfers or swipes and wipes are all examples of stains that may require this type of effort.

Macro-photographs or evidence close-up photographs are taken with the lens inches, not feet, away from the stain. The close-up photograph of a pattern should have as much detail as possible. Camera orientation is critical in the close-up. Forget the fashion photography technique where the camera is shot at every conceivable angle. Place the CCD plane parallel with the plane you are photographing, a tripod is helpful in this task to reduce movement. Once that is done, fill the viewfinder with the subject. Both of these concerns are imperative if the analyst wants clear, undistorted images of the patterns and stains.

If the scene involves impact patterns and area of origin (AO) determinations are necessary, then photograph the individual stains used to make that determination. Again, this photograph is taken with the lens inches away from the individual stain. Using this photograph, another analyst can measure the stain and verify its impact angle. This will effectively preclude any major challenge to the AO determination.

Close-up photographs should be exposed with a scale of reference in the photographs. If the scale is placed in line with the long axis of the stain, measuring the stain in the

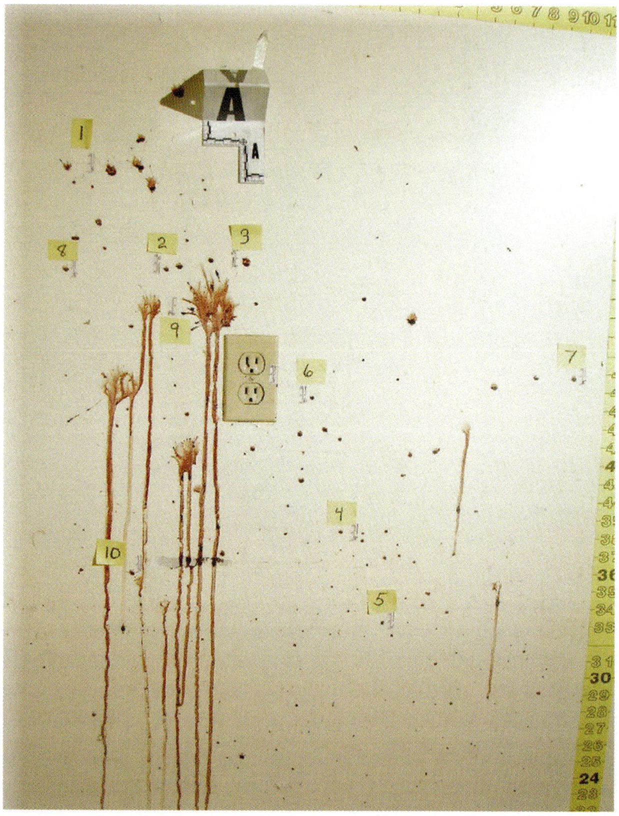

Figure 13.10 Another example of a pattern overall photograph that also serves as the establishing photograph for the close-up photographs. In this instance, we can read the labels of the individual stains in question (Stains 1 through 10) and see their position on the wall.

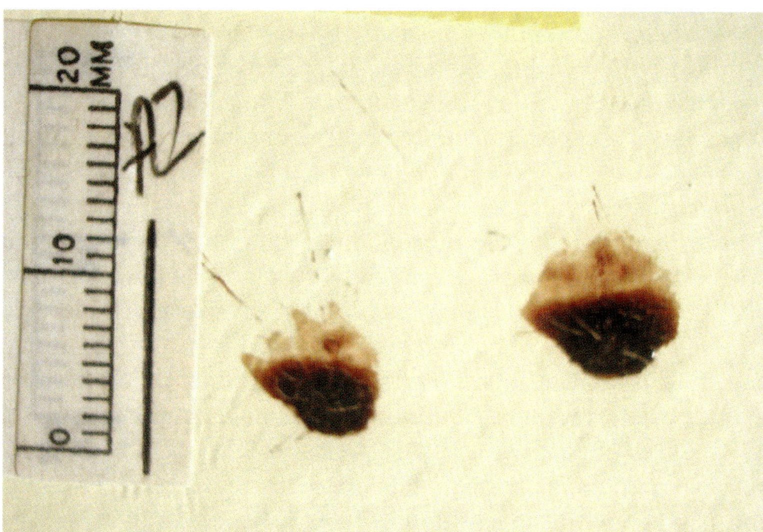

Figure 13.11 A close-up photograph of Stain A2 in Figure 13.10. Without the accompanying pattern overall/establishing photograph, it would be difficult for the viewer to understand this spatter's relationship to the scene.

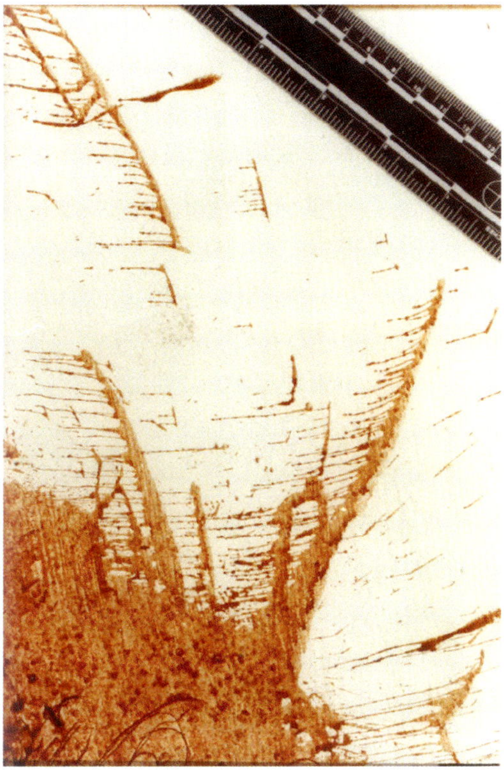

Figure 13.12 An interesting pattern resulting from a beating death. The scale indicates the size of the stains, but where are they found, and what relationship do they hold? Without an accompanying evidence-establishing photograph, Figure 13.13, the viewer has little with which to judge the significance of the pattern. (Photograph courtesy of Donald R. Schuessler, Department of Public Safety, Eugene, OR.)

photograph is simpler, but unnecessary. The use of a scale with both a vertical and horizontal scale is helpful.

Roadmapping is the most effective method of completing photographic documentation of bloodstains. It developed from the efforts of Toby Wolson, a criminalist formerly with the Miami-Dade Police Department, Miami, Florida. Wolson developed this technique based on his scene experience. It later became known as "Roadmapping." Although effective, it is also intrusive. The analyst must insert many "road signs" and scales for reference to map the scene properly. For this reason, the analyst waits until the completion of standard scene processing before completing the Roadmapping technique. Roadmapping is effective for documenting both still and video images.

When using Roadmapping, each major stain group is identified with a unique label, either a letter or number (e.g., Stain Group A). See Figure 13.14. As needed, individual stains or smaller aspects of interest within this larger pattern are given an additional identifier, such as A1 or A2. Labels are taped or placed by each stain, using lettering that allows the label to be read from several feet away.

Adjacent to the patterns, the analyst introduces large reference scales to the surface involved. The most effective approach is to use both a vertical and horizontal scale. These scales are commercially available from many sources. It is important to ensure that such scales retain context to the scene. One does not arbitrarily place these large scales around

Figure 13.13 An evidence-establishing photograph for the Pattern in Figure 13.12. We now understand that the pattern is present on the murder weapon, a shelf from the kitchen. The weapon was used to inflict blunt trauma to the head of the victim. (Photograph courtesy of Donald R. Schuessler, Department of Public Safety, Eugene, OR.)

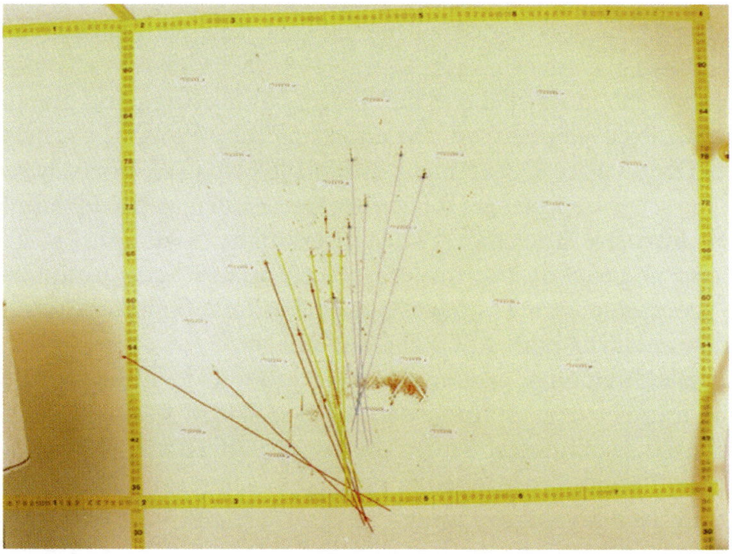

Figure 13.14 An example of the Roadmapping technique. The large yellow scales allow the viewer to know where the various stains are on the wall. The placards are used to identify specific patterns, while the yellow labels identify individual stains of interest within the pattern.

a pattern. For example, if a pattern is present 4 ft up the wall, there is no need to extend the vertical scale from the floor. The analyst should measure to a known point (e.g., 3′ up the wall) and start the scale at that point, marking the scale accordingly. The same is true for the horizontal scale. There may be no need to extend the scale from a corner 7′ from the pattern; the analyst can measure across the wall to an appropriate point and begin the scale placement there.

Once all labels and scales are in place, the actual Roadmapping photography begins. Similar to standard crime scene photography, three basic photographs are needed:

- *Overall Photographs:* These photographs show general orientation of a group of patterns. All primary pattern labels must be readable in this photograph, as this photograph is in effect an evidence-establishing photograph as well.
- *Pattern Overall Photographs*: This photograph captures the entire pattern. For small patterns this will require positioning the camera very close to the surface, while in larger patterns (e.g., a cast-off pattern extending across several surfaces) the camera must be positioned much further back.

  The primary pattern label must be evident in this photograph, as well as a scale. The simplest approach is to place an ABFO scale adjacent to the primary pattern label. If sub-labels were used for any reason (e.g., A1–A2), whereas they may not have been readable in the overall photograph, they must be readable in the pattern overall photograph. Effectively, the pattern overall also acts as the evidence-establishing photograph for the smaller aspects of the pattern. These labels are the "signs" on our roadmap. They preclude us from being lost (see Figures 13.15 and 13.16).

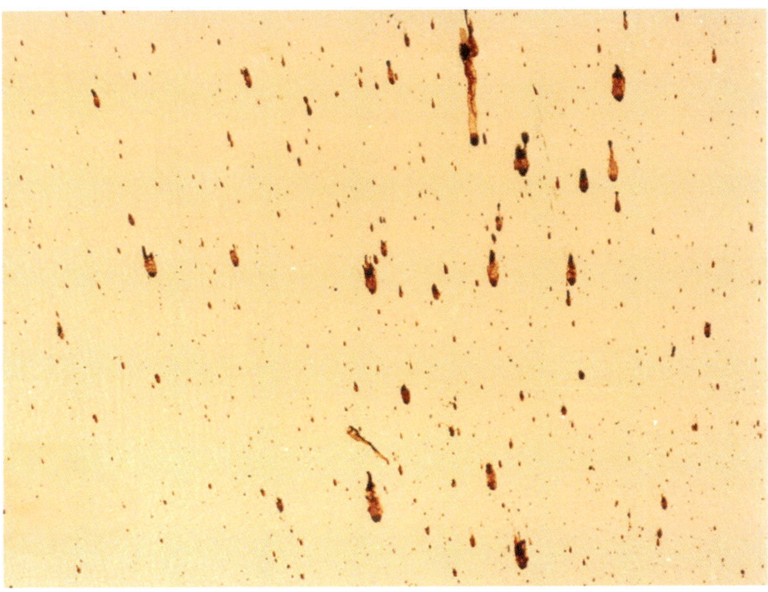

Figure 13.15 The typical close-up photograph found in most documentation files. Although the analyst often exposes such photographs without a scale or label, such photographs do little in and of themselves to define the scene or identify the stain in question. In this instance, we have a spatter pattern, but no evident landmark. In effect, the viewer is "lost" in the scene with no means to associate the picture with its origin. (Photograph courtesy of Toby Wolson, MS, Miami-Dade Police Department, Miami, FL.)

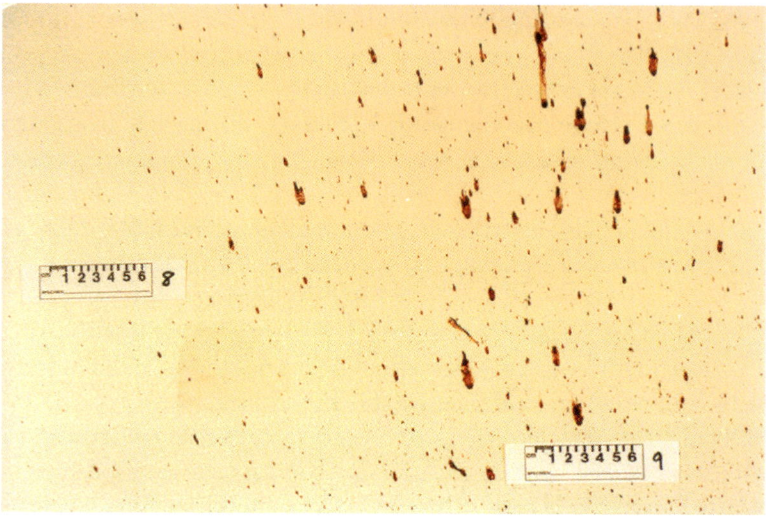

Figure 13.16 A proper close-up photograph of the spatter in Figure 13.15. Using two labels (Numbers 8 and 9) in conjunction with Figure 13.16, the viewer can recognize where the spatter is in the scene. Roadmapping as a method moves the viewer from the overall perspective of the scene to the pattern and then to the detail in the pattern. (Photograph courtesy of Toby Wolson, MS, Miami-Dade Police Department, Miami, FL. From *J. Forensic Identification*, Volume 45, Number 4, August 1995.)

> Another important consideration in the pattern overall is orientation of the CCD plane to the surface plane. If a pattern extends across multiple surfaces (e.g., a wall and floor), the only way to capture the entire pattern is to shoot the area at an angle. This of course violates the CCD plane to surface rule. To resolve this, two additional photographs are taken. The camera is positioned facing the vertical aspect of the pattern and a second photograph taken. Then the camera is repositioned facing the horizontal plane and a third photograph is accomplished. Combined, these three photographs capture all aspects of the pattern, without concern over angular perspective issues.

If necessary, graphic tape or adhesive arrows can be added to show indications of motions in swipes or wipes, or to show points of convergence of spatter and other information that is deemed important.

The rules for Roadmapping are simple:

- No photograph is taken without at least one label fully evident in the photograph.
- Labels must be readable in the appropriate photograph. For overall photographs, this means the primary pattern labels. For the pattern overall photograph, this means all sub-labels.

We cannot overemphasize how effective this method is for conducting subsequent Bloodstain Pattern Analysis. Too often, the relationship of one stain to another is anything but evident in crime scene photographs. Roadmapping takes the viewer from an overall view, to the pattern and then to the individual stains. Every photograph graphically

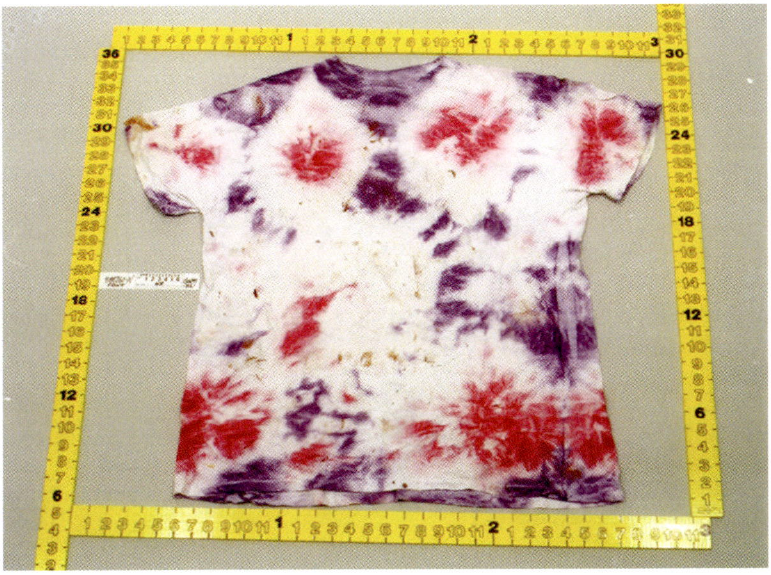

Figure 13.17 Roadmapping is just as effective for documenting items of evidence after collection. In this instance, we have an initial photograph of the overall item, framed by the large scales. (Photograph courtesy of Toby Wolson, MS, Miami-Dade Police Department, Miami, FL. From *J. Forensic Identification*, Volume 45, Number 4, August 1995.)

exhibits its relationship to other stains. It eliminates much of the viewer's confusion and makes for outstanding case documentation.

As effective as Roadmapping is at the scene, it also works effectively for documenting bloodstains on individual items of evidence in the laboratory or office. For instance, Figures 13.17 through 13.19 show documentation of a bloodstained shirt using this method.

Our consideration of photography is also incomplete without discussing video. The video mode of a camera or a video camera can be extremely useful in documenting a bloodstained scene. It certainly serves as a functional method for documenting the scene in situ. A quick, nonintrusive walk-through of the scene can be invaluable later when the analyst is trying to verify the original condition of the scene. In combination with the scene photographs, videography often clears up specific concerns regarding scene integrity issues.

Once the Roadmapping effort is completed, a second videography effort is helpful. Using Wolson's Roadmapping techniques, the scales and "signs" are already in place. The analyst merely "walks" the viewer through the scene, showing the relationship of the specific stains to other items of interest. This video can bring immense clarity to the situation. It provides the viewer with a far better context in which to consider the scene.

One caution is appropriate to video camera usage at the scene. During the video effort, if the analyst or scene technician is simultaneously recording narration, the narration should be limited to a description of the stains observed. Pursuing any "analysis" too soon in a taped medium may come back to haunt the analyst at a later date.

Scene and Pattern Sketches

Narratives, photographs and sketches are basic crime scene documentation tools. Individually, they are helpful; combined they give a more complete understanding of the

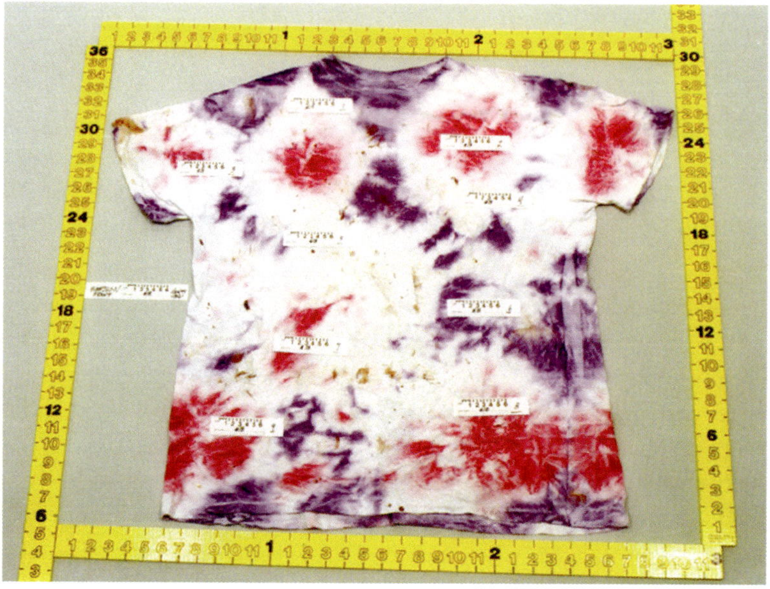

Figure 13.18 In this second photograph of the shirt, an establishing photograph and labels are introduced to identify the specific stained areas on the shirt. (Photograph courtesy of Toby Wolson, MS, Miami-Dade Police Department, Miami, FL. From *J. Forensic Identification*, Volume 45, Number 4, August 1995.)

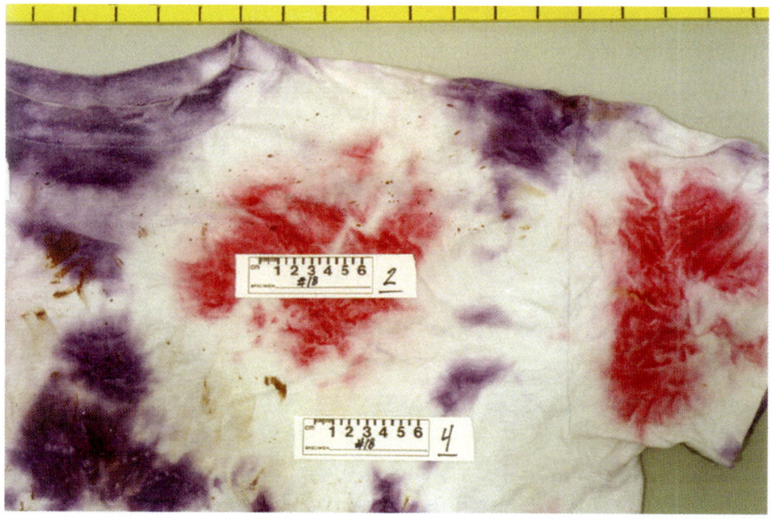

Figure 13.19 An evidence close-up photograph of Stains 2 and 4 on the shirt. By referring to Figures 13.17 and 13.18, the viewer can easily determine the stain locations, even if the collar was not evident in the photograph. (Photograph courtesy of Toby Wolson, MS, Miami-Dade Police Department, Miami, FL. From *J. Forensic Identification*, Volume 45, Number 4, August 1995.)

scene. Crime scene sketches are and will remain an integral part of scene documentation. They specifically help the viewer to "see" the relationship of where items of evidence were and help the court understand the overall circumstances of the crime.

All major stain patterns should probably be included in either the crime scene sketch or a second sketch detailing only the bloodstain patterns. They allow an individual who has never visited the scene to see the relationships of one stain to another or to correlate a stain to its physical surroundings. The level of detail necessary for bloodstains in a sketch depends entirely on the analyst's needs and desires. All major patterns should be addressed in some fashion. This might only require including the pattern label (e.g., Pattern A) at the general location where the pattern was found. However, this does not preclude the analyst from sketching any stain or pattern of significance into the drawing itself.

Oftentimes when considering a bloodstain scene, a cross-projection sketch is necessary to document the scene fully. In the cross-projection sketch, the analyst simply "lays" the surrounding walls down, drawing appropriate stains and evidence onto them. This serves the function of relating one stain pattern to the location of another (see Figure 13.20).

As indicated, the analyst may choose to include only the labels of the stains in the overall crime scene sketch. To provide greater detail, the analyst may wish to create individual sketches of stains and patterns of importance, all in addition to photographs. Such sketches can include freehand drawings of patterns, patterns on clothing, or tracings of a given pattern. A common method currently employed when evaluating bloodstained artifacts (e.g., clothing items) is to take a digital picture and annotate the picture itself. This is far more effective than producing a freehand drawing of the object.

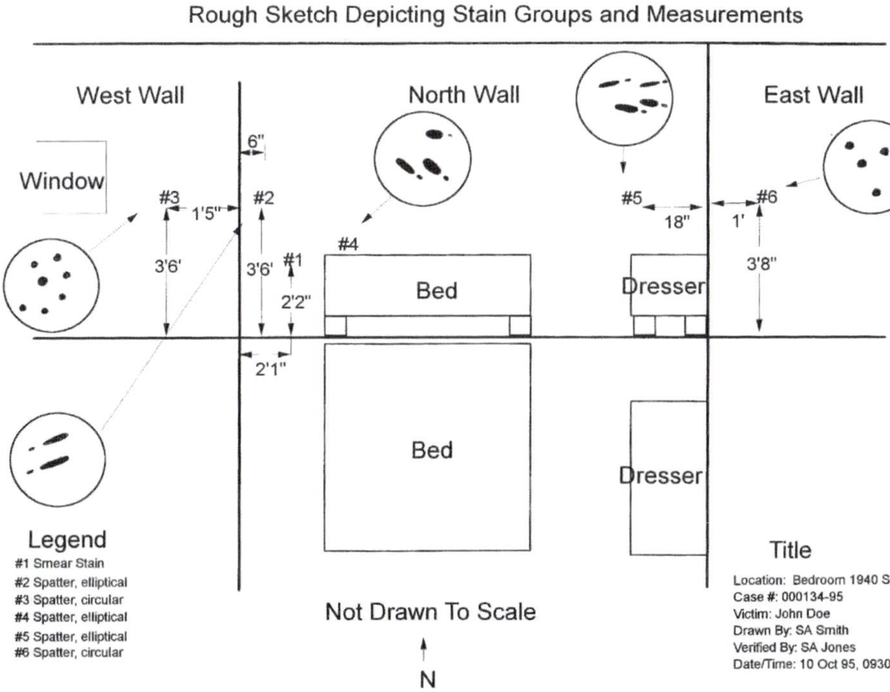

Figure 13.20 The cross-projection sketch. A cross-projection sketch allows the viewer to consider evidence on various surfaces in the scene. For example, Stain Groups 2 and 3 indicate a common source of origin. Without the cross-projection sketch, this relationship might not be immediately evident to anyone reading the report.

Impact patterns demand additional documentation effort, specifically in terms of crime scene mapping. If the analyst intends to conduct any area of origin or area of convergence analysis, each of the individual spatter stains chosen for inclusion must be documented fully. After choosing an adequate number of stains (which is generally defined by the nature of the pattern), the analyst must

- Obtain close-up photographs to allow other analysts to validate each stain's measurements
- Locate each stain in the scene with physical measurements; this will require a minimum of two measurements to place the stain on a given surface depending upon the measurement system employed (e.g., rectangular coordinates, triangulation)
- Sample the pattern for DNA

Written Reports

Photographs, sketches, and videos certainly enhance our overall understanding of the crime scene, but written reports are the foundation of investigative documentation. Therefore, we should not haphazardly approach the written report. The written report is how analysts share much of their knowledge. It should be complete, methodical, and as objective as possible.

The first consideration in the written report should be to describe the information that serves as its basis. The date and time of the scene or evidence viewing should be documented along with a location or an address where the scene or scene photographs were viewed. A list of all documentation considered in the analysis should be listed. The analyst should describe the evidence reviewed and what actions were performed to support the report. Any limitations imposed by the analysis should be discussed. For example, if all observations were made from photographs, the report should state this. If one section of the report is based on photographs and a subsequent section on photographs and on-scene evaluation, the analyst must clarify this. Consider the following example:

> Using photographs provided by S.A. Jones, I evaluated bloodstains present at the scene of a rape/attempted homicide. Additionally, a review was made of all crime scene reports and crime scene sketches.

Major patterns or stain groups are then discussed in an orderly manner (e.g., either in alphanumerical order or perhaps in order of significance). Three different conclusions are included in the analyst's discussion. The first is the classification of the stain/pattern.

> **Stain Group #1:** Impact Pattern – present on the wall adjacent to the headboard.

This classification conclusion is not speculative, it is based on physical characteristics.

The next aspect described is effectively a lay description of the pattern. The pattern should be described in detail, as in the following:

> There are a series of circular and elliptical stains present. The majority of the stains are circular, with very few elliptical stains. The stains in the pattern are primarily 1 to 5 mm in diameter. The majority of spatter present in the pattern are actually less than 2 mm in size.

> The pattern has no obvious linear aspect and no ringlets or vacuoles. There is a radiating dispersion. The directional aspects of the elliptical stains indicate the stains were radiating upward and slightly to the right.
>
> This pattern is slightly offset on the wall above the single bed and slightly left of a large saturation stain on the bedding. The pattern's overall dimensions are 2′ × 3′. The pattern begins at the level of the mattress and extends up the wall for approximately 2′. The pattern extends outward and left along the wall for a distance of 1′ from the left edge of the headboard. The pattern extends across the wall to the right for another 2′. Based upon documentation provided, and the nature of the stains, no specific area of origin determination was possible. The DNA present is consistent with Mr. X.

Specific references that may assist the reader can then be listed. This may include DNA efforts. For example:

> References: Image DSC 0001, 0003, DNA Rpt. dtd. 10 Jan 2022, pg. 3.

The analyst's conclusions follow this section. There are two different conclusions presented in addition to the classification (e.g., Impact Pattern). The first is what the stain or pattern defines alone (without context). Then any conclusions arrived at with context (e.g., scene orientation, DNA). For example:

> Conclusions: This is an impact pattern. A blood source was exposed to some force/impulse forward of this wall. While present at the level of the mattress, and likely while on the mattress, Mr. X received some blow(s).

There are various approaches to how the analyst should present general/overall conclusions. One approach is to present the conclusions at the end of the report in a summary fashion. Presenting a summary of the conclusions at the beginning of the report is just as effective. The only downside to presenting them before the detailed discussion of the stains is the reader has yet to see the details that led to the conclusions. The following is an example of a conclusion sheet:

Stain #	Conclusion
1	Forceful bloodshed occurred near the east wall of the bedroom and within 26″ of the floor
2	Cast-off stain patterns seen on the ceiling describe an object wet with blood being swung in a overhand fashion. These patterns correspond to the forceful incident described in #1
3	The subject was upright and in a seated position when blood flowed from her head wounds
4	The subject was repositioned face down following the infliction of wounds to her head

The written report should give the reader a clear picture of what the analyst observed. Always seek to define specific characteristics that are evident in the stain. Ensure these characteristics are included in the report. The following are more examples of how one might document a bloodstain pattern. Without this type of detail, it is difficult for analysts to support why they reached their conclusions.

A Spatter Pattern Description

Stain Group 2: This is a spatter pattern on a nightstand adjacent to the left side of the bed.

The pattern comprises a series of circular and elliptical-shaped stains, the largest of which is 5 mm in width. Smaller spatter measure down to the 1.5–2 mm range with other smaller spatters present, but they are too small to measure based on photographs alone. Most of the spatter are circular stains on articles present on the nightstand (e.g., a pitcher, cloth, and bottle). Those present on the upper surface of the nightstand itself are elliptical and struck the surface at extreme acute angles. The pattern radiates right to left across the nightstand top.

A void/demarcation line is evident in the stain group. This void is evident on a plastic cup behind the pitcher and is also present within the elliptical stains on the upper surface of the nightstand.

While present on the bed or nightstand, something at the point of the pitcher precluded spatter from occurring on the remaining items.

The directionality of the elliptical stains is consistent with an origin near the large stained area on the bedding.

Using the evident directionality and reverse azimuths of several of the elliptical stains on the nightstand, an area of convergence in two-dimensional space would be approximately 16″–18″ to the right of the edge of the bed mattress. Once again, this correlates with the observed saturation stain area on the bed mattress.

A Blood Pool Description

Stain Group #3: This is a large liquid pool on the floor near the entrance of the cubicle in the room.

The pool is diluted, but the nature of the dilution is not immediately evident.

Lower sections of the stain show spines, indicating the application of force to the liquid. There is little to indicate the nature of this force itself.

The upper right-hand section suggests linear features, which may be more spines. A threefold linear marking is within the lightest section of the stain (in the central area).

This stain is not disturbed except for the spines on the edge facing the bathroom. There are no foot or shoe marks in the stain and the general boundaries are intact.

The blood present within the stain is not the result of drops impacting, either as venous bleeding or spatter. No indication of drips or drip pattern satellites is evident around the stain. The pool received several forceful impacts of some nature, causing the spines. The pool was undisturbed on the side nearest the bathroom, meaning it was deposited and then no further action occurred to that side.

A Pattern Transfer Description

Stain Group #7: This group includes numerous contact stains deposited on the pillowcase. The entire pillowcase is saturated with blood, making many patterns marred or blurred to the point that they are of no use.

However, there are double parallel linear markings within the outer boundary areas of the saturation. Some markings are very short, while others extend for up to 3 cm.

Three general widths are apparent between these double linear boundaries: 4.6 mm, 19.7 mm, and 32.5 mm. These measurements were derived from Xerox copies of the stains, in which a 7% enlargement factor was determined and added.

On the pillowcase, there is one (1) set of lines that closely match the 32.5 mm width, two (2) sets of lines that match the 19.7 mm width, and five (5) sets of lines that match the 4.6 mm width.

These markings suggest a bloody object came in contact with the pillowcase. The orientation of these sets of stains indicates a random application of the object to the pillowcase.

Many more linear features are present on the pillowcase, but none provide specific characteristics beyond indicating an object with a bloody edge was in contact with the pillowcase.

There was one additional contact stain of interest, which lacked further identifying characteristics. The stain is a pattern transfer with a curved feature, perhaps a dual boundary. The curve is not typical of any objects evident in the scene (e.g., a bottle). The nature is insufficient for further identification without a specific object for comparison.

A Complex Pattern Description/Conclusion

Stain Group #8: This complex pattern consists of a saturation stain on the upper sheet of the bedding and includes contact stains evident to the left of the saturation stain and spatter evident to the right.

The contact stains have distinct demarcations, indicating a pattern transfer with dual linear boundaries, similar to those found on the pillowcase. The linear stains range from a few millimeters in length to 11 cm. The application appears less random than that found on the pillowcase, with the majority of the lines angling downward from the saturation stain at about a 30°–40° angle.

The patterns exhibit the same measurements as those on the pillowcase. There are two (2) sets of lines that match with the 4.6 mm width and one (1) set of lines that match with the 32.5 mm width. Impact spatter is evident throughout this stain. The cotton material makes directionality for these limited stains difficult to determine.

The spatter evident to the right of the saturation stain comprises an area of 8″ × 3″. The spatter size range is 0.2–2.4 mm. The definition evident in the stain and the radiating effect evident in the spatter point to a single action as the source of this stain. The force that created this stain projected the droplets at about a 45° angle down and out from the top edge of the sheet.

The same object that came in contact with the pillowcase also came in contact with the sheet. In doing so, it was better defined (e.g., length at least 11 cm).

The nature of the conclusions and the level of certainty described in the body of the report are always a major concern. The detail required in the description is a matter of the nature of the stain or pattern in question. In some instances, a single paragraph may functionally provide all the details required. In instances of complex patterns or pattern transfers the description may require significant effort. As Bunker said, "This is a report which will be examined, reexamined, and dissected by others."[3] If you are not confident that you can support your findings, then do not include them.

Summary

Documentation of bloodstains at the scene is truly an important function. Without quality documentation, the analyst or scene technician may doom any subsequent analysis to failure.

Documentation considerations include the proper collection and preservation of evidence. This requires the analyst understand the needs and desires of the biological scientist. It also requires an understanding of the available chemical detection and enhancement procedures, including their limitations, which were discussed in Chapter 12.

Documentation in the form of sketches and photographs is equally important. They must clearly show stain or pattern detail and define where the stain or pattern was in the scene. Without this detail, the analysts are limited in their conclusions. Without knowing where in the scene the stain was, it is difficult, at best, to make any supportable decisions about the overall Bloodstain Pattern Analysis. Written documentation is the true backbone of the overall documentation effort. The stain and pattern documentation must be clear, concise, and detailed. Full descriptions of what the analyst observed are imperative. Each stain should be fully described logically, leading the reader through the scene.

No matter how well-intentioned an investigation may be, in the end, it will be judged on the analyst's documentation. The analyst's efforts at documentation cannot be haphazard.

Notes

1. Lee, H.C. et al., The Effect of Presumptive Test, Latent Fingerprint and Other Reagents and Materials on Subsequent Serological Identification, Genetic Marker and DNA Testing in Bloodstains, *Journal of Forensic Identification,* Volume 39, Number 6, Nov.–Dec. 1989, pp. 339–358.
2. Wolson, T.L., Bloodstain Pattern Documentation Workshop, presentation to the International Association of Bloodstain Pattern Analysts, Colorado Springs, CO, September 26, 1992.
3. Bunker, J., Documenting and Reconstructing Blood Scenes, Presentation to the Rocky Mountain Association of Bloodstain Pattern Analysts, Advanced Bloodstain Pattern course, Jefferson County, CO, April 22, 1991.

An Introduction to Crime Scene Reconstruction and Analysis

14

A basic reality is that bloodstain pattern analysts are reconstructing crimes. They may concentrate on only one aspect of crime scene reconstruction. Still, it is impossible to act as a bloodstain pattern analyst and not foray into the world of crime scene analysis. For this reason, bloodstain pattern analysts must understand the tenets of the discipline of crime scene analysis and reconstruction. They may not choose to pursue a full-scale crime scene analysis, but they should understand how a full-scale analysis is accomplished.

Although many individuals with varying mindsets utilize the words "crime scene analysis" and "crime scene reconstruction," these terms have specific meanings and contexts. The Association for Crime Scene Reconstruction defines reconstruction as " Using the scientific method to gain explicit knowledge of the series of events surrounding the commission of a crime through deductive reasoning, inductive reasoning, physical evidence in its contextual setting, and their interrelationships."[1] For purposes of this text, one may consider crime scene analysis and crime scene reconstruction as synonymous. Later in this chapter, we will introduce a methodology for crime scene reconstruction known as *Event Analysis*. All three terms relate to the application of the scientific method to the scene data to achieve three basic objectives:

- To identify as many objective actions as possible, an action refers to a change in state of some object (animate or inanimate).
- To establish the temporal order of the identified actions, to whatever level is possible.
- To evaluate specific investigative questions in an attempt to resolve them.

Whereas crime scene analysis and crime scene reconstruction are general terms describing the discipline, Event Analysis is simply one methodology by which one can accomplish crime scene reconstruction.

These "actions" are identified by considering the physical evidence at the scene and the context in which it is found. The analyst identifies what happened during the incident by identifying specific actions that occurred. Just as important to us, examining the scene can lead to objective statements about *what could not have occurred*. But defining the *what* of the crime is not the only consideration of crime scene analysis; wherever and to whatever level possible, crime scene analysis defines the order in which the crime happened. The end product of crime scene analysis is an objective statement of specific actions that occurred, placed in some level of order in which they occurred.

Thus, what happened and in what order it happened is the focus of any crime scene analysis. Interestingly, an obvious issue, the "why" it happened is not considered in CSR. The "why" of crime may be a significant factor in a criminal investigation. However, *why* something happened is never on the table for the crime scene analyst to consider. Physical

DOI: 10.4324/9781003408734-14

evidence cannot answer why someone was stabbed 15 times (unless someone left a confession, and even then, the confession must be viewed for what it is: subjective testimonial evidence). The *why* of crime is the crime scene analyst's event horizon, a point beyond which we cannot see. Physical evidence will tell us much about what happened (e.g., where and how the victim was stabbed). Still, the intent of an individual's mind is not transferred into or viewable through physical evidence, at least not objectively. Leave such opinions to the psychologists or behavioral analysts. Absolute claims by a crime scene analyst about why someone did something are nothing more than speculation.

The end product of crime scene analysis, however, is quite useful. Suppose we can effectively establish that certain things did not occur or that certain things did occur and we are confident of the general order in which these actions occurred. Such knowledge can be considered against any known statements or theories in that case. The analysis is contrasted against these theories (e.g., investigative, prosecution, or defense theories) or statements by involved parties. When compared against the reconstruction, either the statement/theory will be supported, or some part or all of it will be refuted. Wherever the crime scene analysis is found to be out of sync with a given statement or theory, that aspect of the theory must be revisited. If the analysis was conducted properly and all known information considered, the theory or statement must be incorrect in that aspect that conflicts with the data. Data define the conclusion. One does not define the data used based on the conclusion. This is a classic failure of the law profession, a form of confirmation bias. One cannot decide arbitrarily what data and facts they will include in their consideration and ignore that which doesn't support their beliefs.

This value of CSR, evaluating theories, is important before the court and jury. Judges and jurors are asked to evaluate various theories, hear various testimonies, and decide which they believe is accurate. They must listen to and ultimately place a value judgment on all of the testimonial evidence presented (e.g., Do they believe the witness or not?). Obviously, the analysts cannot invade the jury's providence, but through CSR testimony, they can highlight the "what" of the incident in question. This will allow the jury to evaluate better what they are being told. Crime scene reconstruction provides the jury with objective information on which to judge the more subjective information presented to them. This is exactly what the analysis is used for in any unique phenomena investigation. The analysis becomes a benchmark against which any statements or claims can be tested.

Crime Scene Analysis and the Archeologist's Dilemma

Crime scene analysis is often discussed in a jigsaw puzzle analogy, which we have referred to as the archeologist's dilemma. The authors once heard an archeologist explain his professional pursuit as something like putting together a jigsaw puzzle. As he described, his dilemma was as if someone walked up to him and threw down several handfuls of the puzzle pieces from the puzzle box. Then watched as they threw the box top and remaining puzzle pieces away. Using only the pieces of the puzzle provided, the archeologist was still expected to answer, "What does this picture look like?"[2] This puzzle analogy is true for the crime scene analyst. The crime scene analyst shares exceptional correlations with the archeologist in their pursuit, so much so that we will apply basic principles from archeology to crime scene analysis. No doubt, some may ask if this is a fair association.

Consider that archeologists examine historical sites in detail. They document the scene context using detailed notes, photographs, and sketches. Artifacts are carefully recovered from the scene to preserve their value and context. Archeology expects much from the scene documentation. This includes the following:

- The documentation must clearly and accurately describe all features that will disappear as a function of the excavation.
- The documentation must allow for the organization and identification of each artifact removed from the scene.
- The documentation must allow for the correct reconstruction of the strata and artifacts recovered.[3]

An axiom in archeology states: "To excavate is to destroy"; thus, documentation of the scene is considered critical to any subsequent understanding.[4]

The physical artifacts recovered from a site are then examined in depth, oftentimes by other scientific experts, to understand what they are and what they define. The archeologist uses all of this data to identify specific conclusions regarding what happened and in what order things happened (perhaps over decades or centuries) at the site involved.

The crime scene analyst and crime scene technician follow similar paths, with similar expectations. Crime scenes are examined in depth. Scene context is documented using notes, photographs, and sketches, with the recognition that any action taken in the scene affects that context. Artifacts (evidence) are recovered and examined to define what they can tell us. When needed, the evidence is examined by other experts to better understand its nature. Then, using all of this information, the crime scene analyst attempts to identify what happened and in what order it happened. The only real difference between the tasks of the archeologist and crime scene analyst is found in the age of the sites involved. The crime scene analyst deals with far more recent history; thus, we typically have more artifacts and, as a rule, a better overall scene context to consider them against. Archeologists are not crime scene analysts any more than crime scene analysts are archeologists, but our point should be clear: We share similarities in tasks. Therefore, crime scene analysis can and does draw on established principles of archeology to assist us in our pursuit.

Like the archeologist, each artifact discovered at the crime scene becomes a piece of our jigsaw puzzle. No single piece answers all of our questions, but each piece tells us something. Each piece becomes a data item that in combination with other data items may allow us to state with some level of certainty that something did or did not occur. Seeking causal and temporal relationships between the various pieces of our puzzle allows us to create a picture of the incident, our "box top." How complete that picture is or is not, is very much a product of the evidence available. Even though the theory of crime scene analysis is that "nothing just happens," the reality is that we are still limited by technology. Not every action leads to a recognizable artifact, something the analyst can observe and recover through current technology. Like archeology, the greatest limitation in crime scene analysis is there are only so many pieces of the puzzle available in our scene to find and use.

Adding to this dilemma is a concern for recognizing what pieces actually belong to our puzzle. Not only must the analyst understand how the various pieces of the puzzle fit together, they must also work through and recognize which pieces are from some other puzzle. Remember that no scene is ever really a virgin scene, a scene in which no prior activity has ever occurred. Every site has some history. There may be artifacts present that are associated with some prior incident, something that was simply present when our incident occurred. The analyst arrives after the fact to find these pre-incident artifacts. Not knowing what they are looking for, the analyst initially considers them part of the puzzle being examined. Post-incident artifacts may be present, created by police, suspects, EMS, and witnesses. These items have nothing to do with the incident in question, but are present nonetheless when the scene is processed. The analyst must consider all of these concerns to know what pieces of the puzzle they should be trying to put together.

Crime scene analysis is akin to a giant jigsaw puzzle, a daunting task that demands the application of specific methodologies. It demands objective analysis of all information, in which counter-theories are considered and evaluated with the same level of effort as the pet theory of the hour. The analyst must recognize that no matter how smoothly the analysis proceeds, in the end, what we know from the scene is sometimes less than what we do not know. In part, this limitation is based on the available evidence. To accomplish the reconstruction, we have only the evidence discovered, documented, and evaluated, and that evidence may provide but a mere glimpse of the actions and events that encompass the crime. If this glimpse is accomplished using a valid methodology, the end product is still invaluable to everyone's understanding of what actually happened.

How limiting is this task? Consider the Rodney King incident that occurred in Los Angeles in 1989. In the subsequent trial, jurors had to decide if the police officers acted appropriately during the apprehension. In this instance, the jury had an advantage: a videotape of many of the critical portions of the incident. The problem seemed simple: watch the tape and decide who was right or wrong. For example, was Mr. King struck full force in the face by a baton? The opposing counsels presented convincing evidence claiming opposite positions.[5] Even with the tape, decisions of whether this action transpired were anything but simple.

Investigators rarely have the luxury of a videotaped beating, shooting, or murder from which to develop a concise picture of the actual actions involved. Lacking such evidence, the court still expects us to shed light on what did or did not transpire during any given incident. We attempt to do this by using whatever evidence is available. In viewing our scene to define these actions, the evidence encountered will likely fall into one of three broad categories. Without alluding to legal definitions, we might name these categories direct, circumstantial, and peripheral.

Direct evidence is that which gives us clear direction and focus relevant to some action. These kinds of facts are indisputable by anyone. For example, a contact gunshot wound indicates the weapon was in contact with the victim. Thus, whoever operated the weapon must also have been in an appropriate proximity. A latent fingerprint tells us that the depositor of the latent had contact with the item on which it was found. This type of evidence speaks clearly and concisely to everyone.

Following direct evidence is circumstantial evidence, which provides direction but lacks distinct focus. How the information is used in the analysis may be arguable from several positions. Consider a swipe pattern made with the victim's blood. Obviously, it occurred after the victim received some bleeding injury, yet the specifics of who made the swipe and how it was made may still be lacking. This evidence is important, but it demands that we carefully consider any conclusions we draw from it.

On the heels of this second category comes peripheral evidence. Consider a single hair or fiber found at the scene. Can we relate it to the incident, or is it perhaps some accidental deposit totally unrelated to the situation being investigated? This type of evidence may or may not assist in the reconstruction. The problem is determining where and how it fits into the overall picture. Nevertheless, it could be critical to our understanding of the event. More often than not, this type of evidence creates major distractions, not only during the investigation, but also in the subsequent judicial proceeding. As we are not always clear how it relates to the scene, wild claims by counsel are often made regarding that relationship.

In pursuing reconstruction, just like archeologists, we seek to define physical events we were not present to observe. Therefore, we must always remember the following:

- These actions are dynamic. This is to say that any number of similar actions may produce a result, that we observe. These similar actions, however, may deviate in slight but important ways.
- The nature of the evidence to support our decisions on these actions, even in the best of circumstances, will provide only a glimpse of this past.
- Rarely will we have a standard to which we can objectively compare our conclusions. No investigative box top is available for the analyst to refer to!

Therefore, neither we, nor the court can be absolutely certain as to <u>all</u> the specifics encompassing an incident. We can define individual moments with near-absolute clarity, but ultimately, our overall picture is simply the best explanation given the evidence available to us. For this reason, we should never attempt to define the actions surrounding an incident too narrowly. Our responsibility is to make this an objective picture, inferring as little as possible.

A History of Crime Scene Analysis

Although the authors are working to focus on the practice of crime scene analysis, the concepts behind it are not new. Just as Bloodstain Pattern Analysis has a rich history, there is a significant history in forensics relating to reconstructing criminal events using physical evidence. Although the methods were not well documented in some of the earliest references, investigators have clearly applied and continue to apply formal methodologies to this task.

The basic methods associated with crime scene analysis have been described in investigative texts for over 120 years. In 1900, Hans Gross spoke of the necessity of reconstructing the crime by meticulously examining and collecting facts. He warned the investigating officer that by heaping testimony upon testimony, they will "almost always be led astray and found wandering from the goal [the truth]."[6] Although Gross described his beliefs across his entire text and did not present them as a specific methodology, he discussed every aspect of our current beliefs. This included the necessity of making detailed crime scene observations, scientific examination of physical evidence to obtain as much objective data as possible, and applying this information to specific questions regarding the crime in order to "reconstruct the occurrence, built up by hard labor a theory fitted in and coordinated."[7]

Arthur Carey was another author who described this reconstruction process. Carey headed the NYPD Homicide Bureau for nearly 25 years, between 1903 and 1928. In his book *Memoirs of a Murder Man*, Carey commented on the value of CSR:

> If you know by examination of the evidence what happened, then you are far more effective in the interview and interrogation process.[8]

Carey's basic belief carries over into today's modern criminal investigations.

In 1933, Luke May wrote *Scientific Murder Investigation*. May outlined a series of questions that should be answered in the investigative process, but he described the investigation process by leading his readers to the questions. May used an analogy other than the jigsaw puzzle. He likened the investigation to building a structure, stating that without plan and direction, only the simplest structures were possible. He said that a murder investigation was like building a skyscraper; without planned effort and an underlying superstructure, successful completion was impossible. May clearly describes that investigators must "develop other facts, correlating and interlocking to make a whole from apparently disassociated separate units."[9] He also warned against allowing subjective personal theories to guide the investigation and the all-too-common habit of trying to force the pieces of the investigation into such theories. May stated the true mark of the scientific investigator was one who could "work untiringly, obtaining facts upon which to predicate theories, changing his theories as the facts developed warrant."[10] May did not spell out a specific crime scene analysis methodology, but his beliefs set a clear foundation for such a methodology.

Another pioneer in crime scene analysis was Edward Oscar Heinrich, known as the "Wizard of Berkeley." Heinrich offered one of the first true methodologies. He felt that one must first analyze the method of the crime before one could properly understand its purpose or hope to identify the criminal. He described reconstruction as "visualizing the habits and actions of the criminal. I do this by using the debris [evidence] that the criminal leaves behind and relocating it with respect to the criminal episode."[11] Heinrich likened his analysis to "a mosaic…every fact must be evaluated before it can be fit into the pattern. In that way, every fact as it is developed and equated becomes a clue." Heinrich's methodology was basic and included defining what happened, where it happened, and when (in what order) it happened. Heinrich felt the information developed from the crime scene analysis would ultimately lead to answering the more subjective questions of why it happened and who did it. Heinrich can also be credited as one of the first authors to effectively define the foundational theory of crime scene reconstruction. In his words, he described this as follows:

> Rarely are other than ordinary phenomena involved in the commission of a crime. One is confronted with scrambled effects, all of which are attributable to causes separately. The tracing of the relationship between isolated points of fact, the completion of the chain of circumstances between cause and effect, are the highest functions of reason.[12]

In 1933, Henry T.F. Rhodes made the case that crime scene analysis was a specific scientific process. In his text, *Clues and Crime*, he indicated that the objective of the crime scene evaluation was to decide specifically how the crime was committed and in what order the events occurred. Rhodes also clarified that the scientific method was the underlying foundation of any such analysis.

Charles O'Hara published *Fundamentals of Criminal Investigation* in 1956. For years, this text was a standard in many academic criminal justice programs. O'Hara, like Gross, placed a significant emphasis on the objective value of physical evidence and its scientific evaluation. He also spelled out a methodology for crime scene analysis using the scientific method, stating it involved:

- A painstaking and comprehensive collection of data
- Arrangement and correlation of that data
- Defining issues and investigative questions
- The development of a hypothesis along the lines of the available data and subsequent resolution of any hypothesis
- Testing of the hypothesis and elimination, when possible, of contradicting hypotheses
- Testing of the final hypothesis before acceptance[13]

O'Hara offered these ideas as a "representative approach" to the entire investigation, including subjective aspects such as testimonial evidence. Nevertheless, like others before him, he was clear that physical evidence and the crime scene shouldered a distinct responsibility in solving crime. In addition to his representative methodology, O'Hara made the case for conducting a separate crime scene reconstruction.[14] In his fourth chapter, O'Hara discussed reconstructing crimes, indicating that reconstruction was accomplished through physical and mental methods. He stated:

> After a search of the scene, an attempt should be made to determine what occurred from the appearance of the place and its objects, particularly the movement and methods of the criminal… The process of ascertaining the circumstances of a crime is known as reconstructing the crime… From a study of the evidence in this manner, it is often possible to make useful inferences that may be synthesized into a reasonable theory… In reconstructing the actions of the criminal, the investigator should test his theory for logic and consistency. A theory should not be rejected merely because the investigator might not, under the circumstances, behave in a similar manner… No assumptions should be made concerning actions which are not supported by evidence.[15]

James W. Osterburg wrote his text, *Criminal Investigation, A Method for Reconstructing the Past* in 1992. Osterburg made it clear that the investigative process must follow the scientific method. Although he did not define a specific reconstruction methodology, he discussed the importance of using physical evidence to reconstruct events.

The first detailed methodology for reconstructing crime was presented by Rynearson and Chisum in 1989, as part of their work *Evidence and Crime Scene Reconstruction*. Their ideas and concepts were a significant step forward. Of particular interest was their statement that the true value of evidence is a function of time and surroundings. This statement effectively means that context has greater value than mere content. For instance, finding a fingerprint in the scene may be important, but the context in which we find the fingerprint is more important. Context is critical in crime scene analysis, and Rynearson and Chisum understood this. Their described methodology used what they called a storyboard approach, a means of identifying specific events and their sequence or order.[16]

Dr. Henry Lee wrote his text, *Crime Scene Investigation*, in 1994. Lee included a specific chapter on reconstruction, commenting on the importance of using all forensic disciplines as input to any reconstruction. Lacking in Lee's work was a discussion of his methodology. He offered that the scientific method was the underlying reconstruction process, but failed to give a practical understanding of how one achieves the reconstruction with the scientific method.[17]

What we find through an examination of these authors' works are certain common themes regarding analysis. These themes can be stated as follows:

- Data defines the conclusion.
- Objective data is found at the crime scene in objects and scene context, whereas human testimony should always be viewed cautiously.
- Effective forensic evaluation of evidence leads to refined data, which in effect leads to refined conclusions.
- What happened is not the only question of crime scene reconstruction; the order in which they happened is a critical component of any crime scene reconstruction attempt.
- Crime scene analysis uses reductionism, a bottom-up approach, where the analysis is reverse-engineered from the physical evidence and its associated context.

Whatever methodology one chooses for crime scene analysis, these themes must be incorporated into it.

Chisum subsequently contributed a chapter to Brent Turvey's book, *Criminal Profiling*, entitled "An Introduction to Crime Scene Reconstruction," and in 2007 co-authored *Crime Reconstruction* with Turvey. Unfortunately, they did not take the opportunity to expound on previous beliefs, and in the latter reference, Chisum appears to have abandoned his previous methodology for a concept he calls Evidence Dynamics. This effectively is a laundry list of things to ask regarding how the evidence was handled after collection. Although they suggest this as a form of crime scene analysis, it is, in our opinion, severely lacking. Evidence dynamics is skewed far more to what cannot be done, than what can be done. Its focus appears to be on what the police fail to do, rather than what happened. In this work, Chisum offered a simple but somewhat misleading statement regarding CSR:

> When the reconstructionist identifies a discreet event, it is placed where it fits within the elements of the crime. This provides the foundation for the sequence of events and keeps them in order. In this fashion, the *timeline* [italics added] expands from a sequence of general elements to a sequence of discrete events.[18]

Crime scene analysis recognizes the distinct difference between time (absolute chronology) and sequence (relative chronology). In crime scene data, time associations are often few. A timeline is a specific investigative product, which is certainly associated with an event sequence, but they are different. Timelines are linear. One of the greatest failings in crime scene analysis is attempting to do exactly what Turvey and Chisum suggest. Identified actions (in Turvey and Chisum's words, "discrete events") can rarely, if ever, be time associated with every other action and put into a specific linear timeline. Even if one presumes that Turvey (a criminal profiler) was attempting to describe sequencing these actions, sequencing actions to every other action is also rare. As we will see, flowcharts describing the sequence of actions are often convoluted documents. They routinely have many branches. One branch may describe a sequence for a series of related actions, but show only a general sequential relationship to other branches. A direct relationship between every action on the various branches is rarely possible. Granted, Turvey's background in criminal profiling likely influenced some of his understanding of the practical application of crime scene analysis. The development of criminal profiling methods in the

1980s served as a catalyst for bringing crime scene reconstruction back into the forefront of investigative literature. Nevertheless, this association has created a significant misunderstanding of what crime scene analysis really entails.

The Correlation of Crime Scene Analysis to Behavioral Analysis

Criminal profiling provides a service in serial and violent crimes, which may assist the investigator in narrowing the search for a suspect. This evaluation, which considers the crime scene, the victim's actions and background, and the apparent actions of the suspect, provides information that may define the individual as a suspect. The information derived from the process can be valuable. It is also a subjective process, and a good profiler usually explains to the prospective client that these methods are not without fault and are anything but completely accurate. In one case known to the authors, the profile identified 12 characteristics of the probable perpetrator. When the suspect was caught, none of the 12 matched. The primary reason for this disparity was that the case involved a female serial killer who decapitated her victims, a circumstance that failed to fit the statistical data used in developing such profiles.

Our discussion of crime scene analysis concentrates on evaluating the scene and physical evidence, considering specific facts and any information we may safely define from those facts.

Criminal profiling is an important function in the criminal investigation, which is inseparably linked to crime scene analysis. The evolution of criminal profiling techniques has been significant over a relatively short period, but as Geberth noted, psychological profiles are nothing new to criminal investigations.[19] In early criminal investigations, most profiles were completed after the suspect was in custody. More often than not, these methods applied standard psychiatric approaches to establish the sanity of the individual involved. Heinrich, as reported by Chisum, was the first criminalist to go beyond scientific analysis of evidence in his efforts and develop an early form of profiling. Heinrich's methods included consideration of the scene as well as victimology.[20]

Another notable exception to the post-arrest psychiatric approach to profiling was the effort of Dr. James A. Brussel during the 1940s to 1950s. Dr. Brussel used his psychiatric training in an attempt to identify the personality characteristics of the individual known as the "Mad Bomber." This individual set off more than 30 bombs over 16 years. As reported by Geberth, Brussel's evaluation after reviewing investigative efforts spanning those 16 years said:

> "Look for a heavy man—middle age. Foreign born. Roman Catholic. Single. Lives with a brother or sister." He [Brussel] added, "When you find him, he'll be wearing a double-breasted suit. Buttoned." [The suspect] was exactly as described by Dr. Brussel. When taken into custody, he was even wearing a double-breasted suit.[21]

Brussel applied these techniques in other investigations, including the infamous Boston Strangler case.

Regarding its relationship to crime scene analysis and reconstruction, the most significant evolution of profiling came about with the work of Howard Teten. Influenced by previous efforts, but with a greater understanding and appreciation of the role of all forensic disciplines, Teten developed the initial FBI profiling techniques. Over the next 20 years, those techniques were tweaked and enhanced, creating a process that is now, as Turvey says, a "multi-disciplinary skill."[22]

Modern criminal profiling, often called behavioral analysis, generally consists of three stages. These are:

1. *Collecting Profiling Inputs:* These include the crime scene, victimology, forensic information, police reports, background information, and photos.
2. *Creating Decision Process Models:* Defining the homicide style, primary intent, victim risk, offender risk, location, and escalation.
3. *Conducting a Crime Assessment:* Reconstruction of the crime, crime classification, staging issues, motivation, and crime scene dynamics.[23]

As indicated, crime assessment involves reconstructing the crime. Ressler, in describing the assessment phase, stated:

> The crime assessment stage in generating a criminal profile involves reconstructing the *sequence of events* [italics added] and the behavior of both the offender and the victim. Based on the various decisions of the previous stage [the decision process model], this reconstruction of how things happened, how people behaved, and how they planned and organized the encounter provides information about specific characteristics to be generated for the criminal profile.[24]

A viable method of crime scene reconstruction is necessary to attempt any criminal profile. However, such a technique must be objective. This is true because the reconstruction helps ensure objectivity in performing what can only be described as a subjective process. As Turvey himself reported, "Forensic analysis is of paramount importance because it helps to preserve the criminal profiler's objectivity by protecting them from investigative assumptions."[25] An integral and unmistakable part of the third step of criminal profiling, crime assessment, is reconstructing the crime in an objective manner. How one achieves objectivity is through the application of the scientific method.

The Application of the Scientific Method in the Reconstruction Process

As previously described, the scientific method is how humans resolve complex and abstract problems (see Figure 14.1). This is accomplished by asking and answering specific questions and following a defined, but hopefully simple methodology. As was the case for Bloodstain Pattern Analysis, crime scene reconstruction must follow the scientific method if the analyst hopes to achieve defensible conclusions.

Remember that any data used in crime scene analysis is the physical evidence and its context in the scene. These data elements are objective. If considered properly and within any limits set by associated disciplines, the physical evidence will lead the analyst to valid defined statements (actions) about what did or did not occur within the scene. Subjective data is ignored in crime scene analysis; interjecting it weakens the conclusions. The analyst does not simply dismiss subjective data, but recognizes that the crime scene analysis must be developed independently and later considered along with any subjective data. Therefore,

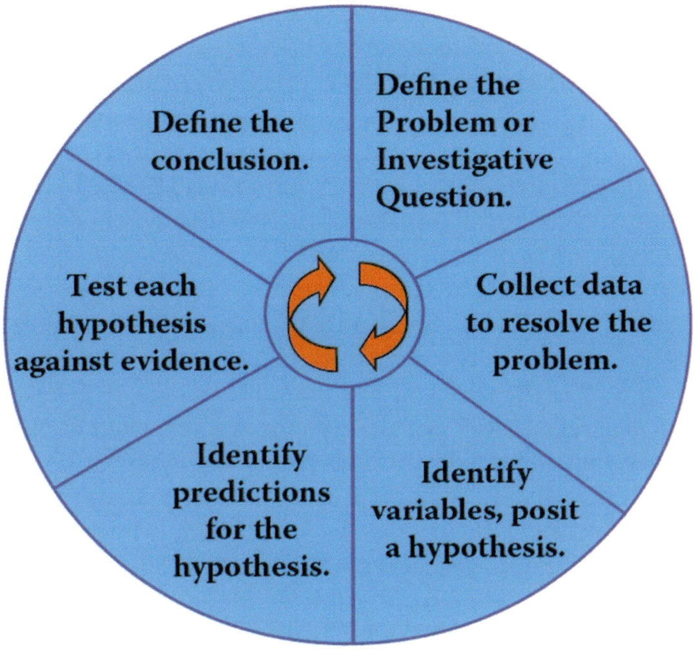

Figure 14.1 The scientific method follows a six-step process that begins with a question and ultimately ends with some answer. However, in a search for true understanding, each new answer usually begs another question. The result of the scientific method is an ever-expanding, self-correcting body of knowledge.

testimonial evidence is rarely included as data in a crime scene reconstruction. To apply the tenets of crime scene analysis, it is important to understand its underlying theory, the supporting principles, and the questions posed about any evidence item. As there are several defined methodologies, the specific methodology used by any given analyst may differ from the next analyst. This chapter will discuss the methodology known as Event Analysis.

Theory and Principles of Crime Scene Analysis

Today, one can absolutely count on lawyers attacking even the most basic scientific concept presented before the court. In such an environment, the crime scene analyst must come to court prepared to do battle. This demands an understanding of the underlying theory of crime scene analysis and the supporting principles the analyst uses. Analysts can expect to be taken to task on the "scientific" aspects of their analysis and must be capable of articulating these ideas to the court. In 1997, we introduced a simple concept in the first edition of this book, which said: *Nothing just happens.* This is the foundational theory in its simplest form. The more articulated description of this theory is:

> Given a series of Actions associated with a specific Incident, any given Action has a unique causal and temporal relationship with every other associated Action.[26]

This idea has always been and remains the basis of all crime scene analysis. Every physical system has components that interact and produce effects. Causes lead to effects, and that defines order. If we know the system's condition at any one point, then we can predict the past and future states of the system to some level. Thus, every action and event that occurs has a set of circumstances that preceded it, and every action and event will produce some subsequent condition. Call it what you will, but if we revisit our archeological correlation, it is nothing less than history. This theory is inviolable; no counsel can ever prove it wrong, no matter how eloquent. No instance in human oral or recorded history has anything that "just happened." Some preceding circumstances existed before any given moment in time, and some set of circumstances or results existed after it. There is no other possibility. This is a basic restatement of the nature of science, causality, associating cause and effect.

This theory that nothing just happens guides and drives everything the crime scene analyst does. However, as true as the theory is, the crime scene analyst is still limited in their ability to recognize these pre- and post-conditions. The archeologist's dilemma is very much in effect, although it does not alter the theory. For instance, 30 years ago, investigators may have been unable to recover dust prints or DNA, but dust prints and DNA were present in the scenes. The theory held true, we simply did not know what to look for or how to capture it. Therefore, the limiting factor in crime scene analysis is recognizing and capturing from the scene and artifacts available, observable cause-and-effect relationships. The crime scene analyst will never be able to define and recognize all of them, but those that are evident tell us much about what happened and in what order it happened.

Recognizing observable cause-and-effect relationships is significant in applying crime scene analysis to specific case conclusions. In science, a conclusion is only viable if it can be tested. Let's consider case-specific conclusions about what happened. If there are no observable or testable predictions that will support or refute the conclusion, it is nothing more than an opinion without foundation (whether one wishes to support or refute the specific conclusion).

As an example, consider a situation in which there is a perforating gunshot injury to a victim found in a kneeling position. A significant blood pool developed after the victim bled out, marring all underlying surfaces on the floor. The terminal ballistic information tells us something about the path of the bullet through the victim, but there is no evidence of a subsequent strike. Therefore, we do not know what path the bullet took after exiting the head. Additionally, no spatter is present in the scene on any surface. The analyst might believe that the head was downturned and that the resulting forward spatter was projected onto the floor and is simply beneath the blood pool. As possible as this scenario may be, it is not a valid conclusion. No observable or testable prediction can be used to refute or corroborate the claim. It is based on an independent assumption that there are spatter beneath the blood pool. In crime scene reconstruction, there must be some observable or testable foundation on which any conclusion is based. Nothing just happens, but we cannot assume the specifics of what happened without a solid foundation (e.g., founded in physical evidence).

Supporting this basic theory of "nothing just happens" are various principles that help us recognize relationships between our data. The obvious one is causality; causes lead to effects. Another is Locard's Principle of Exchange, an underlying principle of forensics for nearly 80 years. Two are taken from archeology; they include Nicolas Steno's Principles of Superposition and Lateral Continuity. A fifth principle is a shared concept in crime scene analysis and archeology, the concept of chronology. The last is spatial and spatial-temporal constraint. All of these principles, in effect, guide the analyst as they consider the scene and context presented.

Causality

At the heart of CSR is the ability to define causal actions and, thus, causal chains. Causes must precede effects, which brings order (relative chronology) to the actions. What is the specific cause for an observed effect? In many instances, understanding cause and effect is simple enough. Examples: A bullet is fired and a wound is created. A wound is created, and blood is deposited. There will always be obvious causal connections present in the scene. However, the analyst is likely to encounter more ambiguous causal connections, and these can't simply be ignored. But contextual bias, even internal bias on the part of the analyst, can always creep into an analysis, particularly when making these more ambiguous causal connections. Confronted with these ambiguous causal connections requires some manner in which to evaluate them more effectively. In this regard, the analyst can utilize a version of what is referred to as Hill's Criteria. Sir Austin Bradford Hill presented his discussion of his criteria for evaluating causal connection in 1965, in the context of epidemiology.[27] They included considering the following issues:

- Strength
- Consistency
- Specificity
- Plausibility
- Coherence
- Temporality
- Experiment
- Analogy

While not all of Hill's described criteria apply to crime scene analysis, from them, we have developed a series of questions that offer a more robust manner to evaluate any causal connections:[28] These questions include:

- How specific is the association?
- How predictable is the association?
- Does the association comport with known scientific principles?
- Does the association comport with experimental effort?
- Does the association violate any aspect of the required causal chain?
- Is the association plausible?

In deference to Hill, we would offer that the ultimate "strength" of any causal connection is defined by considering his subsequent issues. These six questions allow the analyst to better evaluate their own causal connections, particularly when presented with somewhat ambiguous ones. These questions are also useful during any subsequent peer review or when evaluating an opposing analyst's conclusions.

Locard's Principle of Exchange

Edmund Locard set forth his Principle of Exchange in the early 1900s. In describing his beliefs, he said:

> The principle is this one. Any action of an individual, and obviously, the violent action constituting a crime, cannot occur without leaving a mark. What is admirable is the variety of these marks. Sometimes, there will be prints, sometimes simple traces, and sometimes stains.[29]

In effect, this is an extension of the concept of causality. One can only wonder if Locard himself understood just how correct he was, for as forensic technology advanced, so too did the application of his principle. Today, it is succinctly stated: *Every contact leaves its trace.* Every interaction in the crime scene leaves some form of evidence. Fingerprints, DNA, heat signatures, fibers, hairs, soils, and assorted other traces of materials – the list of evidence that may be left behind is nearly endless. Locard's Principle is the basis of all trace evidence examinations; it permeates every area of forensics and serves as the foundation of the forensic linkage triangle (see Figure 14.2). In its common form, Locard's Principle is associated with the world of trace material. Still, if one considers it from a macro-perspective, it restates our theory that nothing happens.

Nicolas Steno's Law of Superposition

Nicolas Steno was a preeminent force in founding modern geology and its subdiscipline, archeology. He defined basic principles for understanding geologic strata, which have since been carried over to evaluating archeological strata, and associated artifacts. One of these principles was the Law of Superposition. Steno's Law of Superposition states that geologic strata (and associated artifacts) are deposited in a time order, oldest to youngest, unless otherwise disturbed. Thus, if three layers of strata and artifacts are uncovered, the one at the bottom is the oldest, and the one in the middle is younger than the bottom layer, but older than the layer above it. Each layer of strata is followed in age by each subsequent overlying stratum. Artifacts in crime scenes are often found in layers. The nature of these layers is not quite geologic strata, but they are layers nonetheless. These layers help us put order to events and actions (see Figure 14.3). Of course, as Steno pointed out, our understanding of the sequence of deposits for these layers can always be affected by some subsequent alteration.

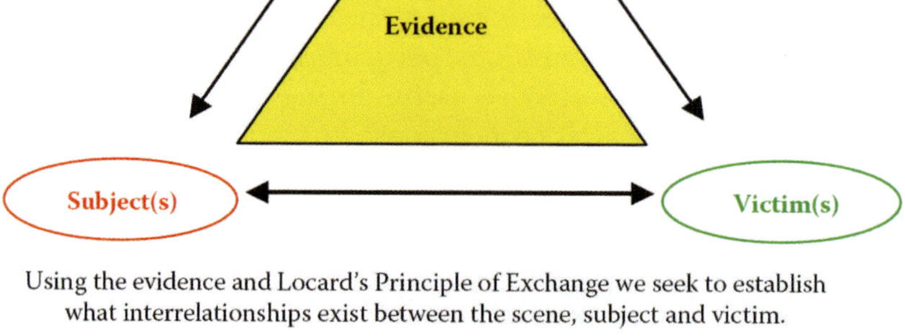

Figure 14.2 The forensic linkage triangle is a basic concept based on Locard's Principle of Exchange.

Figure 14.3 A crime scene example of applying Steno's Principle of Superposition. Considering the victim's hair, the plastic bottle, cell phone, and purse, it is apparent that the bottle is beneath the hair and the hair is beneath the cell phone. Unless altered in some fashion, this establishes a sequential order for their deposit. Depending upon what we find when we lift the purse, it too may be sequenced with these objects.

Nicolas Steno's Law of Lateral Continuity

Another of Steno's principles is Lateral Continuity. It states that disassociated but similar strata can be considered from the same depositional period. Steno believed geologic strata were continuous for extended areas and did not end abruptly. As applied to excavations, this means that if a layer in a dig contains a similar type of soil and similar artifacts to another layer located nearby, but one not in direct association, so long as the similarities are sufficient (nature of the soil or rock, nature of the artifacts), the two layers are considered to be from the same depositional time period. Their disassociation was simply the result of some subsequent geologic alteration. Applying the Principle of Lateral Continuity in simpler terms is a "contextual association." It applies to the crime scene, but in a less direct manner. Disassociated strata are not as common in the crime scene as layering of artifacts, but they do occur in a crime scene context. For example, recognizing a void in a bloodstain pattern is an application of the lateral continuity mindset. When we encounter a void, we recognize the pattern of blood did not simply stop. Something interrupted it, or some portion of the pattern was altered (e.g., removed). Thus, finding an article consistent with the void shape and containing similar bloodstains, we make an association. Another type of disassociated strata/artifact often encountered is spatter (see Figure 14.4). Consider when small spatter are found on both the weapon and surfaces in the crime scene. Suppose a similar spatter is located on a suspect's clothing. In that case, if the similarities of the spatter are sufficient (e.g., DNA, size, and dispersion characteristics) and there is evidence of only one spatter event (e.g., a single gunshot), they represent disassociated strata. However, using this idea of lateral continuity, they can be considered a result of the same deposit.

CSR also expands Steno's idea of continuity to a broader mindset. Continuity, in general, can provide significant data to the analysis. Either a layer is disturbed, or it is not. Consider a claim that arrival and departure from a scene were achieved through an

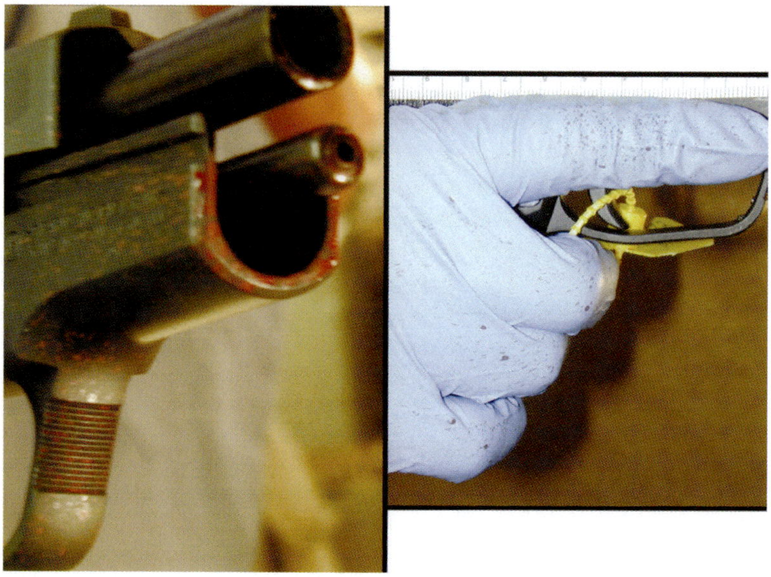

Figure 14.4 A crime scene example suggested by Steno's Principle of Lateral Continuity (contextual association). The event that deposited spatter on the weapon also simultaneously deposited spatter on the gloved hand. If we find these deposits disassociated from one another and sufficient similarities exist (e.g., DNA, size, dispersion), we may consider these artifacts the result of a single deposit.

open window. The window screen, however, is in place, which is explained by an involved party as the suspect simply replaced it to its original condition. Looking at the window, we observe that cobwebs are in place between the screen and the frame. The continuity of this layer is intact and functionally refutes the claim that the window was the entry/exit point. Expanding that belief further, this party claims that the suspect came in through this window and departed across a section of dirt as a means of escape. The window leads directly to uncompacted dirt, which is also muddy. Upon examination, there are no disturbances in the dirt or mud. Once again, the continuity of the layer is intact, further refuting the claim. The concept of continuity may arise in many ways, and the analyst should always be on the lookout for such information. Such data can refute or corroborate claims and investigative theories.

Chronology

Another principle of concern is chronology, the idea of temporal order. This is a bedrock concept in archeology, so much so that it has been stated that any discussion of archeological events without consideration of time and sequence is meaningless. In archeology, there are two forms of chronology – absolute and relative. Absolute chronology relates to time, the ability to date an artifact or event through some mechanism or relationship (e.g., a coin stamped with a date). Relative chronology, conversely, is the ability to define a relative order to a series of artifacts or events (e.g., the layering of artifacts evident through superposition).

In the parlance of criminal investigations, these concepts of chronology are better known as time and sequence. Absolute chronology or timing is established through various mechanisms. Video surveillance, 911 calls, patrol car/body cameras, and similar devices offer the ability to put a specific time to an event. More classic examples exist in the form of time-of-death determinations by the medical examiner or the interruption of daily activities for murder victims. One can never truly predict where timing data will come from. It is not as common as sequencing information, but timing information is becoming more common in the digital age.

Relative chronology or sequencing is the more common evidence encountered in crime scene analysis. It is not concerned with establishing a specific time, but rather with establishing a relative order of events (e.g., Action A preceded Action D). In conjunction with any known timing data, this information will provide an overall time and sequence flow.

Sequence is established by looking for simple relationships between actions and events. In the field of archeology, the evaluation of chronology involves establishing two relationships. Something is *terminus post quem* (a point in time after which a feature was deposited). See Figure 14.5. Or it is *terminus ante quem* (a point in time before a feature could have been deposited). See Figure 14.6. With slight modification, the crime scene analyst applies these relationships to their evaluation of relative chronology. Given the differences between archeology and crime scene analysis, adding a third relationship, *terminus peri quem*, is the only modification required. In crime scene analysis, we deal with a far more recent past. Thus, we often encounter actions that are near-simultaneous to one another. *Terminus peri quem* describes the relationship between near-simultaneous events (see Figure 14.7).

With these three relationships in mind, the crime scene analyst evaluates every action or event, seeking to identify any relationship one may hold with another. For example, for a bloodstain transfer to be produced, there must be an exposed blood source that can be transferred. Thus, in the case of a single wounding, all subsequent pattern transfers in the

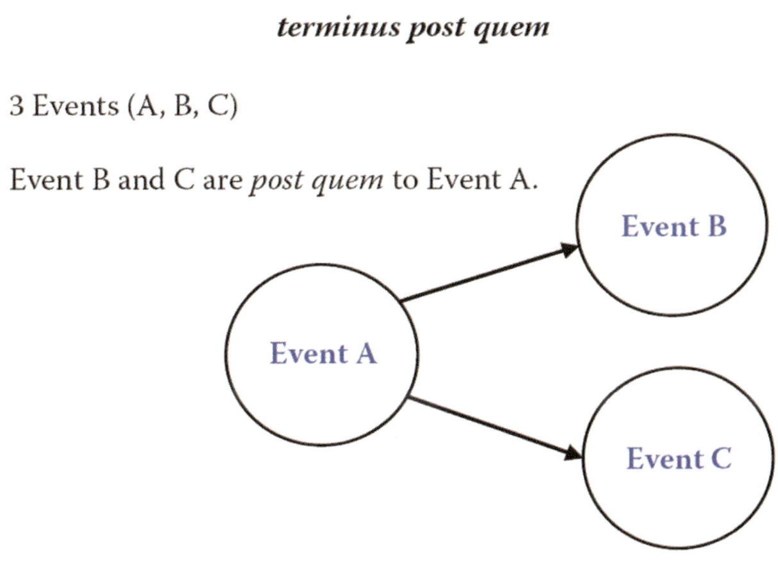

Figure 14.5 The crime scene analyst uses three simple relationships when applying chronology: *terminus post quem*, *terminus ante quem*, and *terminus peri quem*. *Terminus post quem* means that the event or action follows another action.

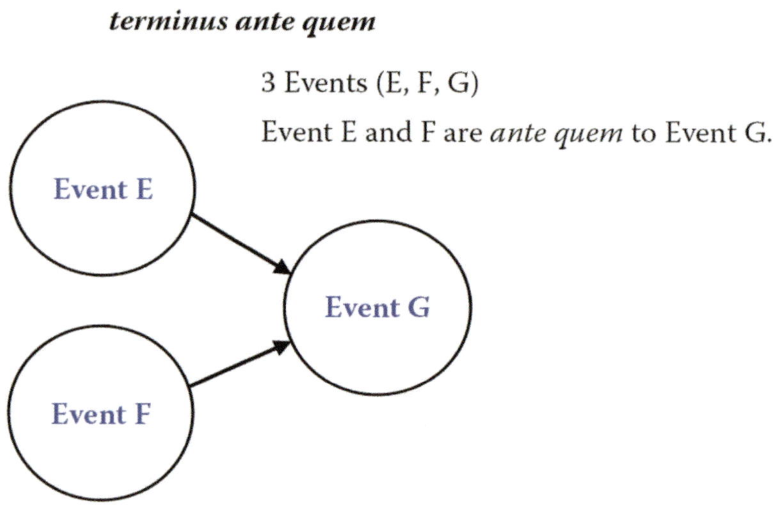

Figure 14.6 *Terminus ante quem* means that the event or action precedes some other action.

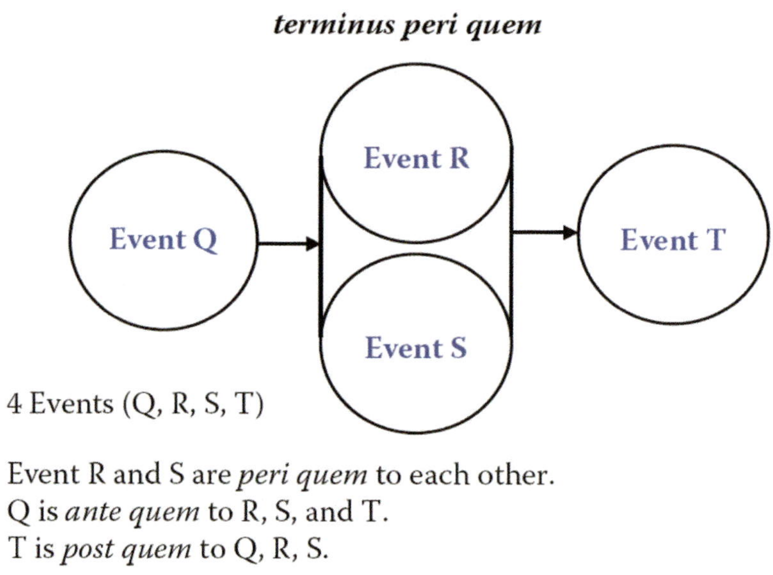

Figure 14.7 *Terminus peri quem* means that in the analyst's eyes, the two events or actions are near-simultaneous.

scene are *terminus post quem* to the wounding. Not every action can or will be associated with every other action, so the ultimate resulting sequence will not be linear. However, the temporal order of portions of the analysis will become evident through the development of simple causal chains of associated actions. The ultimate product will often be convoluted when considered together, with many branches that describe specific events. There will likely be both diverging and converging sections of actions and events (macro-components of the incident). However, some overall sequence hierarchy should be possible using all of the evident relationships. See Figure 14.8.

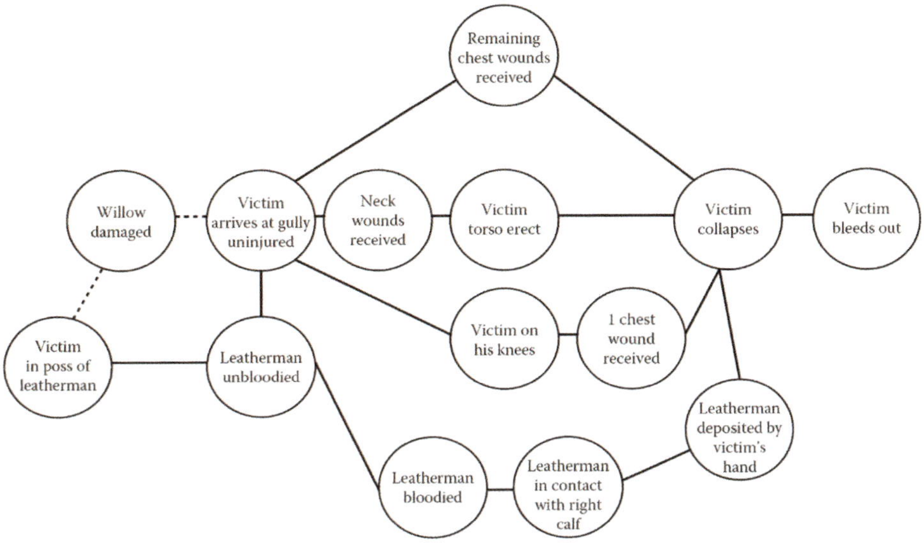

Figure 14.8 A case example of a flowchart. The chart has two specific paths that naturally developed. The upper two-thirds of the chart follows the victim's actions, while the lower third tracks the weapon involved. It is rare that each action identified can be directly associated with every other action, producing a "timeline" for the incident. Nevertheless, the connecting lines help describe how the various actions interrelate sequentially.

Spatial and Spatial-Temporal Constraint

Another principle of concern is that of spatial or spatial-temporal constraint. The latter is perhaps more recognizable to criminal investigators. It involves asking whether the claimed action(s) occurred within the time allotted. It is a common practice when presented with claims that "I was at work, I couldn't possibly have driven from there to the scene in the time you are describing." In such instances, investigators will test the drive distance against the time available. They attempt to match the conditions expected for the time of the drive (e.g., road conditions, traffic) and conduct ad hoc experiments to determine if the claim is corroborated or refuted.

The idea of spatial constraint asks: Will it fit? Can a particular action be accomplished in the area claimed, or can an individual observe something from the position they claim?

Spatial Constraint Example

In a homicide, a woman claims to have shot her estranged husband as he rushed towards her, after entering her home. In the process, she shot him a total of five times. She claims he came directly towards her and never turned away. The shots included one each to the shoulders, one each to the elbows/forearms, and a fatal shot to the head. A series of drips associated with the husband appeared somewhat circular in the scene. This drip "circle" was 8″ from top to bottom and 10″ from left to right.

At trial, the M.E., who fancied himself a bloodstain analyst, told the jury that this "circle" of drips was created as the husband turned in a circle after receiving one of the wounds. The husband was 6′3″, weighed in over 210 pounds, had a size 44 chest, and wore a size 12 shoe. This begs the question, if the blood was deposited before the fatal wound (thus, it came from either an elbow or shoulder wound), how did a man with a 44″ chest and a size 12 shoe turn around in an 8″ × 10″ circle? The "circle" may represent a moment of hesitation. Still, he could not have "turned around" in this limited area.

A spatial constraint decision is often a qualitative conclusion. Whatever might be claimed, we can see clearly it will not fit. Other times, the conclusion may be quantitative using specific measurements. Once again, both spatial and spatial-temporal constraint may allow the analyst to refute or corroborate some claimed action.

These principles are in the analyst's mindset as they look at the scene, artifacts, and associated context. Any one of these principles are likely to come into play in any given scene.

A Methodology for Crime Scene Analysis: Event Analysis

As much as it is true that all science follows the scientific method, the scientific method is a broad-brush concept. Each discipline must have a protocol or methodology to implement the scientific method. Simply stating, "I followed the scientific method," fails to address specifically "how" the analysis was performed. Event Analysis is a specific methodology that allows the analyst to apply scientific methods to crime scene analysis. It incorporates and applies all of the themes discovered in prior authors' work, is based on the theory, applies the associated principles, and provides a practical, proven approach to reconstructing crime scenes.

In Event Analysis, we will use three terms to define a specific context for the reconstruction. These terms are *incident, events,* and *event segments/actions.* We utilized the term event segment when the authors first published the Event Analysis protocol. We now prefer the term action. The overall crime or incident in question is the incident. The incident encompasses all actions related to the occurrence. Using a book analogy, it is the overall story of what happened. There are specific events within the incident. Each event describes a major aspect of the occurrence, a macro-component analogous to a chapter within the book (see Figure 14.9). An event is broken down into segments and the specific actions taken. Using the book analogy, the actions are like the paragraphs and sentences

Figure 14.9 Using an analogy of a book, consider the incident as the entire story of the crime being considered. It is made up of many chapters and events. Each chapter describes a macro-component of the incident.

that make up the chapter (see Figures 14.10 and 14.11). Actions are time snapshots of the incident, moments defined through some physical form. We know that the action occurred based on a set of facts and data, some physical manifestation such as artifacts, or a condition from the crime scene.

A crime scene analysis is reverse-engineered (see Figure 14.12). We use evidence and data to define actions. The more actions we identify, the more we understand the nature of the event. The more fully we can define the events, the more we understand the overall incident. Since this reconstruction method attempts to define the crime or incident by establishing these events and actions, we call the process Event Analysis. The steps taken to create a crime scene reconstruction using Event Analysis are as follows:

- Collect data.
- Establish specific actions (time snapshots).
- Establish which actions are associated.
- Sequence these associated actions, establishing a flow for that event.
- Consider all possible sequences, auditing the evidence when necessary to resolve contradictions.
- Using the same concepts to sequence actions, look for crossover actions in the events and sequence them.
- Flowchart the entire incident and validate the sequence.
- Conduct a peer review.

As an example, consider an assault and rape in a home. The incident likely consists of approaching the house, gaining entry, making contact with the victim, assaulting and raping the victim, and subsequent departure. These events are simply large chunks of our time

The actions are like paragraphs in the chapter.

Figure 14.10 Carrying the analogy further, each chapter comprises paragraphs and sentences containing details that help us understand the story. The actions are like the paragraphs, they describe a specific activity.

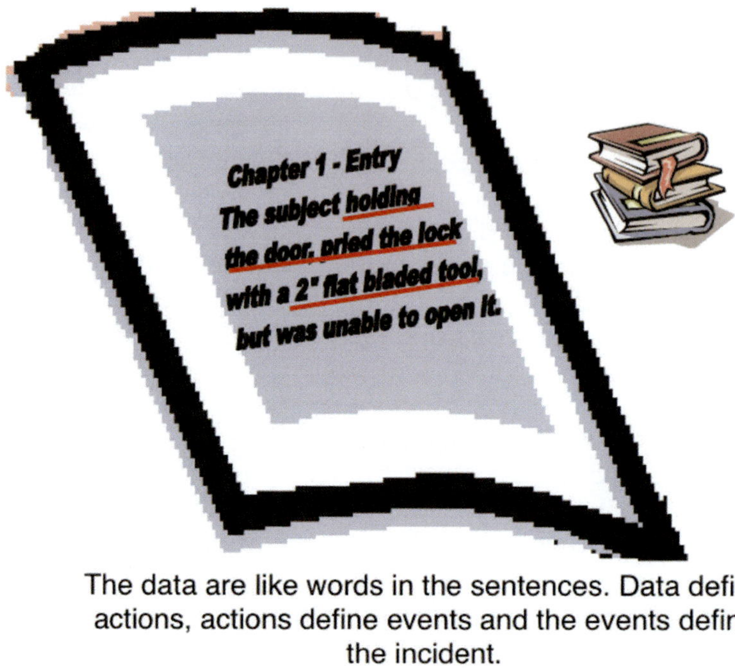

Figure 14.11 Finally, when we look at the detail in the paragraphs, we find specifics (the data) that establish that the action occurred. The more detailed the paragraphs, the more we know about the chapter. The more we know about the chapter, the more we know about the story. In the same way, crime scene data define actions. The actions define the events, and the events tell us the story of the incident.

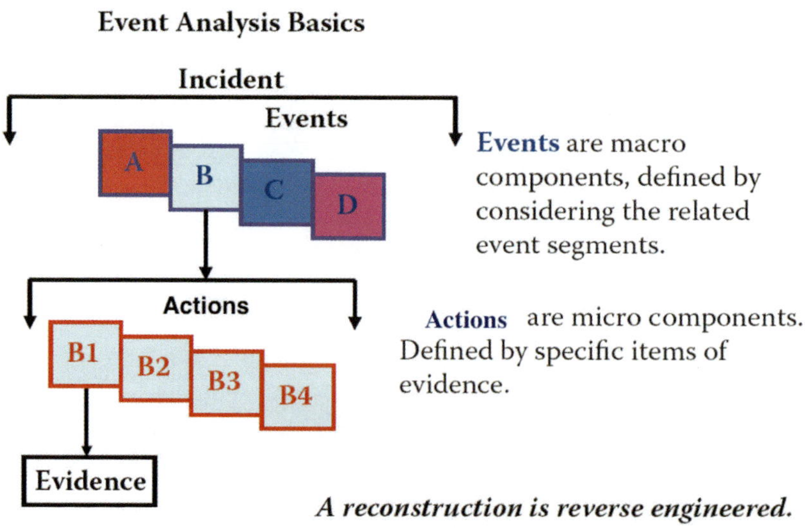

Figure 14.12 A crime scene analysis is reverse-engineered. The evidence leads us to actions. The actions tell us about the events and detail the incident.

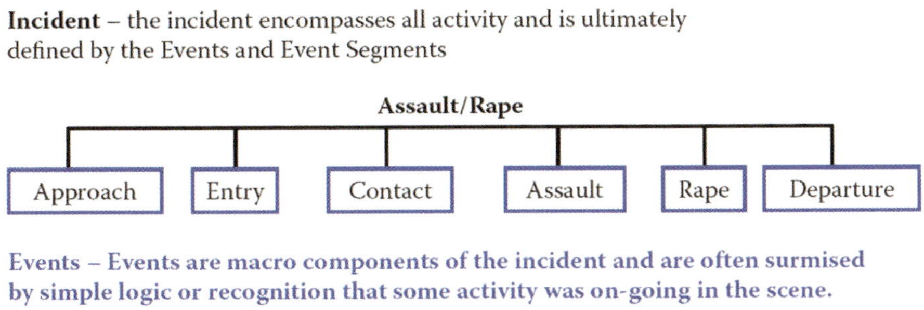

Figure 14.13 Establishing events requires considering what must occur for the incident to transpire. For instance, besides approaching and departing the scene, our subject must gain access to the victim and take control of her to commit the rape. Oftentimes, the analyst will recognize the event or at least that something was ongoing before they understand the details of what was happening.

window, chapters in the book. In the initial scene evaluation, we can often surmise specific events or at least recognize that "something" was happening in an area in the scene. The incident encompasses all these processes or events (see Figure 14.13).

Any event, such as "entry," comprises individual actions (more precisely defined time slots). For example, the rapist may check the door or window in question to see if it is unsecured; then, using a crowbar to break the glass, reach through and unlock the entry and climb through the entry point. Each of these actions results in the creation of items of evidence. Our subsequent discovery and analysis of the evidence at the scene will help us establish that the action occurred. Examples of this evidence might be fingerprints on the outer surfaces of the entry point, glass fractures in the pane of glass, glass fragments on the subject's clothes and shoes, and dust prints or disturbances in the broken glass where the subject stepped. Remember that an action is a statement of some object undergoing change. Evidence is not an "action." A combination of evidence conditions will allow us to state an action occurred.

As the scientific method requires, the analyst seeks to answer smaller questions to resolve some larger ones. This begs the question: What are the questions posed by the CSR analyst? In seeking to identify actions, the analyst asks and answers four recurring questions as they look at the crime scene and the items present in it. The answers to these questions will provide the data used in the analysis. These questions include:

1. What is it?
2. What function did it serve?
3. What interrelationships exist between this item and other items in the scene?
4. What does the item tell us about timing and sequencing?

The first question considers the basic nature of the object. What is it – a gun, a knife, or a fingerprint? This is not a difficult question to answer when presented with obvious objects. Some items, however, may demand significant effort. Examples include fragments (e.g., paint or glass fragments), trace evidence (e.g., a white powder, ignitable liquids), or simply items unfamiliar to the investigative team.

The second question is concerned with how the item was used. An item may be present in the scene, which was not used as intended by its design. For example, a bookend may be used as a blunt force weapon, or stockings may be used as bindings. Not all functional aspects are obvious on first inspection, but the analyst should remain constantly aware of such issues.

The third question looks at interrelationships between articles in the scene. The answer to this question may assist in resolving our understanding of the first two questions. For example, a small metal fragment may be found to be consistent with a deformed bullet, or a liquid at the scene may be identified as fuel consistent with fuel found in a gas can at the scene. Other relationships are also important as well. For example, recognizing a cast-off pattern rising from an impact spatter pattern may aid the analyst in understanding the positioning of parties during the attack, or consideration of a trajectory along with the medical examiner's analysis of the wound may aid in placing the victim in a relatively specific position at the time of wounding. For the analyst, there is no limit to the nature of the interrelationships between objects that may be of interest. The analyst must constantly look for any interrelationships that may be present.

The last question deals with the issue of chronology. What does this item tell the analyst about sequencing different actions or placing an action at a specific point in time? Examples might include the presence of spatter overlaying a shoe mark or the deposition of ceiling debris in a fire scene on top of a victim with no underlying burns. Sequencing information is scattered throughout the crime scene and is critical information to a reconstruction.

By asking and answering these questions, actions are defined in no particular order (see Figure 14.14). Each action is considered against the others, looking for association and relationship. Based upon these relationships, the various actions we believe are associated will begin to provide an understanding of that event (see Figure 15.15). Our function now, is to order these related actions. This ordering gives structure to how the event transpired. Again, all known information must be considered to place these actions in a logical and

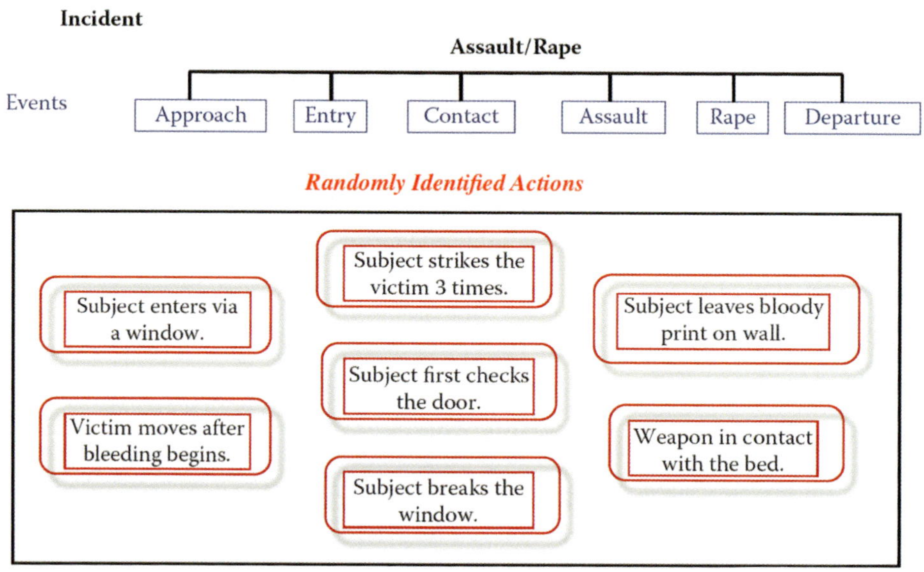

Figure 14.14 Event segments or actions are "snapshots" of the crime. They detail specific activity and are based on our evaluation of specific items of evidence. Initially, we may only know that they occurred, without understanding their sequence or relationship to the events themselves.

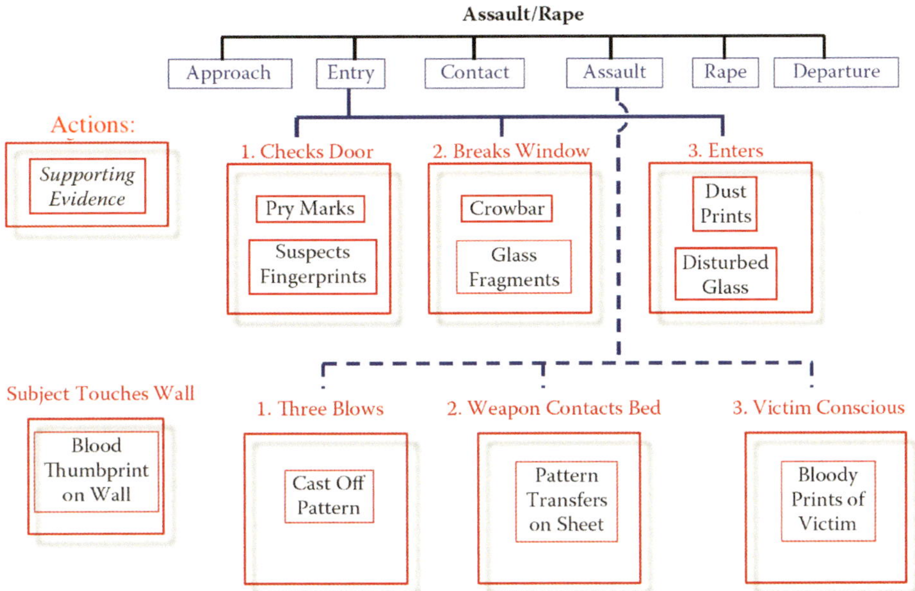

Figure 14.15 In time, we seek to relate the actions to a specific event and then sequence them. This gives us a clearer picture of how the incident transpired. Not every action will have a clear position in the reconstruction. For example, the action related to the subject touching the wall might well belong to the assault, rape, or departure event; unfortunately, there is insufficient evidence to establish where it belongs in the overall reconstruction.

supportable order. Using the three relationships of chronology, each action is considered against every other associated action. Typically, obvious beginning and ending actions are found. Other actions may fit together in a very logical and functional order. There may also be actions that appear to fit the sequence loosely, but perhaps we are uncomfortable with their exact sequence. We are also likely to encounter actions that we believe are related to the event, but for which we do not understand the sequence or order in any fashion. Consider the "Subject Touches Wall" segment in Figure 14.15. We know it happened, but we do not understand when it happened. As simple as sequencing may sound, it is a difficult step (see Figure 14.16). Nevertheless, using the three chronology relationships, we allow an objective order for actions to develop (see Figure 14.17) and then graph the information into a flowchart as in Figure 14.18.

The event flowchart seeks to visualize the reconstruction graphically. As each relationship is defined, it is graphically depicted. Flowcharting begins in the fourth step of EA and continues throughout the analysis effort. This graphic representation is important. Often, the analyst will not recognize a logical flaw in the reconstruction sequence until it is depicted graphically, particularly in complex events. Once drawn out, these flaws are much more evident and easier to recognize. As one might imagine, analysts will go through numerous iterations of the flowchart as they work through sequencing issues.

Remember that every event or action has something that leads to or follows it. Nothing just happens. There is no "Beam me up, Scotty" as in Star Trek; our players must arrive, act, and depart. Consider, for example, the assault segment of "break glass" with a crowbar. Refer back to Figure 14.15. To break the glass with a crowbar, our rapist must first procure one. To procure one, he must either find it on the scene or bring it with him. If found on the

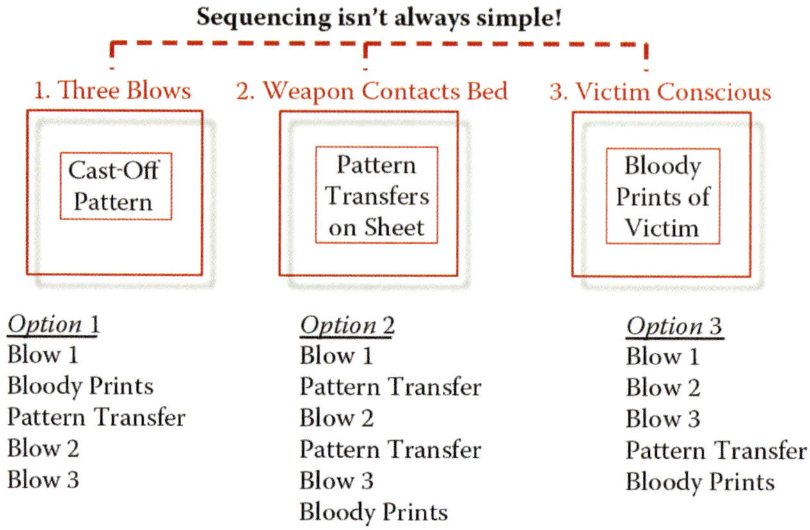

Figure 14.16 Sequencing, although easy enough to understand, is not always simple. We may be presented with a series of actions with little sequencing information available, or a situation in which several permutations are possible.

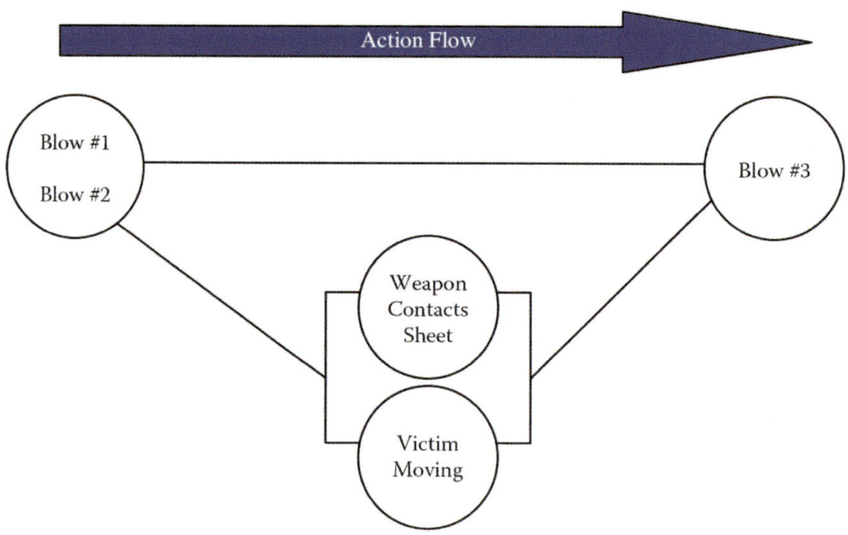

Figure 14.17 Sequencing may not always be simple, but it will ultimately lead the analyst to some conclusion. To get there, the analyst must apply the three basic chronology relationships: something preceded, something followed, or something was simultaneous to some other action. A structure will begin to form, as they tie one event or action to another.

scene, it must have existed there previously or been brought by someone else. If brought by the rapist, he must have brought it from somewhere. Everything has a history. Obviously, these issues become convoluted and lose relevance to the investigation. Still, if for some reason the issue were important, background evidence might well exist to prove its occurrence. This leads us to auditing.

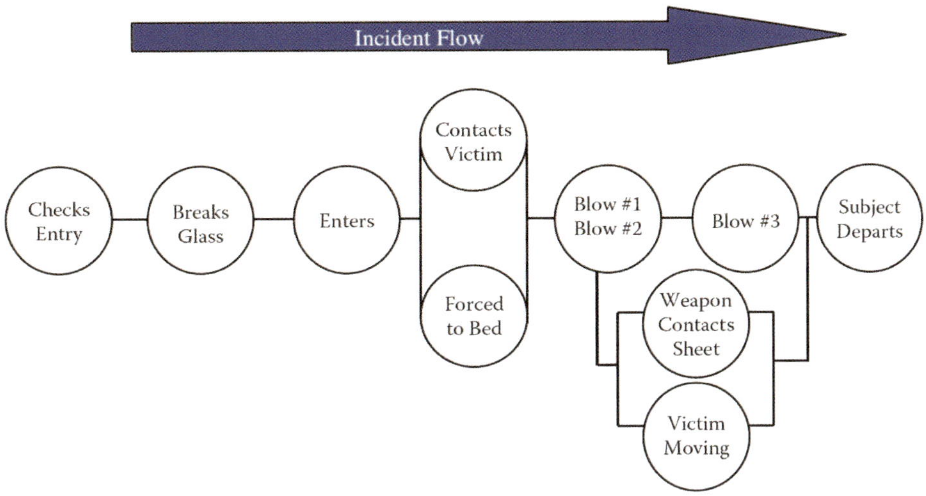

Figure 14.18 Event flowcharting. Using the structure that developed between individual actions, all of the actions are tied together in a flowchart. The application of the chronology relationships provides an objective and supportable sequence. In effect, the flowchart is a graphic representation of the reconstruction.

The investigation can present the analyst with contradictory evidence supporting two or more possible events or action sequences. By looking at the background evidence, this audit can help establish which sequences are more viable.

Consider an example of auditing. Imagine a situation where there are two possible locations in a scene for a shooting. Based on examining the evidence present, either location could conceivably be the location. Testimony by the defendant indicates the shooting occurred at the first location (a small bathroom). He claims he placed the weapon on the floor there and never again moved it. This information supports his overall claims. The barrel of the weapon was bloodied by contact with the victim when it was fired. We also find a bloody pattern transfer consistent with the weapon on a pillow at the second location of interest (see Figure 14.19). In considering the pillow and pattern transfer in relation to the defendant's claim, the following is possible:

- Previously, the pillow was at the location claimed by the defendant. It contacted the weapon and was later moved by someone.
- The shooting occurred as claimed, the weapon was moved and encountered the pillow, and was then returned to its final position.
- The shooting occurred at the second site, where the weapon contacted the pillow. The weapon was then returned to the first location, which was staged to look like the shooting scene.

The analyst's issue becomes one of finding some means of deciding which of these possible explanations is correct. In examining the pattern transfer, we are confident the weapon caused it. The pillow exhibits no other stains. We also find that at the first location, where the defendant claims the shooting occurred, watered-down stains are present from another event. These watered-down stains appear to be related to the splashing of bloody water in the tub. The subject is adamant that these occurred just before the shooting and relate to

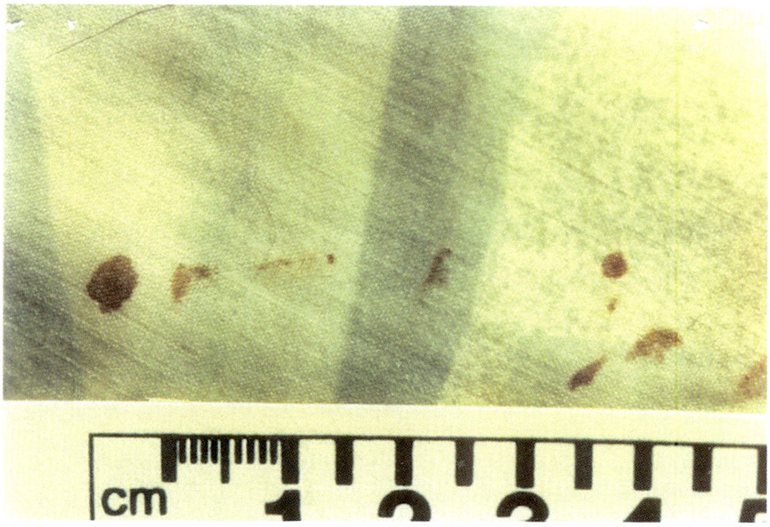

Figure 14.19 This pattern transfer on a pillow in a room disassociated from the homicide presents an auditing situation. Several possible alternative actions and scenarios may explain it. The analyst is left to decide if any of these possibilities can be eliminated to bring more clarity to the situation.

another minor injury. These stains are evident on all surfaces and objects present near the first location.

Given the presence of these splash stains all around the bathtub and the lack of similar stains on the pillow, it is obvious the pillow was not subjected to this event. If the subject is being truthful and the first location is the site of the shooting, then the splash event occurred, causing the stains on the walls and surfaces. The pillow was then moved to this location, the shooting occurred, the weapon encountered it, and the pillow was removed and replaced at the second location. Unfortunately, he neither claims this, nor in his overall description, had sufficient time to take such action. This excludes the pillow's presence at the first location before the shooting. The remaining explanation is: The weapon was moved from the first location, encountered the pillow at the second location, and was then returned, or the second location is the actual shooting site.

Although auditing does not answer absolutely which explanation is *the* explanation, it eliminates one possibility, establishes that the defendant's story, as told, is not completely accurate, and certainly points us back to the second location. In this instance, the watered-down stains, which did not relate directly to the shooting event, provide a means of discriminating between the various possibilities for the shooting. Auditing can be an extremely effective tool for reconstruction, but it demands that we look beyond the immediate issues.

This raises the issue of contingent and non-contingent actions. To achieve some result, whether that be some particular series of actions relating to an event, or perhaps to the end result (e.g., the murder), there are actions that *must* occur. They are a part of the required causal chain to produce the end result. These actions are contingent (necessary) and the end result is not achieved if they do not occur. Contingent actions are the ones most criminal investigators concentrate on; they want to understand how the murder occurred. Other actions are ongoing as well, but are often considered tangents to the actual murder (e.g., a vase is overturned, a contact bloodstain is deposited). To the criminal investigator, such actions tend to have no relevance to the issue at hand.

These noncontingent (effectively unnecessary) actions may well have no immediate probative value in understanding the general aspects of the murder itself, but they often have value to the crime scene analysis. One of the major concerns is in ordering the events themselves. So perhaps we have actions associated with one victim in the bedroom and a second in the kitchen. We are always looking for crossover actions, some action that lets us order, as in this example, something that is happening in the bedroom with something that is happening in the kitchen. Noncontingent actions often lend themselves to such effort. Thus, the analyst is concerned with any and all actions, even when they do not immediately appear probative. They often prove critical in acting as crossover actions.

Sequencing and flowcharting began in the fourth step of Event Analysis and continues throughout the reconstruction process. The sequence developed through individual action relationships and auditing will ultimately provide structure to the relationships between the events as well. As we understand the associated actions sequence, the result often provides cross-over relational data between events. An action related to one event may have a sequential relationship to an action associated with another (see Figure 14.20). For example, the rapist may become bloodied in the altercation, then subsequently leave bloody swipes on drawers when rifling through the home. This information allows the analyst not only to recognize the additional event (rifling of the home), but to validate the sequence of one event to another.

In most instances, the order of the events develops by logically considering the actions in and of themselves, but occasionally there may be lingering issues. For instance, in our example of burglary/rape, the investigative team's initial assessment may suggest that entry was made, followed by rifling of the house, and it was then that the homeowner was raped. However, the relationships established between the actions in Step 4 are more objective than any initial beliefs. Revising the initial beliefs will be necessary if there is a contradiction between the initial assessment and the formal analysis. The presence of the victim's blood on the objects rifled in the scene forces a sequential relationship that demands revising the initial hypothesis.

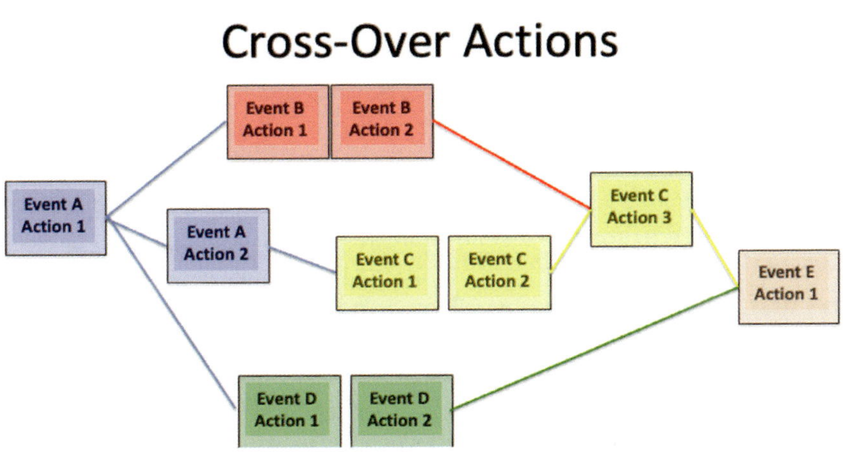

Figure 14.20 In this example, there are five events (A–E). Various actions from one event have a sequential relationship to another event's actions. This cross-over information allows the analyst to put the events themselves in some order objectively.

Once the overall analysis is complete, the analyst must return to their flowchart and validate it. This requires the analyst to question each sequence demonstrated on the flowchart and to look for unrecognized relationships. Once this task is complete, a final form of the flowchart is produced, and the reconstruction is complete unless additional information is forthcoming. In that instance, the additional data must be examined and considered to see if it better defines the overall sequence, or if it suggests that the analyst has incorporated some as yet unrecognized logical error.

At times an attempt to flowchart a scene may fail. The analyst may find the various Events simply do not connect logically. In instances where the scene is known to have been altered (intentionally or not) the result may be an inability to make logical sense of the flow. Should this situation occur and no known scene alteration has been recognized, the analyst should look to determine if some alteration has occurred, and simply not been identified.

Ultimately, some peer review is required of the analyst's ultimate effort. Someone not directly associated with the analysis must review the findings and conclusion, including the sequential aspects, and determine if the conclusions are supported.

In the peer-review, the data and facts used to support any claim of a given action must be reviewed. Is there a foundation for the action? Has that foundation been sufficiently articulated? Is the logic employed flawed in some fashion? Issues of subjective statements or words describing the action may be of concern. Is the foundation for the action subjective, weak, or insufficient to include it in the analysis? Additionally, the sequence presented (via the flowchart) must be explored. Often, analysts will fail to see some simple logical flaw or relationship that might demand reordering the actions or events. The bottom line for the peer review is to determine if, with the same facts and data, would they come to the same conclusion(s).

Putting the Pieces Together

The limitation in the reconstruction task should be evident; the analyst seeks to look back in time. In many instances, no one beyond the subject can provide details regarding what occurred. Even when eyewitnesses exist, the seasoned investigator realizes that witnesses and suspects often provide conflicting testimony for various reasons. In the end, proper analysis of physical evidence shoulders the greatest burden of proof in establishing these past events.

When presented with less than clear circumstances, investigators often depart from objective concerns and turn to subjective issues. Statements like "That's not logical; why would he do that?" begin to appear. It is not that we have no concern with motive or intent, but once we stray to such subjective issues to define the crime, we lose touch with our most authoritative asset. Perhaps Herbert MacDonell put it best in the title of his book, *The Evidence Never Lies*. From Gross to the present, it is recognized that we can best define our incident by considering the physical evidence objectively. From this solid platform, we may always reach out to explore the more subjective concerns.

Given the dynamic nature of our world and its interactions, no one can establish every action related to crime with certainty. Paraphrasing Voltaire, only a charlatan is absolutely certain of everything. We must follow a proven path of evaluating the evidence and context present in the scene, using the scientific method, and applying formal Event Analysis whenever possible. To accomplish this, the crime scene analyst must refine their context of observation, which is done by "assess[ing] the entire scene, noting all competing explanations [and] refusing to be guided by inappropriate preconceived expectations."[30]

By understanding the foundational theory and principles of reconstruction and applying a practical methodology that objectively moves the reconstruction effort forward, the analyst can define specific actions that occurred and the likely order of those actions. Event Analysis is a methodology incorporating the long-standing themes of crime scene analysis. Using Event Analysis, the incident in question begins to emerge as a series of objective moments, which we then link and sequence to produce the best explanation of the events that encompass the incident. Once defined in this fashion, criminal investigators and counsel may always turn to the more subjective issues and, using sound reasoning skills, attempt to define the "why" of the crime.

Notes

1. Association of Crime Scene Reconstruction Bylaws, 2024.
2. Our apologies to the author of these comments; they were observed on an educational program some years ago, the archeologist's name is unavailable.
3. Dever, W.G. and Lance, D.H., *A Manual of Field Excavation Handbook for Field Archeologists*, Hebrew Union College, New York, 1982, p. 74.
4. Dever, W.G. and Lance, D.H., *A Manual of Field Excavation Handbook for Field Archeologists*, Hebrew Union College, New York, 1982, p. 74.
5. Courtroom Television Network, *The Rodney King Case: What the Jury Saw In California V. Powell*, 1992.
6. Gross, H., *System der Kriminalistic,* in the translated and edited version of J. Collyer Adam, *Criminal Investigation A Practical Textbook*, Sweet Maxwell Limited, London, 1924, pp. 37–38.
7. Gross, H., *System der Kriminalistic,* in the translated and edited version of J. Collyer Adam, *Criminal Investigation A Practical Textbook*, Sweet Maxwell Limited, London, 1924, p. 38.
8. Carey, Arthur and McLellan, Howard, *Memoirs of a Murder Man*, Doubleday, Duran and Co, Atlantic City, NY, 1930.
9. May, L.S., *Scientific Murder Investigation*, Institute of Scientific Criminology, Seattle, WA, 1933, p. 9.
10. May, L.S., *Scientific Murder Investigation*, Institute of Scientific Criminology, Seattle, WA, 1933, p. 9.
11. Block, E., *The Wizard of Berkeley*, Coward-McCann Inc., New York, 1958.
12. Block, E., *The Wizard of Berkeley*, Coward-McCann Inc., New York, 1958.
13. O'Hara, C.E., *Fundamentals of Criminal Investigation*, 4th ed., Charles C Thomas Publishing, Springfield, IL, 1976, pp. 20–21.
14. O'Hara, C.E., *Fundamentals of Criminal Investigation*, 4th ed., Charles C Thomas Publishing, Springfield, IL, 1976, p. 58.
15. O'Hara, C.E., *Fundamentals of Criminal Investigation*, 4th ed., Charles C Thomas Publishing, Springfield, IL, 1976, p. 58.
16. Rynearson, J.M. and Chisum, W.J., *Evidence and Crime Scene Reconstruction*, J.M. Rynearson, Shingleton, CA, 1989, pp. 104–105.
17. Lee, H.C., *Crime Scene Investigation*, Central Police University Press, Taiwan, 1994, pp. 198–200.
18. Chisum, W.J. and Turvey, B.E., *Crime Reconstruction*, Elsevier Press, San Diego, CA, 2007, p. 144.
19. Geberth, V.J., *Practical Homicide Investigation*, 3rd ed., CRC Press, Boca Raton, FL, 1996, p. 706.

20. Chisum, W.J., The History of Crime Scene Reconstruction, *ASCR The Scene*, Volume 5, Number 2, April 1998, pp. 14–15.
21. Geberth, V.J., *Practical Homicide Investigation*, 3rd ed., CRC Press, Boca Raton, FL, 1996, pp. 707–708.
22. Turvey, B.E. et al., *Criminal Profiling: An Introduction to Behavioral Evidence Analysis*, Academic Press, San Diego, CA, 1999, p. 11.
23. Ressler, R.K., Burgess, A., and Douglas, J.E., *Sexual Homicide, Patterns and Motives*, Lexington Books, Lexington, MA, 1988, p. 137.
24. Ressler, R.K., Burgess, A., and Douglas, J.E., *Sexual Homicide, Patterns and Motives*, Lexington Books, Lexington, MA, 1988, p. 142.
25. Turvey, B.E. et al., *Criminal Profiling An Introduction to Behavioral Evidence Analysis*, Academic Press, San Diego, CA, 1999, p. 57.
26. Gardner, R., A Qualitative Theory For Crime Scene Reconstruction, *Journal of the Association of Crime Scene Reconstruction*, Volume 20, 2016.
27. Hill, A.B., The Environment and Disease: Association or Causation? *Proceedings of the Royal Society of Medicine*, Volume 58, Number 5, 1965, pp. 295–230.
28. Gardner, R., Applying Hill's Criteria to Determine the Validity of Cause and Effect Associations in Crime Scene Analysis, *Journal of Forensic Identification*, Volume 65, Number. 5, 2015, pp. 803–812.
29. Chisum, W.J. and Turvey, B.E., *Crime Reconstruction*, Elsevier Press, San Diego, CA, 2007, p. 25.
30. Nordby, J.J., Can We Believe What We See, If We See What We Believe, *Journal of Forensic Sciences*, Volume 37, Number 4, July 1992, p. 1122.

Presenting Evidence 15

Preparation for court begins long before any actual court date. Having failed to prepare properly, analysts should not be surprised when a skillful attorney cuts them to shreds on the witness stand. After such a mauling, and usually while nursing a bruised ego, the analyst will likely blame the situation on "that unmerciful attorney." However, if we step into the witness chair unprepared or oblivious to the nature of the court, then perhaps we deserve such treatment. The culmination of any analysis is to demonstrate in court the conclusions drawn from the analysis. In meeting that end, the analyst can never take trial preparation lightly.

One of the more important aspects of presenting evidence as an expert is understanding and meeting the court's expectations. This involves explaining the nature of the discipline and the expert's background.

Understanding the Nature and Content of Daubert or Similar Challenges

Before analysts testify in front of a jury, they often undergo an admissibility hearing called a *voir dire*. In this hearing, the opposing attorney will question the witness on several areas:

- The expert's qualifications in the area the analyst is being proffered as an expert (e.g., the expert's training, education, and experience in Bloodstain Pattern Analysis).
- The science or discipline of Bloodstain Pattern Analysis.
- The accepted methodology and if that methodology was followed.
- Additionally, the analyst may be asked to provide the court with documentation of recent cases in which the expert testified.

In preparing for such a challenge, the analyst should become familiar with the standard that jurisdiction uses as the benchmark for the admissibility hearing. Countries outside the United States will likely have their own rules, which may not remotely follow US rules. Reviewing the following standards may prepare foreign analysts. In US federal court, the Federal Rules of Evidence apply (e.g., Rule 702). In contrast, US state courts follow various *voir dire* standards, including Rule 702, Frye, Daubert, or Kumho. Each of these legal opinions defines a specific standard for expert opinion and, in essence, challenges the expert and the discipline itself. There is overlap between each; however, the analyst should be thoroughly familiar with the standard used by the jurisdiction in which testimony will be offered.

DOI: 10.4324/9781003408734-15

Daubert v. Merrell Dow Pharmaceuticals, 113 S. Ct. 2786 (1993)

A Daubert challenge is one of the more common forms of challenge. It includes the following questions:

1. Has the discipline/methodology been tested?
2. Has the discipline/methodology been subjected to peer review and publication?
3. Is there a known or potential error rate for the discipline?
4. Has the methodology been accepted by the relevant scientific community?

Frye v. United States, 293 F. 1013 (D.C. Cir. 1923)

A Frye hearing includes the following questions:

1. Will the expert testimony assist the jury in understanding the case?
2. Is the testimony based on scientific principles that are accepted in the discipline?
3. Is the witness qualified in this discipline?
4. Is the witness credible?

US Federal Rule 702

A challenge based on US Federal Rule 702 will include the following questions:

1. Is the testimony based on sufficient facts or data?
2. Is the testimony based on scientific principles and methods?
3. Did the witness reliably apply the scientific principles and methods to this case?
4. The court may reject the testimony if the proponent fails to establish admissibility by a preponderance of the evidence.

Responding to *Daubert* or Similar Challenges

To meet and prevail during a *voir dire* challenge, the analyst must be prepared to answer any of the questions posed by the specific challenge. The best way to plan for and respond to a challenge is to build predicate questions. The most basic form of these predicate questions is as follows.

What Is Bloodstain Pattern Analysis?

The analyst must be able to explain to the judge and jury in their own words what Bloodstain Pattern Analysis is. At a minimum, the answer should include the following descriptions and definitions:

- What is the foundational theory and associated principles applied in BPA?
- What is liquid blood?

- What are the purpose and objectives of Bloodstain Pattern Analysis?
- What physical laws and principles apply in Bloodstain Pattern Analysis?
- What information can a study of the bloodstain's physical characteristics identify?
- How was the analysis accomplished, and what accepted methodology was utilized?

In the analyst's own words, they must explain to the jury what an analysis might identify, including the following:

- Identify specific bloodshed events that occurred.
- Identify the location of the attacker and the victim during bloodshed events.
- Identify the sequence of different bloodshed events.
- Identify the movement of people and objects in the scene.

Without intruding on the jury or judge's domain, the analyst should also explain that BPA provides objective information that the trier of fact can use when considering testimonial evidence. Keep in mind, it is not the analyst's domain to tell the jury that "so and so" is lying. They simply provide the jury with factual data that may allow the jury to make that decision. This will typically be done using hypothetical questions posed by counsel.

What Theory or Principles Apply to Bloodstain Pattern Analysis?

In their own words, the analyst should explain how basic principles of mathematics, biology, and physics are used in an analysis. They should describe how energy, force, velocity, surface tension, molecular attraction, and cohesive or disruptive forces affect the patterns as found. This should always start with a brief description of the foundational theory and the three principles associated with BPA.

What Is the Methodology Used in Bloodstain Pattern Analysis?

The analyst should explain the specific methodology utilized in completing the analysis. Areas of interest include:

- In general, that analysis includes gathering information, correlation of data collected using observational skills, and how specific information and evidence were documented.
- How the scientific method was implemented to accomplish the analysis.
- The limits of any conclusion provided should be fully explained, including the fact that Bloodstain Pattern Analysis is a class characteristic evidence.
- If methods such as technical or peer review were used, they should be explained, and if not utilized, the analyst should explain why they were not employed.

Where Has Bloodstain Pattern Analysis Been Accepted in Judicial Settings and within the Scientific Community?

The analyst should explain where the courts and the scientific community have accepted Bloodstain Pattern Analysis.

- The analyst should articulate that Bloodstain Pattern Analysis as a discipline predates most forensic disciplines.
- In its modern form, it has been accepted throughout the United States in municipal, county, state, and federal courts. It has been accepted in many different countries, including the United Kingdom, the Netherlands, Canada, New Zealand, Australia, South Korea, Singapore, Israel, South Africa, and others.
- That BPA is taught in academia at the Associate, Bachelor, Master and even Doctorate level.

What Scientific Studies Have Been Published in Peer-reviewed Journals?

The analyst should be prepared to identify for the court some of the textbooks and studies addressing the scientific validity of Bloodstain Pattern Analysis, particularly those associated with specific case conclusions.

Are There Professional Associations That Recognize Bloodstain Pattern Analysis?

The analyst should be prepared to identify the many professional associations involved in Bloodstain Pattern Analysis, including the International Association of Bloodstain Pattern Analysts (IABPA), the International Association for Identification (IAI), the Association for Crime Scene Reconstruction (ACSR), the American Academy of Forensic Sciences (AAFS), the Scientific Working Group on Bloodstain Pattern Analysis (SWGSTAIN), National Institute of Standards and Technology (NIST), and the Organization of Scientific Area Committees (OSAC).

Is There an Identified Error Rate for Bloodstain Pattern Analysis?

This is one of the more difficult questions the analyst will encounter.

There are certainly error rates associated with specific skills and tasks associated with the discipline. These include the error rates associated with impact angle determinations and area of origin determinations, and classification.

This includes one NIJ funded a black box study by Austin Hicklin et al., at Noblis Research Institute to assess the validity of Bloodstain Pattern Analysis accuracy of 75 experienced analysts identifying 150 lab-created stains. The primary focus was on classification of stains and patterns. The results showed an 11.2% error from the known pattern.

Overall, the supermajority or average responses dropped to 8% for any two examiners contradicting each other. "For classification prompts, responses with a 95% supermajority never contradicted known cause." The majority were always correct when limited to the most consequential prompts.

One of the primary issues related to the NIJ effort was the inclusion of context for classification. In published methodology, classification is accomplished before consideration of context and any ultimate reconstruction opinion regarding the patterns. When subjective context is included in the early stages of classification, various forms of bias can be introduced. Sequential unmasking of context is important, and the study failed in this regard.

The study demonstrated, that even with this subjective context inclusion, classification results were correct in 95% of the results. The study found that often a lack of agreement on the meaning and usage of the pattern terminology caused apparent disagreements between two examiners, contributing to some of the reported errors.[1]

The authors have found throughout the last 20 years that classification error is significantly reduced using the taxonomy criteria. In both BGA basic and advanced courses, students have been presented with both a pre-course and a post-course classification test. Following standard methodology, no context beyond whether the substrate is oriented vertically or horizontally is provided. In basic courses, students routinely achieve about a 65% in pre-course correct responses. Using the taxonomy criteria, post-course responses rise to 97% in correct responses.

Advanced courses routinely occur sometime following a student's basic course. Presented with a pre-course test, students (most of whom have been trained in the taxonomy mindset) achieve a 95% in correct responses. Post-test responses in the advanced courses is similar to that of the basic course results.

What is evident from these results is that using objective and specific criteria, without consideration of subjective context, significantly reduces error in classification.

What cannot be quantified is individual analyst error. No matter the level of competency and knowledge of the analyst, anyone is capable of error in any given situation.

Ultimately, by considering these all of these predicate questions and being prepared to articulate answers for each functionally, the analysts will meet any *voir dire* challenge they encounter. Keep in mind that effectively articulating these answers is important. Just knowing the answer will not do analysts any good. They must be prepared to explain the answer under stress in a contentious arena (i.e., the courtroom) against an adversary who is bent on confusing and preventing the answers from coming out. What is evident from various court decisions, if the court finds the analyst failed to adequately follow accepted methodology, the court may reject the testimony. The analyst can't just talk the talk; they must demonstrate they followed the methodology they described.

General Concerns for Testifying

Analysts should investigate and prepare reports in every case with an eye toward the witness stand. Their own report can often come back to haunt them, so it should not contain inaccuracies, discrepancies, exaggerations, or unsupported theories. When writing the report, the analyst must consider the following questions: How can I substantiate this in court? What is the foundation of this conclusion?

Analysts will find the taxonomy criteria presented in this book particularly functional for setting a classification foundation. When asked why a particular pattern is what the analysts says (for example, a cast-off), rather than indicating the foundation as "experience" or "I've seen it before" and other subjective responses; one can simply state: "This was a pattern of circular and elliptical stains, in a linear distribution, in which the stains were becoming more (or less) elongated on a singular surface."

When presenting an opinion to the court, be prepared to use demonstrative evidence to educate the jury, such as photographs or charts. Computer resources are helpful in this regard. They can effectively illustrate specific conclusions and points related to the analyst's testimony.

The analyst's ability as a teacher is a concern that cannot be overemphasized. An expert opinion that the jury fails to understand is worthless. Conversely, if a conclusion is clear, logical, concise, and understandable, the jury will likely find such testimony extremely valuable. Therefore, analysts should use all appropriate illustrations to let the jury "see" and not simply hear their testimony.

Maintaining Objectivity

It is important to maintain objectivity throughout the trial. Remember, the analyst's role is that of a witness, not an advocate. It is not their job to help the attorney win the case. Such an attitude often leads analysts to include unsubstantiated details in their conclusions. The jury will view this behavior as a bias or prejudice, and rightfully so. In the end, it can destroy the analyst's overall credibility.

Nevertheless, pressures to be "one of the team" will be present. Police organizations employ many bloodstain analysts directly within the structure of the agency. As such, it is easy to identify singularly with the goals of the investigators. Additionally, the analyst may be part of a team with the prosecutor, whose purpose is to build a particular case against the defendant. The analyst is subjected to the "Us versus Them" attitude found in the adversarial court system.

It takes a conscious effort by the analyst to remain impartial, but the discipline demands impartiality. This might seem at odds with our purpose, as the analyst will certainly conclude that something happened. Being adamant in the conclusion is not bad. Still, the analyst becomes adamant only by evaluating all available evidence and considering all possibilities. The analyst's conclusion objectively recognizes inherent strengths and weaknesses. Such scientific impartiality differs from pursuing only evidence that supports a specific theory. Science demands we recognize the difference between the two and pursue the holistic and objective approach.

An important recognition on the analyst's part is that the nature of BPA conclusions is different. Certain conclusions may be arrived at deductively; thus, their certainty is much higher. On the other hand, certain conclusions may be arrived at inductively; thus, it is the best explanation given the available data, but by no means absolute. The best example of this difference in Bloodstain Pattern Analysis is the difference between classification and identifying a source event. If the analyst understands and follows a deductive classification process (using the taxonomy mindset), the classification conclusion arrived at (e.g., this stain is an impact pattern) for most stains will be objective and based on specific physical criteria. On the other hand, identifying a specific source event (e.g., this impact pattern was produced by a gunshot) for the same pattern is not always as certain. Such a conclusion generally involves some form of inductive argument involving the context in which the pattern was found and thus may be open to some level of interpretation. It is imperative that the analyst know the difference in their conclusions and not overstate any part of the analysis. The science is the science; if a given hypothesis cannot be eliminated, it remains a potential explanation.

Analysts often find the legal system intimidating because it has its own unique purpose, values, and logic.[2] The search for truth in the context of the law is simply part of the process by which the goal, "justice," is sought. On the other hand, in analysis, objective fact directing us toward the truth is the goal. The analyst is concerned with being confident that they have identified the best explanation without overstating their conclusions and have done so based on all the evidence available. Of course, the court works in a slightly different fashion. As Peterson explained, "The principal objective of the litigants is to win the case, often at the expense of truth... counsel tries to extract a slanted picture from the witness, and, on cross-examination, opposing counsel seeks to slant the picture the other way."[3]

As Alan Dershowitz, one of O.J. Simpson's defense attorneys, commented, "A criminal trial is anything but a pure search for truth. When defense attorneys represent guilty clients ... their responsibility is to try, by all legal and ethical means, to prevent the truth about their client's guilt from emerging."[4] Truth is very often the unfortunate victim of this wrangling, but analysts cannot afford to play this game. They must present their knowledge factually and objectively, no matter how it may be used. If answering a particular question harms their counsel's position or the even the analyst's conclusion, so be it. The danger for the analyst is defending the indefensible. If an error or some data was not considered, accept the situation. Any attempt to mitigate the lapse creates more difficulties. Truth is truth, the jury or judge must decide if they believe a conclusion and what weight they choose to give it.

Analysts can never allow themselves to forfeit their own or their department's integrity. No matter how perturbing the judicial system may be or how adamant their belief in the guilt of some subject, they can only report what they know. There have been unfortunate cases in which forensic experts have modified or manufactured evidence to support a particular counsel's position. Believing in the guilt or innocence, they attempted to assist counsel inappropriately. There is absolutely no room in forensics for this type of activity. Analysts own their integrity; no one can take it from them. To lose it, you must consciously decide to give it away. Every time you take the stand guard your integrity carefully; the court and society rely on it. Remember, too, that your testimony must never invade the jury's purview. The analyst is not there to define innocence or guilt; this task belongs solely to the jury. The analyst simply gives them knowledge on which to base those decisions.

Settling in and Establishing a First Impression

The first few moments in the witness stand are often the most disconcerting. Unfortunately, these are the same moments when the jury members define their first impressions of the analyst. The following actions may assist the analyst in establishing a positive image.

In preparation for *voir dire* and subsequent testimony, the analyst can assist counsel in preparing qualifying questions. Since analysts know their qualifications better than anyone else, they can create questions that place them in a positive light. Of course, it is imperative that the analyst not embellish their credentials in any fashion. Areas to be covered by these questions include the following:

- Training and years in the discipline
- Position in and years with a given agency
- Articles published or studies conducted
- Schools and classes taught on the subject
- The number of times previously qualified as an expert witness
- Membership in professional organizations related to the field and any honors received

Preparing these questions in advance and providing them to the appropriate attorney ensures two things: (1) the analyst knows what the questions are and knows how best to answer them, and (2) by knowing the nature of the initial questions, the analyst can relax somewhat and will become more comfortable on the stand.

A relaxed attitude is critical, as the jury is evaluating the analyst's appearance and demeanor. This first impression may set the tone for acceptance of the entire testimony. To address this issue, the analyst should always dress conservatively, preferably in business attire. Do not wear items such as sunglasses or anything that might be considered flashy. This has a way of stealing the jury's attention from testimony. If possible, do not testify in a police uniform, as the jury may perceive this as demonstrating partiality to the prosecution.

Demeanor is difficult to maintain, and the analyst constantly walks the "demeanor tightrope." One cannot be "cocky," but just the same, one should not be viewed as "timid." Analysts must appear confident in their knowledge, experience, and conclusions. Hopefully, by acting professionally on the stand, the jury will perceive them as a professional.

To present this professional appearance, analysts must use every action in the courtroom to their advantage. As they walk into the courtroom, all eyes are on them. The analyst should walk in with their head up and shoulders back. When taking the oath, they should stand in a position of attention. Their affirmation should be in a voice that is confident and loud enough that a jury can hear them. The actions and manner of the analyst must say, "I am here to be honest and to tell the truth." If done well, the jury's initial impression of the analyst is that of a serious professional.

Attacks on the analyst come in various ways, but counselors often set the stage and allow analysts to impeach themselves. Trained to bring out any attitude of animosity or bias, the opposing attorneys will seek to do so during cross-examination. If this can be accomplished, the attorneys feel more confident in their own argument, knowing that the jury's opinion of the analyst is diminished. If the analyst refuses to be drawn into such encounters and remains objective and calm, then the jury is unlikely to view the analyst in a derogatory manner.

Remember, objectivity and impartiality are critical. If questioned about a weakness in the case, admit it. It is not the analyst's job to convict, but rather to give whatever testimony can honestly be given and nothing more. The analyst's credibility will suffer if there is any attempt to hide a weakness or subjectively dismiss it. If an error of some type was made in the analysis or the report, this should be brought out on direct examination, so as to show the analyst is not hiding anything from the trier of fact.

When responding to questions that require short answers such as "yes" and "no," look at the attorney asking the questions. Do not look back and forth from the attorney to the jury during a series of such questions. The jury may find the motion of the analyst disconcerting. On longer answers, the analyst should look at the jurors. In doing so, the analyst builds rapport and further establishes an image of credibility for the court and jury. The analyst should not forget to look back at the attorney, particularly when listening to the questions asked.

One often observes that even the most experienced witnesses will eagerly answer the questions of their own attorney and then become immediately defensive or hostile when the opposing attorney asks questions. No matter how aggressive an attorney becomes, the analyst should never get mad or display hostility. If the analyst ignores this aggressiveness and continues to answer with confidence, the jury will tend to view the analyst in a more credible light.

Understanding Cross-examination

For those unfamiliar with court, it is often difficult to understand that a trial has little, if any, relationship to the investigative process. At times, the mantra may be "the best show wins." In the lawyers' mindset, truth has only a limited place in the courtroom. This is best understood by referring to the texts lawyers use to teach other lawyers about science. Contained within one such text is what we like to refer to as the *Smorgasbord Theory of Science*. The following excerpt very effectively explains the underpinnings of the legal profession's mindset:

> Subordination of Scientific Truth to Legal Truth
>
> Forensic scientists recognize that they are but the hired help, and that forensic science is but the handmaiden of the legal system. The validity of facts testified to in a court of law by non-expert witnesses is perpetually subject to challenge. However, when a scientist introduces facts into a court of law, they are subject to the same challenge. Scientific "truths" are established when the validity of a proposition is proven to the satisfaction of a prudent and rational mind. Legal "truths" are not established by the exercise of the scientific method, but by the processes of the adversary system.
>
> The role of physical evidence in the administration of justice is described as follows: Science offers a window through which the law may view the technological advances of our age. Science spreads out a smorgasbord of (hopefully) valid facts and stands back, proudly displaying its wares. The law now picks out the most attractive morsels, applying selection criteria that may or may not have anything to do with science.[5]

It is important to note that this text is by no means some obscure manuscript. Four prominent law professors penned it as the basis for teaching other lawyers and experts about science. It is clear that at least some of the legal system has forgotten that truth is truth. For a legal truth, which is effectively a legal *opinion*, to be valid, it must be based on scientific truths. Logically, it cannot be otherwise. More important to this issue and conveniently forgotten by the legal profession is that arriving at truth of any nature demands consideration of all information and evidence. To base an opinion on only part of the evidence available is a classic example of confirmation bias, a claim often directed at science by the law profession. If we as analysts fail to consider all evidence, lawyers, and judges are quick to chastise and demean us for our failure; yet the phrase "picks out those morsels that appear most attractive" tells the story of the law profession. Lawyers are trained to start with a conclusion and work backward, picking and choosing those facts that support their claim; all others are deemed unworthy. This is not the mindset of any one group of lawyers (i.e., only defense); it is the mindset of the legal profession.

Thus, what counts in court is how each opposing counsel can turn the facts to support their position. Therefore, the analyst must understand and be prepared for the rigors of cross-examination. Counselors use cross-examination to meet several objectives:

- Discrediting the witness
- Discrediting the discipline or methods used
- Discrediting the conclusions or opinions given by the expert
- Misdirecting the jury from important facts or conclusions

Direct questioning in the initial stages of a trial is a walk in the park compared to the antagonistic manner that may be evident during cross-examination. In preparation for this ordeal, analysts must read and know their report. They should review all related reports and any articles or books that have a bearing on the conclusions drawn. Additionally, analysts should know or be cognizant of concepts or theories of other authors that might be counter to their own. Finally, analysts must be able to verbalize under adverse conditions how they reached their conclusions.

During cross-examination, expect the opposing attorney to dwell on those areas for which the analyst is not qualified. An example may be the attorney asking if the analyst has a physics, math, or biology degree. Often, the attorney will attempt to make the analyst answer simple questions, those to which the analyst may know the answer. Using these answers, the attorney will try to further the impression that the analyst claims to be an expert in that area. During cross-examination, these easy little questions have an ugly habit of turning into big, complex questions that ultimately embarrass the analyst. For example, even though the analyst is not a biological scientist, a question from this field will likely be asked in Bloodstain Pattern Analysis. The analyst's best response is to state a lack of expertise in the area and request the attorney to ask the question of someone qualified. If others should answer the question, then let them!

The analyst should never consent to conducting examinations of "new evidence" on the witness stand. It should be evident that analysis demands proper examination in an appropriate setting, with sufficient workspace, lighting, magnification, equipment, and time. Off-the-cuff examinations are exactly that – off-the-cuff. They have no place in the ultimate analysis and do not belong in court.

An associated ploy is to hand the analyst a reproduction from some book, manual, experiment, or case and then ask if the analyst can determine from the excerpt what actually produced the stain. Once again, the analyst should explain that this cannot be done out of context. Obviously, more information is necessary. What was the nature of the surface? Was it vertical or horizontal? What were the associated stains surrounding those presented? What injuries caused the bleeding? What other physical evidence was present which might support or refute any opinions made? An analyst does not make ultimate opinions based on tunnel vision, but rather by utilizing a holistic approach. Hopefully, the judge will preclude an attorney from conducting such "expert testing" procedures. They serve little, if any, function. In the same way, if analysts are forced to answer a question about a specific book or article, they should always request to see the citation and its source before answering.

An attorney usually will rarely use leading questions during direct examination. Conversely, on cross-examination, the attorney wants to control the witness and will likely use leading questions. Generally, we could classify the three types of leading questions as mildly, fairly, and brutally leading. Mildly leading questions often begin with words such as "is," "are," "was," "were," "do," and "did." Fairly leading questions start with words such as "aren't," "weren't," "don't," and "didn't." Brutally leading questions often begin or end with the following phrases: Isn't it true …? Isn't it a fact …? "Won't you admit …? "Won't you concede …?[6] Additionally, the attorney may make a statement instead of asking a question and then ask "Isn't that correct?"

When these words come out of the attorney's mouth, listen to the questions carefully. If you are unclear about what was asked, you should not answer. State clearly that you did not understand the question and ask that it be repeated. If the question is a compound question, take each part separately, answering the first, then the next. If this is impossible, ask the attorney which question they would prefer to have answered first. The analyst must recognize the attack for what it is and respond. The attorney is not looking for an answer; they are trying to confuse the witness, the jury, or both. A good counter to this technique is to pause after each question. Since the analyst is charged with answering the attorney's questions, they also have some control over the pace. Should the attorney interrupt the analyst while talking, the analyst can stop and wait. Once the attorney stops talking, the analyst can ask the court to be allowed to complete the answer if needed. Slow the pace down when possible, and do not become flustered or lose focus by the attorney's actions.

In answering complex questions, be cautious of offering simple "yes" or "no" responses, as the attorney often goes to great lengths to keep you from explaining further. Never start an answer with "Yes, but" or "No, but." By doing so, you have already answered; the only chance to provide the necessary details is to explain the answer first. Instead, turn to the judge and state "I cannot accurately answer the question in a 'yes' or 'no' manner."

Using Demonstrative Aids in Court

Analysts should consider the use of demonstrative aids to assist in explaining their bloodstain pattern testimony to the court. Some of the common aids used include diagrams, charts, and photographs. Additional demonstrative aids include the following:

- 3D computer graphic stills
- 3D computer-generated animation
- PowerPoint or similar presentations

Three-dimensional rendered computer graphic stills can be created for each bloodstain action. These assist the jury in understanding testimony. There are several advantages to their use:

- They represent about one-third of the cost of creating a computer animation.
- If the court rejects one or more of the pictures, they are easily removed. A moving animation is much more time-consuming to edit. Editing, if allowed, must be

accomplished in the time frame directed by the court, and with animation, that is not always possible.
- The 3D renderings direct the jury's attention away from the blood and gore of viewing photographs of the actual victims; thus, they tend to follow the testimony better.

Three-dimensional computer-generated animations are effective but very expensive to produce. The videos allow the court and jury to "see" the analyst's theory of what happened in some specific circumstance. Issues with actual animations include the time required to edit an animation (particularly if directed to by the court) and the subjectivity of animations. The animation has individuals moving throughout, the analysis cannot establish the physical position of all parts of the body at all times. Thus, certain objections by opposing counsel are likely to follow. Care should always be taken to explain to the jury that the computer did not "figure out" what happened, but rather, it simply shows the expert's opinion regarding what occurred.

PowerPoint or similar software packages are excellent for presenting photographs, sketches, and other computer-generated images. One caveat, never make changes after sending a presentation to the court. The process of changing material is so easy that it is hard to resist making last-minute adjustments. However, the presentation before the court must be shown as submitted and shared with opposing counsel. If not, some challenge will ensue. These software packages are very effective because it is easy to redact information in a matter of moments to please the court.

The use of hardware devices such as an Elmo help the analyst present details from photographs or items of evidence. The item to be viewed is placed on the Elmo stand, and the system projects its image to a screen or monitor. These devices allow for zooming in and out on the item with effective magnification. If the analyst is not familiar with the presentation method to be used, the analyst should ask to see and understand the operation before taking the witness stand.

Building Demonstrative Presentations Using Computer Resources

The ability of a jury to "see" the analyst's opinion develop is important in their subsequent consideration of that opinion. If the analyst takes the stand and talks for several hours, they might win over the jury, but more likely than not, the jury will lose focus in following the testimony. The use of demonstrative presentations assists the jury in understanding the interrelationships of the evidence and how that evidence led the examiner to a given conclusion. They also help keep the jury focused on the relevant information.

When instructing, the authors are routinely asked how we prepare some of the demonstrative presentations we share with students. We combine output from various software packages, creating clear, focused, and easily manipulated demonstrative aids. The process requires a computer, three basic software packages, and a scanner. The brand of software purchased is of little importance, but we will describe the process using the brands we utilize.

The first package necessary is a presentation package like Microsoft PowerPoint. This program is the heart of the presentation. It always helps to be familiar with the basic workings of the presentation software. All the other programs will provide input to this package.

The second software package required is a computer-aided/assisted design (CAD) or a home design software program. These programs are commercially available from various software publishers. Examples include Punch Pro and 3D Home Architect. They create architectural-style floor plans and 3D views of the floor plan. Before purchasing the package, check to ensure that it provides the following features:

- A large furniture library
- Editing of the size of the furniture in the software library
- Rotation by degree of furniture items
- Export of 3D rendered screens in a standard graphic file format (e.g., BMP or JPG)

The third package required is a human figure design package, such as Poser. These programs allow the user to pose virtual models in any position desired. The program should have the following features:

- Be easy to learn. You will not work this program all the time, nor will you use it fully. Nevertheless, when needed, you want to be able to use basic processes easily.
- Offer the ability to use different model formats (e.g., life-like, stick-figure, mannequin).
- Import graphic file formats such as JPG or BMP as backgrounds for the program.

Finally, a scanner is required. This item is necessary for converting scene photographs and sketches into graphic files when no digital image is available, for import into the presentation software.

A computer, presentation software, and scanners are usually found in any law enforcement agency. They represent the bulk of the cost associated with purchasing these items. Home design software is relatively inexpensive. Human animation programs may cost several hundred dollars.

Before discussing the process of creating a presentation, there are practical points to identify about designing the presentation. If ignored, they detract from or reduce the effectiveness of the presentation. These points include the following:

- Keep it simple. Fewer presentation slides are better.
- Screens should be orderly and uncluttered, and the lettering should be large. Do not try to jam too much information into a single screen.
- Choose a conservative color scheme and a clear, readable font.
- Whenever possible, eliminate photos that contain gore and bodies. These detract from the presentation and distract the jury.
- Keep conclusions out of the presentation itself. The purpose of the presentation is to demonstrate the evidence and interrelationships. The analyst will testify to the specific conclusion. Failing at this point, the judge may order the redaction of certain information.

The typical presentation method in Bloodstain Pattern Analysis is to walk the jury through the scene, highlighting important stains. To do this effectively, the analyst can create a 3D scene representation.

Using the CAD or home design software, the analyst first creates a 2D sketch of the scene or room in question. The information required by the program from the crime scene sketch and crime scene notes includes ceiling heights, furniture sizes, and general room and opening dimensions. Once the general room layout is complete, the analyst inserts furniture and fixtures from the software library and places them appropriately. It is important to recognize that this model represents the scene; *it is not to exact scale nor* intended to be. The model will not match the scene in every detail, but wherever possible, ensure that it does. Furniture appearance is a major problem. Whenever possible seek items like the actual scene furniture, then size them accordingly. Libraries of a wide range of objects (3D models) can often be found on the Internet.

Once the floor plan and furniture are complete, the analyst can render 3D views from various perspectives. The analyst does this by aiming a virtual camera at specific points in the 2D plan. The software then renders a 3D view of what the viewer would see standing at that location and looking in the given direction. The view or views required by the analyst depend upon the location of items of evidence. The same rules apply to the virtual camera as they do with a normal camera. Keep the lens and primary subject aligned; do not interject odd or inappropriate angles. Only do so, if necessary, as in the case of an item nearly hidden or in an abnormal position. After initial camera placement, almost all programs allow the user to move the virtual camera "on the fly." This allows the user to tweak the virtual camera position and see the immediate result. This will ensure getting the best view possible for the presentation's purpose.

A good starting view for any presentation is an overhead view. Like any crime scene sketch, this view will help orient the jury to the model (see Figure 15.1). Create other views from "within" the room as needed, then export them (see Figure 15.2). Most export methods produce a graphic file, which can be imported into the presentation software. Note that

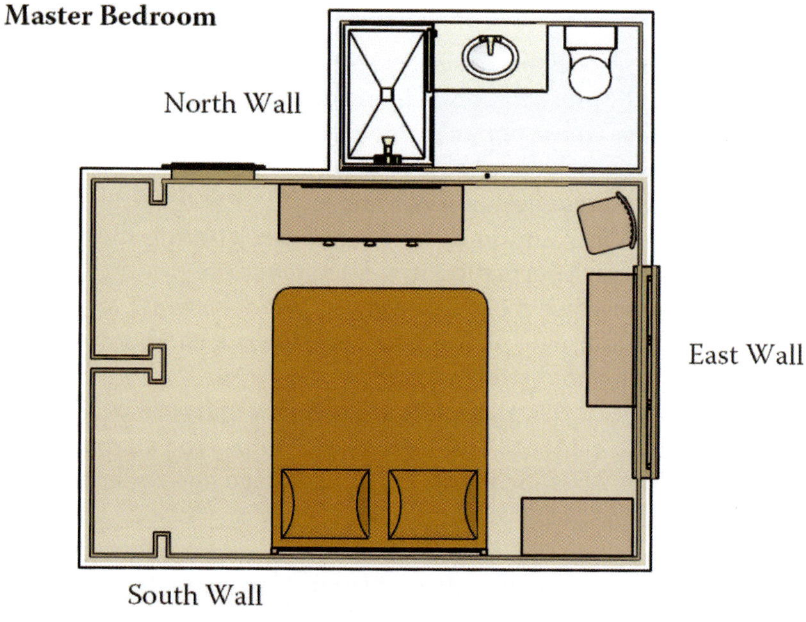

Figure 15.1 An overhead view helps orient the jury to the scene. The jury fully understands where things are, used in conjunction with the crime scene sketch.

Figure 15.2 These programs allow the placement of a camera to create a virtual view from anywhere in the scene. This example shows the view across the room from the doorway. Whatever view is necessary, the software will usually accommodate it. The view is then exported as an image and imported into the presentation software.

most home architect software has various export functions. The most important export function is a particular 3D room view.

Be sure to save the exported files in a temporary folder where you can easily retrieve them. If you allow the software to put them in some default directory on the computer, it is difficult to locate them when needed.

The room views are the raw material for the presentation. Before they can be manipulated, they must be imported into the presentation software. The following instructions apply to PowerPoint, but whatever software is used, the process is not much different.

First, open a new presentation file using the blank slide format. If you create three different room views, you must start with one presentation file with three blank slides.

Browse the directory where you saved the room view files and choose the graphic desired. Using whatever means your programs allow insert this into the presentation. Once inserted the software returns to the edit screen, and the graphic will be displayed on the slide. Size the image as necessary. This view is an image, static and unchangeable except for size in the presentation software. If the graphic must be adjusted, return to the originating software, make the change, and then save and import it back into the presentation software.

Repeat the insert process for each room view you created, but move to a blank slide in the presentation before inserting each new room view. Once created, keep the basic slide intact because you will likely need to make several slides using the same room view. Duplicate it, as needed, before annotating or adding to it. This will save a significant amount of time.

After duplication, the analyst can annotate and add graphics to the slide as required. Annotations, arrows, and circles are used to explain and direct the viewer's attention to specific locations where evidence was found. Be sure to use contrasting colors and a readable font.

Photographs are another integral part of the demonstrative presentation. They may include crime scene photographs or photographs created by the analyst. If necessary, "Auto Balance" or "Enhance" the image. This can lighten a dark image, making it easier to view. Name the image files created so they are easily recognized. Save them to the same temporary directory in which you saved the room views. Combined with 3D views, these photographs will allow the jury to understand better where specific stains were found (see Figures 15.3 and 15.4).

Do not be afraid to crop scene photographs before importing them into the presentation. If you establish from which photograph the cropped image originated (e.g., cropped image of DSC 0001) eliminating possible objections. Cropping allows the analyst to eliminate offending gore, bodies, or to focus on some specific area in the photograph.

Scene or evidence photographs are inserted as with any graphic file. Once inserted the program will return to a slide edit mode, where the picture can be resized as needed (see Figures 15.5 and 15.6).

Human figure animation software is used to create figures in various poses. These are imported into the presentation software in a similar fashion. A significant difficulty exists in combining a 3D room view and a figure. Based on the export method of each image, intermixing the two is nearly impossible. There is one alternative. Programs like Poser allow the analyst to import a scene photograph into the figure software and scale it to match the figure. The scene photograph becomes a background for the Poser figure. Once this is done, the combined image can be exported as with any other Poser image.

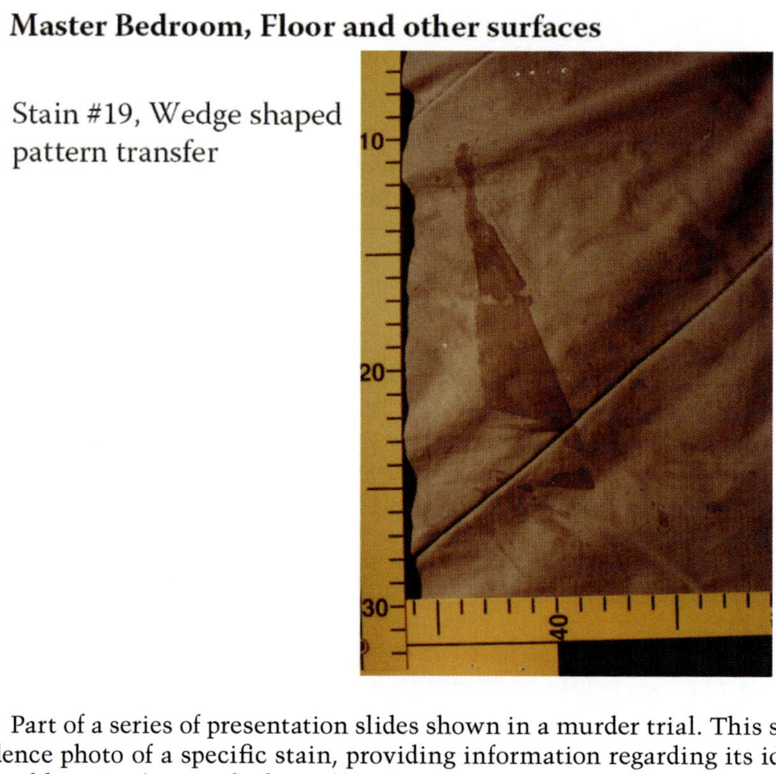

Figure 15.3 Part of a series of presentation slides shown in a murder trial. This slide incorporates an evidence photo of a specific stain, providing information regarding its identity (Stain #19) and general location (master bedroom).

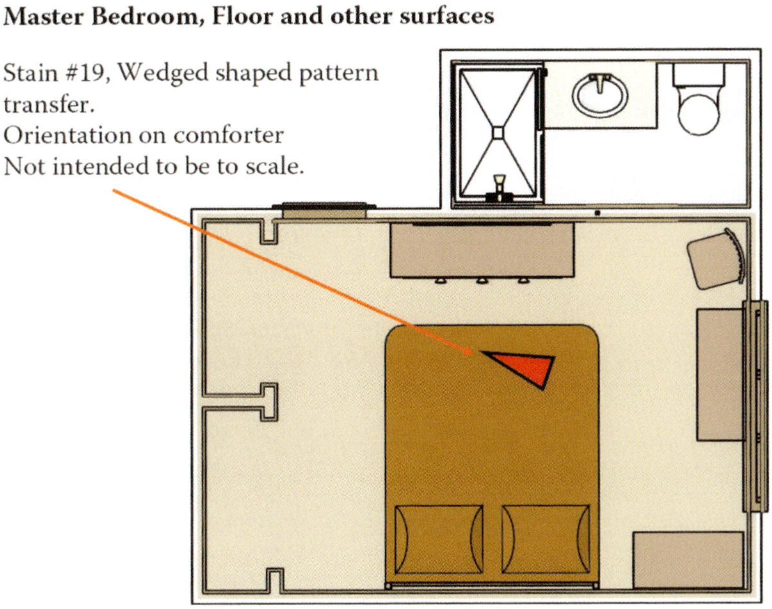

Figure 15.4 This slide combines the image of the overhead view created in the CAD software with arrows and graphics to show the specific location of Stain #19 (from Figure 15.3) in the scene.

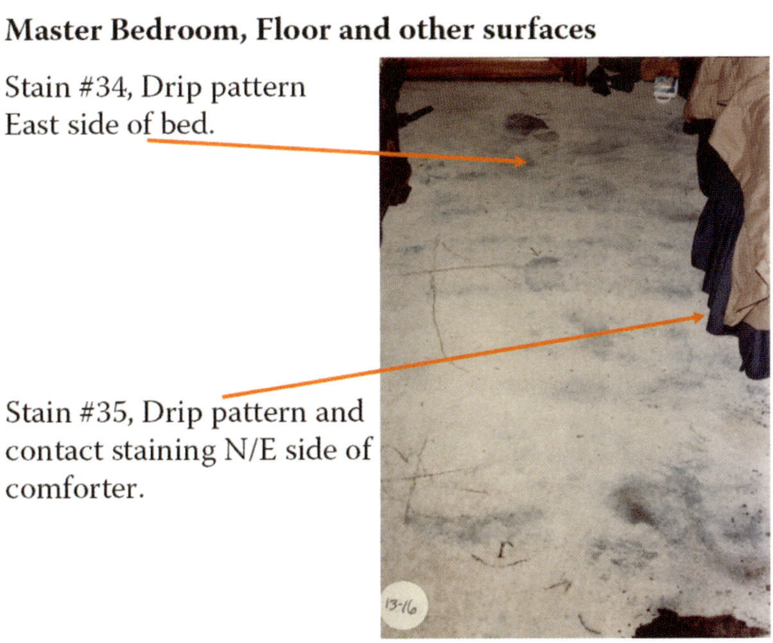

Figure 15.5 This slide shows two stains on the bedroom floor (Stains #34 and #35) using a crime scene photograph. The annotated arrows allow the analyst to help the jury orient where these patterns are located in the photograph. This is effective when multiple stains overlay each other in a photograph.

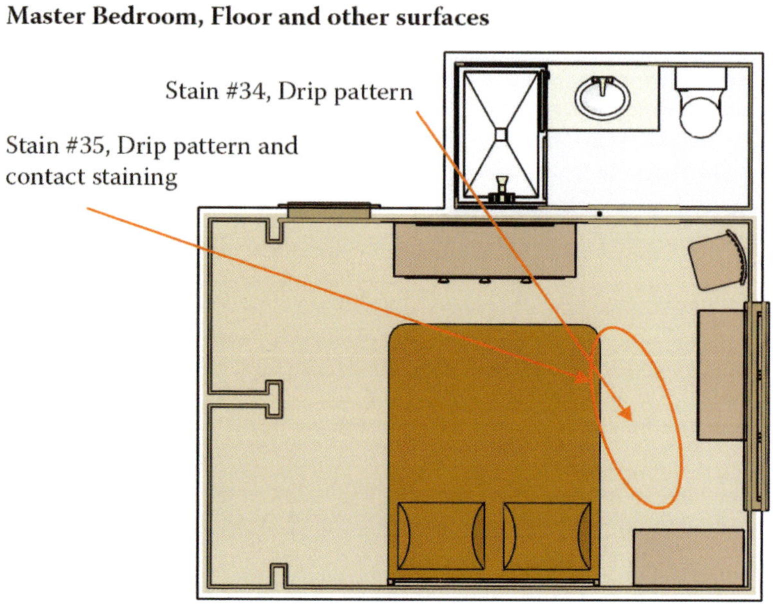

Figure 15.6 Another overhead view orients the position and extent of the two stains shown in Figure 15.5, helping the jury better understand the context and relationship of the patterns.

The human figures are combined in various ways with other pictures to illustrate various points (see Figures 15.7 and 15.8).

Through these software packages, the analyst can create any number of presentation slides to help explain the location and interrelationship of various items of evidence, the results of some experiment conducted, or opinion. The authors prefer a presentation alternating between a scene photograph and a room view. This combination will ensure that the jury can see the stain being discussed and know exactly where it is in the room.

Bloodstain Pattern Analysis Software Applications

Several automation programs may aid the analyst in analyzing and creating demonstrative aids. TRACKS was one such program. Designed by Dr. Alfred Carter of Forensic Computing of Ottawa, TRACKS allowed the user to adjust variables regarding blood droplets (i.e., speed, angle, size) and discover what these changes do to the droplet's flight path. Although primarily intended for students, screenshots taken from TRACKS could be utilized in the courtroom to clarify teaching points to the jury.

Backtrack/Win and Backtrack/Images were software packages that allowed users to evaluate spatter patterns. Using standard equations, the software did the tedious math necessary to define the area of origin. The software then presented the data in both text and graphics. It uses standard equations and physical laws to determine where a droplet may have originated. The programs provide an accurate estimate or representation of the spatter event. Similarly, current programs such as HemoSpat and HemoVision do the same. Although running such programs in court would be inappropriate, screen captures can be inserted effectively into demonstrative presentations.

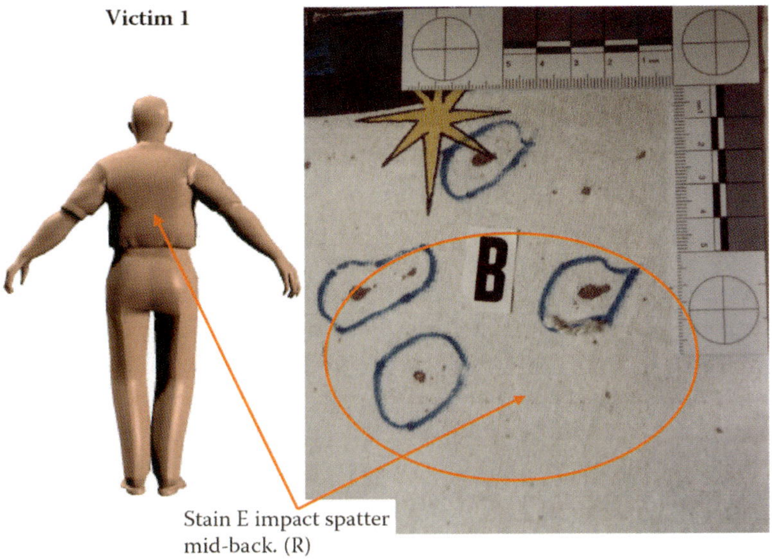

Figure 15.7 In this slide, human animation software created the "representation of the victim." Just as we used the scene views and photographs, this slide allows us to orient the position of Stain E to a specific position on the victim's body.

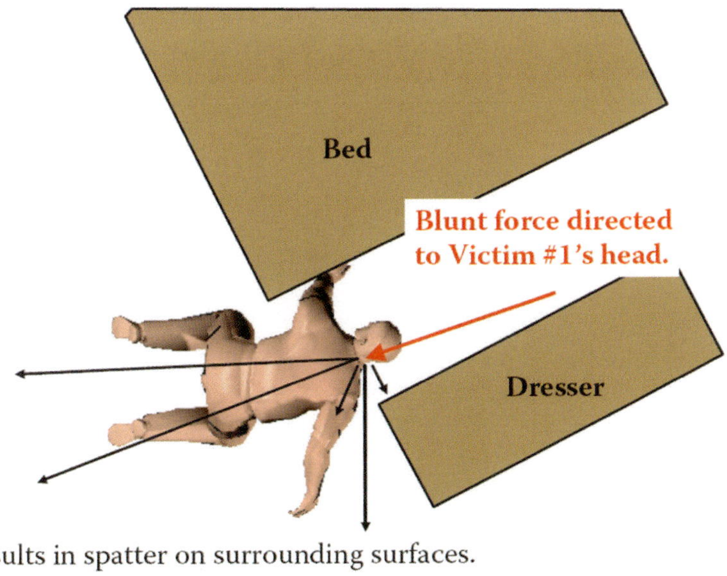

Figure 15.8 After placing the "model" in a position in the virtual scene, we can generate additional graphics to demonstrate the position and relationship of the figure to other objects in the scene.

Another developing trend in computer-aided analysis is the use of imaging software. Often, analysts are presented with poor-quality photographs of evidence and stain patterns. Imaging software allows the analyst to enhance photographs and view details with greater clarity.

Color modification and general enhancement are similar and well-suited to Bloodstain Pattern Analysis, allowing greater clarity in the photograph. For example, a particular color in the image is chosen and modified to stand out more clearly. Color modification works well because the computer distinguishes between colors or shades of colors more effectively than the human eye. The analyst identifies an area in question, and the computer modifies the given pixel color at that location to whatever color the analyst desires. This modification allows better human discrimination by increasing the contrast or shade of all similar pixel colors in the picture. Of note, this method often enhances transfer patterns, allowing better visualization.

A simple form of this process is color reversal. Consider Figures 15.9 and 15.10, which demonstrate this technique in enhancing a spatter and transfer pattern. In the original photograph (Figure 15.9), the dried bloodstains appear dark and do not stand out well against the blue color of the pants. Reversed (Figure 15.10), the pants become red while the patterns become white, making their form and boundaries more evident.

The difference between alteration and enhancement is a major concern in image enhancement and modification. In a presentation to the IABPA, Reeves made the point that enhancement increases the value, while alteration implies a change to the original image. The difference between the two is not that great.[7] The analyst walks a very fine line in the business of image enhancement – one in which the court may or may not approve of the methods. To counter these concerns, the analyst should document each modification process employed and its overall effect, and, wherever possible, create an exhibit

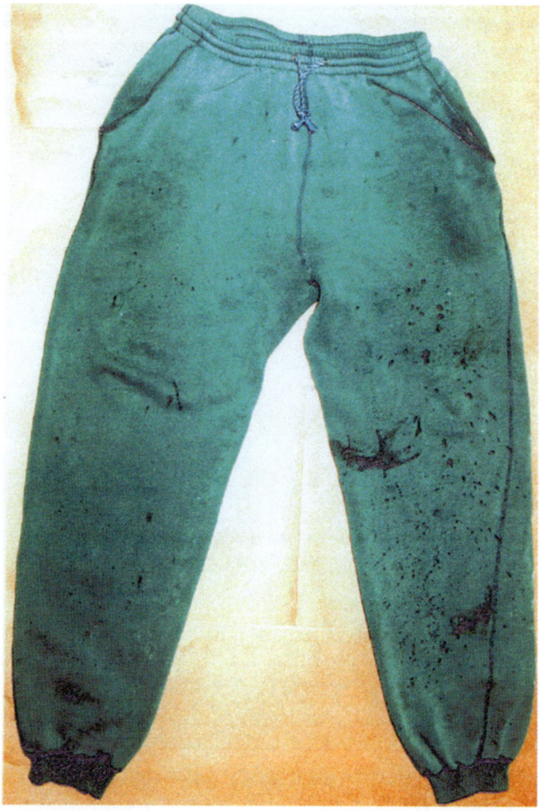

Figure 15.9 An unenhanced computer-generated image from a photograph of a pair of bloodstained pants. Although spatter and stains are evident, the dark-colored blood on the blue background makes viewing difficult. (Photograph courtesy of Norman Reeves.)

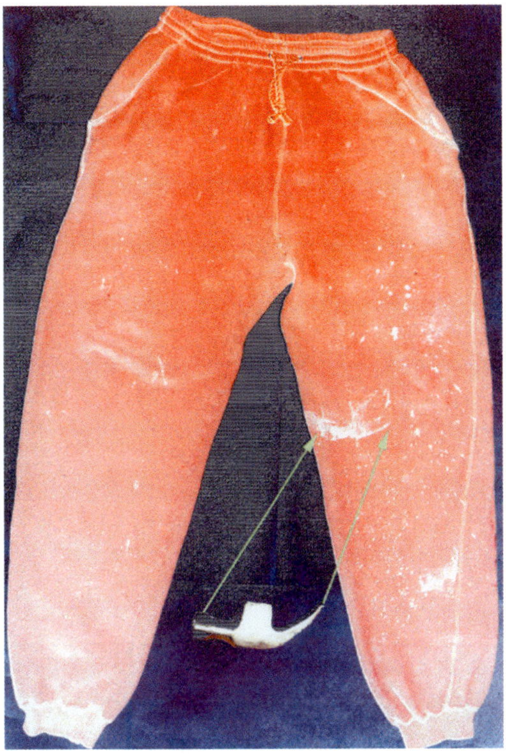

Figure 15.10 An enhanced image of the same photograph in which the analyst inverted the colors. The blue background becomes red, and the dark bloodstains become white. The resulting increased contrast allows for a more functional evaluation. The analyst then superimposed a weapon onto the photograph for comparison to a pattern transfer. (Photograph courtesy of Norman Reeves.)

reflecting each enhancement stage. Since digital cameras have existed and been in use for many years, digital image enhancement is understood by the general public and attacks on digital enhancement are rarer.

Image enhancement is an area that can be exceptionally technical. For example, E-Systems of Dallas developed an image enhancement process clearly suited to both Bloodstain Pattern Analysis and crime scene reconstruction. The method employs techniques and algorithms originally developed for space reconnaissance imaging. E-Systems refers to this process as perspective transformation.

In a case involving one of the authors (Bevel), E-Systems enhanced a photograph of a tile floor with a large bloodstained area (see Figure 15.11). Within the bloodstain is a voided section of specific interest. The area shows indications of the presence of a footwear pattern. The problem associated with the photograph is the oblique angle at which it was taken. This angle introduces obvious distortion to the voided section. Unfortunately, this was the only photograph available of this critical evidence.

Using advanced photogrammetric techniques, E-Systems was able to correct this view. Based on the original photograph's data, the corrected image allowed the viewer to look straight down onto the floor.[8] Although no scale of reference was included in the original photograph, the computer model used the dimensions evident in the floor tiles to build a control grid (see Figure 15.12). This allowed a projection of the original oblique perspective

Figure 15.11 In this crime scene, the photograph is evidence of a footwear impression in the stained area. Unfortunately, the oblique angle of the photograph causes a major perspective distortion, making the photograph of no use to the analyst. Note the lack of a scale of reference. (Photograph courtesy of Henry Muse, Dallas, TX.)

to the vertical perspective. The pattern transfer itself was then enhanced in the image using a mathematical model that corrected imprint details missing from the original image. Figure 15.13 is the enhanced version of the photograph.

These methods go far beyond the simple image enhancements available using off-the-shelf computer software programs. To some extent, E-Systems' methods allow the analyst to correct photographic mistakes during the initial scene processing. These methods are obviously not appropriate for every case, but they are available in instances where a significant enhancement might make the difference in the value of the photograph.

The development of scene scanning systems (e.g., FARO, Leica) represents another source for inclusion in any demonstrative. Three-dimensional scanners produce a point cloud capture of the scene, from the point cloud one can render a complete 3D visual of the area scanned. Virtual cameras within such software allow the user to view any aspect of the scene from any position desired. Similar to the use of architectural software, these captures are an excellent source of graphics. Unlike the architectural software, the view produced is exact to the scene and nearly photographic in nature. Specific captures of a particular view can be included, as well as virtual tours (e.g., a 360° fly-around) to better illustrate the scene to the jury.

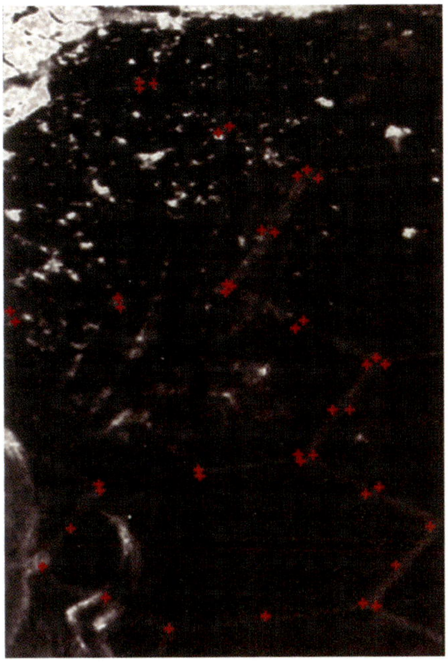

Figure 15.12 Using the tiles as a scale of reference, the computer program created a control grid, which allowed transfer of the angled perspective from the original photograph to a vertical perspective. (Photograph courtesy of Henry Muse, Dallas, TX.)

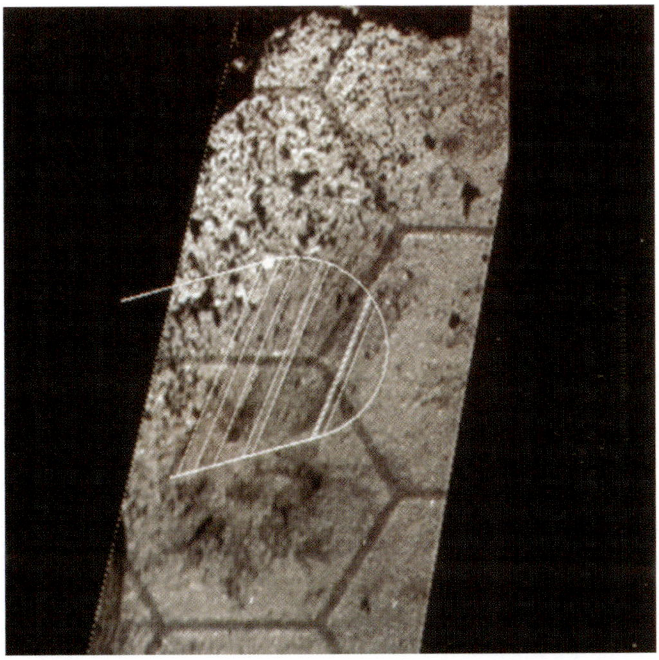

Figure 15.13 E-Systems of Dallas enhanced the original photograph (Figure 15.11). Using a mathematical model, they filled in the missing detail from the pattern transfer and highlighted the entire enhanced pattern in white. Although certainly an "alteration" of the original image, this is an example of a substantive programming technique that serves analysis and reconstruction. (Photograph courtesy of Henry Muse, Dallas, TX.)

Summary

If the reader remembers nothing else from this chapter, remember always that the analyst is a fish out of water in the courtroom. The rules, in effect, have little to do with the analyst's own. It is easy to be drawn in by skillful attorneys who will turn a simple slip of the tongue or minor mistake into a major production. In doing so, they know the evidence will lose center stage as the jury becomes enraptured by the theatrics.

By remaining within comfortable limits in the conclusion itself, by preparing properly, and by applying careful attention to the manner in which they testify, analysts can preclude many opportunities to be party to such courtroom theatrics. Once the evidence is heard, so long as the analyst remains an unbiased and objective expert, the court or jury must ultimately decide to accept or reject the testimony.

A critical concern for the analyst is preparing for Daubert style challenges. Not only must analysts be prepared to defend their own credentials, but they must also come prepared to defend the discipline. This means understanding the foundation and history of the discipline, articulating a concise methodology based on the scientific method, and demonstrating that the analyst utilized such a methodology.

Finally, the analyst must recognize that explaining a conclusion often demands the use of demonstrative evidence. As Bloodstain Pattern Analysis is a graphic discipline, it lends itself to demonstrative presentations far better than many other disciplines. The analyst should use the various automation applications available to build functional demonstrative aids to assist the jury in understanding the scene and the analysis.

Notes

1. Hicklin, R. Austin, Accuracy and Reproducibility of Conclusions by Forensic Bloodstain Pattern Analysts, Elsevier, Forensic Science International, Volume 315, August 2021.
2. Jones, H.W., *Law and the Social Role of Science*, Rockefeller University Press, New York, 1966, p. 124.
3. Peterson, J.L., Ethical Issues in the Collection, Examination and Use of Physical Evidence, in *Forensic Science*, 2nd ed., American Chemical Society, Washington, D.C., 1986, pp. 35–48.
4. Dershowitz, A., As William F. Buckley, Jr. quoted in "Simpson Lawyer Ambiguous on Verdict," United Press Syndicate, June 9, 1996.
5. Faigman, D.L. et al., *Modern Scientific Evidence, The Law and the Science of Expert Testimony*, Thomson West, Eagen, MN, 2005, Volume 4, p. 5.
6. Arizona Prosecuting Attorneys' Advisory Council and Arizona Law Enforcement Officers' Advisory Council, Courtroom Demeanor, Phoenix, AZ, 1982, pp. 21.01–21.15.
7. Reeves, N.H., Computer Digitizing and Enhancements of Photographs, presentation to the joint training conference of the International Association of Bloodstain Pattern Analysts and Association for Crime Scene Reconstruction, Oklahoma City, OK, October 6, 1995.
8. Muse, H. and Bowden, R., E-Systems Forensic Image Processing: A Case Specific Example of Softcopy Exploitation, presentation to the joint training conference of the International Association of Bloodstain Pattern Analysts and Association for Crime Scene Reconstruction, Oklahoma City, OK, October 7, 1995.

Experimentation in Bloodstain Pattern Analysis

16

In unraveling an investigative puzzle, the analyst may have to experiment to clarify a given circumstance. This is particularly true for the bloodstain pattern analyst, who routinely is asked to decide if a particular set of circumstances can create a questioned stain.

When required, the goal of any such experiment is simply to include or exclude some event or set of events as a source of the questioned pattern. With this in mind, we will consider some general guidelines and routinely encountered pitfalls that affect the experimental effort, looking at the basic principles for designing case-specific experiments. The analyst will likely maintain an investigative focus by considering the following factors and guidelines. These guidelines will also help prevent the experimental effort from becoming too subjective. There are five initial considerations:

1. *Define What Is at Issue:* What exactly is the analyst trying to establish or learn? Be clear from the start and maintain focus throughout the effort.
2. *Consider All Viable Explanations for the Result or Pattern:* The analyst cannot go looking for their desired answer. It is data that define the conclusion and not the conclusion that defines the data. As such, it is imperative or essential that the analyst consider all viable possibilities.
3. *Eliminate the Ridiculous:* Particularly when considering bloodstains, given a little time and ingenuity, the analyst can reproduce a particular pattern in various ways. For instance, a misting stain is easily created using a spray bottle, but was a blood-filled spray bottle found in the scene? Certainly, consider all possibilities, but do not waste investigative resources and time on flights of fancy.
4. *Brainstorm with Others:* Analysts cannot always rely on themselves to maintain objectivity. Consider asking investigators, other analysts, attorneys, or individuals not directly involved for input.
5. *Concentrate on Elimination:* The analyst should concentrate their effort on eliminating possibilities that are "not consistent." Do not put emphasis on "identifying" the event. It is not the case that the analyst can associate a set of circumstances to a specific event and always absolutely exclude all other possibilities. The scientific method works on the idea of falsification. Simply by eliminating impossible circumstances, a level of clarity is brought to the situation.

Considerations for the Design and Conduct of Experiments in Bloodstain Pattern Analysis

It is almost inevitable that the analyst will be presented with a situation demanding experimentation. Each bloodstain pattern analyst needs a basic understanding of how to design and use experimentation both in support of and as an integral part of the analysis.

DOI: 10.4324/9781003408734-16

This understanding begins in a basic Bloodstain Pattern Analysis course, where practical exercises and experiments are designed to expose the student to the most common mechanisms of bloodstain production and the nature of the resulting stains and patterns. These exercises offer the analyst exposure to experimentation, including the need to identify a specific objective or purpose and the necessity of controlling variables. What is learned there will act as a foundation for subsequent casework. These experiments, however, do not provide a comprehensive exposure to the vast array of bloodstain producing mechanisms, nor to the variations of circumstances that might be encountered in casework. Analysts will often find complex patterns produced through very dynamic interactions or that result from a combination of crime scene variables. When presented with these circumstances, case-specific experimentation may be necessary.

Numerous texts about scientific and research methodology are available, and a single chapter will not address every aspect of experimentation. This is, however, an important area within our discipline and needs to have, at minimum, a cursory discussion.

In Bloodstain Pattern Analysis, experiments are usually pursued to identify the best explanation among the multiple hypotheses developed that might explain the source of a questioned stain or pattern. This is accomplished by eliminating among the multiple hypotheses those that are refuted. These experiments follow the scientific method. The scientific method involves identifying a discrete question, gathering data to resolve the problem, forming a hypothesis to explain the observation(s), making predictions for the hypothesis, testing each hypothesis, and then drawing conclusions to support or refute each hypothesis. Applying the scientific method in the experiment usually demands engaging in multiple tests, as the analyst must control all variables and change only one variable in each test. Without this control, the analyst would not be able to state with any level of certainty how each variable affected the resulting pattern.

In essence, the analyst looks for cause-and-effect relationships between the variables considered in an attempt to explain how these variables might interact to produce the questioned pattern. The hypothesis is an educated best explanation at what this cause-and-effect relationship is.

In the available literature and references, it is immediately evident that no absolute method or agreed upon series of steps universally defines how the scientific method should be carried out and applied. Irrespective of the author, the scientific method has certain common themes. The following is a discussion of these common steps related to conducting scientific experimentation.

Identify the Investigative Question

Before conducting a meaningful experiment, the analyst must first know what question they are trying to answer. Thus, the analyst must form and state a purpose for any experiment. This statement is also known as the investigative question (IQ) in crime scene analysis. This investigative question is specific and narrow in focus. The broader the question, the more likely there will be an increase in variables. The more variables involved, the more difficult it is to design an experiment that can functionally test them independently. When presented with a broad issue, the analyst will find it necessary to break the broader question up into more discrete issues. Each of these smaller issues or questions is asked and answered independently, and the results are then combined to answer the broader question. In Bloodstain Pattern Analysis, the most often posed investigative question is: "Was a particular mechanism the source event for this pattern?"

The question, will this circumstance produce an exact duplicate of the questioned stain, is not the question of concern. Exact duplication of a pattern is not the goal. More often than not, some variable is considered as to whether that particular variable will affect the ultimate outcome. A classic example is in coagulation studies. Rarely is the analyst trying to duplicate the coagulation, and associate it to an exact time. Generally, they are attempting to identify a range of results for a given time. This is a point often lost on counsel. Always know what you are attempting to define, be clear in posing the question and focus on that.

Initial Observation and Information Gathering

The need for experimentation begins with the initial observations of the scene and the analyst's consideration of the various reasonable events or variables that could have produced the bloodstains in question. Given the class characteristic nature of Bloodstain Pattern Analysis, it is not uncommon that there may be several possible source events associated with any given scene. The analyst uses their experience and/or past experiments as an initial evaluation method. If this initial evaluation does not answer the question in the analyst's eyes, then specific experimentation may be necessary.

After gathering all available and applicable investigative reports, laboratory reports, autopsy protocols, photographs, diagrams, and physical evidence, the analyst studies these to become familiar with all associated details. When preparing for the experiment, it is a good idea for the analyst to review recent literature, textbooks, or other professional publications, looking for any information or ideas that might impact the experimental design.

Identify Variables and Form a Hypothesis

Using the information gathered, the analyst develops one or more theories, based on their training, experience, and past experiments, as to what types of actions or variables may have created the stains or patterns in question. When considering these possibilities, a possibility does not need to be tested if it is excluded by the scene or event. Mike Haag refers to this concept as the "limited universe."[1] Limited universe simply means there is no need to include every possibility within the universe that might explain the pattern. Instead, we consider only those actions that appear viable within our limited universe of the crime scene in question.

This step of identifying the possible variables is critical before the analyst can consider moving forward in the experimental phase. Each possibility considered viable becomes a hypothesis that should be tested.

Considering what variables affect each individual hypothesis is of equal importance to the analyst. Examples might include the volume of blood involved, the nature of the impacting force, or the orientation of the weapon to the wound site. The analyst must consider how that variable will be altered during the experiment. While testing these variables, the analyst must control and change the variables one at a time. If more than one variable is changed at each stage of testing it is difficult to understand their independent effects on the experiment. In that case, it is difficult, if not impossible, to know or recognize which altered variable is changing the pattern observed in the experiment. Keep in mind, however, that oftentimes these variables are linked and work together to cause the end result. In the beginning efforts, it is often best to work with those variables the analyst believes are acting independent of each other.

As indicated, the initial hypotheses are based on prior training, knowledge, experience, case facts, and other experiments conducted in the past. Generally, the analyst answers the question of what caused the observed bloodstain pattern by comparing it to previous experiences involving similar patterns. This experience shows the analyst what basic actions can be excluded as a source and what actions might explain the pattern. Suppose the issue surrounding a given hypothesis is not unique. In that case, the analyst's prior experience may be adequate to answer that aspect without creating a specific experiment (e.g., the basic nature of a radiating impact spatter pattern). In more complex instances, however, the hypothesis may involve unique or never-before-encountered issues demanding significant experimental effort and design.

Design a Functional Experiment to Test Your Hypothesis

For an experiment to provide valid results, it must be designed correctly. In part, this means considering the variables that must be tested during the experiment, determining how they will be adjusted, and deciding which variables may require linking. The design generally requires creating a control test. This control is an additional experimental trial, done under the same conditions as the other experiments, except that no variables are changed during the course of the control experiment. The control acts as a neutral "reference point" for comparison. It ensures that some unrecognized environmental factor is not in play in the analyst's experimental procedure and allows the analyst a greater level of confidence that any change observed in the experiment is an effect of the altered variable.

Experiments are typically done several times to guarantee that the observed effect is reproducible. Reproducibility is a crucial requirement. As the reproducibility of the observed effect increases, the chance that the analyst has introduced some experimental error or is observing some random effect is reduced.

Obtain Materials and Equipment

After considering the design of the experiment, the analyst should list the equipment or items needed. These are gathered and prepared in advance. Remember, changing only one variable during each iteration of the experiment is important. Time is often a variable, so if the analyst is unprepared and stops the experiment to obtain or prepare some aspect of the experiment, they are, in effect, introducing a new variable.

A significant consideration for materials is the need for human or animal blood and whether that blood will be freshly drawn or from a blood bank. In spatter pattern creation experiments, there is little or no difference in the results based on the type of blood used. Freshly drawn blood is usually necessary for drying, clotting, or serum separation experiments.

Conduct the Experiment and Record the Data

Experiments are conducted changing one variable at a time. Thus, a series of experimental runs will make up any experiment. During each run, the analyst should record how and in what way, if any, the changed variable affected the stains under study. Documentation of the experiment should be in the form of written notes. These notes will include the nature of the variable, specifically how it was changed, and the observed change in the pattern that resulted from the manipulation of the variable. The nature of the observations may be descriptions of what was noticed or observed, or they may be actual measurements of changing aspects of the stain pattern (e.g., size, volume, or drying time). In addition to maintaining accurate notes, the resulting observations can be captured using photographs. Ultimately, all the observations will be critical when drawing conclusions at the end of the experimental procedure, so clear and precise documentation is needed.

Analyze and Summarize Results

After concluding the experiment, the analyst must consider the interrelationship of the observed effects in the experimental stains and the variables adjusted throughout the experiment. These effects are then considered relative to the questioned stain or pattern from the scene, in an attempt to resolve the investigative question. Experimental design may still be a factor. After review, it may become necessary to redesign the experiment itself to test an observed effect more accurately and resolve minor issues or contradictions between the results and the questioned stain. However, it should be noted that Bloodstain Pattern Analysis is class characteristic evidence; thus, it is doubtful that any experimental procedure will reproduce every aspect of a questioned pattern. There are simply too many variations in the variables from the scene, many of which are and will remain unknown to the analyst. Experiments cannot reproduce an action in every respect, the experimental effort is not designed to produce an exact "copy" of the questioned pattern.

Once all experimentation is complete and correlation of each variable's effect on the pattern is defined, the analyst should summarize their findings. This can be in a written statement or by the use of tables or graphs. Based on the observed results, the analyst then seeks some conclusion that may assist in supporting or refuting the hypothesis in question.

State the Best Explanation

In terms of defining source events, ultimately the analyst is left to consider the viable possibilities and the observed effects created by changing variables, and to then decide if the observed experimental results support or refute any of the possibilities. Obviously, it would be best to support only one hypothesis, but this is not always the case. A hypothesis cannot be ruled out if, through experimentation, the associated mechanisms/variables produce similar types of stains. However, this does not prevent the analyst from offering an opinion of the "best explanation." Wilson commented on this issue, stating, "If two different hypotheses fit the observed facts and if one is clearly simpler than the other, it is customary to accept the simpler until further evidence causes its rejection."[2] This is a simple application of the principle known as "Occam's Razor," which states that the best theory is the one with the fewest independent assumptions.

Whatever the circumstance encountered, the analyst attempts to use the observed trends and results from the experiment to answer the original question. The end result of this analysis is generally achieved by identifying the "best explanation" based on the available data and variables that were tested.

In considering this concept of best explanation, an objective analyst should not claim that any hypothesis, however extensively tested, was a statement of absolute, universal truth. A hypothesis can always be proven false, but when considering a complex issue absolute answers and one hundred percent certainty do not exist. Thus, the conclusion expressed is simply the best explanation given the available data. Under most situations, it is impossible to test all viable conditions and variables; there are simply too many unknowns in the scene. Thus, someone with a slightly different perspective may find a condition under which the analyst's best explanation hypothesis does not hold true.

Maintaining a Reality Check, Comparing against the Crime Scene

A common error that analysts make in experimentation is not comparing their opinions against the known events within the crime scene, autopsy, and laboratory reports, or not applying known variables from the scene to the experiment. If, for example, a pattern is ultimately identified as an arterial spurt, then the autopsy should identify that some artery was compromised. If it does not, it is imperative that the analyst speak with the ME and see why this contradiction exists (e.g., MEs generally only report damage to major arteries, which does not eliminate the existence of smaller damaged arteries). Another example of checking reality is found in expectorate action. The analyst should validate that the victim had blood in their air passage or that the autopsy identified aspirated blood in the respiratory system.

It is just as important to apply Haag's "limited universe" concept to the conclusions offered. Were the mechanisms and variables evident in the crime scene faithfully reproduced as closely as possible? If not, the results of such experimentation are unlikely to assist the analyst in refuting or corroborating a given hypothesis. There have been cases in which analysts (consciously or subconsciously) "tweaked" the experiment to reproduce a questioned pattern. In doing so, they leave the reality of the crime scene conditions behind and introduce variables that were not in play at the scene. Suppose no evidence or information is available to suggest that some action occurred in the scene. In that case, it is a mistake, if not dishonest, to identify or insinuate that such an event is the possible source for the questioned pattern.

Experimental Errors

Wilson commented on the basic nature of science, stating, "All science rests on the idea that similar events occur in similar circumstances."[3] One of the greatest difficulties in Bloodstain Pattern Analysis is the inability to recreate exactly the events from the scene. No one can create identical circumstances found in beatings, shootings, and other bloodstain producing events. These situations are simply too dynamic. No two circumstances are identical; at the very least, they are different in time and place and many other aspects. These difficulties may lead to errors in experimentation.

If all observed results are inconsistent for the various hypotheses after experimentation, this may suggest some form of experimental error has been introduced into the design. The obvious questions become: Have all viable alternatives been examined, and were the possibilities too refined or limited (e.g., a single position for the application of the weapon)? Designing experiments requires a critical eye to evaluate the experimental design, both in the initial design stages, and later to determine if experimental errors were in play in the observed results. When encountered and recognized, such errors demand rethinking the design of the experiment. Two procedures will assist the analyst in identifying and correcting such an error. First, the analyst should review each step of a procedure to find the source of potential errors. Second, having a second analyst review the procedure and experimental design is often helpful. Sometimes the original analyst will miss an obvious error in the experimental design, while a "neutral" analyst using a fresh perspective will easily identify the problem.

If the analysts apply themselves to their design effort, the most common error encountered is simply a random error introduced into the experiment by doing something differently from one experimental run to the next. An example of this would be when the analyst fails to clean some part of the equipment used and, in doing so, alters (i.e., increases) the amount of blood introduced in each iteration of the experimental run. The analyst must faithfully follow the experimental design to eliminate these random errors.

Pitfalls to Experimentation and Reconstruction Attempts

There are always pitfalls in any effort directed at a bloodstain reconstruction attempt, and these issues carry over to the experimentation conducted in support of any reconstruction. The following are some of the more common pitfalls to consider.

- *Groupthink:* The group involved may form a type of tunnel vision and collectively eliminate concerns that individual members clearly recognize as important. In recreations, we often employ a group effort, and "groupthink" concerns are common. Any concerns or criticism by team members of the methods employed should be objectively considered.
- *Designing an Experiment to Prove Your Result:* Since analysts control the variables, it is entirely possible that they may either consciously or subconsciously create a circumstance that defines exactly what they want. For instance, if trying to disprove that a beating caused a stain, the analyst could fail to use enough blood to create spatter. Thus, the weapon involved would obviously not cause spatter.
- *Failing to Take a Holistic Viewpoint:* Tunnel vision in the individual or the group can eliminate objectivity and doom experiments to failure. As described, design the experiment around the case facts. Consider the results with those same facts in mind. If analysts know they are dealing with a contact wound, then recreating a noncontact wound fails to consider the case facts. The analyst must consider all known facts and information then ensure the experimentation accounts for and includes that knowledge.
- *Confirmation Bias:* The simplest way to explain confirmation bias is that the individual starts with a conclusion and then seeks data that supports it. Rather than trying to disprove a possibility, which is the most effective function of the scientific method, the analyst simply seeks data to support their specific theory.

As a final consideration, be cautious in the exclusion of events. If the analyst chooses to exclude some event as possible, they must have a foundation for exclusion. "Gut feelings" and "Because I think so" are inadequate reasons for exclusion in an experiment or reconstruction. When all is said and done, the analyst's conclusion must agree with their experimental results.

Case Example 1: "Painted Fibers"

This example involved an accused who had bloodstains and tissue on the front of his shirt, allegedly from a shooting event involving multiple parties. The critical question was how these stains got on the shirt. Were they contact or backspatter? Keep in mind, the ultimate guilt or innocence of the accused in this particular case was up to a jury. Integral to this question, one particular hypothesis that was presented in the case was a statement that the stains were contact transfers from brushing up against one victim. Supporting the hypothesis that these stains were simply contact, the analysts reported that individual fibers or threads coated with blood were present. This observed effect was referred to as "painted fibers" and were found disassociated from the other stains. The concept of a "painted fibers" had never been introduced, discussed, or studied in any reference, nor had any study been offered in any professional journal or meeting before trial. The concept was presented to the jury without peer review. Nevertheless, it was stated that these painted fibers or threads would not occur as a function of backspatter associated with a shooting event. Thus, as reported to the jury, the presence of the painted fibers or threads on the shirt in question proved the stains had to be a function of contact transfer. Lacking any prior discussion or research on this issue, and because no prior experiments involving back spatter from shooting a firearm into a wet blood source had produced such a result, the painted fibers seemed to support the hypothesis of a contact transfer.

Presented with the situation, the opposing analyst was left to address if any other variables were present that might affect or produce the resulting painted fibers. One obvious variable in this case was the probability that the shirt involved was damp. This factor introduced another variable that required testing, since all prior experiments involving backspatter (done by the author(s) and others) had always used dry clothing. In these instances, the only observed result similar to the painted fibers was the creation of tiny blood spheres on the threads. No known effort had been directed at experiments involving damp or moist fibers or threads. To evaluate this new hypothesis, an experiment was designed in which the fibers or threads were lightly dampened and then back spatter was directed to the surface. The result of this experiment was the small spheres still landed on the fibers, but with the addition of the damp clothing, these tiny blood spheres wicked or migrated along the fiber or thread and produced the "painted" appearance observed in the actual case (see Figure 16.1).

The fact that painted fibers or threads occur from backspatter landing on damp clothing did not disprove the transfer hypothesis presented at trial. It simply showed that the existence of "painted fibers" could not be used as a foundation to claim that contact transfer was the only viable source of the pattern in question. As it occurs in both instances, the painted fiber becomes a moot point in identifying the questioned pattern as either contact or back spatter.

Case Experiment 2: An Odd Impact/Complex Pattern

In this example, a female was found in bed on her back. A .38 revolver was on the bed beside her. There was a hard contact entry wound to the right temple, with a right-to-left and down-to-up trajectory. There was no apparent exit wound (see Figure 16.2).

The primary investigative question was how the bloodstains on the wall (Figures 16.3 and 16.4) could have been created from this shooting occurrence. The stains were

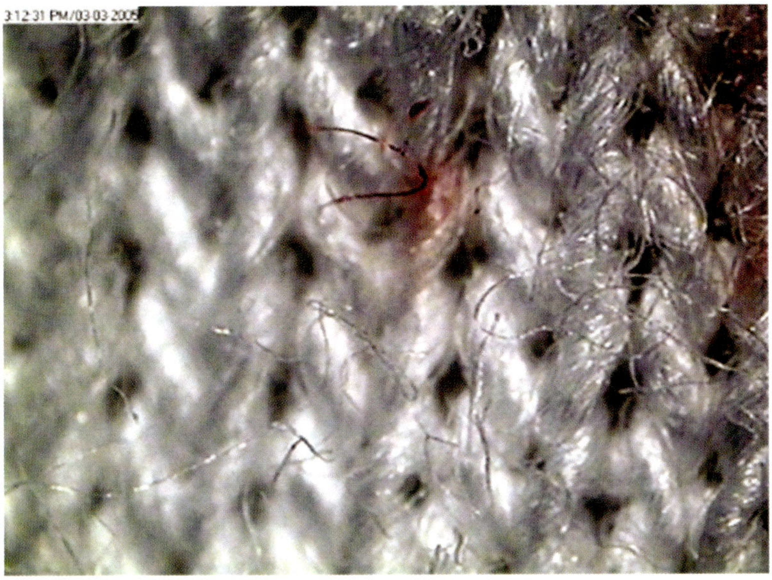

Figure 16.1 A "painted fiber." After appropriate experimentation (e.g., considering and testing all viable possibilities), it was established that "painted fibers" can be produced by both contact and spatter events. In the context of the case, the painted fiber served no function in refuting either one of the potential mechanisms.

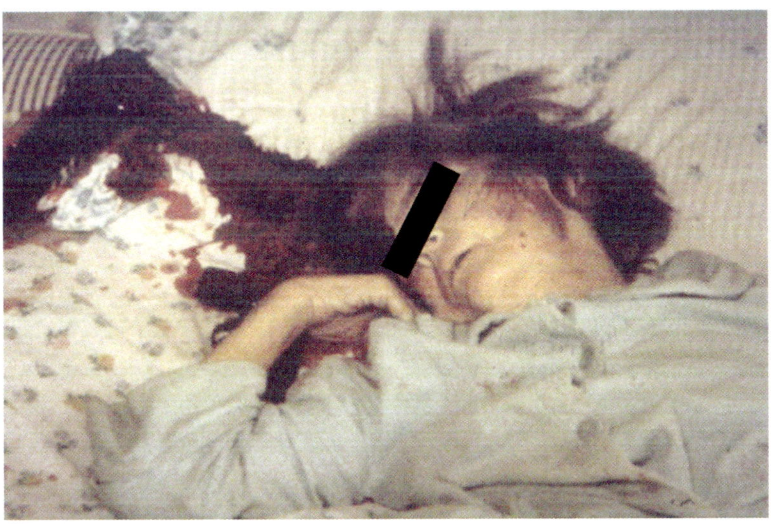

Figure 16.2 The victim of this incident died as a result of a single penetrating gunshot wound to the right temple. There was no exit wound at autopsy.

approximately 6′ away from the victim's head as she lay in her final position on the bed. The pattern indicated a distinct volume of blood was involved, with directionality upward and slightly to the left. There were many submillimeter stains surrounding the main area of pattern. There was also a bullet ricochet to the right edge of the main area of bloodstains. The blood was tested and found to be consistent with the victim.

Figure 16.3 This pattern was present on a nearby wall, over 6′ away from the victim. The blood present in the stain was determined to be consistent with the victim. Given the final position of the victim, the lack of an exit wound, and the atomized nature of the stains in the pattern, it was obvious the pattern could not be reconciled with the known facts.

Figure 16.4 A close-up photograph of the stain in question. The stain reflects distinct features typical of impact spatter associated with gunshot events, but also indicates some volume of blood was involved. Adding to the analyst's concern was the presence of a bullet ricochet in the pattern present on the ceiling above this pattern.

Photographs, a scene diagram, and a synopsis of the facts were sent to two peers. Two different opinions were returned for consideration. One felt that an exit wound was missed at autopsy. To check this possibility, the medical examiner's office was asked to check the X-rays for any overlooked wounds. None were found. The second peer felt that the victim was standing up in bed and the bullet fragmented upon striking the skull. To check this, the bullet extracted at autopsy was weighed and compared to the weight of a whole .38 projectile. The resulting information excluded this possibility.

A round-table discussion was held between the firearms examiner, bloodstain analyst, serologist, detective, and crime scene technicians to consider possible occurrences. One thought offered was that the weapon, while resting on the bed, received a volume of blood down the barrel. The gun was then held up toward the wall and fired a second time by another party.

This possibility was tested using the same weapon and ammunition. Blood was placed into the barrel, and it was immediately fired at a white cardboard target. The weapon was held approximately 18″ away and at a 45° angle. The blood did not produce a similar pattern. In fact, very few bloodstains resulted.

During a subsequent discussion of the experiment and its result, one of the members mentioned that not all variables had been considered. He offered that any passage of time would congeal and thicken the blood in the weapon's barrel. It was felt that congealing might affect the resulting stain pattern. A similar test was conducted, this time allowing the blood to set in the barrel for successive intervals of 30 minutes. After a cumulative period of 3.5 hours, the questioned pattern was replicated, complete with a bullet hole, to the edge of the main pattern area (see Figure 16.5).

Because of the effort directed at the experiment, it was later determined that the husband assisted his wife in her suicide. When placed on the bed after the suicide shot,

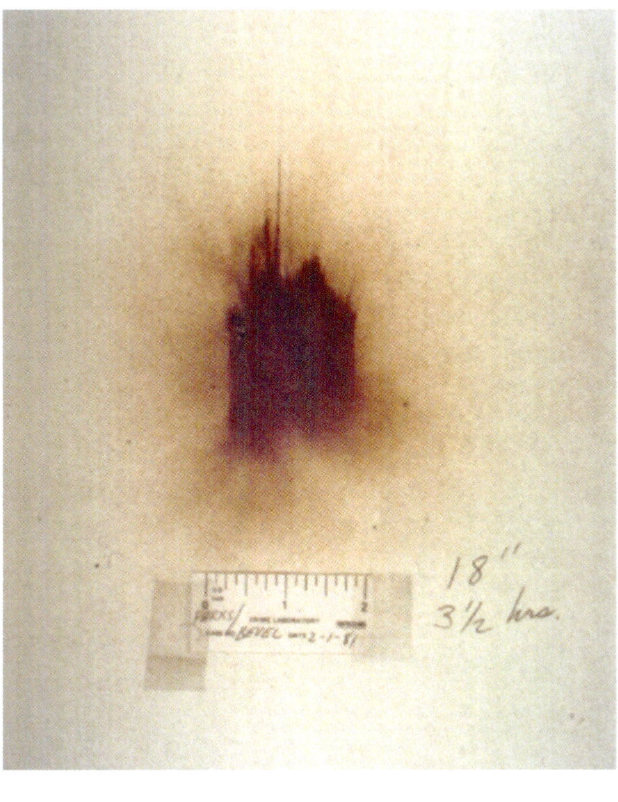

Figure 16.5 A photograph of the stain created during a series of experiments. By allowing blood to sit and congeal in the revolver for a period of ultimately 3.5 hours, the subsequent pattern produced was consistent with the pattern found in the scene.

the barrel filled with blood. This blood then congealed. The gun was left on the bed for 3.5 hours, after which the husband reentered the scene and fired the gun into the wall. The combination of time, congealing, and the second shot were all factors in the ultimate pattern observed.

Case Experiment 3: Spatter or No Spatter

In this example, the following question was posed: Where would blood spatter be found on the clothing of the attacker in a beating death? The involved analyst designed an experiment to test the viable possibilities. A plastic motorcycle helmet with carpet padding affixed to the top and sides of the helmet was used as the model. Whole blood was poured onto the top of the carpet pad. The analyst struck the top of the helmet ten times in succession with a flat-bladed nail bar. The result produced no spatter on the shirt or shorts worn by the analyst. Thus, the experiment was utilized to show that the spatter on the suspect's clothing was not the result of beating the victim, as alleged.

The cardinal rule for all experimentations is: Are the results observable, testable, and repeatable? One of the authors was asked to evaluate the reported results. In retesting this experiment, spatter was created on both the shirt and shorts consistent with the spatter patterns found on the questioned clothing, contrary to the reported results of the first analyst. This begs the question: "Why did the first experiment produce no spatter, while the same experiment conducted by another analyst did?"

This required examining the two experiments to determine which variable had been altered to create the contradicting results. In detailed questioning of the first analyst, it was determined that after the blood was poured on the padding, some period of time passed before the simulated beating began. As a result, much of the blood introduced into the experiment ran down the hard plastic helmet and gravity pooled the rest in the lower areas on the sides of the padding. This left little blood on the top of the pad (the location of the simulated wounding where the impacts were delivered) to create spatter. As one could easily predict, if there is no blood volume at the impact site to spatter, no spatter will be created. The variable involving the limited volume of blood used in the first experiment starkly contrasted with the scene's reality. In the case situation, as the attack continued, the victim's head introduced an ever-increasing volume of blood that was available to be dispersed in the scene.

It could always be argued that, consciously or unconsciously, the experiment was designed to prove the analyst's preexisting opinion. Irrespective of how an error is introduced, it is only through the ideas of the observable, testable, and repeatable practice that we identify problems within the experimental design and recognize confirmation bias rather than objective analysis.

Experiments versus Demonstrations

Sometimes, the analyst may be asked to perform or feel that a demonstration is needed. Demonstrations can be useful, particularly when the analyst is confident of how the stains or pattern occurred, or when a specific set of circumstances is presented as the "explanation." The analyst sets about to demonstrate or show the results of the identified action. In some instances, such a demonstration or reenactment will provide investigative information. Other times it is accomplished merely as a demonstrative aid for the jury's understanding of the event. The actions undertaken to create the desired demonstration are not an "experiment." Although some analysts refer to demonstrations and specific recreations as experiments, to be an experiment, there must be some form of control, and at least one variable changed over the experimental run.

Summary

Experimentation is a task the analyst will likely be called upon to perform at some point. It should not be taken lightly. The analyst should devote significant attention and effort to designing, performing, and documenting the experiments properly. Such experimentation aims to clarify the situation by considering viable hypotheses and seeking to eliminate as many of these as possible. Ultimately, after experimentation, whatever hypotheses remain possible must be considered against the case specifics to establish if some conclusion is possible. As with the application of the scientific method in any circumstance, the analyst recognizes this conclusion is offered as the best explanation given the available data.

Notes

1. Haag, M., Association for Crime Scene Reconstruction Training Conference, *Shooting Incident Reconstruction Workshop*, Albuquerque, NM, February 9–11, 2006.
2. Wilson, Jr., E.B., *An Introduction to Scientific Research*, Dover Science, Mineola, NY, 1990, p. 18.
3. Wilson, Jr., E.B., *An Introduction to Scientific Research*, Dover Science, Mineola, NY, 1990, p. 26.

Appendix A
Weight/Measurement Conversion Table (Approximations)

To Convert	Multiply by	To Get	Symbol
	Linear Measure		
Inches	25.4	Millimeter	mm
Millimeter	.04	Inches	in.
Inches	2.54	Centimeter	cm
Centimeter	.4	Inches	in.
Feet	30.48	Centimeter	cm
Centimeter	.033	Feet	ft
Feet	.33	Meter	m
Meter	3.3	Feet	ft
Yards	.91	Meter	m
Meter	1.09	Yards	yd
	Liquid Measure		
Fluid Ounce	30	Milliliters	ml
Milliliter	.03	Fluid Ounce	fl. oz.
Quart	.95	Liter	l
Liter	1.06	Quart	qt
Gallon	3.8	Liter	l
Liter	.26	Gallon	gal
Milliliter	1000	Microliter	µl
Microliter	.001	Milliliter	ml
	Weight		
Ounce	28	Gram	g
Gram	.035	Ounce	oz
Pound	.45	Kilogram	kg
Kilogram	2.2	Pound	lb
	Metric Unit Equivalents		
Unit	*Number of Units*		
Meter/Gram/Liter	1	one	
Deci	.10	one tenth	
Centi	.01	one hundredth	
Milli	.001	one thousandth	
Micro	.000001	one millionth	

Appendix B
Trigonometric Functions and Their Application in Bloodstain Pattern Analysis

Mathematics by definition is the systematic treatment of magnitude, relationships between figures and forms, and relationships between quantities expressed symbolically.[1] Mathematics allows us to recognize relationships in real objects in order that we may better understand them.

Bloodstain Pattern Analysis draws on mathematics in several areas that were outlined in Chapter 6. In particular, the analyst uses the trigonometric functions. These functions define the relationships between the internal angles that comprise a triangle and the length of its sides. Although the analyst may not fully appreciate these relationships, their specific application to our discipline is relatively easy to understand.

Here we consider only the trigonometric functions related to the right triangle and not those that define the plain triangle. The latter may be of some interest to the analyst, however, as authors have described methods by which the analyst can determine a convergence point using these relationships.[2] These methods are generally used in the algorithms of software designed to determine point of convergence.

To consider properly the trigonometric functions, the analyst must first understand four terms and accept two basic facts true to all triangles. The terms of importance are as follows:

- *Right Triangle:* A triangle in which one of the three angles measures 90°. In Figure B.1, the angle at A is a 90° angle, making the triangle a right triangle.
- *Hypotenuse:* This term designates the side of the triangle opposite the 90° angle. In the figure, the side labeled a is the hypotenuse.
- *Side Opposite:* This term designates the side of a triangle opposite a given angle. The side labeled c is the side opposite angle C.
- *Side Adjacent:* This term designates the side of the triangle adjacent to a given angle, but not the hypotenuse. The side labeled b is the side adjacent to the angle C.

The two basic facts that assist us in understanding the application of the trigonometric functions are as follows:

- For any triangle, the sum of the three internal angles is always 180°.
- For every combination possible for these three angles, there is a distinct ratio between the angles and the lengths of the sides of the triangle.

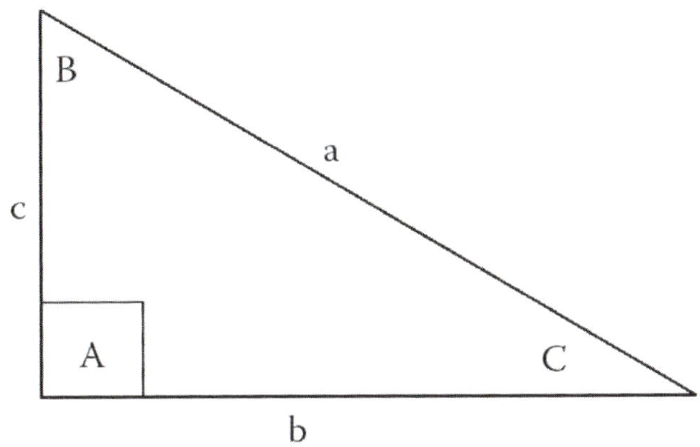

Figure B.1 A right triangle is any triangle in which one angle measures 90°. In this figure, angle A is 90°.

The relationships of concern to the analyst are known as the sine and tangent. Based on these two functions, we know that for any value of angle C in Figure B.1, the following is true:

$$\sin C = \frac{\text{opposite}}{\text{hypotenuse}} \text{ or } \frac{c}{a}$$

$$\tan C = \frac{\text{opposite}}{\text{adjacent}} \text{ or } \frac{c}{b}$$

These ratios are displayed in a trigonometric function table. Generally, such a table shows the ratios for every angle between 0 and 90°, in increments of at least 1°. If you recall the discussion on determining the impact angle, that level of detail is unnecessary. The impact angle determination is generally accurate only to about 3°. Table B.1 is an abbreviated trigonometric function table, in increments of 5°.

Let us begin with the sine function, which helps establish the angle of impact for a given droplet. In Figure B.2, note that the orientation of the right triangle is rotated. The right angle is now at the top of the triangle.

Table B.1 Trigonometric Function Table

Angle	sin	tan	sec	csc	ctn	cos	
0°	0.000	1.000				1.000	90°
5°	0.087	0.087	1.004	11.474	11.430	0.996	85°
10°	0.174	0.176	1.015	5.759	5.671	0.985	80°
15°	0.259	0.268	1.035	3.864	3.732	0.966	75°
20°	0.342	0.364	1.064	2.924	2.747	0.940	70°
25°	0.423	0.466	1.103	2.366	2.145	0.906	65°
30°	0.500	0.577	1.155	2.000	1.732	0.866	60°
35°	0.574	0.700	1.221	1.743	1.428	0.819	55°
40°	0.643	0.839	1.305	1.556	1.192	0.766	50°
45°	0.707	1.000	1.414	1.414	1.000	0.707	45°
	cos	ctn	csc	sec	tan	sin	Angle

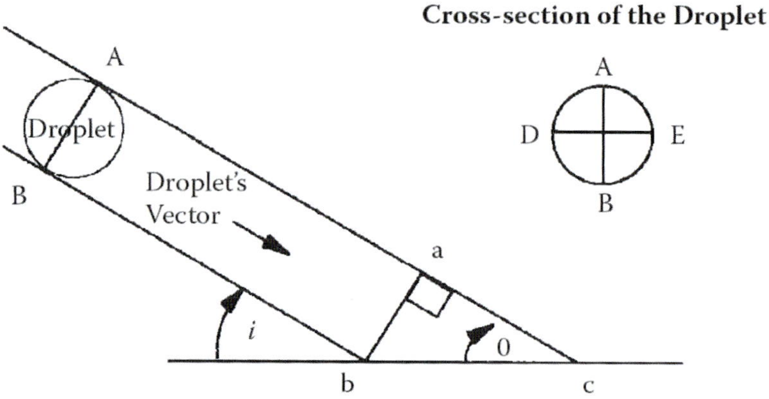

Figure B.2 The trajectory a droplet strikes the target defines the right triangle abc. Using this triangle, we can draw an analogy between the dimensions of the resulting stain and the triangle, which allows the analyst to define the impact angle.

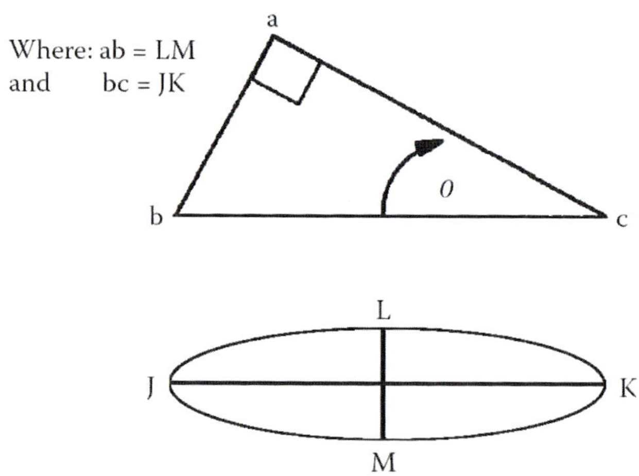

Figure B.3 The relationships drawn between the resulting stain and the right triangle. The side opposite (line ab) is analogous to the width of the stain (line LM). The hypotenuse (line bc) is analogous to the length (line JK).

Given our discussions in Chapter 4 and considering Figure B.2, we should feel confident that:

- The droplet in flight is generally spheroid in shape.
- Therefore, any measurement of the diameter of the droplet will be equal. Thus, in the figure, AB = DE.

In Figure B.2, the triangle we will use to solve the problem is formed by the vertical dimension of our droplet (line ab), the path of the droplet (line ac), and the area on the target surface between the point where the droplet first touches and the termination of the path (line bc). Also note that when viewing Figure B.2, angle *i* is the same as angle 0. Angle *i* defines the impact angle and is what we seek to determine.

In Figure B.3, we transpose the triangle and now compare it to the resulting stain. We can draw an analogy between the length of the hypotenuse (bc) and the length of our stain

(JK), and another between the width element of the stain (LM) and the side adjacent (ab). Using the measurements of JK and LM in the stain, we can define the angle 0, where

$$\sin 0 = \frac{\text{opposite}}{\text{hypotenuse}} \quad \text{or} \quad \frac{ab}{bc} \quad \text{or} \quad \frac{lm}{jk}$$

The result of this division is a ratio. The analyst then finds this ratio in the trigonometric function table and identifies the closest corresponding angle. If the analyst has a scientific calculator, the inverse sine of 0 or arc sine function (ASN), which converts this ratio to the angle, can also be used.

One important note: in viewing the diagrams, it may appear that a 1:1 relationship exists between line LM and line ab or between line JK and line bc. This isn't true. As the droplet impacts, the liquid laterally displaces outward. So the droplet's diameter in flight is much smaller than the resulting stain width; however, because this displacement occurs in both the length and width axes of the resulting stain, it has no effect on the application of the trigonometric functions.

Our application of the tangent function is a little more direct. (Refer to Figure B.4.) Given several droplets that impact a surface as a result of a single event, we may want to define the point of their origin (B). To do this we need to determine the length of the side opposite the two angles c and d, which is line AB. In the figure, line AC is the side adjacent to the angle c and line AD is the side adjacent to the angle d. Because AB is simply a straight line projected from the unknown point of origin to the target surface, the angle at A for both triangles is a right angle.

If we wish to solve for the angle c in Figure B.4, the tangent function tells us that for every right triangle:

$$\tan C = \frac{\text{opposite}}{\text{adjacent}} \quad \text{or} \quad \frac{AB}{AC}$$

We can easily determine the two impact angles (c or d) using the sine function. By measuring from the base of each stain to the point where their paths intersect, we can also determine the length of the adjacent side of each triangle. Using these two known values, we solve for the unknown length of the side opposite (line AB) by balancing the original equation. For the triangle ABC and the angle c, that means

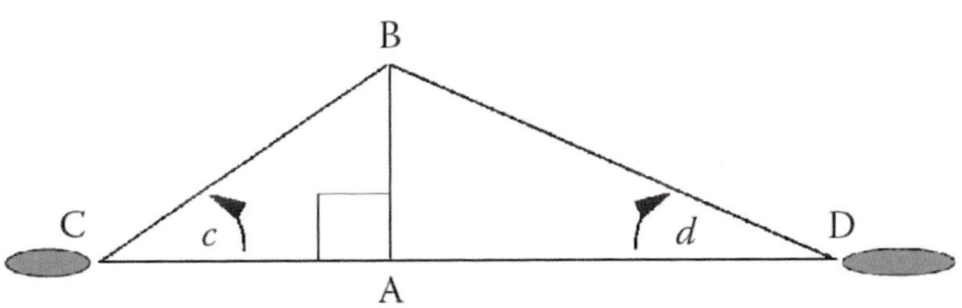

Figure B.4 The relationships of a surface in the scene to another right triangle when using the tangent formula. The probable point above the target where the stain originated is defined by the line AB. The line AC is equal to the distance from the base of Stain C to the convergence (A). The angle c is the impact angle for Stain C.

$$\text{side opposite} = \tan C \cdot \text{side adjacent}$$

or

$$AB = \tan C \cdot AC$$

Considering the triangle created by each droplet's impact, we can establish a general distance above the target surface for Point B. If all of the droplets are from the same origin and event, then this distance should be the same. To understand the limitations of the tangent relationship in defining the point of origin, review the discussion in Chapter 6.

Accuracy, Precision, and Significant Digits

No matter what we choose to measure – bloodstains, football fields, or our own height – there is always a level of uncertainty in any measurement process. The nature of the uncertainty is affected by two concepts: accuracy and precision. Accuracy in a measurement is a statement as to the level of certainty the measurer has for the final measurement. The greater the level of uncertainty in the measurement, the lower the accuracy. Precision, on the other hand, describes the ability of the analyst to repeat the measurement. The more likely it is that the measurement can be repeated, achieving the same result, the higher the level of precision. Accuracy is affected most by the nature of the thing being measured, whereas precision is affected most by the method of measurement employed.

In considering the determination of droplet impact angles using the formulas discussed, accuracy and precision both affect the level of trust the analyst has in the resulting impact angle. In terms of accuracy, it should be self-evident that because we apply straight-line geometry to define an angle that is created by a parabola, the level of accuracy is not great. It is not so much the precision of the measurements used by the analyst, but rather the manner of measurement and the nature of the thing (the impact angle) being measured. We generally accept that impact angles are accurate to $\pm 3°$ for drops impacting at 60° or less.

Precision in measuring the bloodstain is also a factor. Remember, precision speaks to the ability to repeat the measurement and achieve the same result (the same measurement). The manner of measuring the stain is critical in defining precision. For instance, if the analyst uses a ruler scaled to 1 mm to measure small bloodstains, then precision will likely suffer. In viewing such a scale in relation to the stain, the analyst is estimating the final digit of the measurement down to the nearest 0.5 mm. On the other hand, should the analyst use a micrometer, scaled to the nearest 0.1 mm, precision of the measurement increases. Now, the analyst is estimating the final digit for the measurement down to the 0.05 mm level. Just as important to the precision of the measurement is the knowledge or skill of the measurer. As discussed in Chapter 6, if the analyst inappropriately includes portions of the tail or scallops in the measurement, precision of the measurement will suffer.

The result of any impact angle or point of origin determination will ultimately be given as a measurement, either as an angle of impact (e.g., 65°) or a distance from a target (e.g., 2.5 ft). In accepting this measurement, the analyst must recognize the significant digits in the result. This significant digit is always the estimated digit. For instance, given an impact angle determination of 64.5°, the analyst cannot assume the significant digit is 5. The uncertainty present in the accuracy component ($\pm 3°$) tells us the answer is estimated to the second digit (e.g., 4). Recognition of the significant digit in an equation's answer is recognition of the overall uncertainty of the answer itself. To be objective, analysts must keep in mind the effects of accuracy and precision on the measurements, and then represent their investigative findings appropriately. This means keeping an eye on the significant digit, and not alluding to a level of certainty that does not exist.

Notes

1. Stein, J., *The Random House Dictionary of the English Language*, Random House, New York, 1980, p. 825.
2. Carter, A.L. and Podworny, E.J., Bloodstain Pattern Analysis with a Scientific Calculator, *J. Can. Soc. Forensic Sci.*, March 1991.

Appendix C
Dealing with the Risk of Bloodborne Pathogens

Analysts must recognize the risks involved in processing bloodstained scenes. The very nature of the medium the bloodstain pattern analyst works presents a specific health hazard. Unfortunately, law enforcement agencies have been slow to develop programs to deal with this particular issue.

In this chapter, we will discuss the fundamental issues and provide some limited guidelines that may assist the individual analyst or department in establishing methods for operating and managing bloodstained crime scenes.

Bloodborne Diseases

Biohazards concern all crime scene personnel, but the bloodstain pattern analyst is particularly susceptible to this issue. In addition to the presence of blood, death scenes often contain urine, fecal matter, sputum, or vomit. The primary biohazards associated with such situations are HIV (human immunodeficiency virus) and hepatitis. HIV is responsible for Acquired Immune Deficiency Syndrome (AIDS). HIV attacks white blood cells in the host, reducing the ability of the host to fight off other infections. For those exposed to the virus, it very often results in AIDS.[1] The virus destroys the host's ability to fight off other microorganisms, allowing the host to develop various other diseases. These are most often referred to as "Opportunistic Infections" (OIs) which are illnesses that occur more frequently and are more severe in people with HIV due to the damaged immune system.[2]

Current research indicates that the primary exposure avenues are anal or vaginal sex, followed by sharing needles or other drug injection equipment.[3] In terms of accidental exposure (other than through consumption of breast milk or sexual contact), the nature of the exposure required is not great (e.g., contact with an open wound, needle stick injuries). Accidental exposure at the crime scene remains a concern.

Of specific concern to the analyst is the virus' ability to survive. HIV can survive outside the human body for up to 15 days in a liquid state. Dried, the virus survives for 3–13 days, with more conservative estimates at about 3–5 days.[4] Temperature directly affects survivability: refrigeration delays the degradation of the virus, while heat speeds up its demise. HIV on a surface loses most of its ability to infect within several hours. Thus, surface exposure poses only a very limited risk.

The HIV transmission rate for most crime scene exposures appears relatively low. Studies suggest an infection rate of less than 1% for open wounds in contact with infected blood. Mucous membrane exposure has an even lower infection rate, less than 0.1%. These same studies, however, suggest that the risk increases as the volume of infected blood involved in the exposure increases.[5]

Hepatitis is an additional threat present at the crime scene and of specific concern to the bloodstain pattern analyst. Hepatitis occurs in three forms: Hepatitis A (HAV),

Hepatitis B (HBV), and Hepatitis C (HCV). HAV and HBV are both less severe than the HCV form. HAV is transmitted primarily through oral or fecal routes. HBV can be transmitted through direct or secondary contact with infected blood and other body fluids. This includes needle stick injuries or other minor injuries caused by infected articles at the crime scene. In terms of the risk of infection for HBV, infected blood is more likely to result in infection than any other body fluid. HBV can survive in dried blood at room temperature for up to one week.[6] Outside the body, HBV has survival rates as long as seven days.[7]

Both HAV and HBV result in a variety of symptoms that affect the individual, including fever, malaise, nausea, jaundice, and abdominal discomfort. If left unchecked, either form can lead to complications that are more significant or even death.[8] There are, however, preventative inoculations for both HAV and HBV. In the United States, OSHA regulations (OSHA 1910.1030) direct employers of at-risk employees to pay for such immunizations when requested by the employee, or when required by the employer.

HCV is the most severe form of hepatitis. HCV attacks the liver and leads to serious liver problems or death. There are no preventative treatments for HCV. Exposure routes for the disease are similar to HBV, with direct or indirect contact with infected blood the primary exposure risk at the crime scene. Recent studies show a less than 2% infection rate after exposure of an open wound to HCV-infected blood, and there is an even lower incidence rate for exposure through mucous membranes. As of the date of these studies, there have been no reported cases of infection from exposure of intact skin to HCV-infected fluids.[9]

Crime Scene Considerations

As indicated, the survival of these pathogens at the scene is not uncommon. Therefore, those working within the scene must consider possible exposure risks and take effective measures to reduce them.

A significant concern for the police or lab supervisor is who should be allowed on the scene. Scene integrity considerations require limiting the number of people who enter and move about the crime scene. The health risks such pathogens impose are an additional consideration that demands that unnecessary personnel are excluded. With regard to exclusion, the supervisor must consider not only an individual's function at the scene, but also limiting the exposure. This concern is significant if the scene involves the potential for airborne particles of blood is temporarily high.

Additionally, supervisors and analysts must consider personal injuries or wounds, as both represent the most significant source of exposure risk. Exposed cuts, abrasions, or sores offer potential access routes for infection. Individuals with such injuries should approach the scene with extreme caution and, when practical, should avoid the scene if possible. As a minimum protective measure, the individual should be using double-layer protective clothing.

The analyst should be concerned with preventing accidental injuries on scene. Crime scene processing and searches are filled with potential sources of injury. The analyst should move cautiously to avoid accidental needlestick, as well as injuries, cuts, or punctures from weapons, or sharp-edged articles. Any lapse in concentration can open the analyst to an immediate exposure risk.

Protective clothing is an absolute necessity for the analyst to protect against these types of accidents. The purpose of the protective clothing in a DNA world is threefold: protection from accidental exposure at the scene, subsequent incidental contact with blood products tracked out of the scene, as well as preventing contamination of the scene. The minimum requirements for personal protective equipment (PPE), generally include disposable booties, outer garments, and gloves. Additionally, the use of goggles and facemasks should be considered.

Fortunately, scene investigators rarely encounter a situation where significant dried blood may be airborne. In such instances, extra care should be taken to prevent exposure through mucous membranes.

Eating, drinking, smoking, and the application of cosmetics while in the actual crime scene are considered risky behaviors and should be prohibited.[10]

A final concern of exposure risk is contamination of crime scene equipment and clothing during post-processing activities. The use of PPE while in the scene, including shoe booties, is critical. Booties prevent analysts from carrying contaminants to their homes, offices, or cars on their personal footwear. Durable personal items such as the analyst's favorite pen or notebook, should not be utilized inside the scene. Use disposable pens and paper pads. These items are subsequently disposed of when processing is complete, eliminating potential contamination issues. Durable crime scene equipment exposed to biohazards in the scene must be decontaminated. Durable equipment should be washed with a 1:10 dilution of bleach and water. Items like plastic photo tents or scales can be cleaned with less caustic commercial alcohol solutions.

Supervisors should always establish a biohazard collection point at the working perimeter of the actual scene. The collection point should have large plastic trash bags taped open. This helps eliminate excessive handling of contaminated items by crime scene technicians. New PPE should be present here, making it easy for technicians to change gloves and booties when exiting and reentering the scene. Larger law enforcement agencies, especially those having a forensic laboratory section, are equipped to dispose of bagged biohazard materials. In other instances, a local hospital may deal with the small amount of biohazard materials collected during crime scene processing. We suggest coordinating support for the destruction of biohazard residues, whatever their nature before the need arises.

Dealing with Accidental Exposures

No one can guarantee a method of dealing with accidental exposure that effectively eliminates all risks involved. The following guidelines are offered as a primary response to accidental exposure.

Crime scene supervisors should always consider exposure prevention as their primary strategy for protecting workers. The next step in prevention is appropriate vaccination strategies. Any technician who works in biohazard scenes should be vaccinated. If prevention fails, the exposed individual should seek medical assistance to obtain treatment. CDC research indicates that timely reporting is essential, and treatment should be sought in hours, not days. Seek a health care provider or local health department immediately. Ask if a "HBIG" (hepatitis B immune globulin) shot is right for you. Post-exposure measures can prevent infection if treated ideally within 24 hours after the exposure is believed to have occurred.[11]

Current guidelines for on-scene first aid include the following:

- For simple contact, wash the area with soap and water or an antimicrobial wash.
- Exposed mucus membrane tissues should be flushed with water.
- Although prior guidelines suggest allowing needle stick injuries and other wounds to bleed freely for a moment, there is no evidence to indicate this action will reduce the probability of infection.[12] If possible, wash and rinse the wound with soap and water or an antimicrobial solution. The injured party should seek immediate medical assistance and report the incident to supervisors.

Packaging Biohazard Evidence

Proper packaging of biological evidence ensures that no evidence is lost and effectively eliminates subsequent contamination risks by those who must handle the evidence, as well as contaminating the evidence itself.

Packaging methodologies present a contradiction. Clothing and other bloodstained items are generally packaged in paper products to ensure the samples do not degrade. However, when wet or highly saturated and packaged in this fashion at the crime scene, blood can seep through the paper before completely drying.

Because of concerns over exposure to biohazard materials, one consideration is to package the wet items in paper at the scene. This should be done in a manner that prevents interaction between different surfaces on the item. This paper package can then be containerized in plastic until it can be transported to an appropriate facility. Remember that plastic increases the possibility of degrading biological samples, particularly when the plastic is closed or sealed and even more so when exposed to heat (e.g., a car trunk in summer). Some agencies do authorize the use of plastic alone at the scene, with the same concern of transporting as soon as possible. Such items must be transported and placed into appropriate packaging at the earliest opportunity. Once at the agency facility, the item must be air-dried and repackaged. Ultimately, the analyst or law enforcement agency must consider both the evidential considerations as well as the exposure risks and then develop a packaging policy that best fits their organizational needs.

Whatever packaging methods are used, ensure that containers allow some level of assurance to those opening them at the crime lab that the exposure risk is minimal. Double packaging of inner containers and the proper labeling of outer containers prevent surprises for those who will handle the evidence later. This generally includes affixing "Biohazard" labels or clearly labeling "Biohazard" on the outer container.

Exposure Risks in Training and Experimentation

Exposure risks are not limited to crime scenes. Those involved in training, those attending training classes, and those conducting experimentation should avoid unnecessary exposure. Although most blood used for this purpose comes from blood banks, the analyst should apply universal precautions and consider all products contaminated.

An approach in training is the use of animal blood as a substitute for human blood during training courses. Christman presented an interesting study to the International

Association of Bloodstain Pattern Analysts in which he compared different types of animal blood. Neither the HIV nor HBV virus is transmitted through animals such as horses, cows, and sheep. Although such animals may harbor other pathogens (e.g., brucellosis), these pathogens do not represent the risk associated with human blood. For this reason, Christman felt these sources may be very appropriate for use in training police and others in the discipline of Bloodstain Pattern Analysis.[13]

Other Sources of Information on Managing Bloodborne Pathogen Risks

In this appendix, the problems and issues related to bloodborne pathogens are discussed in a very general way. In seeking to establish a comprehensive risk management policy, several documents provide a more in-depth discussion of the subject. We suggest seeking out all of them for further guidance. One of the best documents available for the individual analyst is David Bigbee's manual entitled *The Law Enforcement Officer and AIDS,* which is available through the US Government Printing Office. The CDC offers many articles regarding guidance when dealing with biohazards. Other documents that may be of assistance in establishing a pathogen risk management policy include the following:

A Curriculum Guide for Public Safety and Emergency Response Workers, published by the Centers For Disease Control
AIDS and the Law Enforcement Officer: Concerns and Policy Responses, published by the National Institute of Justice
OSHA Regulation 1910.1030 – Bloodborne Pathogens[14]

Summary

The danger from bloodstained scenes faced by law enforcement personnel and analysts will always be present. Processing the scene demands that someone come in contact with the evidence, and contact means an exposure risk. It is imperative, then, that both analysts and supervisors recognize this danger and introduce steps to reduce the risk. The only rational means of dealing with this problem is to take proactive measures as part of a comprehensive risk management policy; business as usual at the crime scene is not a viable strategy.

Research on treating individuals exposed to the various bloodborne pathogens is ongoing. As a result of this effort, methods and guidelines are likely to change. The best advice is to remain vigilant in reducing accidental exposure risks. Simple steps, such as education and purchasing basic protective gear, will eliminate most unintentional exposures. The prevention of exposure is always preferred over treating individuals for exposure, not only in terms of the well-being of employees, but also from a financial standpoint. When and if prevention fails, pattern any treatment plan on the advice of health care professionals.

Notes

1. Bigbee, D., *The Law Enforcement Officer and AIDS*, 2nd ed., U.S. Government Printing Office, Washington, D.C., November 1988, p. 2.
2. Https://www.cdc.gov/hiv/basics/index.html; April 2021.

3. Https://www.cdc.gov/hiv/basics/index.html; April 2021.
4. Bigbee, D., *The Law Enforcement Officer and AIDS*, 2nd ed., U.S. Government Printing Office, Washington, DC, November 1988, p. 2.
5. U.S. Department of Health and Human Services, CDC, Updated U.S. Public Health Service Guidelines for the Management of Occupational Exposures to HBV, HCV and HIV and Recommendations for Postexposure Prophylaxis, *Morbidity and Mortality Weekly Report*, Volume 50, Number RR-11, June 29, 2001, pp. 7–8.
6. U.S. Department of Health and Human Services, CDC, Updated U.S. Public Health Service Guidelines for the Management of Occupational Exposures to HBV, HCV and HIV and Recommendations for Postexposure Prophylaxis, *Morbidity and Mortality Weekly Report*, Volume 50, Number RR-11, June 29, 2001, p. 3.
7. https://www.cdc.gov./hepatitis/hbv/bfaq.html March 2023.
8. GlaxoSmithKline Biologicals, Twinrix Hepatitis A and B Twin Threats Twin Defense, April 2002, Rizensart, Belgium.
9. U.S. Department of Health and Human Services, CDC, Updated U.S. Public Health Service Guidelines for the Management of Occupational Exposures to HBV, HCV and HIV and Recommendations for Postexposure Prophylaxis, *Morbidity and Mortality Weekly Report*, Volume 50, Number RR-11, June 29, 2001, p. 6.
10. Drayton, L., Guidelines for Personnel Handling Potentially Contaminated Evidence, HQS USACIDC Policy Message 015-87, June 1987.
11. https://www.cdc.gov/hepatitis/hbv/bfaq.html, 2024.
12. Drayton, L., Guidelines for Personnel Handling Potentially Contaminated Evidence, HQS USACIDC Policy Message 015-87, June 1987.
13. Christman, D., A Study To Compare and Contrast Animal and Human Blood Products, International Association of Bloodstain Pattern Analysts and Association of Crime Scene Reconstruction Joint Training Conference, Oklahoma City, OK, October 5, 1995.
14. Stewart, J.D., Bloodborne Diseases: Developing a Training Curriculum, Compuserve Law Enforcement Forum, May 1993.

Appendix D
Using the Bloodstain Pattern Analysis Decision Map

Organization

The organization of the BGA D/map consists of blue diamonds, colored rectangles, ellipsoids, and the associated lines interconnecting them. The diamonds represent the question to be posed. The lines related to the Y(es) or N(o) responses to that question direct the analyst to the next question. Rectangles represent the associated criteria already considered and found up to that point. The ellipsoids represent the most defensible classification, given the questions posed and answered. As BPA operates within the construct of class characteristic evidence, the D/Map is based primarily on excluding pattern types. This ultimately leads to limiting classification(s) to the remaining potential explanations.

The D/Map has a spatter and non-spatter side. The first question asked is: Are the stains only circles and ellipses with or without flow? A "Yes answer leads the user to the spatter side. See Figure D.1. A "No" answer leads the user to the non-spatter side. See Figure D.2.

D/Map Rules

The rules for applying the D/Map are simple. The answer to any question posed is either Yes, No, or I'm unsure. If a yes or no answer is forthcoming, the associated Y and N line directs the analyst to the next question. If the question cannot be answered, the analyst stops at the last point of agreement (the ellipsoid leading to the question being considered). Whatever level this may be, it is regarded as the most objective classification possible.

Questions as presented on the D/Map cannot be skipped, nor can one randomly choose which questions to ask or in what order. One must begin at the top (Q: Are the stain(s) only small circles and ellipses with or without flow?) and then follow the appropriate answer lines to the next question. Ignoring the order of the questions will invalidate the entire document and process. One obvious question not included on the D/Map is: "Is this blood?" The D/Map assumes the analyst has already considered this question.

Understanding the Questions

Understanding what is being asked might seem simple enough, but there must be a clear understanding of exactly what the D/Map characteristics entail. Certain definitions (inferred or otherwise) by the user might lead to a response unintended by the D/Map. Whereas most questions are easily answered, a few questions posed by the D/Map require clarity on the user's part:

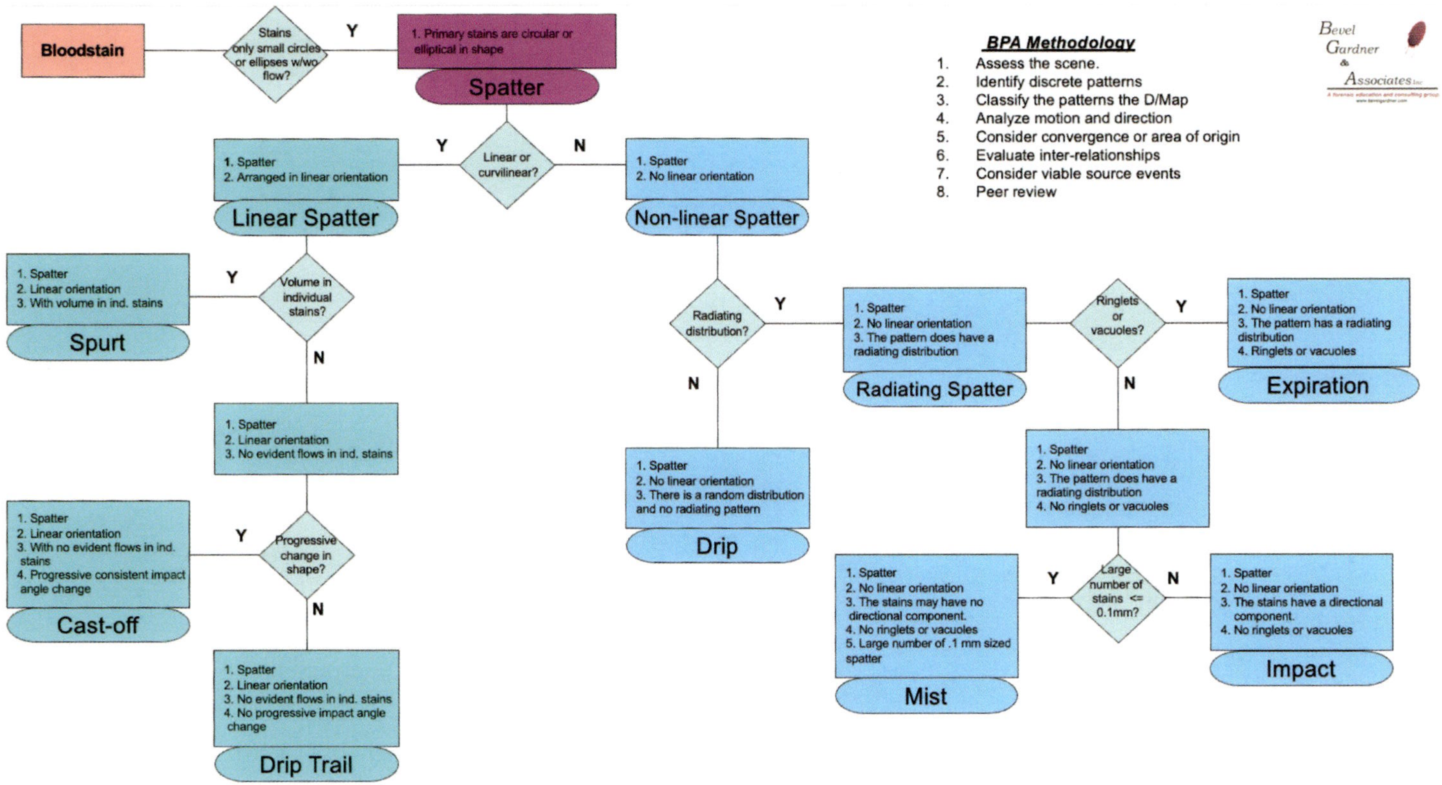

Figure D.1 The first question of the Decision Map leads the user to either the spatter or non-spatter side. A "Yes" answer leads to the spatter side.

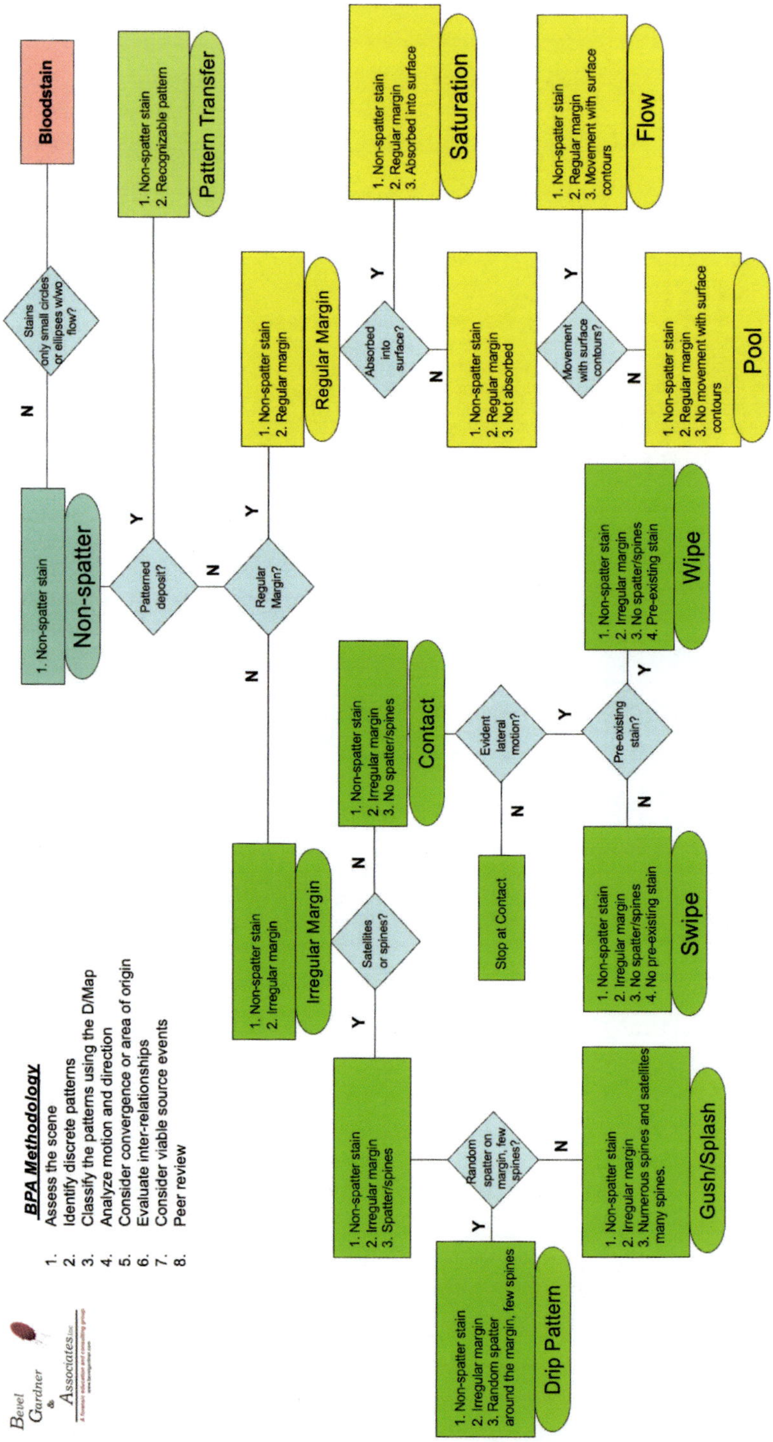

Figure D.2 A "No" answer to the first question leads the user to the non-spatter side.

Regular Margin? A regular margin is that which is consistent with fluid flow. If fluid were poured onto a surface, the margin would be generally straight and smooth with no significant disruptions. Anything else is considered irregular.

Random Spatter on the Margin, Few Spines? The randomness of the spatter is a function of both shape and, if present, directionality associated with any ellipses. The shapes are not singularly circles or ellipses. When ellipses are present, the long axes (the directional angles) point in varying directions that may not be consistent with one another. Spines are linear features associated with the stain. This is one aspect where the analyst must make a call. What are a few? A single or two spines may not be enough for the analyst to decide the question, and that's okay. This question is important as it separates Drip Patterns from Gush/Splash. There may be ambiguous patterns where this cannot be differentiated.

Preexisting Stain? There is evidence of a stain that was then disrupted. That evidence would be singularly a perimeter stain (the dried outer boundary of the preexisting stain). Without evidence of the perimeter stain, one cannot have confidence another stain was present and thus disrupted.

Patterned Deposit? Some aspect of the stain demonstrates distinct angular features, an outline of an image of an object (e.g., hands, knives, shoes), or some repetitive characteristics (e.g., a series of lines spaced equally).

Linear or Curvilinear? This does not mean the stains must align one after the other in a single line. The stains may do so, but they may also be broadly spaced across the pattern (e.g., a 4" width in the pattern). The lines can be straight, arced, or serpentine.

Volume in Individual Stains? The volume aspect is dealt with as a singular aspect. Is there a flow emanating from one or more individual spatter stains? The volume of the droplet was not contained within the spatter; gravity took hold and produced a minor flow. Volume cannot be effectively assessed by size or any other appearance of the spatter stain; evidence of the flow is the only effective method. This means that horizontal surface volume cannot be assessed, which can be problematic.

Ringlets or Vacuoles? Ringlets or vacuoles are small generally, circular voids within individual spatter, caused when an associated air bubble bursts. When answering the question, look for multiple examples not a single example.

Is There Evident Lateral Motion? This pertains to contact stains, where displacement of blood from the original stain boundary, striations, or feathering are present, indicating some type of lateral motion.

Large Number of Stains ≤ (less than or equal to) *0.1 mm?* Small spatter (e.g., 0.2 mm and smaller) are challenging to evaluate visually. The most effective approach is to measure the stains under magnification. Additionally, remember that if one can visualize the ellipse or circle, the stain is not 0.1 mm or smaller. Most human eyes cannot visualize stains of this size; they appear only as a pink or reddish hue.

The D/Map and Complex Patterns

As discussed in Chapter 4, not every pattern encountered can be placed into a specific classification. Complex patterns are recognized as occurring with some frequency. A complex pattern is one that has multiple characteristics that can span across both Spatter and Non-Spatter groups. These patterns don't fit neatly into any singular class of patterns anywhere. They occur from various recognized situations (e.g., drip trail with a drip pattern), as well as in some situations involving highly dynamic events. What is essential is the recognition that they are complex to begin with. Classification of complex patterns is problematic no matter what classification system is employed. This is true with the D/Map and Taxonomy system as well. Most attempts to utilize the D/Map on a complex pattern will fail. One will either be left with a highly ambiguous result (e.g., irregular non-spatter), or forced to ignore some apparent characteristics present. For example, one of the most common complex patterns results when a boot/foot stomps into a blood volume. Using the D/Map, the analyst would generally end up at Gush/Splash, but the contact characteristics will be present and impossible to ignore. Accept the existence of the complex pattern as described in the Taxonomy chapter, and accept that in such instances the D/Map is unlikely to give a functional result.

If it apparent that the pattern is complex, do not attempt to employ the D/Map. Occasionally, after employing the D/Map, it will become evident the pattern is complex. In either instance, deal with the complex pattern as described in Chapter 4, simply document that it is complex and describe any characteristics that may be present.

D/Map Utilization Examples

The following figures show how the D/Map will lead the user to a classification result.

- *Example 1: Figure D.3* The questions posed are as follows:
  - Q: Are the stains only circles or ellipses with or without flow?
  - A: Yes. This leads to the blue side of the D/Map.
    - Q: Are the stains deposited in a linear or curvilinear orientation?
    - A: Yes. This leads to Linear Spatter.
      - Q: Is there evident volume in the individual stains?
      - A: Yes. There are obvious flows emanating from the parent stains. This leads to Spurt.

  The most objective classification for this stain is a Spurt pattern.

- *Example 2: Figure D.4* The questions posed are as follows:
  - Q: Are the stains only circles or ellipses with or without flow?
  - A: Yes. This leads to the blue side of the D/Map.
    - Q: Are the stains deposited in a linear or curvilinear orientation?
    - A: Yes. This leads to Linear Spatter.
      - Q: Is there evident volume in the individual stains?
      - A: No.
        - Q: Is there a progressive change in the shape of the stains?
        - A: Yes. Right to left, the stains are becoming more elliptical. This leads to Cast-off.

  The most objective classification for this stain is a Cast-off pattern.

- *Example 3: Figure D.5* The questions posed are as follows:
  - Q: Are the stains only circles or ellipses with or without flow?

Figure D.3 Example 1 using the Decision Map. The most objective classification for this pattern is Spurt.

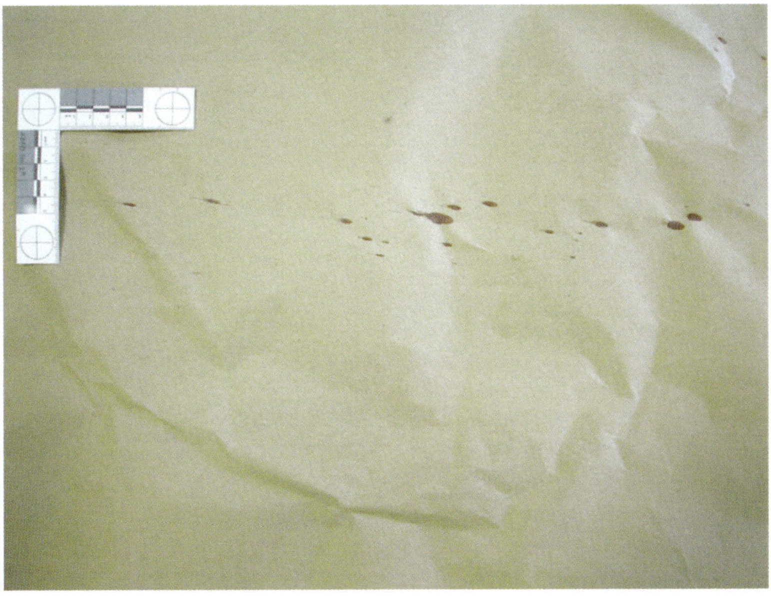

Figure D.4 Example 2 using the Decision Map. The most objective classification for this pattern is Cast-off.

Figure D.5 Example 3 using the Decision Map. The most objective classification for this pattern is Impact.

- A: Yes. This leads to the blue side of the D/Map.
  - Q: Are the stains deposited in a linear or curvilinear orientation?
  - A: No. This leads to Non-Linear Spatter
    - Q: Is there a radiating distribution?
    - A: Yes. This leads to some form of radiating spatter.
      - Q: Are there ringlets or vacuoles in the spatter?
      - A: No.
        - Q: Are a large number of the stains smaller or equal to 0.1 mm?
        - A: No. This leads to Impact.

The most objective classification for this stain is Impact pattern.

- *Example 4: Figure D.6* The questions posed are as follows:
  - Q: Are the stains only circles or ellipses with or without flow?
  - A: No. So, we are directed to the green/yellow side of the D/Map
    - Q: Is there a patterned deposit?
    - A: No. This leads to the margin question.
      - Q: Does the stain have a regular margin (consistent with fluid flow)?
      - A: No. This leads to some form of irregular non-spatter.
        - Q: Are there satellites or spines?
        - A: Yes.

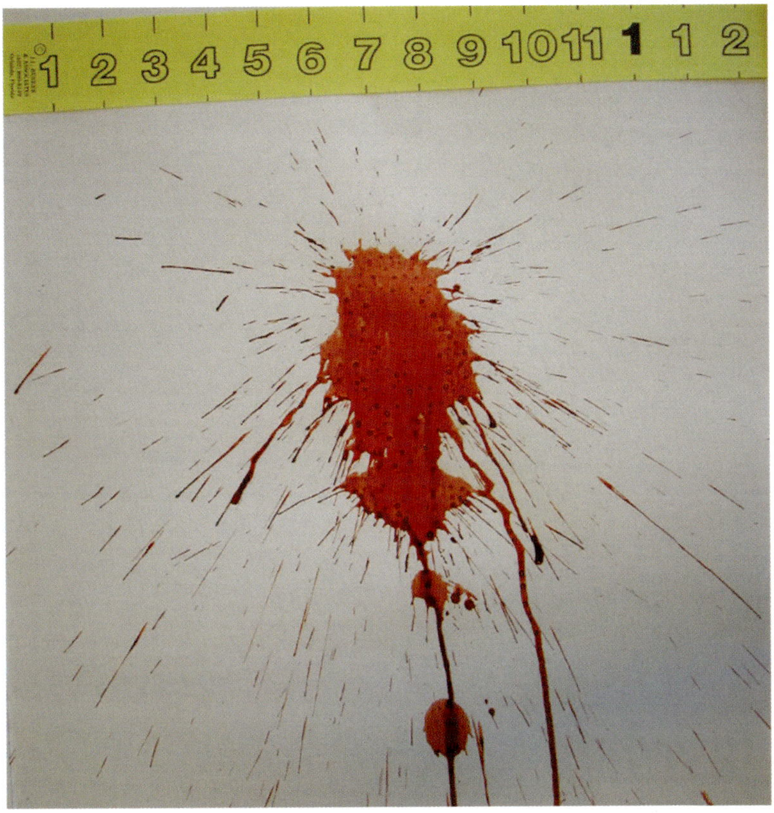

Figure D.6 Example 4 using the Decision Map. The most objective classification for this pattern is Gush/Splash.

        - Q: Are there random spatter (shapes other than just typical circles and ellipses) with random directionalities on the margins with few spines?
        - A: No. This leads to Gush/Splash.

  The most objective classification for this stain is Gush/Splash pattern.

- *Example 5: Figure D.7* The questions posed are as follows:
  - Q: Are the stains only circles or ellipses with or without flow?
  - A: No. This leads to the green/yellow side of the D/Map.
    - Q: Is there a patterned deposit?
    - A: No. This leads to the margin question.
      - Q: Does the stain have a regular margin (consistent with fluid flow)?
      - A: No. This leads to some form of irregular non-spatter.
        - Q: Are there satellites or spines?
        - A: Yes.
          - Q: Are there random spatter (shapes other than just typical circles and ellipses) with random directionalities on the margins with few spines?
          - A: Yes. This leads to Blood Into Blood/Drip Pattern.

Figure D.7 Example 5 using the Decision Map. The most objective classification for this pattern is Drip Pattern/Blood Into Blood.

The most objective classification for this stain is Blood Into Blood pattern.

- *Example 6: Figure D.8* The questions posed are as follows:
  - Q: Are the stains only circles or ellipses with or without flow?
  - A: No. This leads to the green/yellow side of the D/Map.
    - Q: Is there a patterned deposit?
    - A: No. This leads to the margin question.
      - Q: Does the stain have a regular margin (consistent with fluid flow)?
      - A: No. This leads to some form of irregular non-spatter.
        - Q: Are there satellites or spines?
        - A: No. This leads to Contact.
        - Q: Is there evident lateral motion?
          - A. Yes
        - Q: Is there evidence of a preexisting stain?
        - A: Yes. This leads to Wipe.

  The most objective classification for this stain a Wipe pattern.

This pattern represents one of the few potentially confusing aspects of the taxonomy system. Recall from the Pattern Diversity chapter that alterations to patterns are considered adjectives when discussing patterns. This is true with one single exception: the wipe. The wipe is an alteration of another pattern, but right or wrong; it has always been discussed as a distinct pattern type.

Thus, nearly any wipe can technically be considered a complex pattern. In this example, a pool was then disturbed by the wipe. Granted, the pool is no longer present, but its perimeter stain remains. This would be true of individual spatter (drips) that were wiped, a flow that was wiped, etc. In any instance where sufficient perimeter stain exists to conclude wipe, the pattern can always be identified as Complex. Whether reported as complex or simply a wipe, this effectively becomes a decision for the individual analyst.

Figure D.8 Example 6 using the Decision Map. The most objective classification for this pattern is Wipe.

Example 7: Figure D.9 The questions posed are as follows:

Q: Are the stains only circles or ellipses with or without flow?

A: No. This leads to the green/yellow side of the D/Map.

- Q: Is there a patterned deposit?
- A: No. This leads to the margin question.
  - Q: Does the stain have a regular margin (consistent with fluid flow)?
  - A: No. This leads to some form of irregular non-spatter.
    - Q: Are there satellites or spines?
    - A: No. This leads to Contact.
      - Q: Is there evident lateral motion?
      - A: Yes.
        - Q: Is there a preexisting stain?
        - A: No. There is no evidence of perimeter stains present.

This leads to a Swipe.

The most objective classification for this stain is Swipe pattern.

Example 8: Figure D.10 The questions posed are as follows:

Q: Are the stains only circles or ellipses with or without flow?

A: No. Even given a rough surface, we observe this is not singularly spatter. This leads to the green/yellow side of the D/Map.

- Q: Is there a patterned deposit?
- A: No. This leads to the margin question.
  - Q: Does the stain have a regular margin (consistent with fluid flow)?
  - A: No. This leads to some form of irregular non-spatter.
    - Q: Are there satellites or spines?
    - A: Yes.

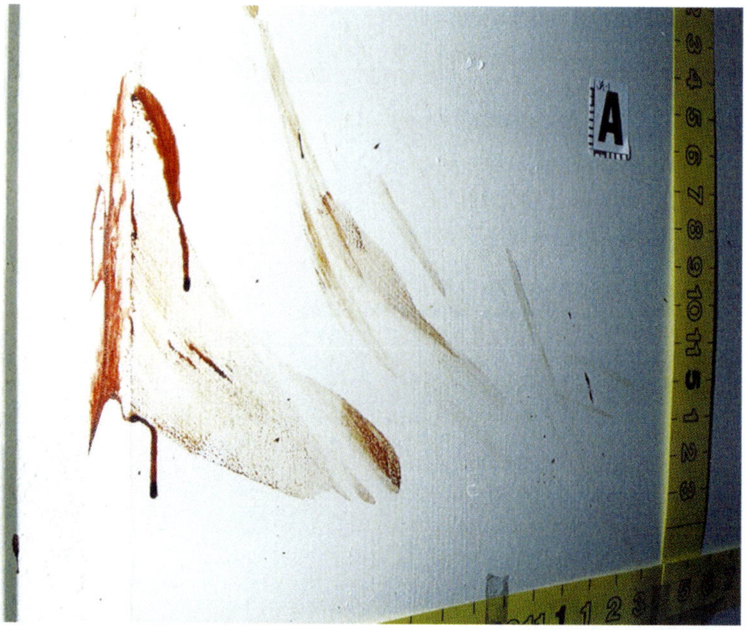

Figure D.9 Example 7 using the Decision Map. The most objective classification for this pattern is Swipe.

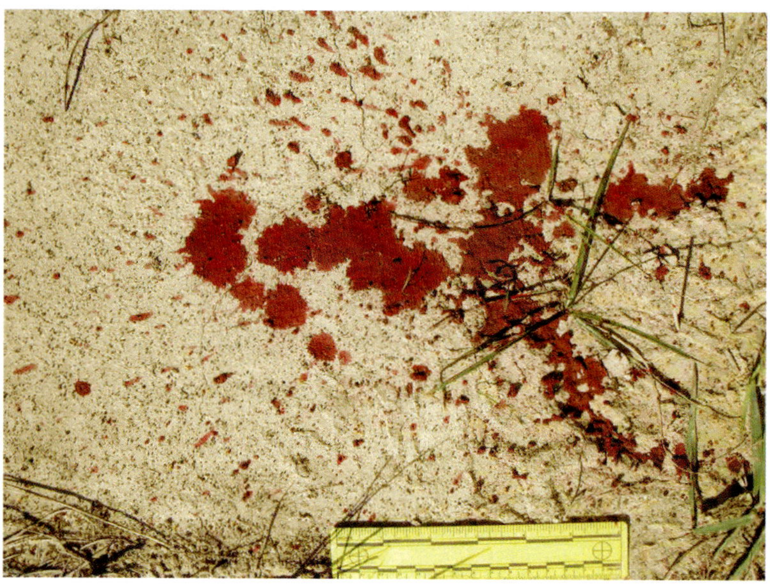

Figure D.10 Example 8 using the Decision Map. The most objective classification for this pattern is an Irregular Margin stain as the characteristics on this substrate are muted.

- Q: Are there random spatter (shapes other than just typical circles and ellipses) with random directionalities on the margins with few spines?
- A: The answer here is "I don't know." At this point, the substrate becomes a distractor. All the satellites are irregular based on the surface characteristics. This demands the classification process stop at Irregular Margin.

The most objective classification for this stain is some form of Irregular Margin pattern. This decision point still includes the potential of both Blood Into Blood or Gush/Splash.

The D/Map and Ambiguous Patterns

Example 8 demonstrates one of the values of the D/Map. When presented with an ambiguous pattern (e.g., muted or absent characteristics for whatever reason), one must pursue the classification, but recognize that the resulting classification may be limited. But even this limited classification as with all class characteristic evidence is probative. In Example 8, Contact is eliminated based on the Satellites/Spines question, leaving only Blood Into Blood and Gush/Splash as viable classifications. Based on the classification Irregular Non-Spatter, neither is inappropriately eliminated. The analyst may arrive at a more refined conclusion at some later point based on context, but the classification remains the same and does not force that issue.

Any classification beyond Bloodstain eliminates possibilities, even when presented with highly ambiguous patterns. For example, a classification that stops at Linear Spatter, effectively eliminates all Non-Spatter, all Radiating Spatter, etc. The three forms of Linear Spatter remain possible explanations. Just the same, a classification of Non-Spatter will eliminate all of the Spatter classifications. Once again recall, that as with all class characteristics evidence, *elimination* of potential answers is key.

Remember always that the D/Map is a tool. It cannot replace the trained analyst. Any ultimate classification arrived at should still be challenged by the user, considering all associated taxa. But when used correctly, it will lead the analyst to the most effective classification. It also provides an easily articulated and objective foundation for any classification opinion. New analysts will find the D/Map a particularly effective learning tool. It ensures they ask the right questions in the proper order. With time and experience, the questions posed will become second nature.

Appendix E
A Review of the OSAC Bloodstain Classification Process Map

In 2019, the NIST OSAC BPA Committee published the OSAC Bloodstain Classification Process Map.[1] See Figure E.1. In reviewing the OSAC Process Map, there are a number of significant questions that arise. The function of a Process Chart or a Decision Map is to act as an Expert system. Thus, any similar tool should be concise, based on current standards/ideas and hopefully not confuse the user. The authors recognize that the Bloodstain Classification Process Map is a work in progress, but it was released to the public in 2019. No subsequent changes are known and the assumed addendum (a porous surface chart) has yet to be released five years later. It is our hope that the OSAC BPA subcommittee will find the comments and criticisms we offer here helpful should future revisions be forthcoming. They are based on years of dedicated effort trying to resolve the classification issue and take into account the myriad of comments, crtitcisms, and suggestions BGA has received from practitioners utilizing the D/Map.

Bloodstain patterns are a form of class characteristic evidence. As such they must be approached as such. Class characteristic evidence is best suited for eliminating possibilities and isolating what remains. The parent–child relationship is critical to a functional system. Every child class of anything (e.g., fiber, paint) must share the characteristics of the parent and then have discerning characteristics that make it different from its siblings.

Thus, in evaluating class characteristic evidence, an expert system should identify specific decision points that eliminate other evidence classes. As the authors have learned over the last 17 years, this requires recognizing not only what questions should be asked, but also deciding whether the question be posed positively or negatively and specifically the order in which the questions should be posed. Any attempt at an expert system without detailed consideration of such issues produces a convoluted and unconcise product.

There are substantive issues in what OSAC published as their version of a Process Map. These include the use of ambiguous terminology, requiring the use of context in classification, questionable or, at the very least, confusing decision points, and a failure to consider in what order decision points should be imposed.

Context Issues

Following published methodology articles (Saviano, Gardner), context is considered in the ultimate conclusion offered by any BPA Analyst. But context only follows more basic considerations, specifically objective classification. The D/Map tool was designed to free the user of context in the classification phase. The OSAC Process chart requires context at various decision points.

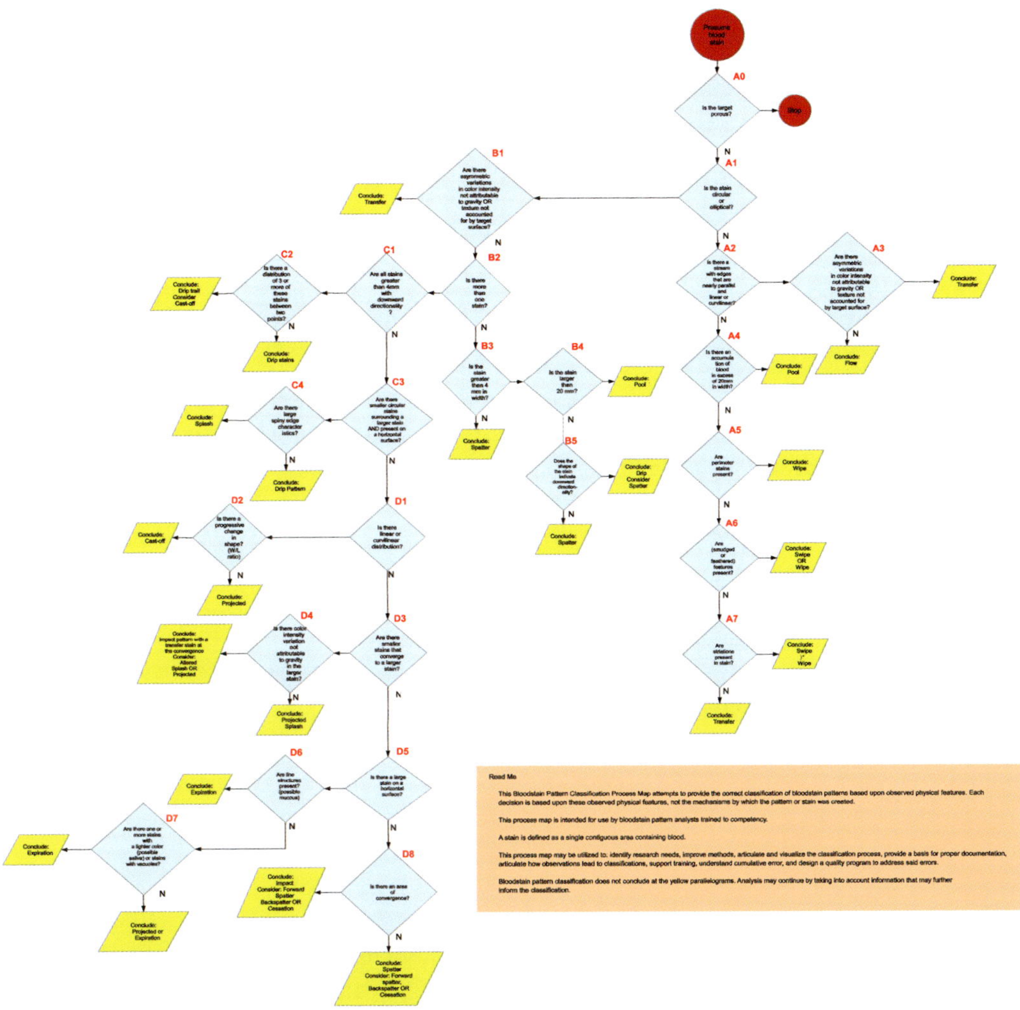

Figure E.1 The OSAC Bloodstain Classification Process Map as published in 2019.

D5 asks about horizontal surface. Can such stains only be deposited on horizontal surfaces? Patterns of every nature can be deposited on many surfaces and effectively recognized. Surface orientations can always be altered between deposition and final position.

C3 In part asks, "AND present on a horizontal surface?" A No to this aspect eliminates all drips. Drips can and do occur on vertical and other than horizontal oriented surfaces. As described above, does one simply eliminate drip because a particular surface is no longer oriented horizontally?

Ambiguous Terminology

The terminology in the OSAC Process chart presents many instances of ambiguous wording.

The words larger and smaller are used in several instances, whereas quantitative numbers are provided elsewhere. Larger than what? Smaller than what? Such terminology must be used sparingly and only when absolutely demanded (e.g., no absolute range of quantification is possible).

A2 The term stream is employed. Stream has a standard definition that is purely mechanistic. Stains do not have streams. While still fluid, there may have been streaming, but as stains, they have size, shape, and margins.

In B1 and A3, the phrase "Asymmetric variations in color not attributable to gravity" is used as the characteristic to be considered. What does this objectively entail? One might assume the answer is striations, feathering and less intense coloration from disruption of the blood, but that is an assumption. What is this asymmetrical variation, and how and where is it defined?

D4 "Intensity variation in color not attributable to gravity." Is this intended to be the same characteristic as asymmetric variation in color? Or, if different, will this color intensity of the associated stains always be present? How does one quantify this intensity variation? Humans do not visualize colors like a computer. Individual biological capabilities, lighting, and background all affect a person's visualization of hue and intensity.

D4 The associated conclusion to an answer of Yes for this color intensity variation issue states: "Conclude Impact pattern with a transfer at the convergence." What factor limits a transfer's location only at the convergence? Can't this color variation caused by a transfer occur elsewhere in the pattern, not simply at the convergence?

D3 The word "Converge" is introduced. The question asks: Are there smaller stains that converge to a larger stain? A Yes answer leads to Projected, Splash, or Impact. That begs the question: What "larger" stain exists in the standard expectation of Impact? And is this "larger stain" meant as a single stain or group of stains? Additionally, recognizing that when BPA talks about directionality (defined by stain shape), convergence occurs by evaluating the reverse azimuth of the stains. Droplets radiate, only their reverse azimuths converge. Radiating is the most appropriate term and characteristic to apply here.

D3-D5-D8 "Area of convergence" is introduced. Again, one presumes the intended question is about a radiating pattern.

A1 "Is the stain circular or elliptical"? Answered No, this leads to a conclusion of Pool. But answering Yes can lead to the path of B1, B2, B3, B4, and once again, arrive at a Pool. Circular or elliptical is a poor choice of words to describe a pool. The circular/elliptical description in general BPA parlance refers to spatter stains, so using it for a pool is questionable. Pools may be somewhat circular in nature, but then again, they can be any shape (e.g., conforming to surface contours). That the decision point allows for a pool using either response, in and of itself, also makes this a questionable decision point.

Questionable Decision Points (Ambiguous or Inappropriate Exclusions or Inclusions)

A0 "Is the target porous?" If the answer is Yes, the response is Stop. We are aware that OSAC intends to produce a porous target Process Map. As published however, lawyers will effectively use this statement to suggest that on a porous surface, classification cannot occur. It doesn't say proceed to the Porous Process Map; it says Stop.

As published, the implication is that *all* BPA classification on *any* fabric (furniture, bedding, clothing) is impossible. This obviously is inconsistent with current beliefs in BPA.

A1 "Is the stain circular or elliptical?"

This question is an appropriate characteristic, but the greater question is: One, several, or all? This ambiguity effectively forces gush, splash, and drip patterns in with typical radiating spatter patterns. Another concern is that satellite stains occur across the gamut of patterns observed, which can then lead to additional confusion.

A2 "Is there a stream with nearly parallel edges or curvilinear?"

The physical characteristic in question for the stain is the margin (e.g., regular margins).

A Yes answer for A2 ultimately offers "Transfer" as a conclusion. A Yes here and then a No at A3 (the asymmetric variation in color characteristic) gets to the obvious Flow. What characteristics do a flow and a transfer share? How and why are these two different patterns associated with the same decision point?

The A2-A3 offers one avenue to some form of Transfer. While the A2-A4-A5-A6, which are offline from that A2-A3 decision, offers a completely different route to Wipe and Swipe. Wipes and Swipes are in effect transfers of blood through contact. These decision points are convoluted and not concise.

A4 "Is there an accumulation of blood over 20 mm in width?" An answer of Yes concludes pool. This issue is particularly difficult for everyone involved. Can a specific volume accumulation be defined for a pool? As much as OSAC attempts to define it by using this 2 cm size; 20 mm fails to account for the obvious. BPA literature has multiple examples of stains (drips) falling to a surface and having widths over 20 mm (e.g., some as wide as 25–27 mm). If the drip occurs on a particularly smooth surface, there will be no edge characteristics; it will simply be a circular stain of more than 20 mm. On an irregular surface it may appear as an asymmetrical stain over 20 mm.

What happens when two large drips fall to the same area and coalesce in a shape other than circular or elliptical? Or when larger droplets associated with a streaming ejection lose energy and fall with gravity to a horizontal surface? The non-elliptical answer at A1 may eliminate many of these stains, but not all. Do these situations now constitute a Pool?

We appreciate OSAC's attempt to define a quantitative value as a characteristic for pool, but how functional is that quantification in practice? There are clear examples of simple spatter associated with drips and projected stains that are easily above this 20 mm size.

A4 and B4 are completely different decision paths, yet both ask the same question (20 mm in width) and both lead to a pool. Again, it is clearly a poor decision point for pool. Any effective decision point leads to the inclusion or exclusion of some class of evidence; an effective expert system should not allow multiple paths to the same class of stain.

A5, A6, and A7. Any answer for these decision points, leads to some form of contact, and the path includes multiple disparate paths (based on either Yes or No) that lead to Wipe. Of particular interest, lacking evidence of a preexisting stain (no perimeter stain) how does one end up objectively with a Wipe? Also, the Yes answers at A6 and A7 lead to "Conclude: Swipe or Wipe." As written, it seems to suggest or allow an analyst to report either swipe or wipe. Was this the intent? Wipe and Swipe are distinct pattern types.

A1, B1, B2, C1, C3, C4 path leads to a splash. However, the C3 decision point answer requires a horizontal target. Is this to suggest a splash (by any definition) cannot occur on a vertical target? The OSAC Splash pattern definition is a bloodstain pattern created from a large volume of liquid blood falling onto a surface. It does not limit the target surface to being horizontal.

D2 "Is there a progressive change in shape?" An answer of Yes concludes a Cast-off, and a No excludes Projected. Many projected patterns demonstrate a change in the shape of the associated stains. It is strictly a matter of how the resulting droplets/masses strike the target. There is no question this is a pertinent decision point when considering cast-off. The

D/Map uses this same consideration, but does so only *AFTER* eliminating the Projected classification.

D3 and D8 present disparate and convoluted paths to Impact pattern. The D3 convergence answer of Yes leads to one form of impact (a complex pattern involving both impact and transfer). The No answer at D3 leads to a second immediate question about a large stain and then a second question about convergence, resulting in what one would generally consider typical Impact (e.g., no larger stain). What is the source of this larger stain in the D3 Yes answer? The D3 Yes answer also fails to consider multiple impacts onto the same surface. The complex pattern created is effectively exactly as their conclusion describes, but there is no transfer involved.

D5 "Is there a large stain on a horizontal surface"? An answer of Yes leads to the only expiration conclusion possible (D5-D6 and D5-D7, regardless of how it is answered). How are expiration stains (small circles and ellipses) now considered large? Once again, the questions suggest expiration can only occur on a horizontal surface.

D8 This decision point lies at the end of a convoluted series of decision points (A1, B1, B2, C1, C3, D1, D3, D5, D8) before asking the most obvious question associated with impact spatter and expiration. It is asked as a convergence, but it is obviously intended to ask if the pattern is radiating. The history of BPA would suggest that a radiating nature would be one of the early and critical decision points, not a secondary or tertiary issue.

Again, instead of asking about the radiating nature of the stains, D8 asks for evidence of an area of convergence. Spatter (e.g., impact, expiration) striking a target perpendicularly may not have an area of convergence. One could argumentatively evaluate the circular stains produced for the Area of Convergence and thus, perhaps identify a general area. But even that is unlikely, given stains impacting at or very near perpendicularly rarely have their reverse azimuths come together effectively. The true issue remains: Do the stains radiate? Radiating is at times recognized by the distribution of the stains, this certainly includes mist where the stains are 0.1 mm in size. Based on their Weber numbers, such small stains always appear as circles and have no directional component. The distribution aspect is the only evidence of their radiating distribution; convergence is not the characteristic in question for these stains.

Bloodstain patterns are a class characteristic evidence. They should be used as such. Any questions asked about critical characteristics must allow for the elimination of groups (classes of patterns). Elimination of non-viable groups leads to isolating potential groups to which the pattern *may* belong. As such, any decision point used in classification must accomplish this exclusion or inclusion effectively and lead one to a singular path for any given classification.

Because BPA is a class characteristic evidence and knowing full well that not every pattern will be absolute (no questionable issues about a characteristic, no situation where characteristics are muted or absent), the OSAC Process Map fails in one major fashion. There are no waypoints; no higher-level classes of patterns that share broad characteristics. There are no clear parent–child relationships, where children share characteristics of the parent. Using the biology analogy (e.g., Domain, Kingdom, Phylum, Class, Order, Family, Genus, Species), the OSAC Process Map assumes that one can only arrive at a bottom-line "Species" decision, or apparently none.

We have demonstrated clearly that bloodstain patterns fall into broader categories (e.g., spatter, linear spatter, radiating spatter, regular margin, irregular margin). This is not a question facing the discipline. Failing to recognize these broader-level classification groups is very problematic. Without them, analysts, when presented with an ambiguous pattern/stain, are forced to either assume or simply not conclude. Both actions are dangerous and lead to subjective analysis. Even the broadest of classifications (beyond "bloodstain") has probative value, it eliminates potential answers. The OSAC Process Map fails in this aspect.

Whereas the OSAC Process Map is, like the BGA Decision Map, at the very least a proper direction for the discipline, it lacks clarity and requires a greater evaluation of the utilized decision points. What is asked and in what order it is asked are critical aspects of any effective expert system. Many of the OSAC decision points used are certainly critical in classifying BPA, but the order in which OSAC presents them leads to confusion. The BGA Decision Map has been published and utilized for over 16 years with success. We would humbly suggest there are lessons present in the D/Map that could effectively assist OSAC in this pursuit.

Note

1. NIST/OSAC BPA Subcommittee Bloodstain Classification Process Map, https://www.nist.gov/system/files/documents/2020/05/19/BPA%20Process%20Map_Dec2019.pd, 24 April 2024.

Index

For Product Safety Concerns and Information,
please contact our EU representative GPSR@taylorandfrancis.com
Taylor & Francis Verlag GmbH, Kaufingerstraße 24,
80331 München, Germany

Printed by Integrated Books International,
United States of America